W9-CRV-124

MONOGRAPHS ON THE PHYSICS AND CHEMISTRY OF MATERIALS

General Editors

C. E. H. BAWN, H. FRÖHLICH,
P. B. HIRSCH, N. F. MOTT

THE FERMI SURFACE

ITS CONCEPT, DETERMINATION, AND USE IN THE PHYSICS OF METALS

BY

A. P. CRACKNELL

CARNEGIE LABORATORY OF PHYSICS, UNIVERSITY OF DUNDEE

AND

K. C. WONG

DEPARTMENT OF PHYSICS, UNIVERSITY OF SINGAPORE

CLARENDON PRESS · OXFORD

1973

PHYSICS

Oxford University Press, Ely House, London W.1

GLASGOW NEW YORK TORONTO MELBOURNE WELLINGTON
CAPE TOWN IBADAN NAIROBI DAR ES SALAAM LUSAKA ADDIS ABABA
DELHI BOMBAY CALCUTTA MADRAS KARACHI LAHORE DACCA
KUALA LUMPUR SINGAPORE HONG KONG TOKYO

ISBN 0 19 851330 5

© OXFORD UNIVERSITY PRESS 1973

73-173504

PRINTED IN NORTHERN IRELAND AT THE UNIVERSITIES PRESS, BELFAST

TO

MARGARET *and* GRACE

Q C
176
.8
E4
C681
PHYS

PREFACE

> Everyone knows what a metal is and can describe many of its characteristics. It is safe to say, however, that few people would define a metal as "a solid with a Fermi surface". This may nevertheless be the most meaningful definition of a metal that one can give today; it represents a profound advance in the understanding of why metals behave as they do. The concept of the Fermi surface, developed by quantum physics, provides a precise explanation of the main physical properties of metals: their conduction of electricity and heat, their hardness and ductility, their lustrous appearance and so on.
>
> A. R. MACKINTOSH (*1963*)

THIS MONOGRAPH is intended for readers with a background of solid-state physics, but not necessarily with any specialist knowledge of the physics of metals. We have tried to steer a middle course between the experimental and the theoretical approaches; thus we have not described experiments in detail, nor have we employed the various sophistications of many-body theory (except in parts of Chapter 6). Group theory, though useful in the calculation of band structures and Fermi surfaces, is not essential and so our discussion of this topic is all contained in one section (Section 1.4) which the reader with no previous knowledge of group theory may wish to omit.

Most of the material in this book is dealt with at much greater length in various more specialized books and review articles which we have referred to in detail for some of the more esoteric aspects of the subject, while being much less careful about giving references for many of the more well-known parts of the subject. This may seem curious but it is deliberate; though we were not writing a definitive treatise, we did feel that it was appropriate to indicate where one

10964

could find further information about points of interest to the specialized reader.

While we describe in general terms the Fermi surface of each of the metallic elements in Chapters 4 and 5, we do not give a complete catalogue of the extensive work that has been done on these metals; for a more extended account see Cracknell (1971*b*). For other more complete catalogues we would refer the reader to the works of Johnson (1965), Grigsby, Johnson, Neuberger, and Welles (1967) and the table given by Visscher and Falicov (1972) based on the unpublished, but widely circulated, bibliography compiled by Professor A. R. Mackintosh of the Technical University, Lyngby, Denmark.

There are a few areas in which we are particularly aware of the limitations of this monograph. First, in Chapter 6 we said little on the subject of **k**-dependent relaxation times and mean free paths. Our excuse for this is that it is a difficult and complex subject on which much work still remains to be done before any really coherent picture can emerge (see, for example, Professor Sondheimer's concluding remarks at the 1968 conference on *Electron mean free paths in metals* (Sondheimer 1969)). Then, our treatments of cohesion in Section 6.1 and of alloys in Section 7.2 are only perfunctory; to do justice to these two topics would require a separate volume. To those who would wish to read further in this connection we would recommend starting with the article by Heine and Weaire (1970). Finally the subject of the metal-insulator transition, which is discussed in Section 7.3, is one on which a considerable amount of research is currently in progress and what we have written on this topic is not only brief but may also quickly become out of date.

We are grateful to all the various authors, editors, and publishers who have granted permission for the reproduction of copyright material, the sources of which are indicated appropriately. In this connection the following publishers have specifically requested to be identified as publishers of the journals indicated: Academic Press Inc. (*Solid State*

Physics), the National Research Council of Canada (*Canadian Journal of Physics*), Springer-Verlag (*Physik der Kondensierten Materie*), and Taylor and Francis Ltd. (*Advances in Physics* and *The Philosophical Magazine*). We are grateful to Taylor and Francis Ltd. for allowing the material originally published in two review articles in *Advances in Physics* and subsequently reprinted as "The Fermi surfaces of metals" (Taylor and Francis Monographs in Physics Series, A. P. Cracknell 1971) to be reproduced in a condensed version in chapters 4 and 5 of this book. We would also like to thank our various colleagues, research students, and other friends for their helpful discussions. We are grateful to Professor Sir Nevill Mott who at various stages made some useful suggestions about the scope and contents of this book and also Professor I. Waller and Dr. J. Tilley who have been kind enough to read and criticize certain parts of Chapter 6 for us. K. C. W. is grateful to Professor A. Rajaratnam, University of Singapore, for his encouragement, to Professor A. Salam and Professor P. Budini, as well as the IAEA and UNESCO, for hospitality at the International Centre for Theoretical Physics, Trieste, during the period March—May 1971, and to Professor K. J. Standley for the hospitality of the Physics Department, University of Dundee, during June 1971.

Dundee and Singapore
October 1971

A. P. C.
K. C. W.

CONTENTS

1

BAND STRUCTURE, THE FERMI SURFACE, AND THE FREE-ELECTRON MODEL

1.1. Introduction

IN DISCUSSING the behaviour of the electrons in a solid material there are some simplifications that arise if we choose to consider materials that are crystalline rather than amorphous or glassy. In this book we shall be concerned with the particular collection of crystalline solids that are metals and, of these, we shall be principally concerned with those metals that are chemical elements.

A simple-minded view of the metallic state of an element of atomic number Z and valence n can be obtained by assuming that each atom loses its outermost n electrons and becomes ionized. Each positively charged ion contains a nucleus of charge $+Z$ together with $(Z-n)$ electrons in various atomic orbitals surrounding the nucleus, so that the net positive charge of the ion is $+n$; these ions are regarded as being fixed at appropriate sites in the crystal structure. The electrons within the ion cores do not contribute significantly to most of the properties of a metal. The remaining n electrons contributed by each atom are not localized on the individual ions but are free to wander throughout the crystal and act as a negatively charged 'glue' that holds together the array of positively charged ion-cores. Very many properties of a metal can be understood from a detailed study of the behaviour of the conduction electrons and, therefore, it is the conduction electrons that we shall be primarily concerned with in this book.

The simplifications referred to in the first paragraph are connected with the fact that because of the regular periodic structure of a crystal there exists a vector quantity **k** that can be used as a good quantum number, to distinguish between the various solutions of Schrödinger's equation for the electrons in a crystal. It is not enough just to obtain the states corresponding to solutions of Schrödinger's equation; it is also necessary to know which of them are occupied, and it is in this connection that the concept of the Fermi surface is useful.

1.2. Bloch's theorem

The atoms or molecules of which a perfect crystalline solid is constituted are arranged in a regular manner. In the language of mathematical crystallography the solid is said to be based on one of the *Bravais lattices*. A lattice is a collection of points arranged in such a way that each lattice point has the same environment in the same orientation. If the crystal were viewed, on a microscopic scale, from any one of these lattice points it would appear exactly the same as if it were viewed from any other lattice point. To obtain complete similarity for the environments of all the lattice points the lattice must, of course, be infinite in extent. In three dimensions there are fourteen essentially different Bravais lattices and they are illustrated in Fig. 1.1. It should be noted that in a crystalline solid there is not necessarily one Bravais lattice point for each atom, neither is a Bravais lattice point necessarily occupied by any atom at all.

One important feature of a Bravais lattice, and therefore also of an ideal crystal, is that it possesses *translational symmetry*. Suppose that we draw three vectors $\mathbf{t}_1$, $\mathbf{t}_2$, and $\mathbf{t}_3$ joining a given point of the Bravais lattice to three of its nearest-neighbour lattice points. Then a bodily movement, or translation, of the whole lattice through any combination of integer multiples of $\mathbf{t}_1$, $\mathbf{t}_2$, and $\mathbf{t}_3$ is a symmetry operation of the lattice, that is, it leaves the lattice in a position that is indistinguishable from its starting position. Thus, if $\mathbf{t}_1$, $\mathbf{t}_2$, and $\mathbf{t}_3$ are given for a particular Bravais lattice, the position vector of any other lattice point of that Bravais lattice can be written as

$$\mathbf{T}_{n_1 n_2 n_3} = n_1\mathbf{t}_1 + n_2\mathbf{t}_2 + n_3\mathbf{t}_3, \tag{1.2.1}$$

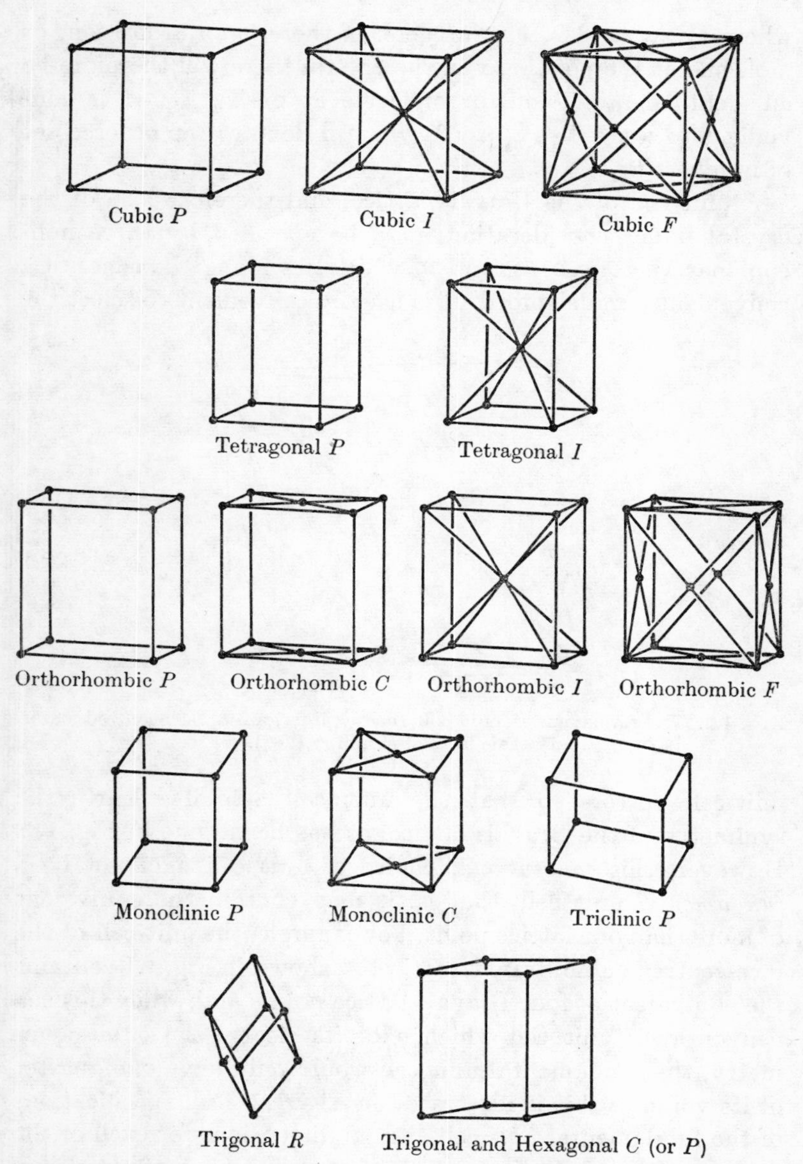

FIG. 1.1. The conventional unit cells of the fourteen Bravais lattices (Phillips 1963).

where n_1, n_2, and n_3 are integers. If there is an atom, ion, or molecule at the position $\mathbf{r}$ in the crystal there will therefore be an identical atom, ion, or molecule at $\mathbf{r}+\mathbf{T}_{n_1n_2n_3}$. It is convenient to write $\mathbf{T}_{n_1n_2n_3}$ as $\mathbf{T}_\mathbf{n}$ where $\mathbf{n}$ denotes the ordered set of numbers (n_1, n_2, n_3).

A unit cell of the Bravais lattice, and therefore also of the crystal under consideration, may be specified by three non-coplanar vectors, $\mathbf{a}$, $\mathbf{b}$, and $\mathbf{c}$, which are along the edges of a convenient parallelepiped. It is usually convenient to select the

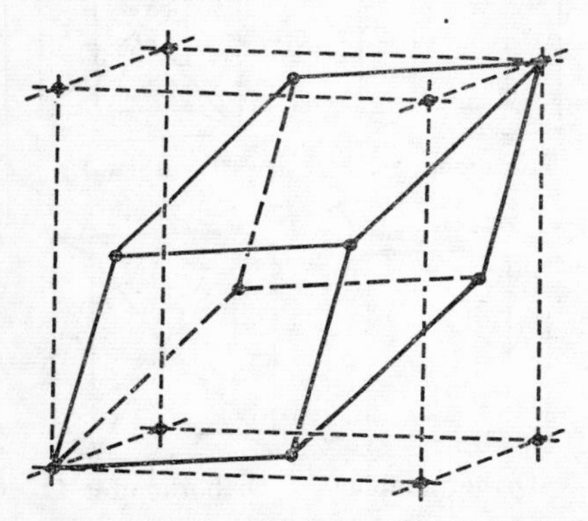

FIG. 1.2. The fundamental unit cell (heavy lines) of a face-centred cubic Bravais lattice. (Phillips 1963).

unit-cell vectors so that the unit cell exhibits clearly the symmetry of the Bravais lattice, as has been done in Fig. 1.1. However, this *conventional* choice of unit cell may not be a *fundamental* unit cell, that is, it may contain the equivalent of more than one lattice point. For example, the unit cell of the face-centred cubic Bravais lattice, F, shown in Fig. 1.1 contains the equivalent of four Bravais lattice points and, while this is a conventional unit cell which exhibits clearly the cubic symmetry, the fundamental unit cell would only have one quarter of its volume; this is illustrated in Fig. 1.2. Full specifications of the fundamental unit cell and its dimensions for each of the fourteen Bravais lattices will be found in Vol. 1 of the *International tables for X-ray crystallography* (Henry and Lonsdale 1965).

A Bravais lattice has been defined as an infinite array of points such that each point has the same environment in the same orientation. We now wish to consider the symmetry operations of a real crystal in which the actual positions of the atoms or molecules are considered. If the fundamental unit cell only contains one atom we may choose the positions of the atoms as the Bravais lattice points. However, if the fundamental unit cell contains more than one atom we must associate a similar configuration of atoms or molecules with each Bravais lattice point, and this configuration of atoms or molecules must possess the symmetry of one of the point groups that is compatible with the point-group symmetry of the Bravais lattice. The crystal will therefore contain both point-group symmetry operations that leave one particular Bravais lattice point unmoved, and also the translational symmetry operations of the Bravais lattice that move all the lattice points through the same displacement. Thus, a general symmetry operation of the crystal will be a compound operation consisting of a point-group operation, which we may call R, followed by a translation through a vector $\mathbf{T}_n$ (see eqn (1.2.1)) of the Bravais lattice. Operations of this type make up one kind of space group called a *symmorphic space group*. This is an example of a set of symmetry operations that satisfy the conditions of the mathematical definition of a group; in this case the group arises by associating with each point of a Bravais lattice one of the point groups compatible with that Bravais lattice. In all there are seventy-three such symmorphic space groups in three dimensions and they can be identified from Vol. 1 of the *International tables for X-ray crystallography* (Henry and Lonsdale 1965).

In a symmorphic space group the point-group operations and the translation operations of the Bravais lattice are essentially separable. In a crystal, however, there may be other more complicated kinds of symmetry elements, namely glide–reflection planes, or screw–rotation axes of symmetry. A glide–reflection operation is a compound operation which consists of a reflection in some plane, m, followed by some translation, specified by a vector **v**. The allowed translations **v** are determined by the condition that two successive performances of such a glide–reflection operation must be equivalent to the effect of a vector $\mathbf{T_n}$ that describes a translation of the Bravais lattice of the crystal. The symmetry operation of a screw–rotation axis is also a compound operation and it consists of a rotation through $(2\pi/f)$ $(f = 1, 2, 3, 4,$ or $6)$ about an axis, followed by a translation **v**. The allowed values of **v** are determined by the fact that if such an operation is performed f times in succession, the result must be equivalent to some translation $\mathbf{T_n}$ of the Bravais lattice of the crystal. A space group that contains at least one glide–reflection plane of symmetry or one screw–rotation axis of symmetry is said to be a *non-symmorphic space group* (or an *asymmorphic space group*). By starting with a symmorphic space group and replacing some or all of the ordinary reflection planes of symmetry by glide–reflection planes and some or all of the ordinary rotation axes of symmetry by screw–rotation axes it is usually possible to derive several non-symmorphic space groups. There are 157 non-symmorphic space groups and their specifications are given in great detail in Vol. 1 of the *International tables for X-ray crystallography* (Henry and Lonsdale 1965).

In studying the behaviour of the electrons in a metal it is necessary sooner or later to make use of quantum mechanics. Leaving aside for the moment any many-body considerations about *quasi-particles* (see Chapter 6), this means that we have to try to solve Schrödinger's equation for each of the conduction electrons in the metal; we then try to obtain a picture of the behaviour of the system of all the conduction electrons in the metal from a synthesis of the behaviour of the individual electrons. This is sometimes called the '*independent particle approximation*'. That is, we assume that it is not only meaningful, but also profitable to describe the behaviour of the system of the conduction electrons in a metal in terms of the behaviour of all the individual electrons under their interactions both with one another and also with other entities in the metal. However, even if one separates the Schrödinger equation for the whole system of N electrons into N equations, one for each of the electrons, the determination of the solutions of these equations will still be a complicated problem in self-consistency; this is because the potential $V(\mathbf{r})$ experienced by a given electron will include contributions from the interactions of that electron with all the other electrons in the crystal. However, if we assume that, in principle at least, $V(\mathbf{r})$ is a physically observable quantity, then we can invoke *Neumann's principle* (Nye 1957; Birss 1964) and assume that $V(\mathbf{r})$ possesses the full translational symmetry of the Bravais lattice of the crystal, that is

$$V(\mathbf{r}) = V(\mathbf{r}+\mathbf{T}_{\mathbf{n}}), \tag{1.2.2}$$

where $\mathbf{T}_{\mathbf{n}}$ is any translation of the Bravais lattice given by eqn (1.2.1). Therefore, in Schrödinger's equation,

$$-\frac{\hbar^2}{2m}\nabla^2\psi(\mathbf{r})+V(\mathbf{r})\psi(\mathbf{r}) = E\psi(\mathbf{r}), \tag{1.2.3}$$

the periodic nature of the potential $V(\mathbf{r})$ will impose conditions on the form of the solutions, $\psi(\mathbf{r})$, of this equation. It is possible to show that the allowed solutions of eqn (1.2.3) take the form

$$\psi_{\mathbf{k}}(\mathbf{r}) = \exp(i\mathbf{k}\,.\,\mathbf{r})u_{\mathbf{k}}(\mathbf{r}), \tag{1.2.4}$$

where $\mathbf{k}$ is a vector that is determined by the basic vectors $\mathbf{t}_1$, $\mathbf{t}_2$, and $\mathbf{t}_3$ of the Bravais lattice of the crystal (see eqn (1.2.1)). $u_{\mathbf{k}}(\mathbf{r})$ is a periodic function with the same periodic properties as $V(\mathbf{r})$, i.e.

$$u_{\mathbf{k}}(\mathbf{r}) = u_{\mathbf{k}}(\mathbf{r}+\mathbf{T}_{\mathbf{n}}). \tag{1.2.5}$$

For each wave function, $\psi_{\mathbf{k}}(\mathbf{r})$, there will be a corresponding energy eigenvalue, $E(\mathbf{k})$, determined by eqn (1.2.3). If $\mathbf{k}$ is regarded as a continuous variable it is possible to show that the corresponding energy $E(\mathbf{k})$ will be a continuous function of $\mathbf{k}$; this statement is sometimes known as *Koopmans' theorem* (Koopmans 1934, Phillips 1961). Suppose that for some wave vector $\mathbf{k}$ there is an energy eigenvalue $E(\mathbf{k})$ that is n-fold degenerate, and that the corresponding n degenerate wave functions are $\psi_1(\mathbf{r})$, $\psi_2(\mathbf{r})$,..., $\psi_n(\mathbf{r})$. If κ_x, κ_y, and κ_z are the components of a small vector $\boldsymbol{\kappa}$ then $\mathbf{k}+\boldsymbol{\kappa}$ will be in the neighbourhood of **k**. Each of the wave functions $\psi_1(\mathbf{r})$, $\psi_2(\mathbf{r}),\ldots,$ $\psi_n(\mathbf{r})$ is a solution of Schrödinger's equation

$$-\frac{\hbar^2}{2m}\nabla^2\psi(\mathbf{r})+V(\mathbf{r})\psi(\mathbf{r}) = E\psi(\mathbf{r}). \tag{1.2.3}$$

If we write

$$\psi_1'(\mathbf{r}) = \exp(\mathrm{i}\boldsymbol{\varkappa}\,.\,\mathbf{r})\psi_1(\mathbf{r}) \tag{1.2.6}$$

then $\psi_1'(\mathbf{r})$ satisfies

$$-\frac{\hbar^2}{2m}\nabla^2\psi_1'(\mathbf{r})+2\frac{\hbar^2}{2m}\mathrm{i}\boldsymbol{\varkappa}\,.\,\boldsymbol{\nabla}\psi_1'(\mathbf{r})+V(\mathbf{r})\psi_1'(\mathbf{r}) = \left(E-\frac{\hbar^2\kappa^2}{2m}\right)\psi_1'(\mathbf{r}). \tag{1.2.7}$$

Since $\boldsymbol{\varkappa}$ is small the second term on the left-hand side of this equation is small and it may therefore be regarded as a perturbation of the original Hamiltonian. If a perturbation calculation is performed one obtains an eigenvalue of the Hamiltonian which is close to $E(\mathbf{k})$ and for which the corresponding wave function has the same translational symmetry as $\psi_1'(\mathbf{r})$, because the whole of the perturbed Hamiltonian in eqn (1.2.7) has the translational symmetry of the Bravais lattice. This shows that the energy $E(\mathbf{k})$ is a continuous function of **k**. Strictly speaking **k** is a discrete variable but the values of **k** are equally spaced and the spacing is very small (see below) so that, for practical purposes, **k** may be regarded as a continuous variable. The dispersion curves of $E(\mathbf{k})$, plotted as a function of **k**, for a given crystal, are referred to as the *band structure* of that crystal.

The statement of eqn (1.2.4) is commonly known as *Bloch's theorem* and its proof is discussed in most reliable books on quantum mechanics. The presence of the $\exp(\mathrm{i}\mathbf{k}\,.\,\mathbf{r})$ factor means that the solution $\psi_{\mathbf{k}}(\mathbf{r})$ has the form of a plane wave that is modulated by the periodic function $u_{\mathbf{k}}(\mathbf{r})$; this is illustrated in Fig. 1.3. **k** is therefore commonly referred to as the wave vector of the solution $\psi_{\mathbf{k}}(\mathbf{r})$. The form of **k** is determined not only by the periodic behaviour of the potential $V(\mathbf{r})$ but also by the boundary conditions that one assumes when solving eqn (1.2.3). Mathematically speaking a Bravais lattice is of

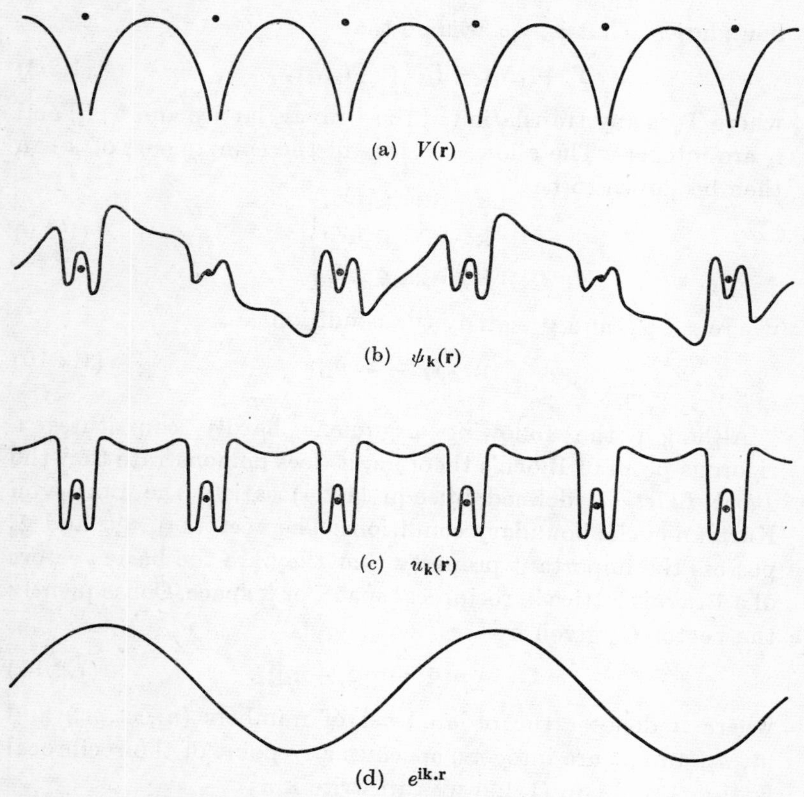

FIG. 1.3. A schematic representation of electronic eigenstates in a crystal. (a) The potential plotted along a row of atoms. (b) A sample eigenstate; the state itself is complex but only the real part is shown. This state can be split into a Bloch function (c), which has the periodicity of the lattice, and (d) a plane wave, the real part of which is shown. (Harrison 1970).

infinite extent in three dimensions, but a real crystal is necessarily only of finite extent. Suppose that the crystal is in the form of a parallelepiped with adjacent edges specified by $N_1\mathbf{t}_1$, $N_2\mathbf{t}_2$, and $N_3\mathbf{t}_3$ where N_1, N_2, and N_3 are (large) integers. It is then possible to apply cyclic (or Born–von Kármán) boundary conditions by assuming that the crystal is surrounded by a regular array of similar parallelepipeds, and requiring that the wave function and its derivative be continuous across the boundaries. That is, we assume that

$$\mathbf{T}_\mathbf{n}+l_1N_1\mathbf{t}_1+l_2N_2\mathbf{t}_2+l_3N_3\mathbf{t}_3 \equiv \mathbf{T}_\mathbf{n} \tag{1.2.8}$$

where $\mathbf{T}_\mathbf{n}$ is any translation of the Bravais lattice and l_1, l_2, and l_3 are integers. The allowed values of the components of $\mathbf{k}$ can then be shown to be

$$\left.\begin{aligned} \mathbf{k}_1 &= m_1\mathbf{g}_1/N_1, \\ \mathbf{k}_2 &= m_2\mathbf{g}_2/N_2, \\ \mathbf{k}_3 &= m_3\mathbf{g}_3/N_3, \end{aligned}\right\} \tag{1.2.9}$$

where $\mathbf{g}_1$, $\mathbf{g}_2$, and $\mathbf{g}_3$ satisfy the condition

$$\mathbf{g}_i \,.\, \mathbf{t}_j = 2\pi\delta_{ij}; \tag{1.2.10}$$

$i, j = 1, 2, 3$.

Although the following argument hardly constitutes a rigorous proof of Bloch's theorem, it does demonstrate that the *Bloch functions* defined by eqn (1.2.4) satisfy the Born–von Kármán cyclic boundary conditions. The vectors $\mathbf{g}_1$, $\mathbf{g}_2$, and $\mathbf{g}_3$ possess the important property that they are the basic vectors of a Bravais lattice in reciprocal space, or $\mathbf{k}$ space. Consequently the vector $\mathbf{G}_\mathbf{n}$ given by

$$\mathbf{G}_\mathbf{n} = n_1\mathbf{g}_1+n_2\mathbf{g}_2+n_3\mathbf{g}_3, \tag{1.2.11}$$

where $\mathbf{n}$ denotes the ordered set of numbers (n_1, n_2, n_3) and n_1, n_2, and n_3 are integers, specifies any point of the reciprocal lattice. From eqn (1.2.9) we can write $\mathbf{k}$ as

$$\mathbf{k} = m_1\mathbf{g}_1/N_1+m_2\mathbf{g}_2/N_2+m_3\mathbf{g}_3/N_3. \tag{1.2.12}$$

The Born–von Kármán cyclic boundary conditions require that if $\mathbf{T}_0$ is given by

$$\mathbf{T}_0 = N_1\mathbf{t}_1+N_2\mathbf{t}_2+N_3\mathbf{t}_3, \tag{1.2.13}$$

then

$$\psi_\mathbf{k}(\mathbf{r}+\mathbf{T}_0) = \psi_\mathbf{k}(\mathbf{r}). \tag{1.2.14}$$

In eqn (1.2.4) we therefore have

$$\begin{aligned} \psi_\mathbf{k}(\mathbf{r}+\mathbf{T}_0) &= \exp\{\mathrm{i}\mathbf{k}\,.\,(\mathbf{r}+\mathbf{T}_0)\}u_\mathbf{k}(\mathbf{r}) \\ &= \exp(\mathrm{i}\mathbf{k}\,.\,\mathbf{r})u_\mathbf{k}(\mathbf{r})\exp\{\mathrm{i}(m_1\mathbf{g}_1/N_1+m_2\mathbf{g}_2/N_2+m_3\mathbf{g}_3/N_3)\times \\ &\quad \times(N_1\mathbf{t}_1+N_2\mathbf{t}_2+N_3\mathbf{t}_3)\} \\ &= \exp(\mathrm{i}\mathbf{k}\,.\,\mathbf{r})u_\mathbf{k}(\mathbf{r})\exp\{\mathrm{i}2\pi(m_1+m_2+m_3)\} \\ &= \exp(\mathrm{i}\mathbf{k}\,.\,\mathbf{r})u_\mathbf{k}(\mathbf{r}), \end{aligned} \tag{1.2.15}$$

since m_1, m_2, and m_3 are integers. Therefore the Bloch functions do satisfy the condition

$$\psi_{\mathbf{k}}(\mathbf{r}+\mathbf{T}_0) = \psi_{\mathbf{k}}(\mathbf{r}), \tag{1.2.16}$$

which is required by the Born–von Kármán cyclic boundary conditions.

1.3. Brillouin zones, band structures, and Fermi surfaces

In the previous section we have noted that in a crystal in which the potential $V(\mathbf{r})$ must be periodic with the periodicity of the Bravais lattice, the wave function of an electron takes the form

$$\psi_{\mathbf{k}}(\mathbf{r}) = \exp(\mathrm{i}\mathbf{k}\cdot\mathbf{r})u_{\mathbf{k}}(\mathbf{r}), \tag{1.2.4}$$

where $u_{\mathbf{k}}(\mathbf{r})$ is a periodic function with the same periodicity as $V(\mathbf{r})$ and $\mathbf{k}$ is a wave vector with components defined by eqn (1.2.9). The basic vectors of the reciprocal space itself are defined by eqn (1.2.10). In this section we shall consider in a little more detail the properties of this reciprocal space.

From eqn (1.2.10) it is possible to obtain explicit expressions for $\mathbf{g}_1$, $\mathbf{g}_2$, and $\mathbf{g}_3$:

$$\left.\begin{aligned}\mathbf{g}_1 &= \frac{2\pi(\mathbf{t}_2 \wedge \mathbf{t}_3)}{\mathbf{t}_1 \cdot (\mathbf{t}_2 \wedge \mathbf{t}_3)},\\ \mathbf{g}_2 &= \frac{2\pi(\mathbf{t}_3 \wedge \mathbf{t}_1)}{\mathbf{t}_2 \cdot (\mathbf{t}_3 \wedge \mathbf{t}_1)},\\ \mathbf{g}_3 &= \frac{2\pi(\mathbf{t}_1 \wedge \mathbf{t}_2)}{\mathbf{t}_3 \cdot (\mathbf{t}_1 \wedge \mathbf{t}_2)},\end{aligned}\right\} \tag{1.3.1}$$

and we have already mentioned that $\mathbf{g}_1$, $\mathbf{g}_2$, and $\mathbf{g}_3$ are the basic vectors of a Bravais lattice. In most cases the Bravais lattice defined by $\mathbf{g}_1$, $\mathbf{g}_2$, and $\mathbf{g}_3$ will be of the same type as the Bravais lattice defined by $\mathbf{t}_1$, $\mathbf{t}_2$, and $\mathbf{t}_3$. Thus if $\mathbf{t}_1$, $\mathbf{t}_2$, and $\mathbf{t}_3$ are the basic vectors of a simple cubic Bravais lattice, it is easy to show that $\mathbf{g}_1$, $\mathbf{g}_2$, and $\mathbf{g}_3$, derived from eqn (1.3.1), form the basic vectors of a simple cubic Bravais lattice in reciprocal space.

The exceptions are that if $\mathbf{t}_1$, $\mathbf{t}_2$, and $\mathbf{t}_3$ describe a face-centred cubic Bravais lattice (F), $\mathbf{g}_1$, $\mathbf{g}_2$, and $\mathbf{g}_3$ will describe a body-centred cubic lattice (I), and vice versa, while if $\mathbf{t}_1$, $\mathbf{t}_2$, and $\mathbf{t}_3$ describe a face-centred orthorhombic Bravais lattice (F), $\mathbf{g}_1$, $\mathbf{g}_2$, and $\mathbf{g}_3$ will describe a body-centred orthorhombic Bravais lattice (I), and vice versa. The reciprocal lattice defined by eqn (1.2.10) or (1.3.1) is identical, apart from a numerical factor, with the reciprocal lattice used by crystallographers in the interpretation of diffraction photographs.

For a crystalline specimen of a metal in the form of a parallelepiped with sides $N_1\mathbf{t}_1$, $N_2\mathbf{t}_2$, and $N_3\mathbf{t}_3$ there are $N_1N_2N_3$ different wave vectors given by equation (1.2.12). All these wave vectors terminate at points in reciprocal space that are contained within the fundamental unit cell of reciprocal space with sides $\mathbf{g}_1$, $\mathbf{g}_2$, and $\mathbf{g}_3$; we may describe this unit cell as the (first) *Brillouin zone*. Suppose we now consider the wave function of an electron with wave vector $\mathbf{k}+\mathbf{G}_n$, where $\mathbf{k}$ is within the Brillouin zone. The wave vector $\mathbf{k}+\mathbf{G}_n$ is then outside the Brillouin zone and $\psi_{\mathbf{k}+\mathbf{G}_n}(\mathbf{r})$ is given by

$$\psi_{\mathbf{k}+\mathbf{G}_n}(\mathbf{r}) = \exp(\mathrm{i}(\mathbf{k}+\mathbf{G}_n)\,.\,\mathbf{r})u_{\mathbf{k}+\mathbf{G}_n}(\mathbf{r}). \tag{1.3.2}$$

By replacing $\mathbf{r}$ by $\mathbf{r}+\mathbf{T}_0$ in this equation we obtain

$$\begin{aligned}\psi_{\mathbf{k}+\mathbf{G}_n}(\mathbf{r}+\mathbf{T}_0) &= \exp\{\mathrm{i}(\mathbf{k}+\mathbf{G}_n)\,.\,(\mathbf{r}+\mathbf{T}_0)\}u_{\mathbf{k}+\mathbf{G}_n}(\mathbf{r})\\ &= \exp\{\mathrm{i}(\mathbf{k}+\mathbf{G}_n)\,.\,\mathbf{r}\}u_{\mathbf{k}+\mathbf{G}_n}(\mathbf{r}),\end{aligned} \tag{1.3.3}$$

that is,

$$\psi_{\mathbf{k}+\mathbf{G}_n}(\mathbf{r}+\mathbf{T}_0) = \psi_{\mathbf{k}+\mathbf{G}_n}(\mathbf{r}), \tag{1.3.4}$$

which means that $\psi_{\mathbf{k}+\mathbf{G}_n}(\mathbf{r})$ also satisfies the Born–von Kármán cyclic boundary conditions. We could rewrite eqn (1.3.2) as

$$\psi_{\mathbf{k}+\mathbf{G}_n}(\mathbf{r}) = \exp(\mathrm{i}\mathbf{k}\,.\,\mathbf{r})\exp(\mathrm{i}\mathbf{G}_n\,.\,\mathbf{r})u_{\mathbf{k}+\mathbf{G}_n}(\mathbf{r}) \tag{1.3.5}$$

or

$$\psi_{\mathbf{k}}^{\mathbf{G}_n}(\mathbf{r}) = \exp(\mathrm{i}\mathbf{k}\,.\,\mathbf{r})u_{\mathbf{k}}^{\mathbf{G}_n}(\mathbf{r}), \tag{1.3.6}$$

where

$$u_{\mathbf{k}}^{\mathbf{G}_n}(\mathbf{r}) = \exp(\mathrm{i}\mathbf{G}_n\,.\,\mathbf{r})u_{\mathbf{k}+\mathbf{G}_n}(\mathbf{r}). \tag{1.3.7}$$

The function $u_{\mathbf{k}}^{\mathbf{G}_n}(\mathbf{r})$ possesses the periodicity of the Bravais lattice since

$$\begin{aligned} u_{\mathbf{k}}^{\mathbf{G}_n}(\mathbf{r}+\mathbf{T}_{n'}) &= \exp\{i\mathbf{G}_n \cdot (\mathbf{r}+\mathbf{T}_{n'})\}u_{\mathbf{k}+\mathbf{G}_n}(\mathbf{r}+\mathbf{T}_{n'}) \\ &= \exp(i\mathbf{G}_n \cdot \mathbf{r})u_{\mathbf{k}+\mathbf{G}_n}(\mathbf{r}) \\ &= u_{\mathbf{k}}^{\mathbf{G}_n}(\mathbf{r}). \end{aligned} \tag{1.3.8}$$

We may therefore regard $\psi_{\mathbf{k}}^{\mathbf{G}_n}(\mathbf{r})$ not as the wave function with wave vector $\mathbf{k}+\mathbf{G}_n$, but as a second solution of Schrödinger's equation with wave vector $\mathbf{k}$, since $\psi_{\mathbf{k}}^{\mathbf{G}_n}(\mathbf{r})$ has, from eqn (1.3.6), exactly the same symmetry properties as $\psi_{\mathbf{k}}(\mathbf{r})$ with regard to the translation group of the Bravais lattice of the crystal. $\psi_{\mathbf{k}}^{\mathbf{G}_n}(\mathbf{r})$ will lead to a second energy eigenvalue that will in general be different from the eigenvalue corresponding to $\psi_{\mathbf{k}}(\mathbf{r})$. It is not necessary therefore to allow an electron's wave vector to take all values in reciprocal space, because all the solutions of Schrödinger's equation (1.2.3) can be found by considering only those wave vectors $\mathbf{k}$ within the Brillouin zone which is just one unit cell of the reciprocal lattice. There will be infinitely many solutions corresponding to each wave vector $\mathbf{k}$ within the Brillouin zone, as we shall see later for example, when discussing the free electron energy bands in Section 1.5.

For any given Bravais lattice, or for any reciprocal lattice, there are many different ways of choosing a unit cell. Even if one restricts oneself to fundamental unit cells, that is to unit cells that contain only the volume associated with one lattice point, there are many different choices of shape for the unit cell. One particularly special unit cell is the *Wigner–Seitz unit cell* which has its centre at one of the points of the reciprocal lattice and which is bounded by planes that bisect the vectors joining this lattice point to its nearest-neighbour lattice points and, possibly, its next-nearest-neighbour lattice points too. The equation of the plane that is the perpendicular bisector of $\mathbf{G}_n$ is

$$\mathbf{k} \cdot \mathbf{G}_n = \tfrac{1}{2}\,|\mathbf{G}_n|^2. \tag{1.3.9}$$

Each face of the Brillouin zone is thus characterized by a reciprocal lattice vector $\mathbf{G}_n$ that is normal to the face. The reason for the choice of the Wigner–Seitz unit cell, rather than some other unit cell of reciprocal space, is that it exhibits the full point-group symmetry of the reciprocal lattice. The (*first*) *Brillouin zone* can be defined as

The set of all **k** *vectors with one of the reciprocal lattice points as origin and having the property that no vector of shorter length can be reached from any of them by adding translation vectors* $\mathbf{G}_n$ *of the reciprocal lattice.*

This means that if any wave vector **k** is within the Brillouin zone, then

$$|\mathbf{k}| \leqslant |\mathbf{k}+\mathbf{G}_n|, \tag{1.3.10}$$

where $\mathbf{G}_n$ is any vector of the reciprocal lattice (see eqn (1.2.11)). The volume of the Brillouin zone of a crystal is simply the volume of a unit cell of the reciprocal lattice, namely

$$\mathbf{g}_1 \cdot (\mathbf{g}_2 \wedge \mathbf{g}_3),$$

which is, of course, the same as the volume of the parallelepiped unit cell with adjacent sides $\mathbf{g}_1$, $\mathbf{g}_2$, and $\mathbf{g}_3$. It is only a matter of convention and convenience that the Wigner–Seitz unit cell of reciprocal space is chosen as the Brillouin zone of a crystal; we should emphasize that it is possible to use any unit cell of the reciprocal lattice so long as its volume is equal to $\mathbf{g}_1 \cdot (\mathbf{g}_2 \wedge \mathbf{g}_3)$ (see, for example, Bradley and Cracknell 1972). Indeed, for crystals of very low symmetry, particularly monoclinic and triclinic crystals, eqn (1.3.9) is very difficult to apply in practice for drawing the Brillouin zone, and even when the Brillouin zone has been drawn on a piece of paper it is very difficult to visualize in three dimensions. Consequently, to define the Brillouin zone for monoclinic and triclinic crystals it is better simply to use the fundamental unit cell of reciprocal space, which is the parallelepiped with edges $\mathbf{g}_1$, $\mathbf{g}_2$, and $\mathbf{g}_3$. The Brillouin zones for all the fourteen Bravais lattices are illustrated in Fig. 1.4.

From eqn (1.2.12) we can see that the number of different vectors **k** within the parallelepiped with edges $\mathbf{g}_1$, $\mathbf{g}_2$, and $\mathbf{g}_3$ (and therefore within the Brillouin zone) is $N_1N_2N_3$, which is just the number of unit cells within the sample of the crystal that we have considered and to which we applied the Born–von

Kármán cyclic boundary conditions. Therefore, the larger our sample of a metallic crystal, the more available states there are, because of the larger number of allowed wave vectors **k** that exist within the Brillouin zone. However, there will be correspondingly more electrons available to occupy these states, so that the fraction of the volume of the Brillouin zone that corresponds to states that are actually occupied by electrons is independent of the size of the sample that we choose to consider. We can study this a little more formally. Suppose that **k** is a given wave vector within the Brillouin zone and that $\mathbf{G}_n$ is some vector of the reciprocal lattice (see eqn (1.2.11)). At the beginning of this section we showed that if $\psi_{\mathbf{k}}(\mathbf{r})$ is a solution of Schrödinger's equation with wave vector **k** then $\psi_{\mathbf{k}}^{\mathbf{G}_n}(\mathbf{r})$, as defined by eqn (1.3.6), can be regarded as a second solution of Schrödinger's equation with wave vector **k**. $\psi_{\mathbf{k}}^{\mathbf{G}_n}(\mathbf{r})$ will lead to an energy eigenvalue that, in general, will be different from that corresponding to $\psi_{\mathbf{k}}(\mathbf{r})$. Therefore, by considering all possible reciprocal lattice vectors $\mathbf{G}_n$ and a given value of **k**, an infinite set of energies will be obtained. If one considers different values of **k** the positions of these energy levels will vary and a set of bands will be obtained. An energy band is therefore an energy level $E(\mathbf{k})$ which, if **k** is regarded as a continuous variable, is a continuous function of **k**. At any given value of **k** in the Brillouin zone there will be a large number of such energy levels and as **k** varies, the relative positions of these energy levels will change too. It is customary to label the energy bands that have energies $E_1(\mathbf{k}), E_2(\mathbf{k}), E_3(\mathbf{k}),\ldots$, with the integers 1, 2, 3,..., where the labels are ascribed in ascending order as the energy increases, i.e. if $E_i(\mathbf{k}) < E_j(\mathbf{k})$ then $i < j$. As **k** varies then it is possible for the energy bands to cross, but when two bands cross we still preserve this convention on both sides of the crossing.

Since **k** is a vector in a three-dimensional space the complete pictorial representation of the band structure of a crystal would require the use of a four-dimensional space. There are two procedures that may be adopted to overcome this difficulty. One can select particular lines in reciprocal space and for each

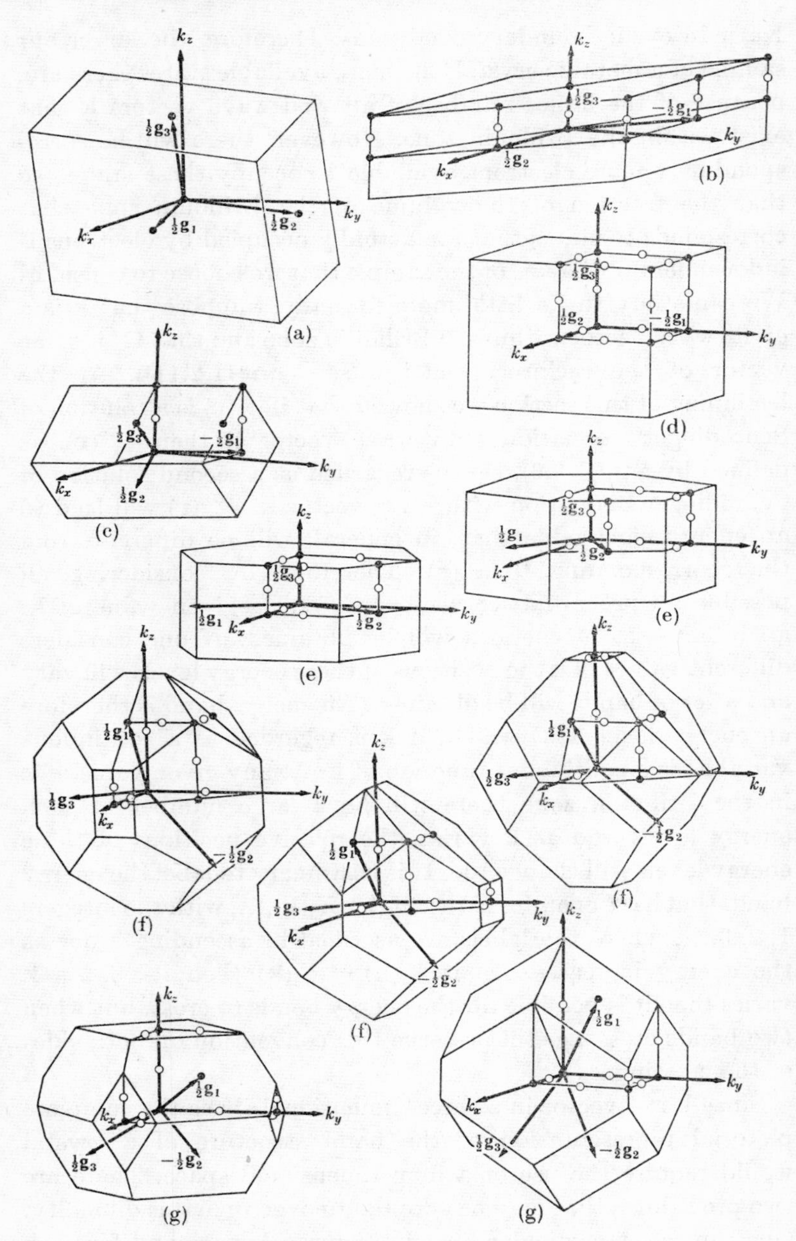

k_z
$\frac{1}{2}\mathbf{g}_3$
k_x
$\frac{1}{2}\mathbf{g}_1$
$\frac{1}{2}\mathbf{g}_2$
k_y
(a)
k_z
$\frac{1}{2}\mathbf{g}_3$
$-\frac{1}{2}\mathbf{g}_1$
k_y
k_x
$\frac{1}{2}\mathbf{g}_2$
(b)
k_z
$\frac{1}{2}\mathbf{g}_3$
$-\frac{1}{2}\mathbf{g}_1$
k_y
k_x
$\frac{1}{2}\mathbf{g}_2$
(c)
k_z
$\frac{1}{2}\mathbf{g}_3$
$\frac{1}{2}\mathbf{g}_2$
$-\frac{1}{2}\mathbf{g}_1$
k_y
k_x
(d)
k_z
$\frac{1}{2}\mathbf{g}_3$
$\frac{1}{2}\mathbf{g}_1$
$\frac{1}{2}\mathbf{g}_2$
k_y
k_x
(e)
k_z
$\frac{1}{2}\mathbf{g}_3$
$\frac{1}{2}\mathbf{g}_1$
k_x
$\frac{1}{2}\mathbf{g}_2$
k_y
(e)
k_z
$\frac{1}{2}\mathbf{g}_1$
$\frac{1}{2}\mathbf{g}_3$
k_x
k_y
$-\frac{1}{2}\mathbf{g}_2$
(f)
k_z
$\frac{1}{2}\mathbf{g}_1$
$\frac{1}{2}\mathbf{g}_3$
k_x
k_y
$-\frac{1}{2}\mathbf{g}_2$
(f)
k_z
$\frac{1}{2}\mathbf{g}_1$
$\frac{1}{2}\mathbf{g}_3$
k_x
k_y
$-\frac{1}{2}\mathbf{g}_2$
(f)
k_z
$\frac{1}{2}\mathbf{g}_1$
k_x
k_y
$\frac{1}{2}\mathbf{g}_3$
$-\frac{1}{2}\mathbf{g}_2$
(g)
k_z
$\frac{1}{2}\mathbf{g}_1$
k_y
k_x
$\frac{1}{2}\mathbf{g}_3$
$-\frac{1}{2}\mathbf{g}_2$
(g)

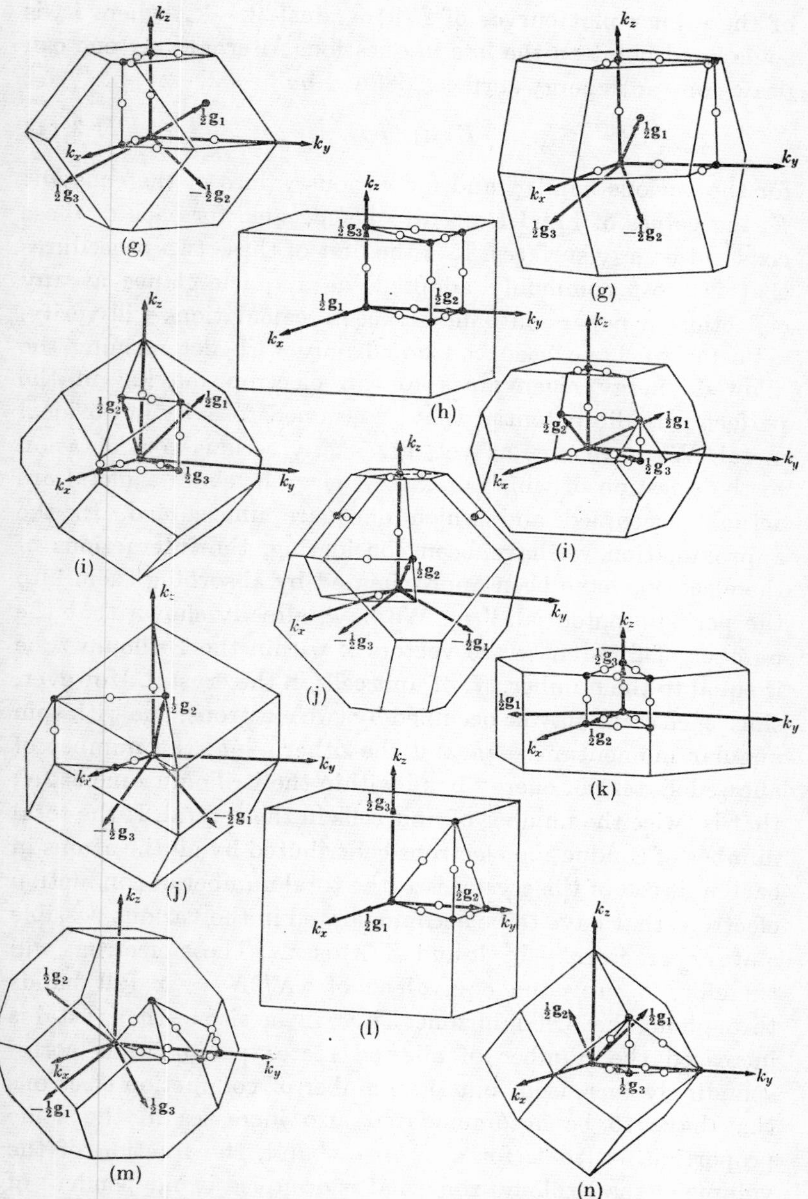

FIG. 1.4. The Brillouin zones for the fourteen crystal systems: (a) triclinic P, (b) monoclinic, P, (c) monoclinic, B or C, (d) orthorhombic, P, (e) orthorhombic, C, (f) orthorhombic, I, (g) orthorhombic, F, (h) tetragonal, P, (i) tetragonal, I, (j) trigonal, R, (k) hexagonal, P, (l) cubic, P, (m) cubic, F, (n) cubic, I. (After Bradley and Cracknell 1972.)

of these lines plot curves of $E(\mathbf{k})$ against $|\mathbf{k}-\mathbf{k}_0|$ where $\mathbf{k}_0$ is some fixed point on the line in question. Alternatively one can draw constant energy surfaces defined by

$$E_j(\mathbf{k}) = E, \tag{1.3.11}$$

for the various bands j and for various values of the constant E; the values of j and E need to be specified for each of these constant energy surfaces. It is the first of these two procedures that is most commonly adopted, as a quick glance at any collection of papers on band structure calculations will verify.

So far we have been concerned only with determining the allowed energy eigenvalues for an electron moving in the perfect periodic potential that is assumed to exist in an ideal metal. We have not, as yet, given any serious consideration to the question of which of these energy levels in a metal are actually occupied and which ones are unoccupied. In the approximation we have been considering, the interactions of the electrons have been approximated by absorbing them into the periodic potential $V(\mathbf{r})$. We have already shown that the number of different wave vectors $\mathbf{k}$ within the Brillouin zone is equal to the number, N, of unit cells in the crystal. However, since each state may be occupied by two electrons, one with spin angular momentum $+\frac{1}{2}\hbar$ and the other $-\frac{1}{2}\hbar$, the number of allowed states per energy band within the Brillouin zone is $2N$, that is twice the number of unit cells in the crystal. If the total number of conduction electrons contributed by all the atoms in each unit cell of the crystal is x, the total number of conduction electrons that have to be accommodated in the various possible states characterized by $\mathbf{k}$ and $E_j(\mathbf{k})$ is xN. These electrons will therefore occupy the equivalent of $xN/2N = \frac{1}{2}x$ full bands throughout the Brillouin zone. Thus, if the size of the crystal is increased the number of allowed states per band is correspondingly increased, but the number of conduction electrons that have to be accommodated also increases in the same proportion, so that, for any given band, the fraction of the volume of the Brillouin zone that is occupied is independent of the size of the crystal.

Of the $2N$ states that exist in any given band for a metal, the ones that will actually be occupied will be determined by the Fermi–Dirac distribution. The probability that a state with wave vector $\mathbf{k}$ in the band $E_j(\mathbf{k})$ will be occupied at temperature T is given by

$$f(E_j(\mathbf{k})) = [\exp\{(E_j(\mathbf{k})-E_\mathrm{F})/kT\}+1]^{-1} \qquad (1.3.12)$$

where E_F is the *Fermi energy* and k is Boltzmann's constant. We have already mentioned that we can regard $\mathbf{k}$ as a continuous variable for practical purposes, and that $E_j(\mathbf{k})$ is a continuous function of $\mathbf{k}$. If the temperature were reduced to absolute zero all those parts of the band $E_j(\mathbf{k})$ for which $E_j(\mathbf{k}) < E_\mathrm{F}$ will be occupied, and all those parts for which $E_j(\mathbf{k}) > E_\mathrm{F}$ will be unoccupied. For non-zero values of T there will be some overspill of electrons from states with energies less than E_F into states with energies greater than E_F; however, E_F still, roughly speaking, marks the division between occupied and unoccupied states, because typical values of E_F in a metal ($\sim$5 eV) are considerably greater than typical values of kT, which would be, for example, $\sim\frac{1}{40}$ eV at room temperature.

If one thinks in terms of constant energy surfaces (see eqn (1.3.11)) the particular constant energy surface corresponding to

$$E_j(\mathbf{k}) = E_\mathrm{F} \qquad (1.3.13)$$

can be regarded as the surface that marks the boundary between the occupied and unoccupied parts of the band j. Although this statement is only rigorously true at $T = 0$ it is still a very good approximation in practice for non-zero values of T because, as we have just mentioned, it is generally the case that $kT \ll E_\mathrm{F}$. The constant energy surface defined by eqn (1.3.13) is called the *Fermi surface*, and we see that its importance lies in the fact that it indicates which wave vectors $\mathbf{k}$ in the Brillouin zone correspond to occupied states and which wave vectors correspond to unoccupied states. We should emphasize that the Fermi surface is not a surface in real space but in an abstract space, reciprocal space. In a real three-dimensional crystal with a complicated band structure there may be several bands that are only partially full so that in such a case there will be different sheets of the Fermi surface for each of the partially filled bands. The Fermi surface in any given band j may consist of a single closed surface or of several isolated closed surfaces; in certain cases these closed surfaces may take a particularly simple form such as a sphere or an ellipsoid. However, the

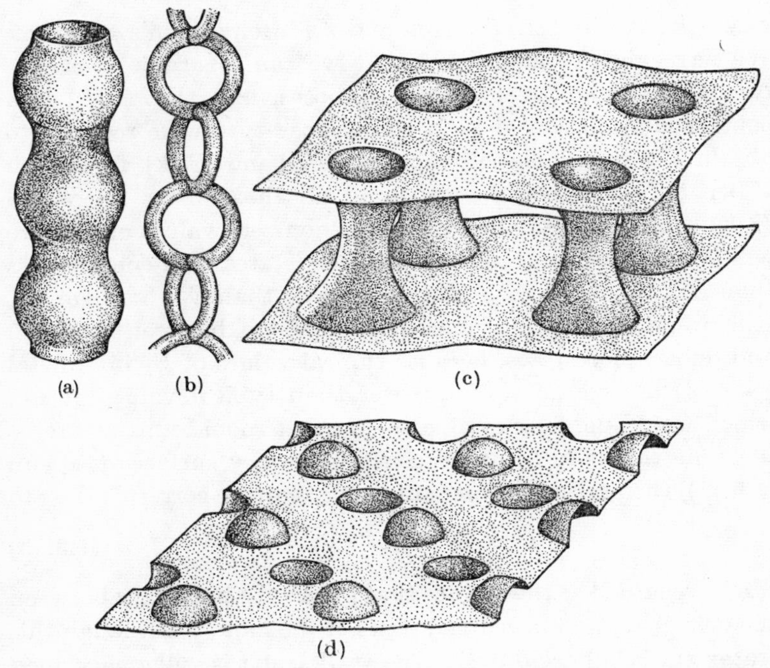

Fig. 1.5. Illustrations of surfaces with different types of connectivity (Lifshitz and Kaganov 1959).

Fermi surface need not necessarily be a closed surface. Because the Brillouin zone is only one unit cell of reciprocal space, each closed sheet of Fermi surface in the Brillouin zone will be repeated in all the other unit cells of reciprocal space. If the Fermi surface intersects the Brillouin zone boundary it may not be a closed surface at all, but some multiply-connected surface that passes through an infinite succession of unit cells in the reciprocal lattice (see Fig. 1.5). We shall see in Chapter 3 that the behaviour of certain of the physical properties of a metal are very substantially affected by the connectivity of the Fermi surface. It is important in studying the topology of the Fermi surface of a given metal to ensure that the correct crystal structure is being considered; this is because the shape of the Brillouin zone depends on the crystal structure of the metal. For various experimental reasons it is usually necessary to use low temperatures in experiments for determining the shape of the Fermi surface (see Chapter 3). However, the importance of the Fermi surface lies in helping to explain the various properties of a metal, not only at low temperatures but very often also at high temperatures as well, that is, so long as kT remains smaller than E_{F}. If the metal undergoes a phase transition it may be impossible to determine directly by experiment the shape of the Fermi surface for the high temperature phase. This has proved to be a serious difficulty in the way of determining the shape of the Fermi surface for one or two metals, for example lithium, which has a phase transition at 78 K. Throughout this book we shall nearly always assume that, provided no phase transition occurs, the band structure $E_j(\mathbf{k})$ of a metal is independent of temperature, and that any effects connected with variations in temperature are associated with changes in the occupation of these bands, according to the Fermi–Dirac distribution. Direct calculations of the effect of temperature on the band structure and Fermi surface of beryllium showed that the changes in the dimensions of the Fermi surface of beryllium are indeed very small in the range 0–300 K (Tripp 1970). It is only when the Fermi surface encloses very small pockets of carriers that any temperature dependence of the band structure, apart from simple scaling due to changes in the lattice constants as a result of thermal expansion, becomes significant (see for example, the case of zinc, p. 69).

If an electron in free space is subjected to a force such as, for example, that due to an electric field, its motion can be described by using Newton's second law of motion,

$$F = m\mathrm{d}^2x/\mathrm{d}t^2.$$

However, the response of an electron in a metal to an external force is modified because of the interactions between that electron and the rest of the metal. It is shown in many general textbooks on solid state physics that it is possible to retain the use of Newton's second law of motion for a conduction electron under the influence of an external force provided that the free-electron mass is replaced by an effective mass tensor $m_{ij}(\mathbf{k})$ defined by

$$\frac{1}{m_{ij}(\mathbf{k})} = \frac{1}{\hbar^2} \frac{\partial^2 E_p(\mathbf{k})}{\partial k_i \, \partial k_j} \tag{1.3.14}$$

for a given band p. The physical reason for this form of the definition can be seen by studying the form of the second-order term

$$\tfrac{1}{2} \sum_i \sum_j (\mathbf{k}_i - \mathbf{k}_{i0}) \cdot (\mathbf{k}_j - \mathbf{k}_{j0}) \frac{\partial^2 E_p(\mathbf{k})}{\partial k_i \, \partial k_j}$$

$$= \tfrac{1}{2} \hbar^2 \sum_i \sum_j \frac{1}{m_{ij}(\mathbf{k})} (\mathbf{k}_i - \mathbf{k}_{i0}) \cdot (\mathbf{k}_j - \mathbf{k}_{j0}) \tag{1.3.15}$$

in the Taylor expansion of $E_p(\mathbf{k})$ about a fixed point $\mathbf{k}_0$, which reduces to the expression $\hbar^2 \, |\mathbf{k}|^2/2m$ for the kinetic energy of a moving electron in the free-electron case.

Since $E_p(\mathbf{k})$ generally is not a simple function of $\mathbf{k}$, it is often difficult to determine a precise expression for $m_{ij}(\mathbf{k})$ which will also, of course, be a function of $\mathbf{k}$. It is possible to relate $m_{ij}(\mathbf{k})$ to the curvature of the energy bands $E_p(\mathbf{k})$, or to the curvature of the constant-energy surfaces defined by $E_p(\mathbf{k}) = E$. Suppose that we consider the variation of the energy $E_p(\mathbf{k})$ as a function of $\mathbf{k}$ along a straight line whose direction is defined by a unit vector $\mathbf{m}$, so that $\mathbf{k} = \mathbf{k}_0 + \mathbf{m}k$. The curvature of the band $E_p(\mathbf{k}_0 + \mathbf{m}k)$ is given by

$$\frac{1}{\rho(\mathbf{k})} = \frac{\hbar^2}{m_{\text{av}}(\mathbf{k})} \cdot \frac{1}{(1 + \hbar^2 \mathbf{v}(\mathbf{k})^2 \cos^2\alpha)}, \tag{1.3.16}$$

where $\mathbf{v}(\mathbf{k})$ is the electron velocity defined by

$$\mathbf{v}(\mathbf{k}) = (1/\hbar) \nabla_{\mathbf{k}} E_p(\mathbf{k}); \tag{1.3.17}$$

$m_{av}(\mathbf{k})$ is defined by

$$1/m_{av}(\mathbf{k}) = \mathbf{m}_i\mathbf{m}_j/m_{ij}(\mathbf{k}), \qquad (1.3.18)$$

and α is the angle between $\mathbf{v}(\mathbf{k})$ and $\mathbf{m}$. The relationship between $m_{ij}(\mathbf{k})$ and the curvature of the Fermi surface is rather more complicated (for details see Jan 1968). The effective mass defined by eqn (1.3.14) is usually called the *dynamical effective mass*. However, there are other effective masses which one may encounter; these include, for example, the *cyclotron mass* (see Section 3.6) and the *thermal effective mass* (see Section 3.2.2). The term 'effective mass' is often used somewhat carelessly in the literature and one has to remember that, for a given metal, different kinds of effective masses defined from different phenomena are not necessarily equivalent.

It may be regarded as self-evident that the conduction electrons in a metal obey Fermi–Dirac statistics, as we have assumed in eqn (1.3.12). However, since there exist phenomena such as superconductivity in which simple ideas of the electrons in a metal as a collection of independent fermions do not apply, it is perhaps appropriate to enquire more closely into the question of the relevance of the Fermi–Dirac distribution. Of course, although it is perhaps only rather indirect evidence, it is the existence of the vast body of experimental work that has been successfully described in terms of the concept of the Fermi surface, some of which will be discussed later in this book, which is perhaps the most convincing evidence for the relevance of the Fermi–Dirac distribution to the conduction electrons in a metal. The possibility of obtaining a rather more direct experimental verification of the Fermi–Dirac distribution of the conduction electrons in a metal has been explored by Shapira and Neuringer (1967) in their study of the line shapes of the 'giant' quantum oscillations in the ultrasonic attenuation in gallium.

The concept of the Fermi surface was introduced in the very early days after the introduction of quantum mechanics to the study of the behaviour of electrons in metals; however, its full significance was not appreciated for a long time, principally because it proved to be difficult to determine the Fermi surface, either theoretically or experimentally, for most metals. It was shown by Jones (1934) that for the semi-metal bismuth, one band is very nearly full with only very small pockets of holes, while the next band is almost empty and contains only very small pockets of electrons. The unit cell of bismuth contains an even number of electrons so that the material is *compensated*, that is it has equal numbers of electrons and holes in the small pockets of carriers in the two partially-full bands (see Section 4.6). Jones was able to estimate the sizes of the pockets of carriers in bismuth by studying the electrical resistivity as a function of the concentration of tin as an impurity. During the following twenty-five years a little information was obtained about the Fermi surfaces of a few other metals (mainly the alkali metals and bismuth) from experimental work on quantum oscillations in the diamagnetic susceptibility—the *de Haas–van Alphen effect*—and in the electrical resistance—the *Schubnikow–de Haas effect*—and from one or two of the other experimental methods that will be described in Chapter 3. One of the chief difficulties experienced on the experimental side was in the preparation of single crystal specimens with a sufficiently high degree of purity to enable the required 'effects' to be observed. On the theoretical side, the size and speed of the computing machinery that was available was completely inadequate for the performance of band structure calculations that would be sufficiently accurate to make sensible predictions of the Fermi surfaces of most metals. Towards the end of the 1950s both the experimental and the theoretical situation improved significantly and by 1960, when an important conference on the subject was held (Harrison and Webb 1960), some definite facts about the Fermi surfaces of a number of metals had begun to emerge. The determination of the Fermi surface of copper by Pippard (1957*b*), using the anomalous skin effect (see Sections 3.4 and 5.2), marked the beginning of this period of expansion. Because the anomalous skin effect does not enable one to make direct deductions about features of the shape of the Fermi surface,

Pippard had to guess a form for the Fermi surface of copper and calculate the anisotropy of the surface impedance for comparison with the results of the experimental measurements. Adjustments were then made to the assumed Fermi surface until the best fit was obtained. Although this procedure was successful for copper it is unlikely that it would be equally successful for a metal with a more complicated Fermi surface. Since about 1960, as will be seen from Chapters 4 and 5, a very large amount of experimental and theoretical work has been devoted to the problem of the determination of the Fermi surfaces of the metallic elements, with considerable success. However, it has to be remembered that the determination of the shape of the Fermi surface of a metal is not, or should not be, an end in itself. The determination of the shape of the Fermi surface of a given metal should be seen as one important step towards obtaining a better understanding of all those properties of the metal which depend in some way on the behaviour of the conduction electrons in the metal (see Chapters 3 and 6).

The discussion that we have given has been concerned entirely with electrons moving in crystalline solids, but the argument that was based on assuming that the potential $V(\mathbf{r})$ in Schrödinger's equation possesses the periodicity of the Bravais lattice, can be adapted to various other particles or quasi-particles in a crystal. It can be shown in a similar manner that a phonon, which is a quantum of lattice-vibrational energy in a crystal, or a magnon, which is a quantum of spin-wave energy in a magnetic crystal, will have a wave function that is characterized by a wave vector in reciprocal space defined by eqn (1.2.10). Once again, all possible solutions can be obtained by considering the wave vectors that lie within the first Brillouin zone of the crystal. Some care may be necessary when considering magnons because where a crystal can exist in a magnetic form and in a non-magnetic form, the Bravais lattice may be different in the two cases, even when there has been no actual displacements of any of the atoms in the crystal during the transition to the magnetic phase. However, phonons and magnons obey Bose–Einstein statistics rather than Fermi–Dirac statistics and the concept of the Fermi surface does not apply to such particles or quasi-particles. These particles do not have to satisfy the Pauli exclusion principle and the number of phonons or magnons present in a crystal will vary with temperature.

Before considering, in the next two chapters, the various theoretical and experimental methods that are available for the determination of the Fermi surface of a metal, we shall devote the rest of this chapter to one or two other general matters. These include *group theory*, which is capable of simplifying the calculation of the band structure and Fermi surface of a metal, and the so-called *free-electron model* (or sometimes the *empty-lattice model*) which is capable, for some metals at least, of giving a good first approximation to the band structure and Fermi surface of a real metal.

1.4. Group theory

Group theory has been very widely used in connection with the study of band structures and Fermi surfaces. It is, therefore, appropriate that we should, at this juncture, describe the nature of the information that can be obtained from the application of group theory to the study of the conduction electrons in a metal without, however, describing too many of the details. We assume that the reader is familiar with the basic concepts of the theory of finite groups and their representation by groups of unitary matrices, on which subject there are numerous books available (Heine 1960; Tinkham 1964; Cracknell 1968; Streitwolf 1971). The use of group theory is not essential for the study of energy bands and Fermi surfaces in metals but it has been widely used because of the simplifications it introduces in many numerical calculations (Koster 1957; Nussbaum 1966). The labels that are often used for the various electron energy bands in a metal are group-theoretical in origin. However, it is possible for the reader to omit this section without any very grave loss of comprehension of the remainder of this book.

There are two principal uses of group theory in connection with the study of band structures and Fermi surfaces. First, it is possible to identify certain essential degeneracies in the band structure of a given metal that exist as a consequence of the symmetry of the structure of the metal irrespective of the value of the potential $V(\mathbf{r})$. It is also possible to predict any lifting of these degeneracies that may occur if the effect of spin–orbit coupling is included. Secondly, for many years it was not possible, with the computers that were then available, to handle the large series expansions and determinants involved in calculations of $E(\mathbf{k})$ for arbitrary wave vectors $\mathbf{k}$ in the Brillouin zone. However, these calculations can be simplified group-theoretically for wave vectors $\mathbf{k}$ that occupy certain special positions in the Brillouin zone.

The investigation of the symmetry properties of the wave function of an electron (or of a phonon or magnon or any other particle or quasi-particle) in a crystal was initiated by Bouckaert, Smoluchowski, and Wigner (1936). The basic theorem behind the application of group theory to quantum mechanics, which is commonly called *Wigner's theorem*, states that

if the Hamiltonian operator, $\mathscr{H}$, of a quantum-mechanical system possesses a group, **G**, *of symmetry operations, then each of the eigenfunctions ψ of $\mathscr{H}$ must belong to (that is, must transform according to) one of the irreducible representations of* **G**.

For an atom or molecule the eigenvalues of $\mathscr{H}$ are well separated, but for a crystal, as we have seen in Sections 1.2 and 1.3, the eigenvalues are so close together that they form a quasi-continuous function of $\mathbf{k}$. This quasi-continuous function $E_j(\mathbf{k})$, the energy eigenvalue as a function of $\mathbf{k}$, is what we have called an *energy band*. We have in fact already ensured that the wave function $\psi_{\mathbf{k}}(\mathbf{r})$ belongs to one of the representations of the translational group of the Bravais lattice of the crystal by taking $\psi_{\mathbf{k}}(\mathbf{r})$ to be of the form

$$\psi_{\mathbf{k}}(\mathbf{r}) = \exp(i\mathbf{k} \cdot \mathbf{r})u_{\mathbf{k}}(\mathbf{r}). \qquad (1.2.4)$$

However, we have not so far taken any steps to ensure that

$\psi_{\mathbf{k}}(\mathbf{r})$ has the correct transformation properties under the rotation and reflection symmetry operations of the crystal, so that it belongs to one of the irreducible representations of the space group, **G**, of the crystal.

A symmetry operation R of a point group can be represented by a 3 by 3 matrix, **R**, giving the action of that operation on the vector $\mathbf{r}(=(x, y, z))$ where the symmetry operation moves the point specified by $\mathbf{r}$ to another position specified by $\mathbf{r}'$, where

$$\mathbf{r}' = \mathbf{R}\mathbf{r}. \tag{1.4.1}$$

Since a translation by a vector $\mathbf{v}$ sends the point specified by $\mathbf{r}$ to the point $\mathbf{r}+\mathbf{v}$ a general symmetry operation of any space group may be represented by R and a vector $\mathbf{v}$, where $\mathbf{v}$ may be either a vector of a glide–reflection or screw–rotation operation or a vector $\mathbf{T}_{\mathbf{n}}$ of the translational group of the Bravais lattice. It is convenient to write these in one symbol $\{R \mid \mathbf{v}\}$ which is called a *Seitz space-group symbol* (Seitz 1936). It is then fairly easy to write down the rule which is the rule for the group multiplication of these symbols. If $\{R_1 \mid \mathbf{v}_1\}$ acts on $\mathbf{r}$ to produce $\mathbf{r}'$ then

$$\{R_1 \mid \mathbf{v}_1\}\mathbf{r} = \mathbf{r}' = \mathbf{R}_1\mathbf{r}+\mathbf{v}_1 \tag{1.4.2}$$

and therefore

$$\begin{aligned}\{R_2 \mid \mathbf{v}_2\}\{R_1 \mid \mathbf{v}_1\}\mathbf{r} &= \{R_2 \mid \mathbf{v}_2\}\mathbf{r}' = \mathbf{R}_2\mathbf{r}'+\mathbf{v}_2\\ &= \mathbf{R}_2(\mathbf{R}_1\mathbf{r}+\mathbf{v}_1)+\mathbf{v}_2\\ &= \mathbf{R}_2\mathbf{R}_1\mathbf{r}+(\mathbf{R}_2\mathbf{v}_1+\mathbf{v}_2).\end{aligned} \tag{1.4.3}$$

Thus the multiplication rule is

$$\{R_2 \mid \mathbf{v}_2\}\{R_1 \mid \mathbf{v}_1\} = \{R_2R_1 \mid \mathbf{v}_2+R_2\mathbf{v}_1\}. \tag{1.4.4}$$

Suppose that **T** is the group of the translational symmetry operations of a crystal, then the elements of the group **T** are given by eqn (1.2.1):

$$\mathbf{T}_{n_1n_2n_3} = n_1\mathbf{t}_1+n_2\mathbf{t}_2+n_3\mathbf{t}_3, \tag{1.2.1}$$

where $\mathbf{t}_1$, $\mathbf{t}_2$, and $\mathbf{t}_3$ are basic vectors of the Bravais lattice and we write $\mathbf{T}_{n_1n_2n_3}$ as $\mathbf{T}_{\mathbf{n}}$ for convenience. Then it is possible to express a space group **G** as an expansion in terms of left coset representatives of $\mathbf{T}$,

$$\mathbf{G} = \{E \mid \mathbf{0}\}\mathbf{T} + \{R_1 \mid \mathbf{v}_1\}\mathbf{T} + \{R_2 \mid \mathbf{v}_2\}\mathbf{T} + \ldots + \{R_h \mid \mathbf{v}_h\}\mathbf{T}, \quad (1.4.5)$$

where the coset representatives are Seitz space-group symbols and E is the identity operation. If $\mathbf{G}$ is a symmorphic space group it will be possible to choose the coset representatives so that $\mathbf{v}_1 = \mathbf{v}_2 = \ldots = \mathbf{v}_h = \mathbf{0}$. For each wave vector, $\mathbf{k}$, in the Brillouin zone we now define $\mathbf{G}^{\mathbf{k}}$, *the group of the wave vector* $\mathbf{k}$, which consists of all those symmetry operations $\{R_i \mid \mathbf{v}_i + \mathbf{T}_{\mathbf{n}}\}$ of the space group $\mathbf{G}$ for which R_i, acting in reciprocal space on the wave vector $\mathbf{k}$, sends $\mathbf{k}$ into an equivalent wave vector $\mathbf{k}+\mathbf{G}_{\mathbf{n}'}$, where $\mathbf{G}_{\mathbf{n}'}$ is given by an equation of the form of eqn (1.2.11). $\mathbf{G}^{\mathbf{k}}$ is still a space group and is some, possibly trivial, subgroup of the space group $\mathbf{G}$. For a wave vector $\mathbf{k}$ that does not terminate on the Brillouin zone boundary, there can be no other wave vector within the Brillouin zone that is equivalent to $\mathbf{k}$, so that the point-group operations R_i of elements in $\mathbf{G}^{\mathbf{k}}$ must be those that leave $\mathbf{k}$ invariant. For example, suppose we consider the space group $Pm3m(O_h^1)$ and a wave vector $\mathbf{k}$ that lies along the k_z axis in the Brillouin zone (see Fig. 1.6). The point-group operations that leave $\mathbf{k}$ invariant are those operations of $m3m(O_h)$ that are either pure rotations about the k_z axis or reflection planes passing through the k_z axis; these operations form the point group $4mm(C_{4v})$ so that $\mathbf{G}^{\mathbf{k}}$ is the tetragonal space group $P4mm(C_{4v}^1)$. For a wave vector $\mathbf{k}$ that terminates on the surface of the Brillouin zone the determination of $\mathbf{G}^{\mathbf{k}}$ is slightly more complicated. In this case, in addition to operations that leave $\mathbf{k}$ invariant, $\mathbf{G}^{\mathbf{k}}$ also includes any operations in $\mathbf{G}$ that send $\mathbf{k}$ into a wave vector $\mathbf{k}'$ that differs from $\mathbf{k}$ by a vector of the reciprocal lattice, that is,

$$\mathbf{k}' = \mathbf{k} + \mathbf{G}_{\mathbf{n}'}. \quad (1.4.6)$$

$\mathbf{G}^{\mathbf{k}}$, the group of $\mathbf{k}$, can be written in terms of left coset representatives of $\mathbf{T}$,

$$\mathbf{G}^{\mathbf{k}} = \{E \mid \mathbf{0}\}\mathbf{T} + \{S_1 \mid \mathbf{w}_1\}\mathbf{T} + \{S_2 \mid \mathbf{w}_2\}\mathbf{T} + \ldots + \{S_m \mid \mathbf{w}_m\}\mathbf{T}, \quad (1.4.7)$$

where the coset representatives $\{S_i \mid \mathbf{w}_i\}$ are a subset of the coset representatives in the expansion of **G** in eqn (1.4.5). Clearly for a wave vector that is in a completely general position in the Brillouin zone, $\mathbf{G}^{\mathbf{k}}$ will simply consist of $\{E \mid \mathbf{0}\}\mathbf{T}$ which is the group of the translational symmetry operations of

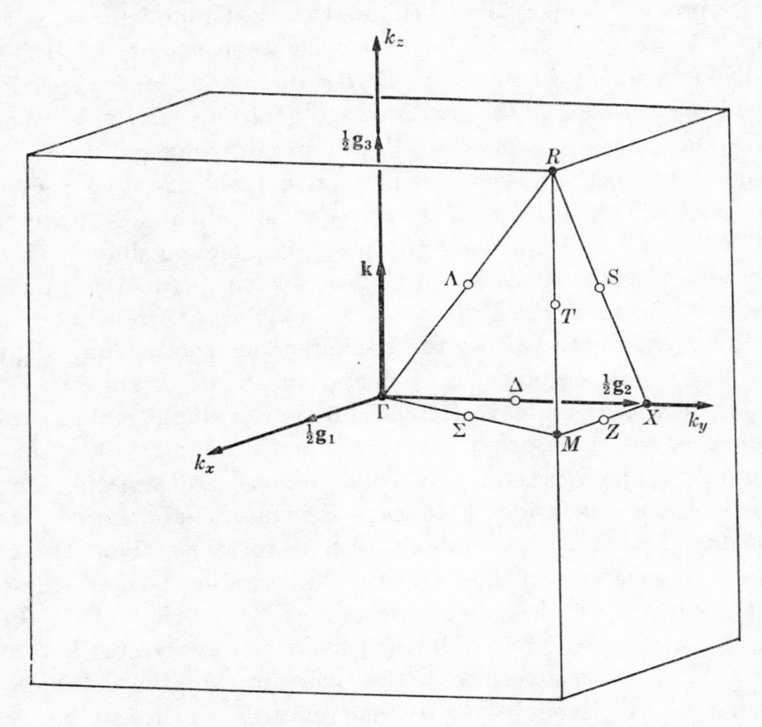

FIG. 1.6. The Brillouin zone for the space group $Pm3m(O_h^1)$ illustrating a wave vector, **k**, directed along a special line of symmetry.

the Bravais lattice. It is only when **k** lies on some axis of symmetry or plane of symmetry that $\mathbf{G}^{\mathbf{k}}$, the group of **k**, is non-trivial and can give extra information about the energy bands and wave functions beyond what is already known from the properties of **T**, the group of the Bravais lattice of the crystal.

The application of group theory in the simplification of the form of the wave function $\psi_{\mathbf{k}}(\mathbf{r})$ depends on the result that

$\psi_{\mathbf{k}}(\mathbf{r})$ can be regarded as belonging to one of the irreducible representations of $\mathbf{G}^{\mathbf{k}}$ (Seitz 1936). The transformation properties of $\psi_{\mathbf{k}}(\mathbf{r})$ under those symmetry operations of the crystal are therefore determined by the appropriate irreducible representation of $\mathbf{G}^{\mathbf{k}}$ and the degeneracies of the energy levels of the electrons at $\mathbf{k}$ are given by the degeneracies of these representations of $\mathbf{G}^{\mathbf{k}}$. There will be certain special points in the Brillouin zone with wave vector $\mathbf{k}$ such that all points $\mathbf{k}+\boldsymbol{\varkappa}$ in the neighbourhood of $\mathbf{k}$ possess less symmetry than $\mathbf{k}$; the point specified by $\mathbf{k}$ is then called a special *point of symmetry*. Such points of symmetry will normally be either at the centre of the Brillouin zone ($\mathbf{k} = 0$) or in some special position on the surface of the Brillouin zone. In a similar way there will be certain lines that are called *lines of symmetry* in the Brillouin zone; if $\mathbf{k}$ ends somewhere on a line of symmetry the group of the wave vector $\mathbf{k}$ is of higher order than the group of any neighbouring wave vector $\mathbf{k}+\boldsymbol{\varkappa}$ that is not on this line, where $\boldsymbol{\varkappa}$ is a small arbitrary vector. Again, there are certain planes in a Brillouin zone that are called *planes of symmetry;* if $\mathbf{k}$ ends somewhere on a plane of symmetry the group of the wave vector $\mathbf{k}$ is of higher order than the group of any neighbouring wave vector $\mathbf{k}+\boldsymbol{\varkappa}$ that is not on this plane, where $\boldsymbol{\varkappa}$ is a small arbitrary wave vector.

The study of points of symmetry, lines of symmetry and planes of symmetry within a Brillouin zone provides a valuable method for investigating the degeneracies of the energy bands $E_j(\mathbf{k})$ for the various wave vectors in the Brillouin zone. We have already shown in Section 1.3 that the energy $E_j(\mathbf{k})$ is a continuous function of $\mathbf{k}$. If we consider a point of symmetry with wave vector $\mathbf{k}$ and choose a small vector $\boldsymbol{\varkappa}$, such that $\mathbf{k}+\boldsymbol{\varkappa}$ is a general wave vector, then the energy eigenvalues at $\mathbf{k}+\boldsymbol{\varkappa}$ will all be non-degenerate while at $\mathbf{k}$ there may be many degeneracies. There is thus a sticking together of the energy eigenvalues as $\boldsymbol{\varkappa}$ is reduced to zero (see Fig. 1.7). If $\mathbf{k}$ is the wave vector of some point on a line (or plane) of symmetry and we choose $\boldsymbol{\varkappa}$ so that $\mathbf{k}+\boldsymbol{\varkappa}$ is on the same line (or plane) of symmetry, then the group of the wave vector $\mathbf{k}+\boldsymbol{\varkappa}$ is the same as the group of the wave vector **k** so that the degeneracies of $E_j(\mathbf{k}+\varkappa)$ are the same as the degeneracies of $E_j(\mathbf{k})$. The sticking together of the energy eigenvalues will therefore be the same all along a line of symmetry or all over a plane of symmetry.

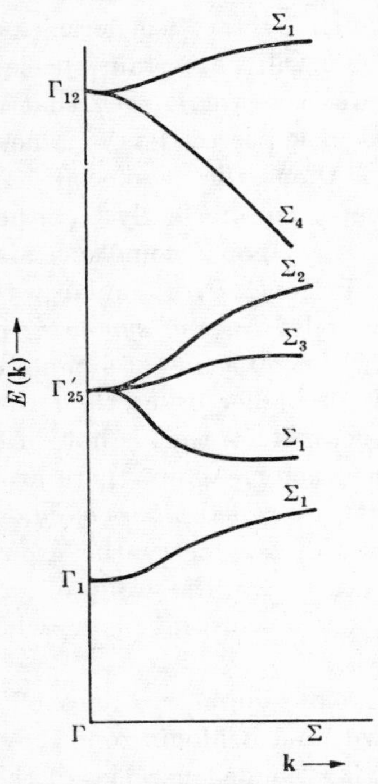

FIG. 1.7. The sticking together of energy bands at a point of symmetry; a schematic diagram for a cubic crystal. Γ and Σ can be identified from Fig. 1.6.

Finally, we may suppose that $G^{\mathbf{k}+\varkappa}$, the group of the wave vector $\mathbf{k}+\varkappa$, is a subgroup of $G^{\mathbf{k}}$, but still contains more elements than just the identity element. Such a situation arises, for instance, if $\mathbf{k} = 0$, which represents the central point of the Brillouin zone, and we choose $\varkappa$ so that $\mathbf{k}+\varkappa$ is along a line of symmetry. Alternatively, **k** may be on some line of symmetry with $\mathbf{k}+\varkappa$ on a plane of symmetry that passes through the line of symmetry. In this situation the energy eigenvalue

$E_j(\mathbf{k})$ will belong to some irreducible representation, Γ_j, of $\mathbf{G}^{\mathbf{k}}$. When the representation Γ_j is restricted to the elements of the group $\mathbf{G}^{\mathbf{k}+\boldsymbol{\kappa}}$, it will form a representation of $\mathbf{G}^{\mathbf{k}+\boldsymbol{\kappa}}$ which may or may not be irreducible. If Γ_j forms an irreducible representation of $\mathbf{G}^{\mathbf{k}+\boldsymbol{\kappa}}$ the degeneracy of the energy eigenvalues corresponding to Γ_j at $\mathbf{k}$ will not be lifted at $\mathbf{k}+\boldsymbol{\kappa}$. If Γ_j forms a reducible representation of $\mathbf{G}^{\mathbf{k}+\boldsymbol{\kappa}}$ the degeneracy of the eigenvalues corresponding to Γ_j at $\mathbf{k}$ will be partially or totally lifted at $\mathbf{k}+\boldsymbol{\kappa}$. The actual representations of $\mathbf{G}^{\mathbf{k}+\boldsymbol{\kappa}}$ which arise from the representation Γ_j of $\mathbf{G}^{\mathbf{k}}$ can be identified by determining the *compatibility relations* between the representations of $\mathbf{G}^{\mathbf{k}}$ and of $\mathbf{G}^{\mathbf{k}+\boldsymbol{\kappa}}$ using the character tables of the two groups. The process of restricting the elements of the group $\mathbf{G}^{\mathbf{k}}$ to those elements that belong to the group $\mathbf{G}^{\mathbf{k}+\boldsymbol{\kappa}}$ is sometimes described as *subduction* of the representations of $\mathbf{G}^{\mathbf{k}}$ onto the group $\mathbf{G}^{\mathbf{k}+\boldsymbol{\kappa}}$.

The degeneracies that have been discussed so far are intrinsic degeneracies necessitated by the symmetry of the Hamiltonian of an electron in the crystal. However, it is also possible for accidental degeneracies to occur at isolated wave vectors $\mathbf{k}$. A considerable number of theorems were proved by Herring (1937*b*) concerning the situations in which such accidental degeneracies might occur. Accidental degeneracies between energy bands that belong to different irreducible representations of $\mathbf{G}^{\mathbf{k}}$ frequently occur. On the other hand, Herring was able to show that in certain circumstances which depend on the crystal in question, an accidental degeneracy between two energy bands that belong to the same irreducible representation of $\mathbf{G}^{\mathbf{k}}$ may be 'vanishingly improbable'. By vanishingly improbable is meant that the degeneracy will disappear if some infinitesimal change is made in the potential $V(\mathbf{r})$ without altering the symmetry of $V(\mathbf{r})$. Of particular importance from among the results of Herring (1937*b*) was a proof that the occurrence of accidental degeneracy between energy bands belonging to the same irreducible representation of $\mathbf{G}^{\mathbf{k}}$ is vanishingly improbable for crystals with an inversion centre of symmetry.

We illustrate the ideas of the group of the wave vector and of the compatibility relations between points, lines, and planes of symmetry which we have discussed above for symmorphic space groups by considering the space group $Fm3m(O_h^5)$ (Bouckaert, Smoluchowski, and Wigner 1936; Altmann and Cracknell 1965). $Fm3m$ is the space group of one of the three structures which are commonly exhibited by metals, namely the face-centred cubic (or cubic close packed) structure. In this structure there is only one metal atom per unit cell and the atoms can be taken to be coincident with the points of the cubic Bravais lattice F shown in Fig. 1.1. The basic vectors $\mathbf{t}_1$, $\mathbf{t}_2$, and $\mathbf{t}_3$, referred to the axes $Oxyz$, may be taken to be

$$\left.\begin{aligned} \mathbf{t}_1 &= \tfrac{1}{2}(0, a, a), \\ \mathbf{t}_2 &= \tfrac{1}{2}(a, 0, a), \\ \mathbf{t}_3 &= \tfrac{1}{2}(a, a, 0), \end{aligned}\right\} \tag{1.4.8}$$

where a is the length of an edge of the conventional unit cell. The basic vectors of the reciprocal lattice for this space group can be found from eqns (1.2.10) and (1.3.1). They are

$$\left.\begin{aligned} \mathbf{g}_1 &= \frac{2\pi}{a}(-1, 1, 1), \\ \mathbf{g}_2 &= \frac{2\pi}{a}(1, -1, 1), \\ \mathbf{g}_3 &= \frac{2\pi}{a}(1, 1, -1), \end{aligned}\right\} \tag{1.4.9}$$

which can be seen to be the basic vectors of a body-centred cubic Bravais lattice in reciprocal space. The Brillouin zone was illustrated in Fig. 1.4(m) and is reproduced in Fig. 1.8. The various points of symmetry and lines of symmetry are indicated in Fig. 1.8. In Table 1.1 the coordinates of each point, line, and plane of symmetry are given, referred to the basic vectors $\mathbf{g}_1$, $\mathbf{g}_2$, and $\mathbf{g}_3$ of the appropriate reciprocal lattice. For each of the points, lines, and planes of symmetry, the group of the wave vector can be identified. To do this we need to adopt a scheme for labelling the point-group symmetry operations of the crystal. The scheme that we use is best

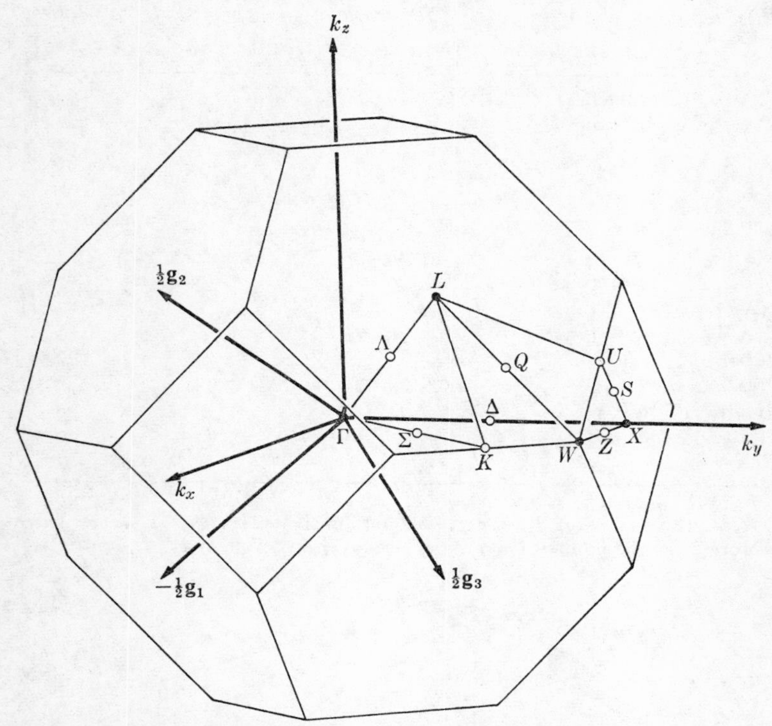

FIG. 1.8. The Brillouin zone for the space group $Fm3m(O_h^5)$. The special points of symmetry are given by: Γ, $\mathbf{k} = 0$; X, $\mathbf{k} = \frac{1}{2}\mathbf{g}_1 + \frac{1}{2}\mathbf{g}_3$; L, $\mathbf{k} = \frac{1}{2}\mathbf{g}_1 + \frac{1}{2}\mathbf{g}_2 + \frac{1}{2}\mathbf{g}_3$; W, $\mathbf{k} = \frac{1}{2}\mathbf{g}_1 + \frac{1}{4}\mathbf{g}_2 + \frac{3}{4}\mathbf{g}_3$.

explained by considering Fig. 1.9. We consider active operations, that is operations that move the points $\mathbf{r} = (x, y, z)$ of space and not the axes of space. C_{nr}^{+} is an anticlockwise rotation of the points of space through $2\pi/n$ radians about the axis labelled by r in Fig. 1.9. C_{nr}^{-} is a clockwise rotation of the points of space through $2\pi/n$ radians about the same axis. E is the identity and I is the inversion. C_{2m} is a rotation through π about the axis labelled by m. σ_m is a reflection in the plane normal to the axis labelled by m $(= x, y, z)$ and σ_{dp} is a reflection in the plane normal to the axis labelled by p $(= a, b, c, d, e, f)$. It should be noted that $IC_{3n}^{+} = S_{6n}^{-}$, $IC_{3n}^{-} = S_{6n}^{+}$ $(n = 1, 2, 3, 4)$. The group of the wave vector $\mathbf{k}$ can then be identified from Table 1.2 which

TABLE 1.1

The group of the wave vector for each of the special wave vectors in the face-centred cubic Brillouin zone.

$\Gamma(0, 0, 0)$	$m3m\ (O_h)$	$E, C_{3j}^{\pm}, C_{2m}, C_{4m}^{\pm}, C_{2p}, I, S_{6j}^{\pm}, \sigma_m, S_{4m}^{\pm}, \sigma_{dp}$
$X(\frac{1}{2}, 0, \frac{1}{2})$	$4/mmm\ (D_{4h})$	$E, C_{2y}, C_{4y}^{\pm}, C_{2x}, C_{2z}, C_{2c}, C_{2e}$ $I, \sigma_y, S_{4y}^{\pm}, \sigma_x, \sigma_z, \sigma_{dc}, \sigma_{de}$
$L(\frac{1}{2}, \frac{1}{2}, \frac{1}{2})$	$\bar{3}m\ (D_{3d})$	$E, C_{31}^{\pm}, C_{2b}, C_{2e}, C_{2f}, I, S_{61}^{\pm}, \sigma_{db}, \sigma_{de}, \sigma_{df}$
$W(\frac{1}{2}, \frac{1}{4}, \frac{3}{4})$	$\bar{4}2m\ (D_{2d})$	$E, C_{2x}, C_{2d}, C_{2f}, \sigma_y, \sigma_z, S_{4x}^{\pm}$
$\Delta(\Gamma X)$	$4mm\ (C_{4v})$	$E, C_{2y}, C_{4y}^{\pm}, \sigma_x, \sigma_z, \sigma_{dc}, \sigma_{de}$
$\Lambda(\Gamma L)$	$3m\ (C_{3v})$	$E, C_{31}^{\pm}, \sigma_{db}, \sigma_{de}, \sigma_{df}$
$\Sigma(\Gamma\Sigma)$	$mm2\ (C_{2v})$	$E, C_{2a}, \sigma_z, \sigma_{db}$
$S(XS)$	$mm2\ (C_{2v})$	$E, C_{2c}, \sigma_{de}, \sigma_y$
$Z(XW)$	$mm2\ (C_{2v})$	$E, C_{2x}, \sigma_y, \sigma_z$
$Q(LW)$	$2\ (C_2)$	E, C_{2f}
$C(\Gamma LK)$	$m(C_{1h})$	E, σ_{db}
$O(\Gamma WX)$	$m\ (C_{1h})$	E, σ_z
$J(\Gamma XL)$	$m\ (C_{1h})$	E, σ_{de}
$B(XUW)$	$m\ (C_{1h})$	E, σ_y

$j = 1, 2, 3, 4$; $m = x, y, z$; $p = a, b, c, d, e, f$.

Note. The planes of symmetry are not illustrated in Fig. 1.8. The wave vectors **k** for the points of symmetry are given in terms of $\mathbf{g}_1$, $\mathbf{g}_2$, and $\mathbf{g}_3$.

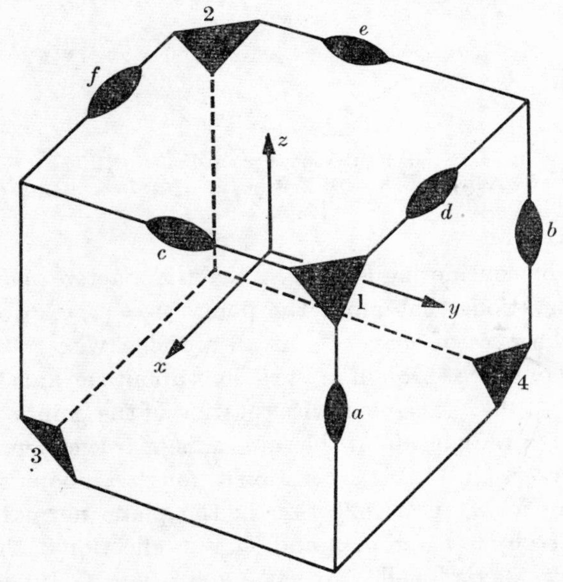

FIG. 1.9. The symmetry operations of the cubic groups. Notice that the σ_{dp} planes ($p = a, b, c, d, e, f$) are perpendicular to the C_{2p} axes. (Altmann and Bradley 1963.)

TABLE 1.2

Character table of the group of the wave vector for all the wave vectors of the cubic group Fm3m (O_h^5)

Γ	E / I	$8C_{3j}^{\pm}$ / $8S_{6j}^{\mp}$	$3C_{2m}$ / $3\sigma_m$	$6C_{4m}^{\pm}$ / $6S_{4m}^{\mp}$	$6C_{2p}$ / $6\sigma_{dp}$
$m3m$ (O_h)					
Γ_1, Γ_1'	1	1	1	1	1
Γ_2, Γ_2'	1	1	1	-1	-1
$\Gamma_{12}, \Gamma_{12}'$	2	-1	2	0	0
$\Gamma_{15}', \Gamma_{15}$	3	0	-1	1	-1
$\Gamma_{25}', \Gamma_{25}$	3	0	-1	-1	1

Σ, K	S, U	Z				
Σ, K			E	C_{2a}	σ_z	σ_{db}
	S, U		E	C_{2c}	σ_{de}	σ_y
		Z	E	C_{2x}	σ_z	σ_y
$mm2$ (C_{2v})						
Σ_1, K_1	S_1, U_1	Z_1	1	1	1	1
Σ_2, K_2	S_2, U_2	Z_2	1	1	-1	-1
Σ_4, K_4	S_3, U_3	Z_3	1	-1	1	-1
Σ_3, K_3	S_4, U_4	Z_4	1	-1	-1	1

Δ	X	W					
Δ			E	C_{2y}	$C_{4y}^{\pm}$	σ_z, σ_x	σ_{de}, σ_{dc}
	X		E	C_{2y}	$C_{4y}^{\pm}$	C_{2x}, C_{2z}	C_{2e}, C_{2c}
			I	σ_y	$S_{4y}^{\mp}$	σ_x, σ_z	σ_{de}, σ_{dc}
		W	E	C_{2x}	$S_{4x}^{\pm}$	C_{2d}, C_{2f}	σ_y, σ_z
$4mm$ (C_{4v})	$4/mmm$ (D_{4h})	$\bar{4}2m$ (D_{2d})					
Δ_1	X_1, X_1'	W_1	1	1	1	1	1
Δ_1'	X_4, X_4'	W_2	1	1	1	-1	-1
Δ_2	X_2, X_2'	W_1'	1	1	-1	1	-1
Δ_2'	X_3, X_3'	W_2'	1	1	-1	-1	1
Δ_5	X_5, X_5'	W_3	2	-2	0	0	0

TABLE 1.2 (*Continued*)

Λ		E	$C_{31}^{\pm}$	$\sigma_{db}, \sigma_{de}, \sigma_{df}$
	L	$\begin{cases} E \\ I \end{cases}$	$C_{31}^{\pm}$ $S_{61}^{\mp}$	C_{2b}, C_{2e}, C_{2f} $\sigma_{db}, \sigma_{de}, \sigma_{df}$
$3m\ (C_{3v})$	$\bar{3}m\ (D_{3d})$			
Λ_1	L_1, L_1'	1	1	1
Λ_2	L_2, L_2'	1	1	-1
Λ_3	L_3, L_3'	2	-1	0

					E	σ_{db}
C					E	σ_{db}
O					E	σ_z
J					E	σ_{de}
B					E	σ_y
				Q	E	C_{2f}
$m(C_{1h})$				$2(C_2)$		
C^+	O^+	J^+	B^+	Q^+	1	1
C^-	O^-	J^-	B^-	Q^-	1	-1

A	E
$1\ (C_1)$	
A	1

Notes to Table 1.2.

1. $j = 1, 2, 3, 4$; $m = x, y, z$; $p = a, b, c, d, e, f$ (see Fig. 1.9).
2. For direct product groups we only give one quarter of the character table. If R is an element for which $\chi(R)$ is given in the table, then $\chi(IR) = +\chi(R)$, where I is the inversion, for representations in the first column for each $\mathbf{k}$ (such as Γ_{12} for example) and $\chi(IR) = -\chi(R)$ for representations given in the second column (such as Γ_{12}' for example).

gives the character table of the group of $\mathbf{k}$ for each point, line, and plane of symmetry. The point-group operations given in Table 1.2 for any particular wave vector are the rotational parts $E, S_1, S_2, \ldots, S_m$ of the coset representatives $\{S_i \mid \mathbf{w}_i\}$ of $\mathbf{T}$ in the expansion of $\mathbf{G}^{\mathbf{k}}$ in eqn (1.4.7). Since the space group $Fm3m$ is a symmorphic space group, $\mathbf{w}_i = 0$, and the left coset representatives actually form a finite group, which we shall call $\bar{\mathbf{G}}^{\mathbf{k}}$ and which is one of the thirty-two point groups. The irreducible representations of the group of $\mathbf{k}$ are labelled in the notation of Bouckaert, Smoluchowski, and Wigner (1936).

We illustrate the concept of the group of the wave vector by considering one example. The point X lies on the k_y axis which is parallel to the y axis of the original crystal, see Fig. 1.8. Therefore point-group operations of symmetry that are rotations about the y axis, namely C_{4y}^+, C_{2y} and C_{4y}^- are operations that leave $\mathbf{k}$ unaltered and therefore belong to the group of $\mathbf{k}$. The effect of the inversion I or the reflection plane σ_y on the wave vector $(2\pi/a)(0, 1, 0)$, the wave vector of the point X, is to produce $(2\pi/a)(0, -1, 0)$, which is related to $(2\pi/a)(0, 1, 0)$ by the reciprocal lattice vector $\mathbf{g}_1+\mathbf{g}_3$. These elements, together with the rotations about the k_y axis generate the point group $4/mmm(D_{4h})$ consisting of the symmetry operations

$$\begin{array}{cccccccc} E, & C_{4y}^+, & C_{2y}, & C_{4y}^-, & C_{2x}, & C_{2z}, & C_{2c}, & C_{2e} \\ I, & S_{4y}^-, & \sigma_y, & S_{4y}^+, & \sigma_x, & \sigma_z, & \sigma_{dc}, & \sigma_{de}. \end{array}$$

The character table of this group, together with the labels used for the irreducible representations of this group can be found from Table 1.2. Similarly, from Table 1.2 we can write down the group of $\mathbf{k}$ for each point, line, or plane of symmetry in the Brillouin zone.

We now consider the compatibility relations that exist between the irreducible representations of the group of $\mathbf{k}$ for various points in the Brillouin zone. For example the group of the wave vector on a line of symmetry must be some sub-group of the group of the wave vector for the points of symmetry at the ends of that line. Thus the irreducible representations of the group of $\mathbf{k}$ for the electron energy levels at a point of symmetry will form a representation, which may be reducible or irreducible, of the group of the wave vector on a line of symmetry connected to that point. Thus, for instance, if the representation Γ_1 of Γ is restricted to the elements of Δ it would form the representation Δ_1 of Δ as can be seen from inspection of Table 1.2. Thus Γ_1 and Δ_1 are said to be *compatible*, and an energy band belonging to Γ_1 can therefore continue along Δ as a Δ_1 band, but it cannot continue along Δ as a Δ_2, Δ_1', Δ_2', or Δ_5 band. Alternatively, the representation Γ_{15}' when restricted to Δ, forms a reducible representation and inspection of Table 1.2 shows that

$$\Gamma'_{15} = \Delta'_1 + \Delta_5. \tag{1.4.10}$$

Thus an energy band belonging to Γ'_{15} and which is therefore threefold degenerate at Γ, splits into one non-degenerate energy

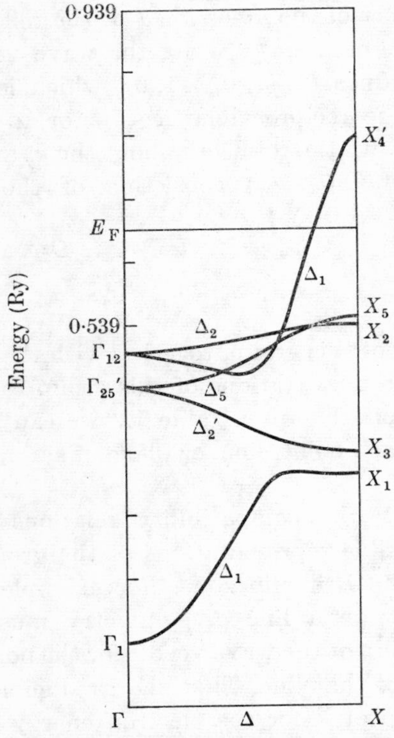

FIG. 1.10. Illustration of the compatibilities along Γ—Δ—X in the band structure of copper (Burdick 1963).

band and one two-fold degenerate energy band along Δ. The compatibilities for the line Γ—Δ—X of the face-centred cubic Brillouin zone, Fig. 1.8, are illustrated in Fig. 1.10. The compatibilities for all the points and lines of symmetry in the Brillouin zone for the space group $Fm3m$ are given in Table 1.3.

The example that we have just considered is simpler than many that could have been chosen because $Fm3m$ (O_h^5) is a symmorphic space group. For any space group $\mathbf{G}$ which is symmorphic, $\mathbf{G}^{\mathbf{k}}$ will also be symmorphic for any choice of the wave vector $\mathbf{k}$. This means that the coset representatives $\{S_i \mid \mathbf{w}_i\}$ in eqn (1.4.7) can all be chosen with $\mathbf{w}_i = \mathbf{0}$, and therefore the irreducible representations of $\mathbf{G}^{\mathbf{k}}$ are simply

TABLE 1.3

Compatibility relations for the f.c.c. structure

(i) Γ: Δ, Λ, Σ.

Γ_1	Γ_2	Γ_{12}	Γ'_{15}	Γ'_{25}
Δ_1	Δ_2	$\Delta_1\Delta_2$	$\Delta'_1\Delta_5$	$\Delta'_2\Delta_5$
Λ_1	Λ_2	Λ_3	$\Lambda_2\Lambda_3$	$\Lambda_1\Lambda_3$
Σ_1	Σ_4	$\Sigma_1\Sigma_4$	$\Sigma_2\Sigma_3\Sigma_4$	$\Sigma_1\Sigma_2\Sigma_3$

Γ'_1	Γ'_2	Γ'_{12}	Γ_{15}	Γ_{25}
Δ'_1	Δ'_2	$\Delta'_1\Delta'_2$	$\Delta_1\Delta_5$	$\Delta_2\Delta_5$
Λ_2	Λ_1	Λ_3	$\Lambda_1\Lambda_3$	$\Lambda_2\Lambda_3$
Σ_2	Σ_3	$\Sigma_2\Sigma_3$	$\Sigma_1\Sigma_3\Sigma_4$	$\Sigma_1\Sigma_2\Sigma_4$

(ii) X: Δ, Z, S.

X_1	X_2	X_3	X_4	X_5	X'_1	X'_2	X'_3	X'_4	X'_5
Δ_1	Δ_2	Δ'_2	Δ'_1	Δ_5	Δ'_1	Δ'_2	Δ_2	Δ_1	Δ_5
Z_1	Z_1	Z_4	Z_4	Z_2Z_3	Z_2	Z_2	Z_3	Z_3	Z_1Z_4
S_1	S_4	S_1	S_4	S_2S_3	S_2	S_3	S_2	S_3	S_1S_4

(iii) L: Λ, Q.

L_1	L_2	L_3	L'_1	L'_2	L'_3
Λ_1	Λ_2	Λ_3	Λ_2	Λ_1	Λ_3
Q^+	Q^-	Q^+Q^-	Q^+	Q^-	Q^+Q^-

(iv) W: Z, Q.

W_1	W_2	W'_1	W'_2	W_3
Z_1	Z_2	Z_2	Z_1	Z_3Z_4
Q^+	Q^-	Q^+	Q^-	Q^+Q^-

TABLE 1.3 (*Continued*)

(v) symmetry planes.

Plane	+	−
O (ΓWX)	Σ_1, Σ_4	Σ_2, Σ_3
	$\Delta_1, \Delta_2, \Delta_5$	$\Delta_1', \Delta_2', \Delta_5$
	Z_1, Z_3	Z_2, Z_4
C (ΓLK)	Σ_1, Σ_3	Σ_2, Σ_4
	Λ_1, Λ_3	Λ_2, Λ_3
J (ΓXL)	Λ_1, Λ_3	Λ_2, Λ_3
	S_1, S_3	S_2, S_4
	$\Delta_1, \Delta_2', \Delta_5$	$\Delta_2, \Delta_1', \Delta_5$
B (XUW)	S_1, S_4	S_2, S_3
	Z_1, Z_4	Z_2, Z_3

related to the irreducible representations of one or other of the thirty-two point groups. However, for the non-symmorphic space groups, the determination of the irreducible representations of $\mathbf{G}^\mathbf{k}$ is slightly more complicated. It was considered for the two important non-symmorphic space groups of diamond $Fd3m(O_h^7)$ and of the close-packed hexagonal structure $P6_3/mmc$ (D_{6h}^4) by Herring (1942). The extra complication that arises in a non-symmorphic space group is that it is not possible to choose the left coset representatives $\{R_i \mid \mathbf{v}_i\}$ so that $\mathbf{v}_i = \mathbf{0}$ for all of them. Then the coset representatives $\{S_i \mid \mathbf{w}_i\}$ in the expansion of $\mathbf{G}^\mathbf{k}$, the group of $\mathbf{k}$, may also not all be able to have $\mathbf{w}_i = \mathbf{0}$; this means that on their own the set of the $\{S_i \mid \mathbf{w}_i\}$ may not form a group. However, it is possible to show that the problem of determining the irreducible representations of the infinite group $\mathbf{G}^\mathbf{k}$ can be simplified to the determination of the irreducible representations of the finite group, which is the factor group $\mathbf{G}^\mathbf{k}/\mathbf{T}^\mathbf{k}$, where $\mathbf{T}^\mathbf{k}$ is the group of all those translations $\mathbf{T}_\mathbf{n}$ of the Bravais lattice for which $\mathbf{k} \cdot \mathbf{T}_\mathbf{n}$ is an integer multiple of 2π. It may still happen for certain points, lines, or planes of symmetry that the factor group $\mathbf{G}^\mathbf{k}/\mathbf{T}^\mathbf{k}$ for a non-symmorphic space group may be isomorphic with one of the thirty-two point groups. Even if $\mathbf{G}^\mathbf{k}/\mathbf{T}^\mathbf{k}$ is not isomorphic with one of the thirty-two point groups it is still a finite group of small order (in all cases of order $\leqslant 96$ or, if we include the double groups, $\leqslant 192$).

One of the three common metallic structures has to be described by a non-symmorphic space group; this is the hexagonal close-packed (h.c.p.) structure which belongs to the space group $P6_3/mmc(D_{6h}^4)$. It would not be appropriate in the present book to consider in any greater detail the problem of the determination of the irreducible representations of the non-symmorphic space groups. The irreducible representations of all the symmorphic and non-symmorphic space groups have been determined by a number of authors and the results tabulated for all the special wave vectors **k** in the Brillouin zone for each space group (Faddeyev 1964; Kovalev 1965; Miller and Love 1967; Zak, Casher, Glück, and Gur 1969; Bradley and Cracknell 1972).

In addition to the spatial symmetry operations which we have been considering, the effect of the operation of time-inversion must also be considered. If we are considering a crystal in which there is no magnetic ordering it will possess time-reversal symmetry in addition to all the spatial symmetry operations of **G**. The complete group of the crystal is then the direct product $\mathbf{G} \otimes (E+\theta)$ where θ is the operation of time inversion; this group is, of course, the grey space group (or type II Shubnikov space group) derived from **G** (see for example Bradley and Cracknell 1972). The addition of the extra symmetry operation θ may cause some extra degeneracies in the energy levels of a particle or quasi-particle with wave vector **k** (Wigner 1932; Herring 1937*a*; Frei 1966; Bradley and Cracknell 1970). If $\Gamma_j^{\mathbf{k}}$ is an irreducible representation of $\mathbf{G}^{\mathbf{k}}$ there are several possibilities;

(a) there is no change in the degeneracy of $\Gamma_j^{\mathbf{k}}$;

(b) the degeneracy of $\Gamma_j^{\mathbf{k}}$ becomes doubled; or in terms of energy levels, two different energy levels, both described by the same representation $\Gamma_j^{\mathbf{k}}$, become degenerate;

and

(c) the degeneracy of $\Gamma_j^{\mathbf{k}}$ becomes doubled but, unlike (b), two different, that is inequivalent, representations $\Gamma_{\mathbf{k}}^{j}$

and $\Gamma^{\mathbf{k}}_{j'}$ of $\mathbf{G^k}$ become degenerate because of the addition of θ to the space group of the crystal.

In addition to the possibility of causing some extra degeneracies in the spectrum of the energy eigenvalues at $\mathbf{k}$, the inclusion of time-reversal symmetry will always cause the spectrum of the energy levels at $\mathbf{k}$ and $-\mathbf{k}$ to be identical if $\mathbf{k}$ and $-\mathbf{k}$ are not already related by one of the operations of the space group $\mathbf{G}$ of the crystal. Of course, if $\mathbf{k}$ and $-\mathbf{k}$ are already related by one of the operations of the space group $\mathbf{G}$, as for example when the crystal has a centre of inversion so that $\{I \mid \mathbf{v}\}$ is in $\mathbf{G}$, the spectrum of the energy eigenvalues at $\mathbf{k}$ is the same as that at $-\mathbf{k}$ even in the absence of time-reversal symmetry. This extra degeneracy between $E_j(\mathbf{k})$ and $E_j(-\mathbf{k})$, which is called a type (x) degeneracy, is very easy to recognize as it occurs when time-reversal symmetry is present and there are no space group elements that transform $\mathbf{k}$ into $-\mathbf{k}$.

Rather than thinking of time-reversal symmetry, it is possible to regard these various extra degeneracies as being due to the fact that the Hamiltonian used in determining the band structure is real, which is only a different way of looking at the same problem (Herring 1937*a*). The behaviour of the conduction electrons in a metal can be determined from a knowledge of the solutions $\psi(\mathbf{r})$ of Schrödinger's equation for the electrons:

$$\frac{-\hbar^2}{2m}\nabla^2\psi(\mathbf{r})+V(\mathbf{r})\psi(\mathbf{r}) = E\psi(\mathbf{r}). \tag{1.2.3}$$

The Hamiltonian operator on the left-hand side of this equation can be assumed to be real. The basic theory of the extra degeneracies that arise because of the reality of the Hamiltonian of a system (in other words because of time-reversal symmetry) was given by Wigner (1932). If the Hamiltonian, $\mathscr{H}$, is real then $\psi_{\mathbf{k}}(\mathbf{r})$ and $\psi_{\mathbf{k}}(\mathbf{r})^*$ are two eigenfunctions of $\mathscr{H}$ with the same eigenvalue $E_i(\mathbf{k})$. Thus if $\psi_{\mathbf{k}}(\mathbf{r})$ and $\psi_{\mathbf{k}}(\mathbf{r})^*$ are linearly independent functions, the energy eigenvalue has an extra double-degeneracy that is not predicted from the analysis of $\mathbf{G^k}$, the group of the wave vector $\mathbf{k}$. On the other hand, if

$\psi_{\mathbf{k}}(\mathbf{r})$ and $\psi_{\mathbf{k}}(\mathbf{r})^*$ are not linearly independent there will be no extra double-degeneracy of the energy band structure due to time-reversal symmetry. The wave function used by one observer to describe some state of the system is transformed, by the operation of complex conjugation of $\psi_{\mathbf{k}}(\mathbf{r})$, into a wave function describing the same physical state as could be used by a second observer with the same x, y, and z, axes, but whose time axis is in the opposite direction. It is then desirable, for ease of manipulation, to turn this condition for the extra degeneracy due to time-reversal from a condition on the wave function $\psi_{\mathbf{k}}(\mathbf{r})$, which in general is probably unknown, to a condition on the space group irreducible representations that are used to classify the wave functions and energy eigenvalues.

Suppose that **D** is a set of matrices, in general complex, that form an irreducible representation of $\mathbf{G}^{\mathbf{k}}$, the group of the wave vector **k**, then **D** will have one of the following set of properties:

(a) all the matrices **D** are real,
(b) the matrices **D** and **D*** (**D** is complex) belong to different irreducible representations of $\mathbf{G}^{\mathbf{k}}$, that is, to irreducible representations that are not equivalent,
(c) **D** and **D*** (**D** complex) are equivalent representations of $\mathbf{G}^{\mathbf{k}}$.

It was shown by Wigner (1932) that, when spin is ignored,

(a) there is no extra degeneracy,
(b) there is an extra degeneracy and the representations **D** and **D*** 'stick together',
(c) there is an extra degeneracy and the representation **D** occurs twice.

This is equivalent to considering the irreducible corepresentations of the grey magnetic space group of the crystal (Bradley and Cracknell 1970, 1972). In the case when spin is included, that is, when spin–orbit coupling is introduced into the Hamiltonian of the system, then the degeneracies in (a) and (c) become interchanged from those given above (Elliott 1954). By using a theorem due to Frobenius and Schur (1906) Herring was able to express the conditions (a), (b), and (c) in terms of the characters of the space-group irreducible representations.

$$\left.\begin{aligned} &\text{(a)} \qquad \sum_{\{R_i|\mathbf{v}_i\}} \chi(\{R_i \mid \mathbf{v}_i\}^2) = h \\ &\text{(b)} \qquad \sum_{\{R_i|\mathbf{v}_i\}} \chi(\{R_i \mid \mathbf{v}_i\}^2) = 0 \\ &\text{(c)} \qquad \sum_{\{R_i|\mathbf{v}_i\}} \chi(\{R_i \mid \mathbf{v}_i\}^2) = -h \end{aligned}\right\} \tag{1.4.11}$$

where h is the order of the group and the summations are over all those elements of the space group that send $\mathbf{k}$ into $-\mathbf{k}+\mathbf{G}_{\mathbf{n}'}$ ($\mathbf{G}_{\mathbf{n}'}$ is given by eqn (1.2.11) and is any vector of the reciprocal lattice). The summations in eqn (1.4.11) can be restricted to elements of the space group that do not differ by an element of $\mathbf{T}^{\mathbf{k}}$, the group of all those translations $\mathbf{T}_{\mathbf{n}}$ of the Bravais lattice for which $\mathbf{k} \cdot \mathbf{T}_{\mathbf{n}}$ is an integer multiple of 2π. When the test in eqn (1.4.11) is applied to the hexagonal close-packed structure, the result is that time-reversal symmetry introduces extra degeneracies at R and at all general points on the top face of the Brillouin zone. This has important practical consequences because it means that, in the absence of spin–orbit coupling, the energy bands $E_j(\mathbf{k})$ must always be at least two-fold degenerate all over the hexagonal face AHL at the top of the Brillouin zone. This affects the connectivity of any sheet of the Fermi surface that intersects this face and leads to the use of the so-called *double-zone scheme* which is often used for hexagonal close-packed metals, see Section 1.6.

At the beginning of our discussion of time-reversal we said that we were excluding any consideration of metals that are magnetically ordered. The problem of the degeneracies of the band structure of a magnetically ordered metal will be considered in Section 5.1.

1.5. Free-electron energy bands

In Sections 1.1 and 1.2 it was shown that the solutions of Schrödinger's equation

$$-\frac{\hbar^2}{2m}\nabla^2\psi(\mathbf{r})+V(\mathbf{r})\psi(\mathbf{r}) = E\psi(\mathbf{r}) \tag{1.2.3}$$

in a periodic potential $V(\mathbf{r})$ take the form

$$\psi_{\mathbf{k}}(\mathbf{r}) = \exp(i\mathbf{k}\,.\,\mathbf{r})u_{\mathbf{k}}(\mathbf{r}) \tag{1.5.1}$$

where $\mathbf{k}$ is given by eqn (1.2.12). It is possible then to substitute $\psi_{\mathbf{k}}(\mathbf{r})$ from eqn (1.5.1) into eqn (1.2.3) to obtain an equation for $u_{\mathbf{k}}(\mathbf{r})$:

$$-\frac{\hbar^2}{2m}\{-\mathbf{k}^2u_{\mathbf{k}}(\mathbf{r})+2i\mathbf{k}\,.\,\nabla u_{\mathbf{k}}(\mathbf{r})+\nabla^2u_{\mathbf{k}}(\mathbf{r})\}+ \\ +V(\mathbf{r})u_{\mathbf{k}}(\mathbf{r}) = E(\mathbf{k})u_{\mathbf{k}}(\mathbf{r}). \tag{1.5.2}$$

Therefore,

$$\nabla^2u_{\mathbf{k}}(\mathbf{r})+2i\mathbf{k}\,.\,\nabla u_{\mathbf{k}}(\mathbf{r})+\frac{2m}{\hbar^2}\left\{E(\mathbf{k})-\frac{\hbar^2\mathbf{k}^2}{2m}-V(\mathbf{r})\right\}u_{\mathbf{k}}(\mathbf{r}) = 0. \tag{1.5.3}$$

In any real metal it is very difficult to determine a reliable expression for $V(\mathbf{r})$ and we shall devote considerable attention to that problem in Chapter 2. However, if one assumes $V(\mathbf{r}) = 0$, it is possible to obtain a set of energy bands for a structure with any given Bravais lattice and this is commonly called the *free-electron approximation*. It is a valuable approach because it illustrates many of the general features of the energy band structures of metals and their use in determining the shape of the Fermi surface. Indeed, as we shall see in Chapter 4, for many of the simpler non-transition and non-rare-earth metals the free-electron model gives a very realistic first approximation to the actual band structure and Fermi surface, see Fig. 1.11. A fairly detailed treatment of free-electron band structures is given by Jones (1960).

It is perhaps as well to list the approximations that we are making in this section and the next:

(a) the electrons move in a perfect crystal having a regular periodic structure with no defects,

(b) the nuclei of the atoms of which the metal is composed are assumed to be at rest (i.e. $T = 0$ K),

(c) the proper many-body treatment is approximated by the individual particle treatment, i.e. eqn (1.2.3) is a one-electron equation,

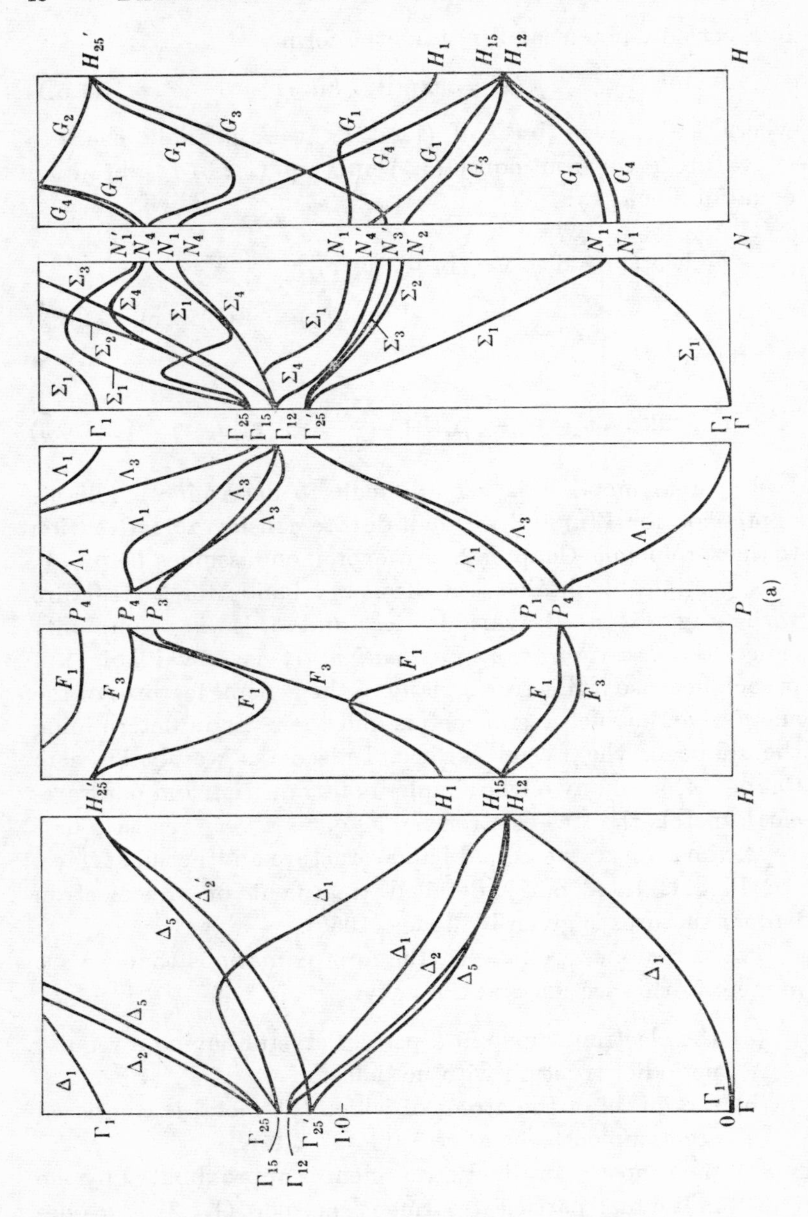

Γ
H
P
N
1·0
0
Γ_1
Γ_{25}
Γ_{15}
Γ_{12}
Γ_{25}'
H_{25}'
H_1
H_{15}
H_{12}
P_4
P_3
P_1
N_1'
N_4
N_1
N_4'
N_3
N_2
Δ_1
Δ_2
Δ_5
Δ_2'
F_1
F_3
Λ_1
Λ_3
Σ_1
Σ_2
Σ_3
Σ_4
G_1
G_2
G_3
G_4

(a)

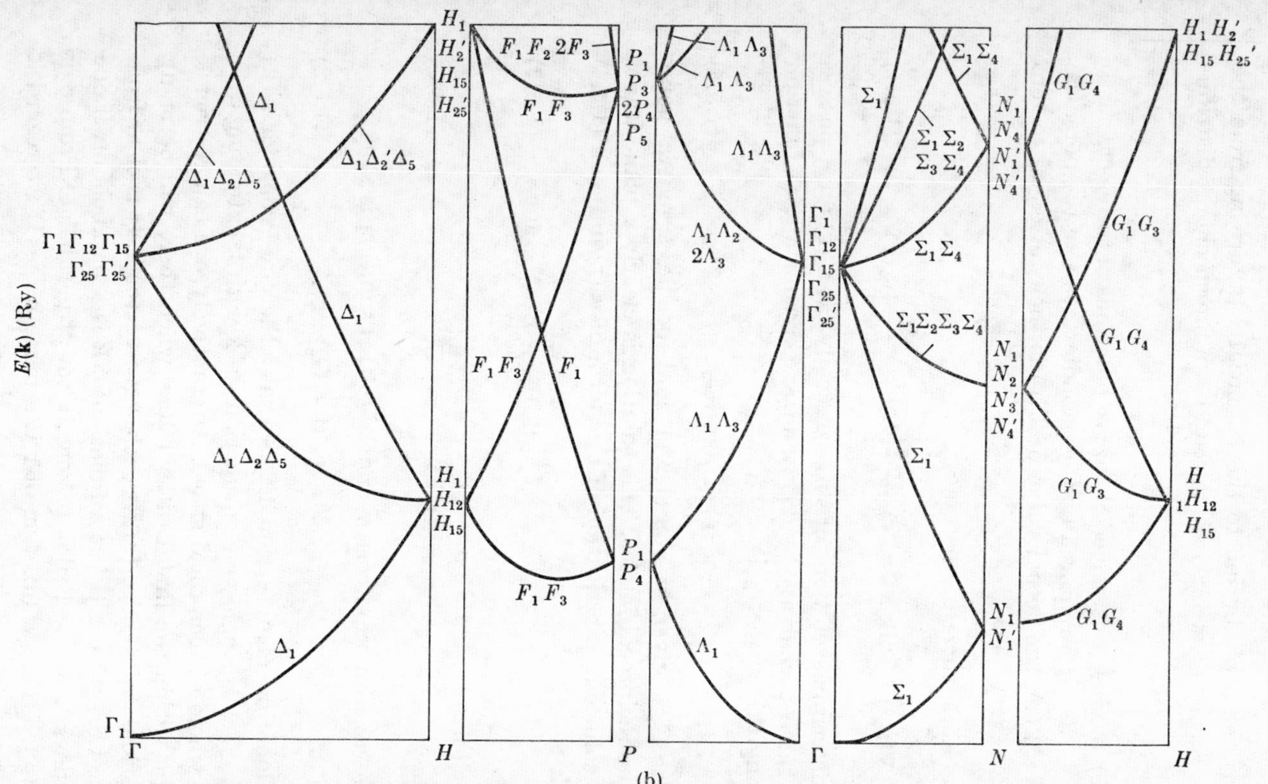

FIG. 1.11. Energy of conduction and higher energy bands of sodium along symmetry directions, calculated by Kenney: (a) actual bands, and (b) free-electron bands. (Slater 1965).

(d) the contribution to the potential $V(\mathbf{r})$ arising from the other electrons is replaced by a uniform charge distribution,

and

(e) having constrained the wave functions of the electrons to the form required by the periodic nature of the potential $V(\mathbf{r})$, the actual value of $V(\mathbf{r})$ is then set equal to zero everywhere.

Eqn (1.5.3) therefore becomes

$$\nabla^2 u_{\mathbf{k}}(\mathbf{r})+2\mathrm{i}\mathbf{k}\cdot\nabla u_{\mathbf{k}}(\mathbf{r})+\frac{2m}{\hbar^2}\left\{E(\mathbf{k})-\frac{\hbar^2\mathbf{k}^2}{2m}\right\}u_{\mathbf{k}}(\mathbf{r}) = 0. \quad (1.5.4)$$

To determine solutions of this we recall that $u_{\mathbf{k}}(\mathbf{r})$ must possess the same periodicity as $V(\mathbf{r})$ and a convenient trial function having this property is

$$u_{\mathbf{k}}(\mathbf{r}) = \exp(\mathrm{i}\mathbf{G}_{\mathbf{n}}\cdot\mathbf{r}) \quad (1.5.5)$$

where $\mathbf{G}_{\mathbf{n}}$ is given by eqn (1.2.11) and specifies a general point of the reciprocal lattice. This solution is clearly a plane-wave-like solution. Substitution of this form of $u_{\mathbf{k}}(\mathbf{r})$ into eqn (1.5.4) leads to

$$-|\mathbf{G}_{\mathbf{n}}|^2-2\mathbf{k}\cdot\mathbf{G}_{\mathbf{n}}-|\mathbf{k}|^2+\frac{2m}{\hbar^2}E(\mathbf{k}) = 0 \quad (1.5.6)$$

which has the solution

$$E(\mathbf{k}) = \frac{\hbar^2}{2m}|\mathbf{k}+\mathbf{G}_{\mathbf{n}}|^2. \quad (1.5.7)$$

The wave function corresponding to this solution is thus

$$\psi_{\mathbf{k}}(\mathbf{r}) = \exp\{\mathrm{i}(\mathbf{k}+\mathbf{G}_{\mathbf{n}})\cdot\mathbf{r}\}. \quad (1.5.8)$$

As mentioned in Section 1.2, we may therefore consider $E(\mathbf{k})$ and $\psi_{\mathbf{k}}(\mathbf{r})$ as single-valued functions of $\mathbf{k}$ with $\mathbf{k}$ extending over the whole of reciprocal space. Alternatively, we can restrict $\mathbf{k}$ to lie within the first Brillouin zone when there will be many solutions $E(\mathbf{k})$ and $\psi_{\mathbf{k}}(\mathbf{r})$ for each wave vector $\mathbf{k}$. It is clear from eqn (1.5.7) that $E(\mathbf{k})$ as a function of $\mathbf{k}$ is a parabola or, if one confines $\mathbf{k}$ to the Brillouin zone, a set of parabolae. From eqn (1.5.8) $\psi_{\mathbf{k}}(\mathbf{r})$ as a function of $\mathbf{r}$ is a plane wave. If we regard $\mathbf{k}$

as being restricted so as to end within the Brillouin zone then for any given $\mathbf{k}$ one simply has to substitute all the various possible values of $\mathbf{G}_n$ into eqn (1.5.7) to determine all the energy eigenvalues at $\mathbf{k}$. These eigenvalues could then be assigned to the appropriate irreducible representation of $\mathbf{G}^{\mathbf{k}}$, the group of the wave vector $\mathbf{k}$, by studying the transformation properties of $\psi_{\mathbf{k}}(\mathbf{r})$ in eqn (1.5.8) under the symmetry operations of $\mathbf{G}^{\mathbf{k}}$. Having determined the symmetries of the wave functions the compatibilities could then be determined.

We now consider briefly an example of the details of the solutions $E(\mathbf{k})$ and $\psi_{\mathbf{k}}(\mathbf{r})$; we shall consider the derivation of the free-electron energy bands for one space group $Fm3m(O_h^5)$, the space group of the face-centred cubic structure. The Brillouin zone for this structure is illustrated in Fig. 1.8 and the values of $\mathbf{k}$ for the points of symmetry are given in Table 1.1. Suppose we write

$$\mathbf{k} = \frac{2\pi}{a}(\xi, \eta, \zeta) \tag{1.5.9}$$

and also from eqns (1.2.11) and (1.3.1),

$$\mathbf{G}_n = n_1\mathbf{g}_1 + n_2\mathbf{g}_2 + n_3\mathbf{g}_3 \tag{1.5.10}$$

that is

$$\mathbf{G}_n = \frac{2\pi}{a}(-n_1, n_1, n_1) + \frac{2\pi}{a}(n_2, -n_2, n_2) + \frac{2\pi}{a}(n_3, n_3, -n_3) \tag{1.5.11}$$

$$= \frac{2\pi}{a}(l_1, l_2, l_3) \tag{1.5.12}$$

where

$$\left.\begin{aligned} l_1 &= -n_1 + n_2 + n_3 \\ l_2 &= n_1 - n_2 + n_3 \\ l_3 &= n_1 + n_2 - n_3 \end{aligned}\right\}, \tag{1.5.13}$$

and n_1, n_2, and n_3 are integers. Therefore, from eqn (1.5.7), the energy $E(\mathbf{k})$ is given by

$$E(\mathbf{k}) = \frac{\hbar^2}{2m}\left(\frac{2\pi}{a}\right)^2\{(\xi + l_1)^2 + (\eta + l_2)^2 + (\zeta + l_3)^2\} \tag{1.5.14}$$

and from eqn (1.5.8) the corresponding wave function, $\psi_{\mathbf{k}}(\mathbf{r})$, is given by

$$\psi_{\mathbf{k}}(\mathbf{r}) = \exp\left[\mathrm{i}\frac{2\pi}{a}\{(\xi+l_1)x+(\eta+l_2)y+(\zeta+l_3)z\}\right]. \quad (1.5.15)$$

For a point of symmetry there are often several vectors $\mathbf{n} = (n_1, n_2, n_3)$ which lead to the same value of $E(\mathbf{k})$ at that point. If $V(\mathbf{r})$ were to be changed slowly from zero to the potential that exists everywhere in a real crystal, many of the energy bands in the free-electron band structure would slowly change in form, the accidental degeneracies would be lifted, and the only one that would remain would be the essential degeneracy due to symmetry.

We consider first the axis Γ—Δ—X, see Fig. 1.12(a). At Γ, $\mathbf{k} = 0$, while along Δ, $\mathbf{k} = (2\pi/a)(0, \eta, 0)$, and at X, $\mathbf{k} = (2\pi/a)(0, 1, 0)$ (see Table 1.1). Therefore the corresponding energy eigenvalues are, from eqn (1.5.14),

$$E_\Gamma = \frac{\hbar^2}{2m}\left(\frac{2\pi}{a}\right)^2(l_1^2+l_2^2+l_3^2) \quad (1.5.16)$$

$$E_\Delta = \frac{\hbar^2}{2m}\left(\frac{2\pi}{a}\right)^2\{l_1^2+(l_2+\eta)^2+l_3^2\} \quad (1.5.17)$$

and

$$E_X = \frac{\hbar^2}{2m}\left(\frac{2\pi}{a}\right)^2\{l_1^2+(l_2+1)^2+l_3^2\}. \quad (1.5.18)$$

Therefore to plot the free-electron energy bands along Γ—Δ—X, one has to evaluate E_Γ, E_Δ, and E_X for all the various sets of values of (l_1, l_2, l_3) that satisfy eqn (1.5.13). If

$$\mathbf{l} = (l_1, l_2, l_3) = (0, 0, 0)$$

then $E_\Gamma = 0$, $E_\Delta = (\hbar^2/2m)(2\pi/a)^2\eta^2$ and $E_X = (\hbar^2/2m)(2\pi/a)^2$. It is convenient to measure the energy $E(\mathbf{k})$ in units of $(\hbar^2/2m)(2\pi/a)^2$ when we have $E_\Gamma = 0$, $E_\Delta = \eta^2$ and $E_X = 1$. This energy band is clearly a parabola and from eqn (1.5.15) the corresponding wave function is

$$\psi_\Gamma(\mathbf{r}) = \exp\left\{\mathrm{i}\frac{2\pi}{a}(0)\right\}, \quad (1.5.19)$$

so that the wave function is a constant and must belong to the irreducible representation Γ_1, see Table 1.2. Along Δ, the wave function is then

$$\psi_\Delta(\mathbf{r}) = \exp\left(\mathrm{i}\frac{2\pi}{a}\eta y\right), \tag{1.5.20}$$

$(0 \leqslant \eta \leqslant 1)$, and at the end point X the wave function is

$$\psi_X(\mathbf{r}) = \exp\left(\mathrm{i}\frac{2\pi}{a}y\right) = u_1(\mathbf{r}). \tag{1.5.21}$$

It happens that the energy level $E_X = 1$ is twofold degenerate and can also be obtained with $\mathbf{n} = (-1, 0, -1)$ so that $\mathbf{l} = (0, -2, 0)$ and

$$\psi_X(\mathbf{r}) = \exp\left(-\mathrm{i}\frac{2\pi}{a}y\right) = u_2(\mathbf{r}). \tag{1.5.22}$$

Thus there are two energy bands connected to the energy level $E_X = 1$. For one of these $\mathbf{l} = (0, 0, 0)$ and this one is connected to the energy level $E_\Gamma = 0$. The second band, connected to $E_X = 1$, for which we found $\mathbf{l} = (0, -2, 0)$, will be connected not to $E_\Gamma = 0$ but to a different level at Γ, in fact to the level $E_\Gamma = 4$. There are several other values of $\mathbf{l}$ that lead to an energy $E_\Gamma = 4$ at Γ, namely $\mathbf{l} = (-2, 0, 0)$, $(0, 0, -2)$, $(2, 0, 0)$, $(0, 2, 0)$ and $(0, 0, 2)$, so that this energy eigenvalue is sixfold degenerate. Since the irreducible representations of $\mathbf{G}^{\mathbf{k}}$ at Γ have dimension of either one, two, or three there is a considerable amount of accidental degeneracy of the energy levels at Γ in the free-electron approximation; this accidental degeneracy will in general be removed by the non-zero potential $V(\mathbf{r})$ existing in a real metal. By inspection of the possible values of $\mathbf{l}$ we find that there is another set of degenerate energy levels at Γ, $E_\Gamma = 3$, and in fact this energy level is eightfold degenerate; once again there must be a considerable amount of accidental degeneracy at $E_\Gamma = 3$. Values of $\mathbf{n}$ and $\mathbf{l}$ for the levels $E_\Gamma = 3$ and $E_\Gamma = 4$ are given in Table 1.4.

The group-theoretical symbols that are used to label the energy bands in Fig. 1.12 can be determined by studying the transformation properties of the wave functions $\psi_{\mathbf{k}}(\mathbf{r})$ obtained with various different values of $\mathbf{l}$ in eqn (1.5.15). To obtain functions with the correct symmetry properties it may be

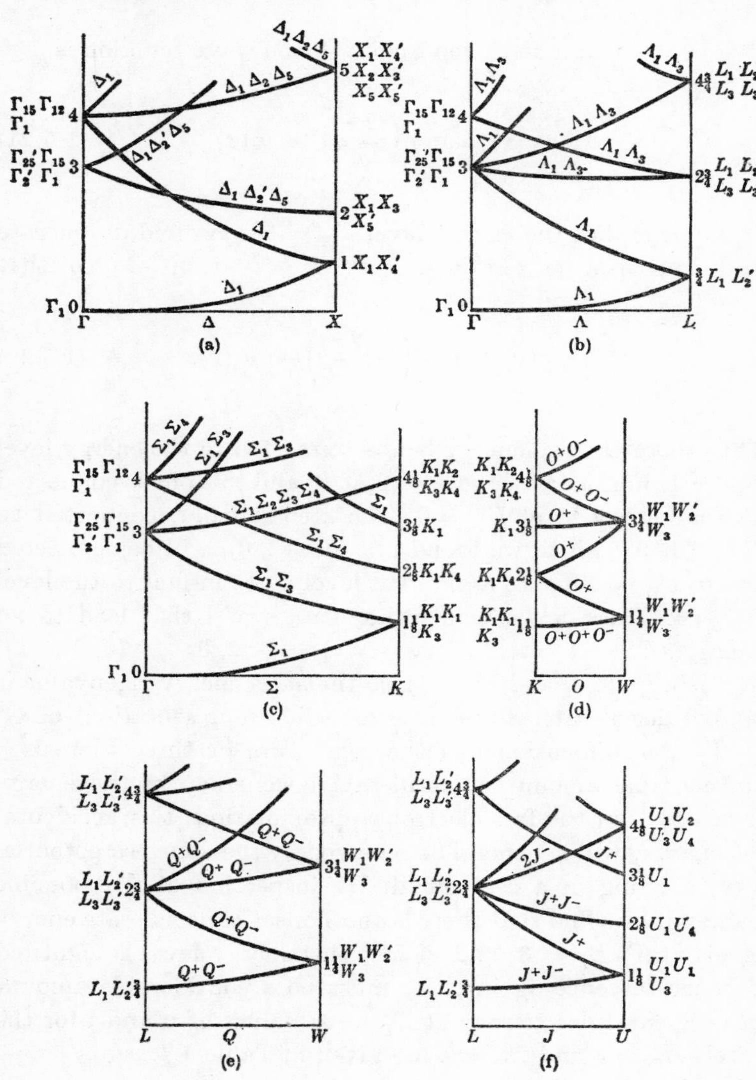

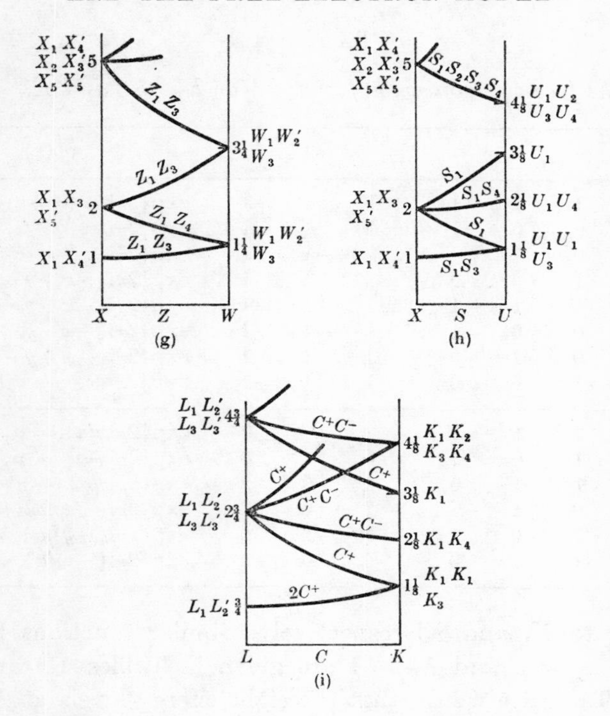

FIG. 1.12. The free-electron energy bands for the face-centred cubic structure. Energies are given in units of $(2\pi/a)^2$Ry.

necessary to take linear combinations of degenerate wave functions (for a detailed discussion see, for example, the book by Jones (1960)). Thus, whereas $\psi_\Gamma(\mathbf{r})$ in eqn (1.5.19) belongs to Γ_1 and $\psi_\Delta(\mathbf{r})$ in eqn (1.5.20) belongs to Δ_1, we now have to take linear combinations

$$\phi_1(\mathbf{r}) = u_1(\mathbf{r})+u_2(\mathbf{r}) = 2\cos\left(\frac{2\pi}{a}y\right), \tag{1.5.23}$$

and

$$\phi_2(\mathbf{r}) = u_1(\mathbf{r})-u_2(\mathbf{r}) = 2\mathrm{i}\sin\left(\frac{2\pi}{a}y\right), \tag{1.5.24}$$

to obtain wave functions that have the proper symmetry properties at X. It is possible to show that $\phi_1(\mathbf{r})$ and $\phi_2(\mathbf{r})$

TABLE 1.4

The free-electron wave function for $E_\Gamma = 3$ and $E_\Gamma = 4$

E	$\mathbf{n}$			$\mathbf{l}$			$\psi_{\mathbf{k}}(\mathbf{r})$
3	1	1	1	1	1	1	$\exp\{2\pi i(x+y+z)/a\}$
	0	0	1	1	1	−1	$\exp\{2\pi i(x+y-z)/a\}$
	0	1	0	1	−1	1	$\exp\{2\pi i(x-y+z)/a\}$
	1	0	0	−1	1	1	$\exp\{2\pi i(-x+y+z)/a\}$
	−1	−1	−1	−1	−1	−1	$\exp\{2\pi i(-x-y-z)/a\}$
	0	0	−1	−1	−1	1	$\exp\{2\pi i(-x-y+z)/a\}$
	0	−1	0	−1	1	−1	$\exp\{2\pi i(-x+y-z)/a\}$
	−1	0	0	1	−1	−1	$\exp\{2\pi i(x-y-z)/a\}$
4	0	1	1	2	0	0	$\exp\{2\pi i x/a\} = u_1$
	1	0	1	0	2	0	$\exp\{2\pi i y/a\} = u_2$
	1	1	0	0	0	2	$\exp\{2\pi i z/a\} = u_3$
	0	−1	−1	−2	0	0	$\exp\{2\pi i(-x)/a\} = u_4$
	−1	0	−1	0	−2	0	$\exp\{2\pi i(-y)/a\} = u_5$
	−1	−1	0	0	0	−2	$\exp\{2\pi i(-z)/a\} = u_6$

belong to X_1 and X_4' respectively. Similar functions for the levels $E_\Gamma = 3$ and $E_\Gamma = 4$ are given in Tables 1.5 and 1.6. From Table 1.5 we see that the eight energy levels at $E_\Gamma = 3$ arise from an accidental degeneracy between two non-degenerate levels, Γ_1 and Γ_2', and two threefold degenerate levels, Γ_{15} and Γ_{25}'. Thus, the inclusion of a more realistic potential $V(\mathbf{r})$ would split these levels, see Fig. 1.10. Similarly, from

TABLE 1.5

Symmetrized functions for $E_\Gamma = 3$

Representation of Γ	*Wave function*
Γ_1	$\cos(2\pi x/a)\cos(2\pi y/a)\cos(2\pi z/a)$
Γ_2'	$\sin(2\pi x/a)\sin(2\pi y/a)\sin(2\pi z/a)$
Γ_{25}'	$\cos(2\pi x/a)\sin(2\pi y/a)\sin(2\pi z/a)$
	$\sin(2\pi x/a)\cos(2\pi y/a)\sin(2\pi z/a)$
	$\sin(2\pi x/a)\sin(2\pi y/a)\cos(2\pi z/a)$
Γ_{15}	$\sin(2\pi x/a)\cos(2\pi y/a)\cos(2\pi z/a)$
	$\cos(2\pi x/a)\sin(2\pi y/a)\cos(2\pi z/a)$
	$\cos(2\pi x/a)\cos(2\pi y/a)\sin(2\pi z/a)$

TABLE 1.6

Symmetrized functions for $E_\Gamma = 4$
(See Table 1.4 for definition of u_1, u_2, etc.)

Representation of Γ	*Wave function*
Γ_1	$u_1+u_2+u_3+u_4+u_5+u_6$
Γ_{12}	$\begin{cases} u_1+u_2-2u_3+u_4+u_5-2u_6 \\ u_1-u_2+u_4-u_5 \end{cases}$
Γ_{15}	$\begin{cases} u_1-u_4 \\ u_2-u_5 \\ u_3-u_6 \end{cases}$

Table 1.6 we see that the six degenerate levels at $E_\Gamma = 4$ arise from an accidental degeneracy between one non-degenerate level, Γ_1, one twofold degenerate level, Γ_{12}, and one threefold degenerate level, Γ_{15}. The energy levels at X can be determined in a similar way to the method used above at Γ, when energy levels will be found at $E_X = 1$, $E_X = 2$, and $E_X = 5$. The process can be repeated again for the wave vectors **k** that terminate on the line of symmetry Δ. It is possible to determine the symmetry of a band that connects Γ and X from the compatibility relations that exist between Γ and Δ at one end and between X and Δ at the other end. Thus, for instance, from Table 1.3 the bands that emerge from the eightfold degenerate level $E_\Gamma = 3$ must contain the following irreducible representations for Δ:

Γ	Δ
Γ'_{25}	Δ'_2, Δ_5
Γ_{15}	Δ_1, Δ_5
Γ'_2	Δ'_2
Γ_1	Δ_1

namely $2\Delta_1$, $2\Delta'_2$, and $2\Delta_5$, see Fig. 1.12(a). In Fig. 1.12 we show the lowest 15 energy bands at Γ and show also how they change as **k** varies over the Brillouin zone.

The discussion that we have given for the face-centred cubic structure can be adapted in a straightforward manner to obtain the free-electron energy bands for a metal with any other structure. We use the term "free-electron energy bands" for any given metal with an arbitrary crystal structure, to mean the set of energy bands $E(\mathbf{k})$ given by eqn (1.5.7), using the relevant vectors $\mathbf{G}_n$ obtained from the basic vectors $\mathbf{t}_1$, $\mathbf{t}_2$, and $\mathbf{t}_3$ of the appropriate space group, and using the actual values of the lattice constants for that metal. Therefore, strictly speaking, even for two metals with a common crystal structure, such as Al and Pb which are both face-centred cubic metals, the free-electron energy bands will not be identical but will differ by a scaling factor that involves the square of the ratio of the lattice constants of the two metals. The free-electron energy bands for the simple cubic and body-centred cubic structures, as well as the face-centred cubic structure, are considered in some detail in the book by Jones (1960). The free-electron model that we have described in this section is sometimes referred to as the "empty lattice model" (Shockley 1937); this is perhaps a more appropriate title since although we are concerned with free electrons in the sense that $V(\mathbf{r})$ is zero, we are still requiring their wave functions to satisfy periodic boundary conditions associated with the existence of a regular lattice structure in the crystal.

1.6. Free-electron Fermi surfaces

In the previous section we showed that for a free electron in a periodic lattice the energy bands are given by

$$E(\mathbf{k}) = \frac{\hbar^2}{2m}|\mathbf{k}+\mathbf{G}_n|^2 \tag{1.5.7}$$

and the corresponding wave functions are given by

$$\psi_k(\mathbf{r}) = \exp\{i(\mathbf{k}+\mathbf{G}_n) \cdot \mathbf{r}\} \tag{1.5.8}$$

where $\mathbf{G}_n$ is a vector of the reciprocal lattice. Eqn (1.5.7) defines a set of constant energy surfaces and the Fermi surface will be the particular constant energy surface corresponding to $E(\mathbf{k}) = E_F$ where E_F is the Fermi energy. The value of E_F will depend on the number of conduction electrons per fundamental unit cell of the crystal. It can be seen that it is only the

Bravais lattice of the crystal, and not the complete space group, that affects the values of $E(\mathbf{k})$ in eqn (1.5.7). When we speak of the "free-electron Fermi surface" for a given metal with an arbitrary crystal structure, we mean the Fermi surface that is obtained by using the free-electron energy bands given in eqn (1.5.7) with the vectors $\mathbf{G}_n$ appropriate to the actual crystal structure of the metal in question. The shape of the free-electron Fermi surface for a given metal in the free-electron approximation therefore depends on the Bravais lattice of the crystal, the lattice spacings a, b, and c, and the number of conduction electrons per unit cell. There are many metals which occur in one of the three common crystal structures: the cubic close-packed (face-centred cubic or f.c.c.), the body-centred cubic (or b.c.c.), and the hexagonal close-packed (or h.c.p.) structures; it is therefore of some general interest to consider the shape of the free-electron Fermi surface for each of these structures and for metals of various valence. However, other examples of free-electron Fermi surfaces only apply either to one particular metal, for example Ga or Sn, or to the metals in one particular group of the periodic table, for example As, Sb, and Bi. Such free-electron Fermi surfaces will be mentioned in the appropriate positions in Chapters 4 and 5.

The shapes of the free-electron Fermi surfaces for the three common metallic structures were illustrated, for metals of various valence, by Harrison (1960*b*). The volume of the Brillouin zone is equal to

$$\mathbf{g}_1 \cdot (\mathbf{g}_2 \wedge \mathbf{g}_3) = 8\pi^3/\{\mathbf{t}_1 \cdot (\mathbf{t}_2 \wedge \mathbf{t}_3)\} = 8\pi^3/V, \qquad (1.6.1)$$

where V is the volume of the fundamental unit cell of the crystal. The Brillouin zone contains $2N_1N_2N_3$ allowed electronic states per band where $N_1N_2N_3$ is the number of unit cells in the specimen of the metal under consideration. If the total number of conduction electrons contributed by all the atoms in any one fundamental unit cell of the metal is x, the volume of $\mathbf{k}$-space that will be occupied corresponds to $\frac{1}{2}x$ bands, allowing for spin, and will therefore be $\frac{1}{2}x(8\pi^3/V)$. If one were to use extended $\mathbf{k}$-space rather than just one particular unit cell of $\mathbf{k}$-space, namely the Brillouin zone, eqn (1.5.7) would simplify to

$$E(\mathbf{k}) = \frac{\hbar^2}{2m} |\mathbf{k}|^2, \tag{1.6.2}$$

so that the constant energy surfaces would be spheres. Consequently the Fermi surface would also be a sphere. A convenient way of discussing the free-electron Fermi surface for a given metal is therefore to use the known number of conduction electrons per unit cell to determine the volume of the spherical free-electron Fermi surface in extended **k**-space and hence k_F, its radius. k_F will be determined by

$$\tfrac{4}{3}\pi k_F^3 = \tfrac{1}{2}x\left(\frac{8\pi^3}{V}\right), \tag{1.6.3}$$

therefore

$$k_F = (3x\pi^2/V)^{\frac{1}{3}}. \tag{1.6.4}$$

For the three common metallic structures it is convenient to express k_F in terms of $(2\pi/a)$, so that

$$k_F = \left(\frac{2\pi}{a}\right)\left(\frac{3xa^3}{8\pi V}\right)^{\frac{1}{3}}. \tag{1.6.5}$$

Correspondingly, the Fermi energy is given by

$$E_F = \frac{\hbar^2}{2m}\left(\frac{2\pi}{a}\right)^2\left(\frac{3xa^3}{8\pi V}\right)^{\frac{2}{3}} = \frac{\hbar^2}{2m}\left(\frac{3x\pi^2}{V}\right)^{\frac{2}{3}}. \tag{1.6.6}$$

The spherical Fermi surface of which we have found the radius in eqn (1.6.5) is in the extended zone scheme, but in practice one does not usually use the extended zone scheme so that it is necessary to investigate what happens to those parts of the sphere of occupied states within the Fermi surface that have **k** vectors which terminate outside the Brillouin zone. For the end point of any such wave vector **k** corresponding to an energy eigenstate that is occupied but is outside the Brillouin zone, it is possible to find some vector $\mathbf{G}_n$ of the reciprocal lattice (given by equation (1.2.11)) which can be added to **k** to give another wave vector which is within the Brillouin zone. In Section 1.3 it was shown that the wave vectors **k** and $\mathbf{k}+\mathbf{G}_n$

are equivalent and have the same set of energy eigenvalues. This means that any piece of the sphere of occupied states can be moved bodily through a suitable reciprocal lattice vector chosen so that these occupied states come within the Brillouin zone. We may illustrate this first of all with a two-dimensional example. Suppose we consider a hypothetical two-dimensional metal in which the metal atoms are situated on the lattice points of a square two-dimensional Bravais lattice with basic

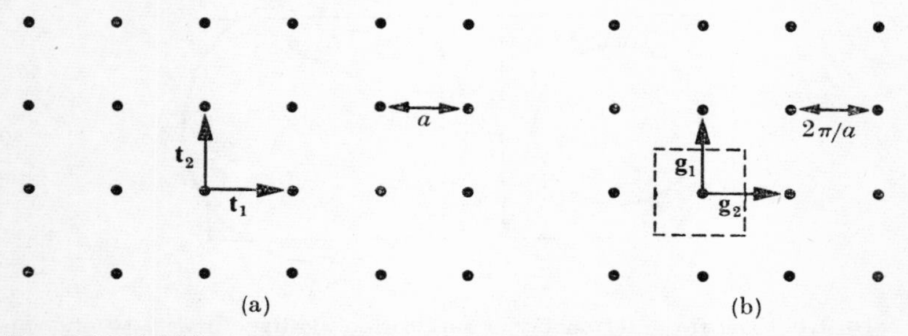

Fig. 1.13. (a) The crystal structure of a hypothetical two-dimensional metal, and (b) the reciprocal lattice and Brillouin zone for the same hypothetical metal.

vectors $\mathbf{t}_1 = a\mathbf{i}$ and $\mathbf{t}_2 = a\mathbf{j}$. The reciprocal lattice is also a square Bravais lattice and the Brillouin zone is a square unit cell of this reciprocal lattice as shown in Fig. 1.13; $\mathbf{g}_1 = (2\pi/a)\mathbf{i}$ and $\mathbf{g}_2 = (2\pi/a)\mathbf{j}$ and the area of the Brillouin zone is $4\pi^2/a^2$. If the valence of the metal is 1, the area occupied by electrons will be half the area of the Brillouin zone, allowing for the two possible spin states for each electron. Therefore

$$\pi k_{\mathrm{F}}^2 = \frac{1}{2}\frac{4\pi^2}{a^2} \tag{1.6.7}$$

so that

$$k_{\mathrm{F}} = \frac{1}{\sqrt{(2\pi)}}\left(\frac{2\pi}{a}\right) \doteqdot 0.399\left(\frac{2\pi}{a}\right). \tag{1.6.8}$$

The free-electron Fermi surface is therefore completely contained within the first Brillouin zone in the extended zone scheme, that is, it is entirely in band 1 in the reduced zone scheme, see Fig. 1.14. If the valence of the metal is 2, the radius, k_F, of the Fermi surface will be given by

$$\pi k_F^2 = \frac{4\pi^2}{a^2} \qquad (1.6.9)$$

so that

$$k_F = \frac{1}{\sqrt{\pi}}\left(\frac{2\pi}{a}\right) \doteqdot 0.564\left(\frac{2\pi}{a}\right), \qquad (1.6.10)$$

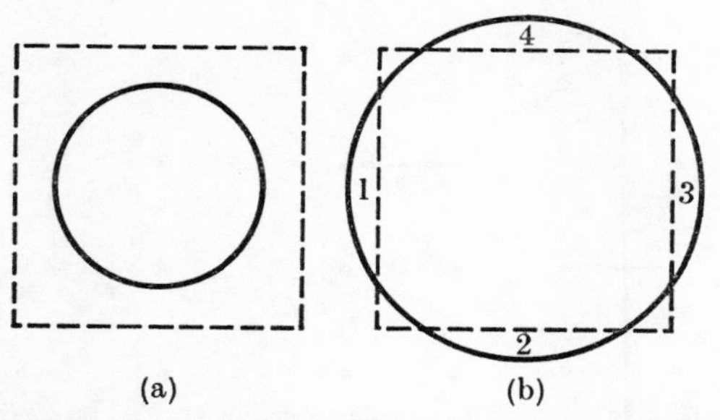

FIG. 1.14. The free-electron Fermi surface for the hypothetical two-dimensional metal shown in Fig. 1.13; (a) valence 1, (b) valence 2, extended zone scheme.

and the Fermi surface is now not completely contained within the first Brillouin zone in the extended zone scheme, see Fig. 1.14(b). The regions of occupied states labelled 1, 2, 3, and 4 in Fig. 1.14(b) which are outside the first Brillouin zone can be moved into the first Brillouin zone by translations through $+\mathbf{g}_2$, $+\mathbf{g}_1$, $-\mathbf{g}_2$, and $-\mathbf{g}_1$, respectively. In the reduced zone scheme there will therefore be pieces of Fermi surface in band 1 and band 2, see Fig. 1.15. The argument could be extended to cases of higher valence.

We now return to the case of three-dimensional real metals. The value of V, the volume of the fundamental unit cell, is given by $\mathbf{t}_1 \cdot (\mathbf{t}_2 \wedge \mathbf{t}_3)$ and using the appropriate values of $\mathbf{t}_1$, $\mathbf{t}_2$, and $\mathbf{t}_3$ we obtain for the face-centred cubic structure

$$V = \tfrac{1}{4}a^3 \qquad (1.6.11)$$

for the body-centred cubic structure

$$V = \tfrac{1}{2}a^3 \qquad (1.6.12)$$

and for the hexagonal close-packed structure

$$V = \tfrac{1}{2}\sqrt{3}\ \rho a^3 \qquad (1.6.13)$$

where $\rho = c/a$. If there is no distortion of the hexagonal structure from perfect close packing, ρ is equal to $\sqrt{(\frac{8}{3})}$, but many metals that are described as having the hexagonal

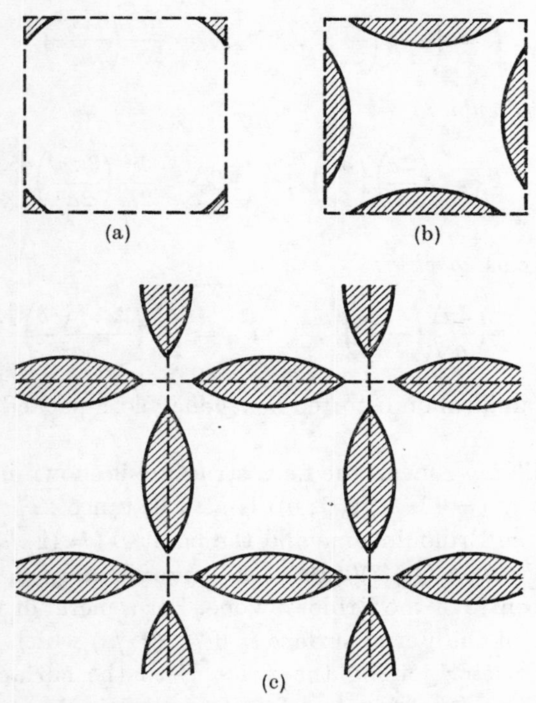

Fig. 1.15. The free-electron Fermi surface for the hypothetical two-dimensional metal of valence 2 shown in Fig. 1.14(b); in the reduced zone scheme, (a) band 1, unoccupied regions shaded, (b) band 2, occupied regions shaded, and (c) in the repeated zone scheme, band 2, occupied regions shaded.

close-packed structure do have a slight distortion and therefore a slightly different value of ρ. By substituting the values of V from equations (1.6.11–13) into eqns (1.6.5) and (1.6.6) we obtain, for each value of x, the values of k_F, the radius of the free-electron Fermi surface in the extended zone scheme, and E_F, the Fermi energy. For the f.c.c. and b.c.c. structures there is only one atom in each fundamental unit cell so that x is just equal to the valence of the metal. For the h.c.p. structure which has two atoms per unit cell, the value of x will be twice the valence of the metal. The values of k_F and E_F for these three structures are then given by

face-centred cubic:

$$k_F = \left(\frac{2\pi}{a}\right)\left(\frac{3x}{2\pi}\right)^{\frac{1}{3}}, \qquad E_F = \frac{\hbar^2}{2m}\left(\frac{3x\pi^2}{4a^3}\right)^{\frac{2}{3}} \tag{1.6.14}$$

body-centred cubic:

$$k_F = \left(\frac{2\pi}{a}\right)\left(\frac{3x}{4\pi}\right)^{\frac{1}{3}}, \qquad E_F = \frac{\hbar^2}{2m}\left(\frac{3x\pi^2}{2a^3}\right)^{\frac{2}{3}} \tag{1.6.15}$$

and

hexagonal close-packed:

$$k_F = \left(\frac{2\pi}{a}\right)\left(\frac{x\sqrt{3}}{4\pi\rho}\right)^{\frac{1}{3}}, \qquad E_F = \frac{\hbar^2}{2m}\left(\frac{2x\pi^2\sqrt{3}}{\rho a^3}\right)^{\frac{2}{3}} \tag{1.6.16}$$

where for an ideal undistorted hexagonal close-packed structure $\rho = \sqrt{(\frac{8}{3})}$.

The Brillouin zone of the f.c.c. structure is shown in Fig. 1.8. The point X $(= (2\pi/a)(0, 1, 0))$ is at a distance $2\pi/a$ from the centre of the Brillouin zone and the point L $(= (2\pi/a)(\frac{1}{2}, \frac{1}{2}, \frac{1}{2}))$ is at a distance $(\frac{1}{2}\sqrt{3})(2\pi/a)$, or approximately $0{\cdot}866(2\pi/a)$, from the centre of the Brillouin zone. For a metal of valence 1, the radius of the Fermi surface is $0{\cdot}782(2\pi/a)$ which is shorter than any vector joining the point Γ to the surface of the Brillouin zone. The Fermi surface for $x = 1$ is thus completely within the first Brillouin zone in the extended zone scheme; this means that it is entirely contained in band 1 in the more conventional reduced zone scheme (see Fig. 1.16). In the later chapters of this book we shall see that in the alkali metals this free-electron Fermi surface is a very close approximation to the actual Fermi surfaces for those metals. For a metal with valence 2, the radius, k_F, of the Fermi surface is $0{\cdot}985(2\pi/a)$, which means that the Fermi surface now intersects the surface of the Brillouin zone and this is best studied by looking at the

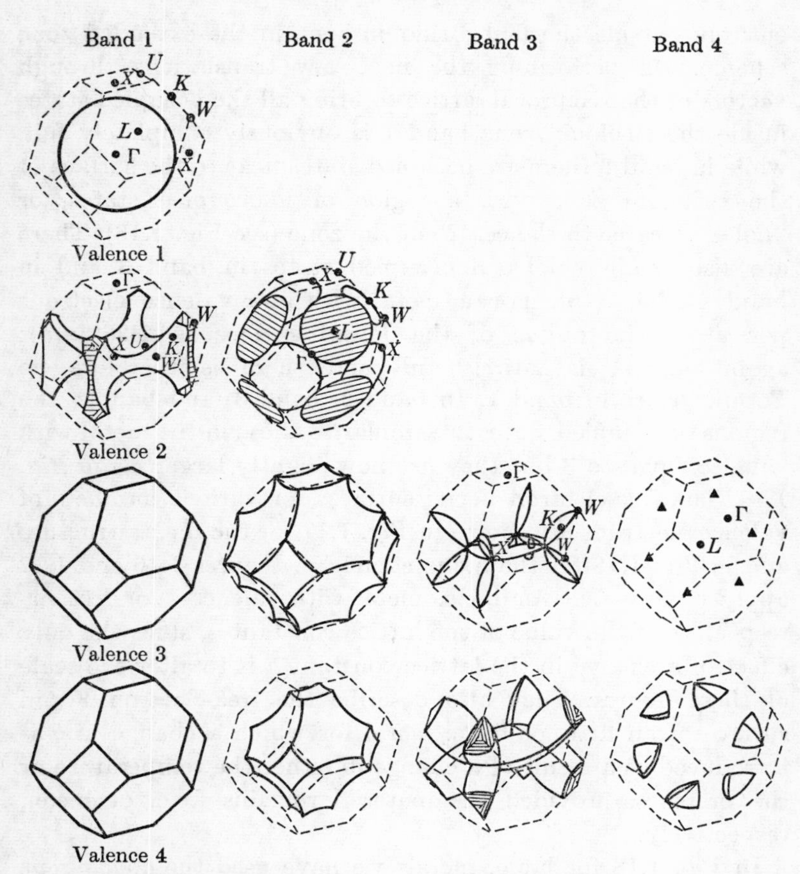

FIG. 1.16. Free-electron Fermi surfaces for f.c.c. metals of various valence (Harrison 1960*b*).

various important cross-sections of the Brillouin zone and Fermi surface. The Fermi surface for $x = 2$ cuts the large hexagonal faces of the Brillouin zone but does not cut the small square faces. Each of the eight spherical caps cut off by the hexagonal faces of the Brillouin zone can be moved back into the Brillouin zone by a translation through the reciprocal lattice vector which is normal to the appropriate hexagonal face (see Fig. 1.16). For an f.c.c. metal with valence 3 the radius, k_F, of the Fermi surface is $1{\cdot}13(2\pi/a)$, and this Fermi surface is entirely outside the first Brillouin zone in the extended zone scheme. On performing the necessary translations through vectors of the reciprocal lattice to bring all the occupied states inside the Brillouin zone band 1 is obviously completely full, while in band 2 there are occupied states near to the surface of the Brillouin zone with a region of unoccupied states, or "holes", nearer to the centre of the zone (see Fig. 1.16). There are also small regions of occupied states in band 3 and in band 4 (Fig. 1.16). For a metal with four valence electrons per atom the radius of the Fermi surface is $1{\cdot}25(2\pi/a)$; again band 1 is entirely full and there is therefore no Fermi surface in band 1. In band 2, band 3, and band 4 the regions of occupied states are similar to those in the metal with valence equal to 3 but they are now slightly larger, as in Fig. 1.16. The free-electron Fermi surfaces for various numbers of valence electrons are shown in Fig. 1.17 for the b.c.c. structure and in Fig. 1.18 for the h.c.p. structure. Figures 1.16 and 1.17 apply to any f.c.c. or b.c.c. metal with valence 4 or less, irrespective of the value of the lattice constant a, since the only effect of a change in the lattice constant a is to alter the scale of these figures. They also describe the free-electron Fermi surface of an f.c.c. or b.c.c. metal for which a change in a is introduced as a result of a change in either the temperature or the pressure, provided the material remains f.c.c. or b.c.c., respectively.

In Fig. 1.18 for h.c.p. metals we have used the *double-zone scheme*. This is useful because of the important group-theoretical result mentioned in Section 1.4 that the energy bands $E_j(\mathbf{k})$ are twofold degenerate all over the large hexagonal face of the Brillouin zone of the h.c.p. structure. Consequently, if the Fermi surface cuts this hexagonal face at all it must do so in pairs of sheets which stick together at this face. It is therefore convenient in helping to visualize the Fermi surface of an h.c.p. metal to use a double zone which has twice the volume of a fundamental unit cell of **k**-space. In this double zone scheme the energy bands $E_j(\mathbf{k})$ for which j is odd are represented in the part of the zone specified by $|\mathbf{k}_z| \leqslant \pi/c$ and bands

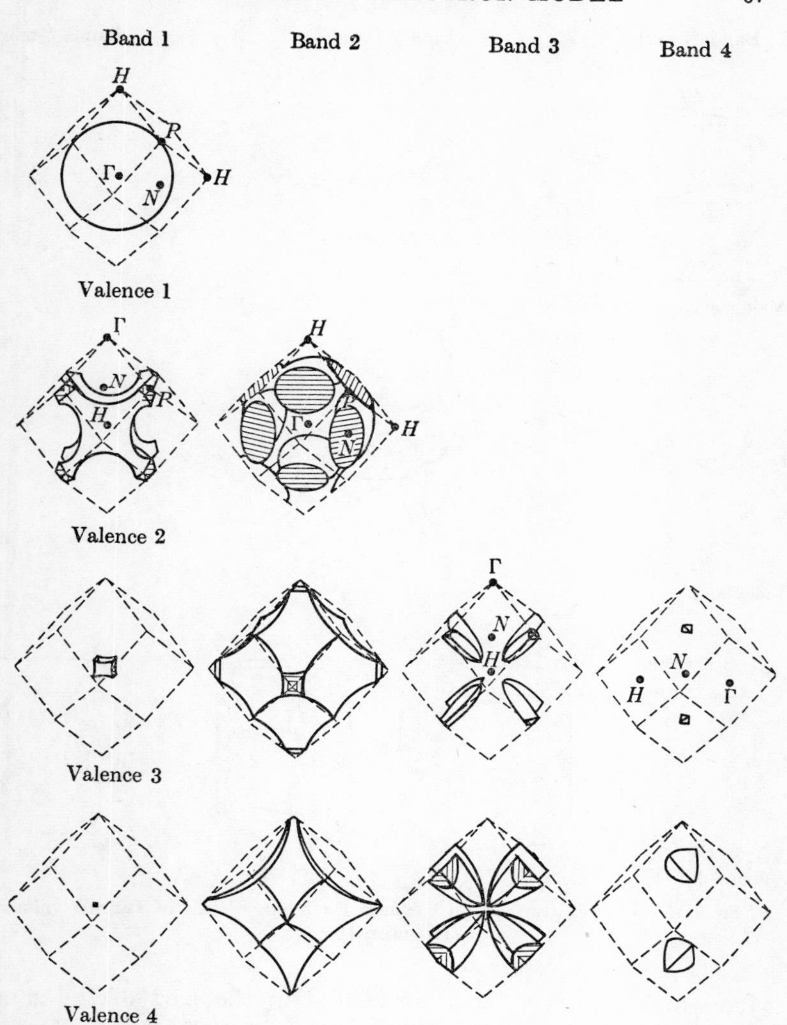

FIG. 1.17. Free-electron Fermi surfaces for b.c.c. metals of various valence (Harrison 1960*b*).

Bands 1 and 2 | Bands 3 and 4 | Bands 5 and 6

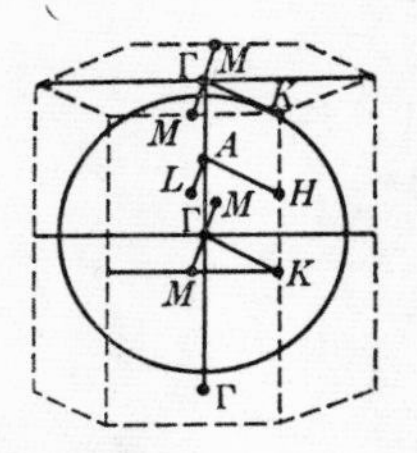

Valence 1

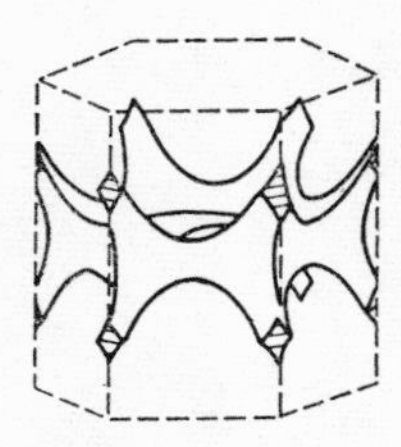

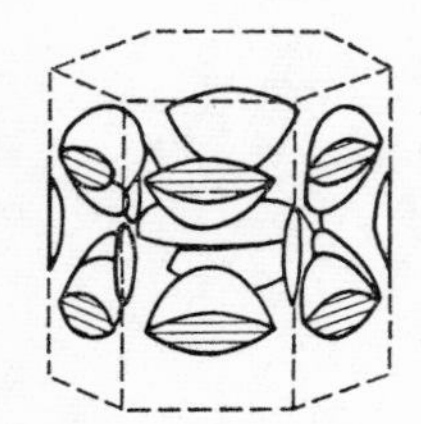

Valence 2

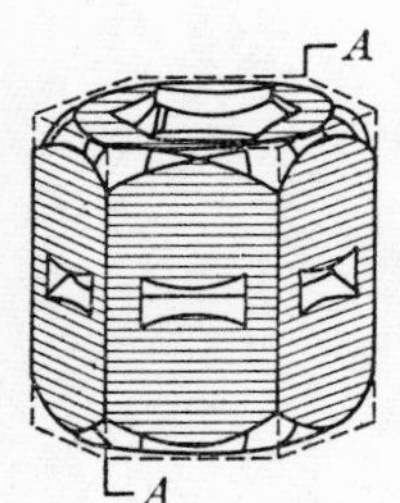

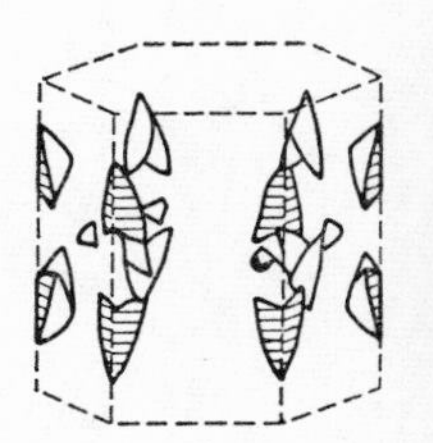

Valence 3 Section A – A

FIG. 1.18. Free-electron Fermi surfaces for h.c.p. metals of various valence (Harrison 1960*b*).

for which j is even are represented in the part of the zone specified by $\pi/c \leqslant |\mathbf{k}_z| \leqslant 2\pi/c$. Of course, the energy bands and the Fermi surface in the region $\pi/c \leqslant |\mathbf{k}_z| \leqslant 2\pi/c$ could all be folded back into the conventional first Brillouin zone and indeed this has to be done once spin–orbit coupling is introduced (see Section 2.11).

The illustrations in Fig. 1.18 are for the value of $\rho = \sqrt{(\frac{8}{3})}$ which is the ideal value for an h.c.p. metal, whereas for many h.c.p. metals ρ departs from this ideal value. Often the effect of the departure of c/a from the ideal value of $\sqrt{(\frac{8}{3})}$ is not particularly important because, although it alters the relative dimensions of the various pieces of Fermi surface, it does not generally cause any alterations of the general features, that is, it does not cause pieces of Fermi surface to appear or disappear and it does not alter the connectivities of the various sheets of the Fermi surface. However, for Cd c/a is sufficiently greater than $\sqrt{(\frac{8}{3})}$ that the vertical needle- or cigar-shaped pockets of electrons in bands 3 and 4 at K do not exist at all in the

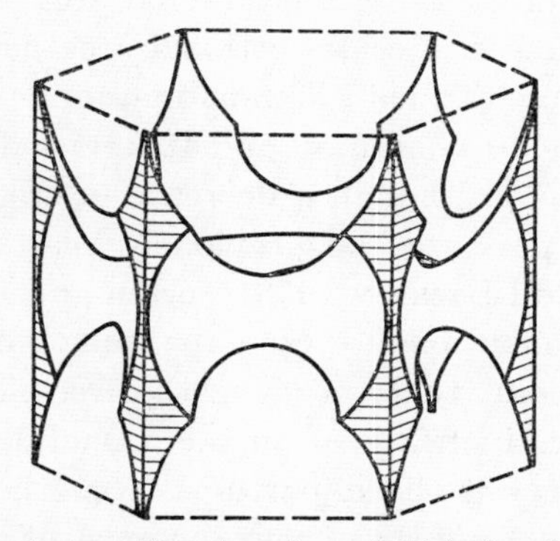

FIG. 1.19. The free-electron Fermi surface of Cd in bands 1 and 2 (Harrison 1960*b*).

free-electron Fermi surface (cf. Fig. 1.18). The critical value of c/a for the disappearance of these needles is 1·8607 (Harrison 1960*b*) which is substantially smaller than the actual value of c/a for Cd (about 1·886). Simultaneously the connectivity of the arms of the monster in bands 1 and 2 becomes altered, see Fig. 1.19. For Zn c/a is smaller than in Cd and is estimated by Harrison (1960*b*) to be 1·8246 at the low temperatures usually used in Fermi surface studies; this is smaller than the critical value of 1·8607 so that in the free-electron model these needles survive for Zn. It is interesting to note that at high pressures c/a for Cd decreases and eventually reaches a value at which, in the free-electron model, these needle-shaped pockets of electrons would be expected to appear. Evidence for the appearance of these needles in Cd has been obtained in magnetoresistance measurements on the metal at pressures up to 15 kbar (Gaĭdukov and Itskevich 1963; Itskevich and Voronovskiĭ 1966). Verkin, Kuzmicheva, and Svechkaryov (1968) were able to observe the disappearance of the needles in pure Zn when the temperature was raised as well as the appearance of the needles in Cd when doped with Mg; at the disappearance or appearance of the needles a maximum occurs in the susceptibility. The principal effect of pressure on Zn is to decrease the axial ratio c/a and this, in the free-electron approximation, has the effect of increasing the horizontal cross-sectional area of the needles in band 3; this increase in cross-sectional area has been observed experimentally by various different methods (Balain, Grenier, and Reynolds 1960; Gĭadukov and Itskevich 1963; O'Sullivan and Schirber 1965) although a decrease in cross-sectional area with increasing pressure had previously been reported (Dmitrenko, Verkin, and Lazarev 1958; Verkin and Dmitrenko 1958, 1959). We have mentioned these experiments on Zn and Cd to emphasize the point that, in using the free-electron model to describe the Fermi surface of an h.c.p. metal, it is important to take into account the departures from the ideal value of the axial ratio $\rho = c/a$ that occur for most of them.

In this section we have been concerned with the deduction of the shape of the Fermi surface in the free-electron approximation. For metals which are made up of atoms with a complicated electronic structure—metals of high atomic number and especially transition and rare-earth metals—this approximation is so far removed from the truth as to be almost useless. However, as we shall see in Chapter 4 there are many simple metals for which the free-electron model gives a very good first approximation to the Fermi surface of the metal.

2

THE CALCULATION OF ELECTRONIC ENERGY BAND STRUCTURES AND FERMI SURFACES

IN CHAPTER 1 we discussed the energy band structure and the Fermi surface of a metal in the free-electron approximation. In Chapter 4 we shall see that this model gives a surprisingly good first approximation to the band structures and Fermi surfaces of many of the simpler non-transition and non-rare-earth metals. In the present chapter we turn to the consideration of the problem of determining the band structure and Fermi surface of a real metal in which some allowance must be made for the fact that $V(\mathbf{r})$ is not zero everywhere. There are two main lines of approach.

First, in the individual particle approximation one can take Schrödinger's equation

$$-\frac{\hbar^2}{2m}\nabla^2\psi(\mathbf{r})+V(\mathbf{r})\psi(\mathbf{r}) = E\psi(\mathbf{r}) \tag{1.2.3}$$

and try to construct *ab initio* some realistic form of the potential $V(\mathbf{r})$ experienced by an electron in the metal. The starting point for the construction of $V(\mathbf{r})$ is usually the potential of a suitably chosen free ion, calculated by solution of the Hartree–Fock equations (Hartree 1957; Herman and Skillman 1963) and then realistic estimates are made of as many additional contributions as possible; these additional contributions include the electrostatic potential due to the other conduction electrons, exchange among the conduction electrons, exchange between conduction electrons and ion-core electrons, correlation among ion-core electrons, among conduction electrons and between ion-core electrons and conduction electrons, spin–orbit coupling and one or two other contributions. Then, having constructed $V(\mathbf{r})$, there are several methods available for solving eqn

(1.2.3) subject to the appropriate boundary conditions that actually exist in the metal, see Sections 2.2–5. We shall see in Chapters 4 and 5 that *ab initio* band structure calculations have now been performed for a large number of different metals; in many cases the accuracy and usefulness of the results have been very much limited by uncertainties in the construction of $V(\mathbf{r})$. Unfortunately the accurate calculation of $V(\mathbf{r})$ is complicated by the fact that the calculation of many of the contributions to $V(\mathbf{r})$ requires a knowledge of the conduction electron wave functions which can only be found by solving eqn (1.2.3). In other words the accurate calculation of $E(\mathbf{k})$ and $\psi_{\mathbf{k}}(\mathbf{r})$ from first principles involves an enormous problem in self-consistency similar to the problem of the solution of the Hartree–Fock equations for an atom or ion, but made very much more complicated by the very much larger number of particles and equations. Self-consistent calculations for the electrons in a metal are only now just beginning to be performed (see, for example, Snow and Waber (1967) for a review of early attempts in this direction).

The second approach, which has been used quite extensively in connection with the calculation of the band structures of simple metals, is by the use of a pseudopotential; this approach owes much of its success to the fact that the free-electron model gives a very good first approximation to the band structures and Fermi surfaces of many simple metals. This means that the net scattering of a conduction electron by the atoms in one of these metals is weak. This does not, however, mean that the actual potential $V(\mathbf{r})$ itself is weak, which would obviously not be true near the nuclei of the atoms in the metal. The weakness of the scattering does, however, mean that a realistic pseudopotential can be expressed in terms of quite a small number of Fourier coefficients $V(\mathbf{k})$ which can then be regarded as adjustable parameters that are varied until the calculated Fermi surface for a given metal gives the best fit to a set of experimental caliper dimensions or cross-sectional areas of the Fermi surface of that metal. One can regard the pseudopotential method as a sophisticated interpolation scheme based on physical rather than mathematical ideas which enables the shape of the entire Fermi surface to be determined numerically from the necessarily rather small number of experimentally determined caliper dimensions or extremal areas of cross section. Having used various direct experimental measurements for a few important directions or planes to determine the parameters that specify the pseudopotential, the effect of the conduction electrons on various other macroscopic static or transport properties can then be expressed in terms of the pseudopotential.

There have been quite a number of books and review articles written on the subject of the methods used in band structure calculations and the reader who is interested in more details than it is possible to give in this chapter is advised to consult these sources: Reitz (1955); Herman (1958); Pincherle (1960, 1971); Blount (1962*b*); Callaway (1964); Heine (1964, 1969); Slater (1965, 1967); Fletcher (1971).

We ought to make some brief mention of the problem of units. In calculations of the wave functions and the energies of the electrons in an atom it is common to use the system of units which are sometimes called *atomic units* or *Hartree units*. In this system (see, for example, Hartree (1957)) we choose:

$$m = 1, \qquad |e| = 1, \qquad \text{and} \qquad \hbar = 1$$

where m is the rest-mass of an electron, $|e|$ is the magnitude of the charge of an electron and $2\pi\hbar$ is Planck's constant. With this choice of units one finds that

(i) the atomic unit of length is a_0, the radius of the first Bohr orbit in an atom of hydrogen, which is approximately 0·529 Å or $5{\cdot}29 \times 10^{-11}$ m,

(ii) the atomic unit of energy is *twice* the ionization energy of an atom of hydrogen in Bohr's theory, that is 2 Ry, or approximately 27·2 eV or $4{\cdot}36 \times 10^{-18}$ J, and

(iii) c, the velocity of light, takes the value $c \risingdotseq 137$ atomic units of velocity.

The factor of 2 which relates the atomic unit of energy to the rydberg is a very fruitful source of numerical errors and great care is needed so as not to confuse the two units in practice.

Calculations of the band structure of a solid are nearly always performed using the atomic system of units because this is the most convenient system to use since the equations become simplified by the disappearance of m, e, and $\hbar$. However, one has to acknowledge that a large amount of experimental data is not analyzed in terms of atomic units. For example, whereas calculated band structures are usually given in terms of the atomic unit of energy ($= 2$ Ry), or in terms of the rydberg, Ry ($= \frac{1}{2}$ atomic unit of energy), experimental results are usually presented in terms of electron volts (eV), which are quite convenient practical units of energy. Moreover, there is a second set of units which is commonly used in work on relativistic problems and which is different from atomic units or Hartree units (see Section 2.10). Therefore, because atomic units are not universally adopted in practice, and to avoid possible confusion when we come to consider relativistic effects, we shall retain the constants m, e, and $\hbar$ in our formulae and equations.

2.1. The crystal potential

In this section we shall give some discussion of the problems involved in constructing the potential $V(\mathbf{r})$ for use in *ab initio* band structure calculations. It is common to write $V(\mathbf{r})$ as a sum of identical atomic-like potentials centred at the positions, $\mathbf{R}_\alpha$, at which the various ions in the structure are located

$$V(\mathbf{r}) = \sum_{\alpha} U(|\mathbf{r}-\mathbf{R}_\alpha|) \tag{2.1.1}$$

where $U(r)$ can be written as

$$U(r) = \frac{-1}{4\pi\epsilon_0 r} Z_p(r)\,|e|. \tag{2.1.2}$$

$Z_p(r)\,|e|$ is an effective charge resulting from the positive charge of $+Z\,|e|$ on the nucleus which is screened to a variable extent depending on r; Z is the atomic number of the metal in question.

As r becomes very small so $Z_p(r) \to +Z$ and as r becomes very large $Z_p(r) \to 0$.

The construction of $V(\mathbf{r})$ is a complicated many-body problem involving all the nuclei and all the electrons of all the atoms in the metal. In practice the positions of the nuclei can be regarded as fixed and any effects due to the lattice vibrations need only be introduced later (see Sections 6.4 and 6.5). It is convenient to separate the electrons into (i) ion-core electrons, which are regarded as being quite firmly localized on the individual ions, and (ii) conduction electrons, which are regarded as being free to wander throughout the specimen of the metal in de-localized states possessing band-like properties, that is, for which $E(\mathbf{k})$ depends quite strongly on $\mathbf{k}$. Although this division between localized ion-core electrons and de-localized or itinerant conduction electrons is quite sharp for many metals—the alkali metals for example—it should not be assumed that this will always automatically be completely rigorous for all metals. Electrons deep inside the ion core experience a potential $V(\mathbf{r})$ not significantly different from the potential they would feel in a single free atom of that element. For these electrons the energy eigenvalues will be $\mathbf{k}$-independent and will still be meaningfully described by the quantum numbers associated with atomic energy levels. However, it is possible for electrons that are less deep within the ion core to be spread out into bands with a definite band structure; even though such bands are separated by a finite gap from the conduction band, it is clearly only an approximation to assume that these electrons can be completely adequately described by energy levels and wave functions obtained from Hartree–Fock calculations on free atoms or ions. This dependence of $E(\mathbf{k})$ on $\mathbf{k}$ for electrons within the ion core of a metal in the fourth row of the periodic table (potassium, calcium, etc.) is illustrated in Fig. 2.1 for the 3s and 3p electrons in f.c.c. solid argon; although this is not a metal, the behaviour of the 3s and 3p electrons in nearby f.c.c. metals in the periodic table is unlikely to be very different from the case of solid argon. However, we do not wish to make too much of this formal difficulty of the division of the electrons

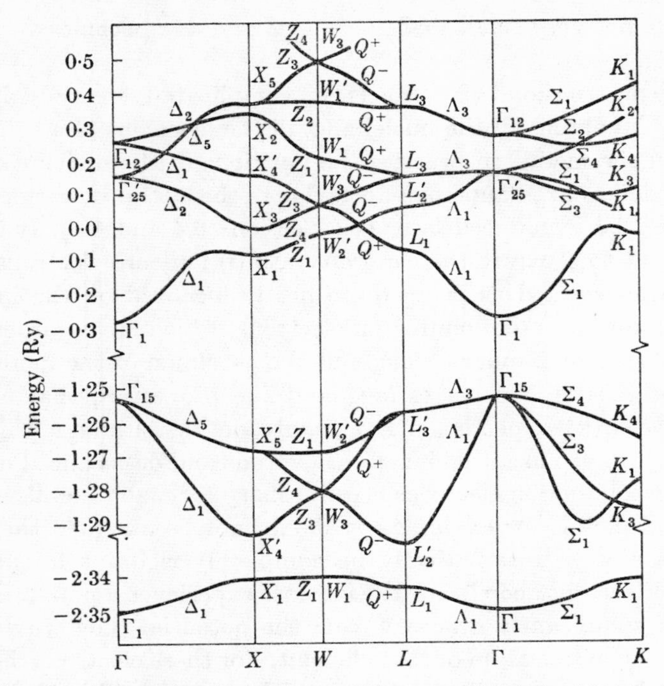

FIG. 2.1. Energy bands for f.c.c. solid argon (Mattheiss 1964*a*); the lowest band arises from the full 3s-states, the next three bands, between −1·25 Ry and −1·29 Ry, arise from the full 3p-states, and the higher bands arise from empty states.

into ion-core and conduction electrons because for practical purposes this division is sufficiently valid to be of considerable use.

Accepting the division into ion-core and conduction electrons the main contributions to the potential $V(\mathbf{r})$ experienced by an electron in a metal (Heine 1957*c*; Falicov 1962) includes:

(i) electrostatic potential due to the ion cores,
(ii) electrostatic potential due to the other conduction electrons,
(iii) exchange among the ion-core electrons,
(iv) correlation among the ion-core electrons,
(v) exchange between ion-core electrons and conduction electrons,

(vi) correlation between ion-core electrons and conduction electrons,
(vii) exchange among the conduction electrons,
(viii) correlation among the conduction electrons,
(ix) spin–orbit coupling, and
(x) deviation from spherical symmetry around the ion core.

The electrostatic potentials (i) and (ii) give the largest contributions to $V(\mathbf{r})$, while the exchange and correlation effects give significant extra contributions to $V(\mathbf{r})$. The magnitude of the spin–orbit coupling contribution depends very much on the metal in question. Generally for metals of low atomic number, Z, it will be very small but it becomes significant for metals of high Z. If the ion-core electronic energy levels and wave functions are assumed not to depart significantly from those in the appropriate free ion, then contributions (i) and (iii) are included in the potential obtained from the atomic (or ionic) Hartree–Fock self-consistent field. Some idea of the relative magnitudes of various contributions to $V(\mathbf{r})$ can be obtained from the potential used by Falicov (1962) for Mg which is given in Table 2.1. This particular potential is chosen by way of illustration because it shows rather nicely the relative significance of the different contributions to $V(\mathbf{r})$ and not because of any particular intrinsic accuracy it might possess; indeed recent experimental evidence suggests that this potential is not particularly accurate and produces a band structure and Fermi surface for Mg which is not sufficiently free-electron-like (see Section 4.3).

In writing the potential in the form given in eqn (2.1.1.) we have not paid very careful attention to the form of the potential in the region in which an electron is approximately equidistant from two different nuclei, that is, in the region near the boundaries of the Wigner–Seitz unit cell of the crystal. For a hypothetical two-dimensional metal with the square Bravais lattice shown in Fig. 1.13(a), it should be apparent that although $U(r)$ is, for small r ($\leqslant \frac{1}{2}a$), spherically symmetrical, this is no longer the case if $\frac{1}{2}a < r < \frac{1}{2}a\sqrt{2}$. In general if $r_{\min}$ and

TABLE 2.1

Various contributions to the crystal potential in Mg
(The potential is expressed in the form given in eqn (2.1.2))

r	$2Z_1$	$2Z_2$	$2Z_3$	$2Z_4$	$2Z_5$	$2Z_P$
0	24·000	0·000	0·000	0·000	0·000	24·000
0·025	22·099	−0·004	0·387	−0·045	0·005	22·442
0·05	20·417	−0·007	0·693	−0·090	0·009	21·022
0·10	17·780	−0·014	1·118	−0·181	0·019	18·723
0·15	15·809	−0·021	1·746	−0·271	0·028	17·292
0·20	14·188	−0·028	0·180	−0·361	0·038	14·017
0·28	11·993	−0·039	1·040	−0·504	0·053	12·542
0·36	10·160	−0·048	1·338	−0·647	0·068	10·870
0·44	8·661	−0·052	1·586	−0·798	0·083	9·488
0·52	7·472	−0·053	1·824	−0·931	0·098	8·409
0·60	6·555	−0·053	2·111	−1·001	0·113	7·654
0·80	5·144	−0·043	3·554	−1·484	0·150	7·389
1·00	4·495	−0·029	−2·026	−1·751	0·188	0·876
1·20	4·211	−0·016	−0·262	−2·072	0·225	2·086
1·60	4·039	−0·004	0·016	−2·665	0·300	1·686
2·00	4·008	−0·001	0·018	−3·175	0·375	1·226
2·60	4·001	0·000	0·006	−3·736	0·443	0·713
3·40	4·000	0·000	0·001	−4·000	0·217	0·218
4·20	4·000	0·000	0·000	−4·000	0·005	0·005
5·00	4·000	0·000	0·000	−4·000	0·000	0·000

r the radius in atomic units.
$2Z_1$ potential due to the ion-core.
$2Z_2$ correlation among ion-core electrons.
$2Z_3$ exchange between core and conduction electrons.
$2Z_4$ potential due to conduction electrons.
$2Z_5$ potential due to the deviation from uniform distribution of conduction electrons around each nucleus.
$2Z_P$ the total potential excluding correlation and exchange among conduction electrons.

(Falicov 1962)

r_{max} are, respectively, the shortest and longest distances from the origin to the surface of the Wigner–Seitz unit cell of any three-dimensional crystal, then $U(r)$ does not possess spherical symmetry if $r_{min} < r \leqslant r_{max}$. The spherical symmetry of $U(r)$ can be restored artificially by using a muffin-tin potential in which $U(r)$ is set equal to some constant value U_0 throughout the entire region near the surfaces of the Wigner–Seitz unit

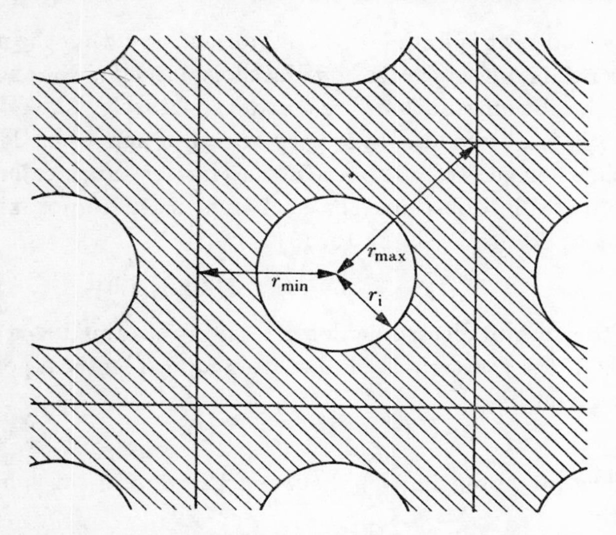

FIG. 2.2. Diagram to illustrate the construction of a muffin-tin potential for a hypothetical two-dimensional metal.

cells. The extent of this region is a little arbitrary but it must cover at least the region indicated in Fig. 2.2 where

$$r_i = r_{min} - (r_{max} - r_{min}) = 2r_{min} - r_{max}. \qquad (2.1.3)$$

r_i will then be smaller than r_0, the radius of the equivalent sphere which has a volume equal to the volume of the Wigner–Seitz unit cell of the crystal. A schematic representation of a muffin-tin potential is shown in Fig. 2.3.

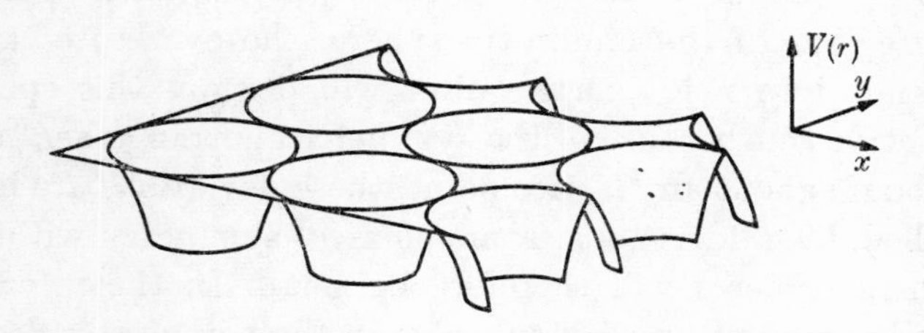

FIG. 2.3. A schematic representation of a 'muffin-tin' potential. Each well drops to infinity at the centre of the atom, but is truncated in the figure (Harrison 1970).

A method that is frequently used for constructing the muffin-tin potential is one suggested by Mattheiss (1964*b*). The exchange and Coulomb contributions are treated separately. The Coulomb part $U_c(r)$ is obtained as a superposition of spherical Coulomb potentials derived from the free atom Hartree–Fock calculation for the metal in question. The exchange term is approximated as a local potential (Slater 1951*a*)

$$U_x(r) = -3(e^2/4\pi\varepsilon_0)[(3/8\pi)\rho(r)]^{\frac{1}{3}} \tag{2.1.4}$$

where the crystal electronic density $\rho(r)$ is obtained as a superposition of the Hartree–Fock densities $\rho_0(r)$. The total potential is then given by

$$U_t(r) = U_c(r) + U_x(r). \tag{2.1.5}$$

The constant potential, U_0, between the muffin-tin spheres is given by

$$U_0 = \frac{3}{(r_0^3 - r_i^3)} \int_{r_i}^{r_0} U_t(r) r^2 \, dr. \tag{2.1.6}$$

If we choose this constant potential in the region between the muffin-tin spheres as a new zero of energy, then the potential within each muffin-tin sphere is given by

$$U(r) = U_t(r) - U_0. \tag{2.1.7}$$

That the muffin-tin potential represents a reasonable approximation to the actual crystal potential has been argued on the following grounds (Segall 1957). In the neighbourhood of the atomic sites $V(\mathbf{r})$ is atomic in character and is, therefore, commonly assumed to be spherically symmetrical. At the outer regions of the muffin-tin sphere, the overlap of atomic potentials into neighbouring sites will destroy this spherical symmetry. But because of the symmetrical arrangement of the neighbours about any lattice point, these deviations are largely cancelled. The deviations from spherical symmetry within the muffin-tin spheres will therefore be small. In the interstitial region between the muffin-tin spheres, the potential is flat, and can be approximated by a constant U_0. At these points the actual crystal potential is small. By moving the zero of the energy scale so that $U_0 = 0$, any deviations from an arbitrarily chosen approximating potential can be accounted for by perturbation theory. Ham and Segall (1961) tested the appropriateness of the approximating potential by comparing the exact eigenvalues calculated from a Mathieu potential for a simple cubic lattice with the approximate eigenvalues calculated from the corresponding muffin-tin potential. They found that the errors were small and could be further reduced by a perturbation calculation.

2.2. The tight-binding method

In Section 1.2 we have seen that, as a result of the regular periodic structure of a crystalline sample of a metal, the wave function $\psi_{\mathbf{k}}(\mathbf{r})$ must satisfy Bloch's theorem, so that we can write

$$\psi_{\mathbf{k}}(\mathbf{r}+\mathbf{T}_{\mathbf{n}}) = e^{i\mathbf{k}.\mathbf{T}_{\mathbf{n}}}\psi_{\mathbf{k}}(\mathbf{r}). \tag{2.2.1}$$

In the rest of Chapter 1 we were concerned with studying the form of $\psi_{\mathbf{k}}(\mathbf{r})$ for the conduction electrons rather than the ion-core electrons. We have already mentioned in this present chapter that the wave functions and energy levels of the ion-core electrons behave very much the same as if they were in corresponding states in a single free atom or ion of the given metallic element. Nevertheless the wave function of an ion-core electron must still be able to be written as a Bloch function in the form given in eqn (2.2.1). It is therefore not un-natural to suppose that for ion-core electrons $\psi_{\mathbf{k}}(\mathbf{r})$ can be constructed from an atomic wave function which is repeated at each of the atomic sites in the metal, in other words

$$\phi_{\mathbf{k},m}(\mathbf{r}) = (N)^{-\frac{1}{2}} \sum_{\alpha} \exp(i\mathbf{k} \,.\, \mathbf{R}_{\alpha}) u_m(\mathbf{r}-\mathbf{R}_{\alpha}) \tag{2.2.2}$$

where N is the number of atoms in the crystal and m refers to one of the s, p, d,... atomic wave functions and the $\mathbf{R}_{\alpha}$ specify the positions of the ions. The tight-binding method involves taking functions of the form of equation (2.2.2) as the wave functions for the conduction electrons as well as for the core electrons. If the crystal potential is represented as a sum of identical atomic-like potentials

$$V(\mathbf{r}) = \sum_{\alpha} U(|\mathbf{r}-\mathbf{R}_\alpha|) \tag{2.1.1}$$

the Hamiltonian $\mathscr{H}(\mathbf{r})$ can be written as

$$\mathscr{H}(\mathbf{r}) = \mathscr{H}_0(\mathbf{r})+\sum_{\alpha\neq 0} U(|\mathbf{r}-\mathbf{R}_\alpha|) \tag{2.2.3}$$

where

$$\mathscr{H}_0(\mathbf{r}) = -\frac{\hbar^2}{2m}\nabla^2+U(\mathbf{r}), \tag{2.2.4}$$

so that, by definition the function $u_m(\mathbf{r})$ satisfies

$$\mathscr{H}_0(\mathbf{r})u_m(\mathbf{r}) = E_{m0}u_m(\mathbf{r}). \tag{2.2.5}$$

The atomic wave functions $u_m(\mathbf{r}-\mathbf{R}_\alpha)$ on a given ion are assumed to be normalized. The energy $E_m(\mathbf{k})$ of an electron with wave vector $\mathbf{k}$ is determined by finding the eigenvalues of $\mathscr{H}(\mathbf{r})$ in eqn (2.2.3) using first-order perturbation theory and the wave function $\phi_{\mathbf{k}.m}(\mathbf{r})$ in eqn (2.2.2). Therefore

$$E_m(\mathbf{k}) = \frac{\int \phi^*_{\mathbf{k}.m}(\mathbf{r})\mathscr{H}(\mathbf{r})\phi_{\mathbf{k}.m}(\mathbf{r})\,\mathrm{d}\mathbf{r}}{\int \phi^*_{\mathbf{k}.m}(\mathbf{r})\phi_{\mathbf{k}.m}(\mathbf{r})\,\mathrm{d}\mathbf{r}}$$

$$= E_{m0}+\frac{\sum\limits_{\alpha}\sum\limits_{\substack{\alpha' \\ \alpha'\neq 0}} \mathrm{e}^{-i\mathbf{k}.\mathbf{R}_\alpha}\int u^*_m(\mathbf{r}-\mathbf{R}_\alpha)U(|\mathbf{r}-\mathbf{R}_{\alpha'}|)u_m(\mathbf{r})\,\mathrm{d}\mathbf{r}}{1+\sum\limits_{\substack{\alpha \\ \alpha\neq 0}} \mathrm{e}^{-i\mathbf{k}.\mathbf{R}_\alpha}\int u^*_m(\mathbf{r}-\mathbf{R}_\alpha)u_m(\mathbf{r})\,\mathrm{d}\mathbf{r}}. \tag{2.2.6}$$

The three-centre integrals in eqn (2.2.6) are difficult to evaluate. In practice it is common to neglect these three-centre integrals, that is, to neglect all terms in the double summation except those for which $\alpha' = \alpha$.

The above argument is based on the use of non-degenerate perturbation theory and therefore it is only appropriate for energy bands that can be assumed to be derived from atomic s states. If $u_m(\mathbf{r})$ describes an atomic wave function with angular momentum quantum number, l, greater than zero, that is $u_m(\mathbf{r})$ is the wave function of an atomic p, d,... electron, then the method has to be modified by the use of degenerate perturbation theory. Instead of using eqn (2.2.6) to determine the perturbed energy eigenvalues $E_j(\mathbf{k})$, they now have to be obtained by solving the secular equation

$$\det |\mathscr{H}_{mm'}(\mathbf{k}) - E_j(\mathbf{k})\mathscr{I}_{mm'}(\mathbf{k})| = 0 \tag{2.2.7}$$

where

$$\mathscr{H}_{mm'}(\mathbf{k}) = (1/N) \sum_{\alpha'} \sum_{\alpha''} \exp\{-i\mathbf{k} \cdot (\mathbf{R}_{\alpha'} - \mathbf{R}_{\alpha''})\} \int u_m^*(\mathbf{r} - \mathbf{R}_{\alpha'}) \mathscr{H}(\mathbf{r}) u_{m'}(\mathbf{r} - \mathbf{R}_{\alpha''})\, d\mathbf{r} \tag{2.2.8}$$

and

$$\mathscr{I}_{mm'}(\mathbf{k}) = \delta_{mm'} + \sum_{\alpha \neq 0} \exp(-i\mathbf{k} \cdot \mathbf{R}_\alpha) \int u_m^*(\mathbf{r} - \mathbf{R}_\alpha) u_{m'}(\mathbf{r})\, d\mathbf{r}. \tag{2.2.9}$$

Many of these matrix elements are also difficult to evaluate in practice and in most applications three-centre integrals, which are of the form $\int u_m^*(\mathbf{r} - \mathbf{R}_\alpha) U(|\mathbf{r} - \mathbf{R}_{\alpha'}|) u_{m'}(\mathbf{r})\, d\mathbf{r}$, and overlap integrals, which are of the form $\int u_m^*(\mathbf{r} - \mathbf{R}_\alpha) u_{m'}(\mathbf{r})\, d\mathbf{r}$, have been neglected. This is a good approximation only if the atomic wave functions are highly localized around each atomic core, with little overlap onto adjacent atoms. Slater and Koster (1954) tabulated the interaction integrals in the two-centre approximation for s, p, and d functions in cubic crystals and this was extended to the hexagonal close-packed structure by Miąsek (1957). It seems unlikely that in practice calculations that neglect three-centre integrals and overlap integrals can be expected to be particularly accurate. An illustration of this was provided in a detailed calculation by Wohlfarth (1953) for a linear chain of hydrogen atoms.

In addition to the errors introduced by the neglect in practice of three-centre integrals and overlap integrals there is another significant source of error in the tight-binding method as we have described it. While a wave function of the form $\phi_{\mathbf{k},m}(\mathbf{r})$ in eqn (2.2.2) gives a good description of the wave function of an ion-core electron, for which an angular momentum quantum number is still well defined, it is much less clear that a wave function of this form is appropriate to the description of conduction-electron wave functions, particularly in view of the considerable success of the free-electron model which we have described extensively in Chapter 1 and in which the conduction-electron wave functions are simply plane waves. In a real metal the conduction electron wave functions for nearly every value of **k** are not pure s, p or d functions so that an expansion of the form of eqn (2.2.2) which involves just one type of atomic wave function is not an expansion in terms of a sufficiently complete set of functions. Calculations based on the tight-binding method have been performed for a number of metals, but as we might expect this tends to be more successful where the bands are still able to be fairly realistically ascribed to a particular s-, p- or d-type behaviour, as for instance in the d bands of transition metals. This shortcoming of the tight-binding method could be overcome by including for each electronic wave function in an expansion of the form of eqn (2.2.2), wave functions $u_m(\mathbf{r})$ appropriate to s, p, d, etc. atomic energy levels rather than just one atomic wave function, that is,

$$\phi_{\mathbf{k},n}(\mathbf{r}) = (1/\sqrt{N}) \sum_m A_m \sum_\alpha \mathrm{e}^{\mathrm{i}\mathbf{k}.\mathbf{R}_\alpha} u_m(\mathbf{r}-\mathbf{R}_\alpha), \qquad (2.2.10)$$

where the A_m are some coefficients that remain to be determined. The tight-binding method can now be seen very clearly as an extension to solids of the LCAO (linear combination of atomic orbitals) method which is used extensively in constructing wave functions for electrons in molecules. One of the uses to which the tight-binding method has been applied has been as the basis of an interpolation scheme that is based on physical rather than mathematical foundations, see Section 2.7.

2.3. The Wigner–Seitz and cellular methods

The cellular method, which was introduced by Slater (1934) and has been used quite extensively in band structure calculations since then, has its origins in the method of Wigner and Seitz (1933). Wigner and Seitz assumed that the electrons in a metal move independently within a regular array of fixed positive ions. Since the wave functions of the electrons obey Bloch's theorem it is only necessary to determine the wave functions in one Wigner–Seitz unit cell of the structure, such as that shown in Fig. 2.4 for the b.c.c. structure. This unit cell was then replaced by a sphere, sometimes called the Wigner–Seitz sphere, with volume equal to the volume of the unit cell, and the potential $V(\mathbf{r})$ was assumed to be spherically symmetrical; all this gross oversimplification has the advantage that it

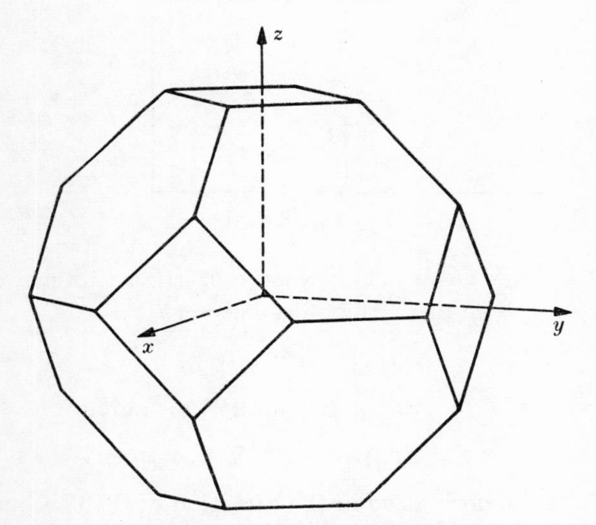

FIG. 2.4. The Wigner–Seitz unit cell for the b.c.c. structure.

effectively reduces the situation to a one-dimensional quantum-mechanical problem. The one-electron wave function $\psi_{\mathbf{k}}(\mathbf{r})$ is then subject to the boundary conditions that $\psi_{\mathbf{k}}(\mathbf{r})$ and $\nabla\psi_{\mathbf{k}}(\mathbf{r})$ are continuous at the surface of the Wigner–Seitz sphere. This treatment gives reasonable results for the point Γ at the centre of the Brillouin zone where $\mathbf{k} = 0$ and was applied by Wigner and Seitz (1933, 1934) to Na. For other points in the Brillouin zone the method is less successful.

In the cellular method proposed by Slater (1934) account is taken of the shape and symmetry of the Wigner–Seitz unit cell instead of approximating to it by a sphere. From Fig. 2.5 the boundary condition on $\psi_{\mathbf{k}}(\mathbf{r})$ becomes, at the interface between cell one and cell two,

$$\psi_1(\mathbf{r}_B) = \psi_2(\mathbf{r}_B), \tag{2.3.1}$$

and $\psi_2(\mathbf{r}_B)$ can be related to the wave function in cell one at A as a result of Bloch's theorem by a phase factor

$$\psi_2(\mathbf{r}_B) = \mathrm{e}^{\mathrm{i}\mathbf{k}.\mathbf{T}}\psi_1(\mathbf{r}_A) \tag{2.3.2}$$

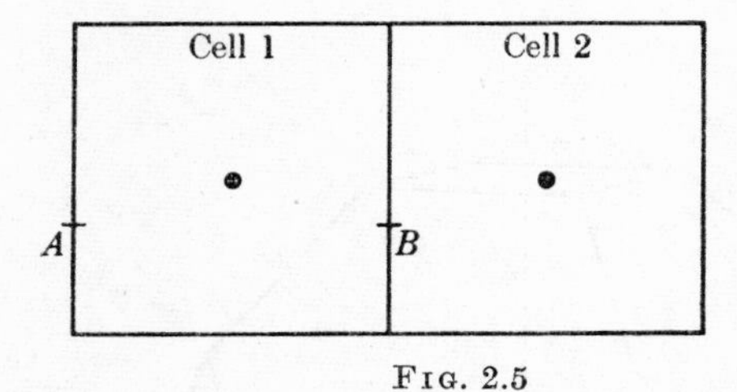

FIG. 2.5

where $\mathbf{T} = \mathbf{r}_B - \mathbf{r}_A$ and is a vector of the reciprocal lattice. The boundary condition therefore becomes

$$\psi_1(\mathbf{r}_B) = \mathrm{e}^{\mathrm{i}\mathbf{k}.\mathbf{T}}\psi_1(\mathbf{r}_A). \tag{2.3.3}$$

Similarly, the boundary condition on $\nabla\psi_{\mathbf{k}}(\mathbf{r})$ gives

$$\nabla_{\mathbf{T}}\psi_1(\mathbf{r}_B) = -\mathrm{e}^{\mathrm{i}\mathbf{k}.\mathbf{T}}\nabla_{\mathbf{T}}\psi_1(\mathbf{r}_A) \tag{2.3.4}$$

where $\nabla_{\mathbf{T}}$ is the derivative in the direction of $\mathbf{T}$. The boundary conditions which relate the wave functions and their derivatives in two adjacent cells have therefore been reduced in eqns (2.3.3) and (2.3.4) to relations between the wave function $\psi_1(\mathbf{r})$ at different points in the same cell. It is often possible to exploit the symmetry of the crystal to simplify the boundary conditions still further (see, for example, Altmann (1958*a*)). It is necessary to assume that the potential $V(\mathbf{r})$ is spherically symmetrical so that $\psi_{\mathbf{k}}(\mathbf{r})$ can be expanded in terms of radial functions $R_l(r)$ and spherical harmonics $Y_l^m(\theta, \phi)$; if

$$\psi_{\mathbf{k}}(\mathbf{r}) = u_{\mathbf{k}}(\mathbf{r})\exp(\mathrm{i}\mathbf{k} \,.\, \mathbf{r})$$

we may write

$$u_{\mathbf{k}}(\mathbf{r}) = \sum_l \sum_m A_{\mathbf{k}lm} R_l(r) Y_l^m(\theta, \phi). \tag{2.3.5}$$

Of course it is only an approximation to assume that the potential $V(\mathbf{r})$ is spherically symmetrical in the outer regions of the cell as we have seen before in Section 2.1. When a wave function $\psi_{\mathbf{k}}(\mathbf{r})$ of this form is substituted into the boundary conditions in eqns (2.3.3) and (2.3.4) these equations become

$$\sum_l \sum_m A_{\mathbf{k}lm} R_l(r_B) Y_l^m(\theta_B, \phi_B) = e^{i\mathbf{k}.\mathbf{T}} \sum_l \sum_m A_{\mathbf{k}lm} R_l(r_A) Y_l^m(\theta_A, \phi_A) \tag{2.3.6}$$

and

$$\nabla_{\mathbf{T}} \sum_l \sum_m A_{\mathbf{k}lm} R_l(r_B) Y_l^m(\theta_B, \phi_B) = -e^{i\mathbf{k}.\mathbf{T}} \nabla_{\mathbf{T}} \sum_l \sum_m A_{\mathbf{k}lm} R_l(r_A) Y_l^m(\theta_A, \phi_A). \tag{2.3.7}$$

Originally it was only possible to handle wave functions at special points of symmetry in the Brillouin zone for which many of the coefficients $A_{\mathbf{k}lm}$ could be simplified or completely eliminated group-theoretically (Von der Lage and Bethe 1947; Bell 1954; Altmann 1958*a*; Altmann and Bradley 1965; Altmann and Cracknell 1965); however, this is not an essential part of the method. By using a number of different pairs of boundary points $\mathbf{r}_A$ and $\mathbf{r}_B$ in eqns (2.3.6) and (2.3.7) a set of linear simultaneous algebraic equations is obtained for the unknown coefficients $A_{\mathbf{k}lm}$. However, the quantities $R_l(r_B)$ and $R_l(r_A)$ depend on the value of $E_j(\mathbf{k})$ and it is, of course, the object of the calculation to determine $E_j(\mathbf{k})$. This means that one has to use a number of different trial values of $E_j(\mathbf{k})$ until an exact fit to the boundary condition equations is obtained. Better results can be obtained by using an excess number of boundary conditions and using a least squares fitting procedure to determine $E(\mathbf{k})$ and the coefficients $A_{\mathbf{k}lm}$ (Altmann 1958*b*). Shockley (1937) showed that if $V(\mathbf{r})$ is set equal to zero the free-electron energy bands given previously in Section 1.5 should be reproduced by the calculation; this is sometimes called the *empty lattice test*. In recent empty-lattice tests of the cellular method Altmann, Davies, and Harford (1968) were able to reproduce free-electron, or empty-lattice, energies at the special points of symmetry in the Brillouin zone to an accuracy of

2×10^{-5} Ry. The cellular method has been applied to the calculation of the band structures of quite a large number of real metals.

2.4. The orthogonalized plane wave and augmented plane wave methods

In Chapter 1 we found it convenient to ignore completely the non-zero values of the atomic potentials within a metal. If the solutions $\psi_{\mathbf{k}}(\mathbf{r})$ of Schrödinger's equation are written as

$$\psi_{\mathbf{k}}(\mathbf{r}) = \exp(i\mathbf{k} \cdot \mathbf{r})u_{\mathbf{k}}(\mathbf{r}), \tag{1.2.4}$$

then we have seen that for the free-electron approximation $u_{\mathbf{k}}(\mathbf{r})$ is itself a plane wave

$$u_{\mathbf{k}}(\mathbf{r}) = \exp(i\mathbf{G}_{\mathbf{n}} \cdot \mathbf{r}), \tag{1.5.5}$$

where $\mathbf{G}_{\mathbf{n}}$ is some vector of the reciprocal lattice of the crystal. For a real metal in which $V(\mathbf{r})$ cannot be set equal to zero the simple expression for $u_{\mathbf{k}}(\mathbf{r})$ given in eqn (1.5.5) is not satisfactory, that is, we have not used a complete set of basis functions in expanding $u_{\mathbf{k}}(\mathbf{r})$. To obtain a complete set of functions for use in the expansion of $u_{\mathbf{k}}(\mathbf{r})$ it is necessary to express $u_{\mathbf{k}}(\mathbf{r})$ as a linear combination of plane waves corresponding to all possible reciprocal lattice vectors $\mathbf{G}_{\mathbf{n}}$, that is, we write

$$u_{\mathbf{k}}(\mathbf{r}) = \sum_{\mathbf{n}} C_{\mathbf{kn}} \exp(i\mathbf{G}_{\mathbf{n}} \cdot \mathbf{r}). \tag{2.4.1}$$

This corresponds to a Fourier analysis of $u_{\mathbf{k}}(\mathbf{r})$ in three dimensions. If $\psi_{\mathbf{k}}(\mathbf{r})$ is to be a solution of the one-electron Schrödinger equation, given for example in eqn (1.2.3) the coefficients $C_{\mathbf{kn}}$ and the corresponding energy eigenvalues $E(\mathbf{k})$ can be determined by the usual method so that, writing $\mathscr{H}(\mathbf{r})\psi_{\mathbf{k}}(\mathbf{r})$ for the left-hand side of eqn (1.2.3) we have

$$\mathscr{H}(\mathbf{r})\psi_{\mathbf{k}}(\mathbf{r}) - E(\mathbf{k})\psi_{\mathbf{k}}(\mathbf{r}) = 0 \tag{2.4.2}$$

then multiplying by $\exp(-i(\mathbf{k}+\mathbf{G}_{\mathbf{n}'}).\mathbf{r})$ and integrating we have

$$\sum_{\mathbf{n}} C_{\mathbf{kn}}\Big\{\int \exp(-i(\mathbf{k}+\mathbf{G}_{\mathbf{n}'}) \cdot \mathbf{r})\mathscr{H}(\mathbf{r})\exp(i(\mathbf{k}+\mathbf{G}_{\mathbf{n}}) \cdot \mathbf{r})\, d\mathbf{r} + \\ -E(\mathbf{k})\int \exp(-i(\mathbf{G}_{\mathbf{n}'}-\mathbf{G}_{\mathbf{n}}) \cdot \mathbf{r})\, d\mathbf{r}\Big\} = 0. \tag{2.4.3}$$

This is an infinite set of simultaneous linear equations from which, in principle, the unknown coefficients $C_{\mathbf{kn}}$ can be determined. The condition for the existence of a solution is that

$$\det|\mathscr{H}_{\mathbf{n'n}}(\mathbf{k})-E(\mathbf{k})\mathscr{I}_{\mathbf{n'n}}(\mathbf{k})| = 0 \tag{2.4.4}$$

where

$$\mathscr{H}_{\mathbf{n'n}}(\mathbf{k}) = \int \exp(-\mathrm{i}(\mathbf{k}+\mathbf{G}_{\mathbf{n'}})\,.\,\mathbf{r})\mathscr{H}(\mathbf{r})\exp(\mathrm{i}(\mathbf{k}+\mathbf{G}_{\mathbf{n}})\,.\,\mathbf{r})\,\mathrm{d}\mathbf{r} \tag{2.4.5}$$

and

$$\mathscr{I}_{\mathbf{n'n}}(\mathbf{k}) = \int \exp(-\mathrm{i}(\mathbf{G}_{\mathbf{n'}}-\mathbf{G}_{\mathbf{n}})\,.\,\mathbf{r})\,\mathrm{d}\mathbf{r}. \tag{2.4.6}$$

Equation (2.4.4) is, of course, similar to eqn (2.2.7) and by terminating the expansion of $\psi_{\mathbf{k}}(\mathbf{r})$ at some convenient value of $\mathbf{n}$ the values of $E(\mathbf{k})$ and the coefficients $C_{\mathbf{kn}}$ can be determined. Unfortunately the convergence of a plane wave expansion of the form given in eqn (2.4.1) is very poor, and the larger the expansions that one tries to handle, the more lengthy and tedious it becomes to solve the secular eqn (2.4.4) to determine $E(\mathbf{k})$.

Once again, as in the case of the cellular method described in Section 2.3, it is possible to use group theory to reduce the number of independent and non-zero coefficients in an expansion of $\psi_{\mathbf{k}}(\mathbf{r})$ in terms of plane waves, (see, for example, Cornwell 1969) however, even if this is done the convergence of the expansion of $\psi_{\mathbf{k}}(\mathbf{r})$ is still too slow for use in real band structure calculations. Two different ways out of this difficulty were proposed by Slater (1937, 1953) and Herring (1940). Although Slater's suggestion, which became known as the A.P.W. (augmented plane wave) method, was made before Herring's suggestion, which became known as the O.P.W. (orthogonalized plane wave) method, it was the O.P.W. method that was more extensively used for a long time until only fairly recently it has been displaced by the A.P.W. method. An extensive discussion of the A.P.W. method is given by Loucks (1967*a*).

We have seen in section 2.2 that the tight-binding wave functions given in eqn (2.2.2)

$$\phi_{\mathbf{k},m}(\mathbf{r}) = (1/\sqrt{N})\sum_{\alpha}\exp(\mathrm{i}\mathbf{k}\,.\,\mathbf{R}_{\alpha})u_m(\mathbf{r}-\mathbf{R}_{\alpha}) \tag{2.4.7}$$

are good approximations to the wave functions of ion-core electrons although they are often inadequate for conduction electrons. Herring (1940) proposed the construction of wave functions for the conduction electrons from plane waves using the fact that these wave functions must be orthogonal to the wave functions of all the ion-core electrons. If we start with a plane wave with wave vector $(\mathbf{k}+\mathbf{G}_\mathbf{n})$ this function can be orthogonalized to the ion-core wave functions by forming the function

$$X_{\mathbf{kn}}(\mathbf{r}) = (NV)^{-\frac{1}{2}} \exp\{\mathrm{i}(\mathbf{k}+\mathbf{G}_\mathbf{n})\cdot\mathbf{r}\} - \sum_m \mu_{\mathbf{k}m}\phi_{\mathbf{k},m}(\mathbf{r}) \quad (2.4.8)$$

where V is the volume of a unit cell of the crystal, and determining the coefficients $\mu_{\mathbf{k}m}$ from the orthogonalization condition

$$\int \phi^*_{\mathbf{k},j}(\mathbf{r})X_{\mathbf{kn}}(\mathbf{r})\mathrm{d}\mathbf{r} = 0$$

we find

$$\mu_{\mathbf{k}m} = (1/\sqrt{(NV)})\int \phi^*_{\mathbf{k},m}(\mathbf{r})\exp\{\mathrm{i}(\mathbf{k}+\mathbf{G}_\mathbf{n})\cdot\mathbf{r}\}\,\mathrm{d}\mathbf{r}. \quad (2.4.9)$$

The orthogonalized plane waves $X_{\mathbf{kn}}(\mathbf{r})$ then form a complete set of basis functions for the expansion of the wave function $\psi_\mathbf{k}(\mathbf{r})$ of a conduction electron so that we write

$$\psi_\mathbf{k}(\mathbf{r}) = \sum_\mathbf{n} C^X_{\mathbf{kn}}X_{\mathbf{kn}}(\mathbf{r}). \quad (2.4.10)$$

In practice it has been found that expansions of $\psi_\mathbf{k}(\mathbf{r})$ of the form of eqn (2.4.10) in terms of the orthogonalized plane waves $X_{\mathbf{kn}}(\mathbf{r})$ converge much more rapidly than simple plane-wave expansions of the form of eqn (2.4.1). The energy eigenvalues $E(\mathbf{k})$ can then be found by substituting an expansion for $\psi_\mathbf{k}(\mathbf{r})$ of the form of eqn (2.4.10) into eqn (2.4.2); this leads to a secular equation similar to eqn (2.4.4)

$$\det|\mathscr{H}^X_{\mathbf{n'n}}(\mathbf{k}) - E(\mathbf{k})\mathscr{I}^X_{\mathbf{n'n}}(\mathbf{k})| = 0 \quad (2.4.11)$$

where

$$\mathscr{H}^X_{\mathbf{n'n}}(\mathbf{k}) = \int X^*_{\mathbf{kn'}}(\mathbf{r})\mathscr{H}(\mathbf{r})X_{\mathbf{kn}}(\mathbf{r})\,\mathrm{d}\mathbf{r} \quad (2.4.12)$$

and

$$\mathscr{I}^X_{\mathbf{n'n}}(\mathbf{k}) = \int X^*_{\mathbf{kn'}}(\mathbf{r})X_{\mathbf{kn}}(\mathbf{r})\,\mathrm{d}\mathbf{r}. \quad (2.4.13)$$

Once $E(\mathbf{k})$ has been determined the coefficients $C^X_{\mathbf{kn}}$ can be obtained, if required, by solution of the set of simultaneous equations associated with the determinant in eqn (2.4.11). Once again, as in the case of the cellular method described in Section 2.3, it was only possible in the early days of band structure calculations to handle wave functions for which $\mathbf{k}$ corresponds to one of the special points or lines of symmetry in the Brillouin zone. This was because it is possible at the points of symmetry to use group theory to construct symmetrized combinations of plane waves that belong to the appropriate group $\mathbf{G}^{\mathbf{k}}$ and thereby reduce the number of unknown coefficients in eqn (2.4.10) by showing that for that particular $\mathbf{k}$ some of the $C^X_{\mathbf{k}n}$ vanish or that some simplifying relationships exist between different $C^X_{\mathbf{k}n}$ (see, for example, Callaway 1964; Slater 1965; Cornwell 1969). The first application of the O.P.W. method was to beryllium by Herring and Hill (1940) and since then, as we shall see in Chapters 4 and 5, it has been applied to a number of other metals.

It is a feature of the O.P.W. method, and incidentally also of the tight-binding method, that the matrix elements in the secular equation can be calculated directly, assuming the ion-core wave functions are known. There is no implicit dependence of the matrix elements on the eigenvalue $E(\mathbf{k})$. The determination of $E(\mathbf{k})$ therefore simply involves the direct solution of eqn (2.4.11) with a set of matrix elements that have definite numerical values. In some other methods, such as the A.P.W. method and the Green's function method, the values of the matrix elements depend on $E(\mathbf{k})$ so that it is not possible to solve the secular equation directly. One then has to choose a value of the energy and evaluate the determinant, and then repeat this whole procedure for various energies until one finds the particular energy that makes the determinant vanish. This is clearly more tedious than the direct solution which is possible in the O.P.W. method. However, there are several disadvantages of the O.P.W. method and in practice its use has generally only been particularly successful for simple metals. Part of the reason for this is that one has to assume the existence of a clear distinction between ion-core electrons and conduction electrons and this may not be very realistic for transition metals or rare earth metals. Moreover, although in principle it would be possible to determine the core functions $u_m(\mathbf{r})$ from a self-consistent calculation for the metallic crystal, this is unrealistic and in practice the $u_m(\mathbf{r})$ are assumed to be the same as in a free atom of the same element. For non-simple metals this is unlikely to be a very good approximation. Even if the ion-core functions $u_m(\mathbf{r})$ are known, the procedure of orthogonalization to construct the functions $X_{\mathbf{kn}}(\mathbf{r})$ becomes more difficult for heavy elements because there will be a larger number of ion-core states to be included.

We now turn to the consideration of the A.P.W. method which was introduced by Slater (1937). When a conduction electron is near the centre of a Wigner–Seitz unit cell of a crystal the potential that it experiences is quite accurately described by a screened Coulomb potential and so the expansions which are used in the cellular method (see eqn 2.3.5) can be expected to converge quite rapidly. On the other hand when a conduction electron is in the region near the boundaries of the Wigner–Seitz unit cell the potential that it experiences is very nearly constant so that in this region one would expect the conduction-electron wave functions to take the form of plane waves. Slater therefore proposed that the conduction-electron wave function could be expanded in terms of functions $\Phi_{\mathbf{kn}}(\mathbf{r})$ which have the property that within a spherical region of radius r_i, $\Phi_{\mathbf{kn}}(\mathbf{r})$ takes the form given by eqn (2.3.5) and outside this spherical region $\Phi_{\mathbf{kn}}(\mathbf{r})$ is a plane wave of the form $\exp(\mathrm{i}(\mathbf{k}+\mathbf{G_n})\,.\,\mathbf{r})$; $\Phi_{\mathbf{kn}}(\mathbf{r})$ can be written in the form

$$\Phi_{\mathbf{kn}}(\mathbf{r}) = A^0_{\mathbf{kn}}\epsilon(r-r_i)\exp(\mathrm{i}(\mathbf{k}+\mathbf{G_n})\,.\,\mathbf{r}) + \\ +\sum_l \sum_m A_{\mathbf{kn}lm}\epsilon(r_i-r)R_l(r)Y^m_l(\theta,\phi), \qquad (2.4.14)$$

where $\epsilon(x)$ is a step function defined by

$$\begin{aligned} \epsilon(x) &= 1 \qquad \text{if } x \geqslant 0 \\ \epsilon(x) &= 0 \qquad \text{if } x < 0. \end{aligned} \qquad (2.4.15)$$

The use of a function of the form of eqn (2.4.14) corresponds to using a muffin-tin potential of the form described in Section

2.1. The function $\Phi_{\mathbf{kn}}(\mathbf{r})$ is then described as an *augmented plane wave*. It is desirable to choose the coefficients $A_{\mathbf{kn}lm}$ so that $\Phi_{\mathbf{kn}}(\mathbf{r})$ is continuous across the spherical surface of radius r_i. Since the plane wave $\exp(\mathrm{i}\mathbf{k}\,.\,\mathbf{r})$ can be expanded as

$$\exp(\mathrm{i}\mathbf{k}\,.\,\mathbf{r}) = 4\pi \sum_{l} \sum_{m} (\mathrm{i})^l j_l(kr) Y_l^m(\theta, \phi) Y_l^{m*}(\theta_{\mathbf{k}}, \phi_{\mathbf{k}}) \tag{2.4.16}$$

the continuity of $\Phi_{\mathbf{kn}}(\mathbf{r})$ can be achieved if

$$A_{\mathbf{kn}lm} = 4\pi(\mathrm{i})^l Y_l^{m*}(\theta_{\mathbf{k}+\mathbf{G}_\mathbf{n}}, \phi_{\mathbf{k}+\mathbf{G}_\mathbf{n}})[j_l(|\mathbf{k}+\mathbf{G}_\mathbf{n}|\, r_i)A^0_{\mathbf{kn}}]/R_l(r_i) \tag{2.4.17}$$

where $j_l(kr)$ is a spherical Bessel function and $\mathbf{k}$ is expressed in spherical polar coordinates as $(k, \theta_{\mathbf{k}}, \phi_{\mathbf{k}})$. It should be noticed that because $R_l(r_i)$ depends on $E(\mathbf{k})$ the value of the coefficient $A_{\mathbf{kn}lm}$ given in eqn (2.4.17) also depends on the value of $E(\mathbf{k})$, but $E(\mathbf{k})$ is, of course, what a band structure calculation aims at determining.

The wave function $\psi_{\mathbf{k}}(\mathbf{r})$ of an electron with wave vector $\mathbf{k}$ can then be expanded in terms of these augmented plane waves

$$\psi_{\mathbf{k}}(\mathbf{r}) = \sum_{\mathbf{n}} C^{\Phi}_{\mathbf{kn}} \Phi_{\mathbf{kn}}(\mathbf{r}), \tag{2.4.18}$$

where the summation is over all the vectors $(\mathbf{k}+\mathbf{G}_\mathbf{n})$ that are related to $\mathbf{k}$ by a translation of the reciprocal lattice. The energy eigenvalues $E(\mathbf{k})$ can then be found in a similar way to that described previously in connection with the O.P.W. method by solution of a secular equation similar to eqn (2.4.11),

$$\det |\mathscr{H}^{\Phi}_{\mathbf{n}'\mathbf{n}}(\mathbf{k}) - E(\mathbf{k})\mathscr{I}^{\Phi}_{\mathbf{n}'\mathbf{n}}(\mathbf{k})| = 0 \tag{2.4.19}$$

where

$$\mathscr{H}^{\Phi}_{\mathbf{n}'\mathbf{n}}(\mathbf{k}) = \int \Phi^*_{\mathbf{kn}'}(\mathbf{r})\mathscr{H}(\mathbf{r})\Phi_{\mathbf{kn}}(\mathbf{r})\, \mathrm{d}\mathbf{r} \tag{2.4.20}$$

and

$$\mathscr{I}^{\Phi}_{\mathbf{n}'\mathbf{n}}(\mathbf{k}) = \int \Phi^*_{\mathbf{kn}'}(\mathbf{r})\Phi_{\mathbf{kn}}(\mathbf{r})\, \mathrm{d}\mathbf{r}. \tag{2.4.21}$$

Once $E(\mathbf{k})$ has been determined, the coefficients $C^{\Phi}_{\mathbf{kn}}$ can be obtained, if required, by solution of the set of simultaneous equations associated with the determinant in eqn (2.4.19). However, there is the complication, which did not arise in the tight-binding and O.P.W. methods, in that the values of the matrix elements $\mathscr{H}^{\Phi}_{\mathbf{n}'\mathbf{n}}(\mathbf{k})$ and $\mathscr{S}^{\Phi}_{\mathbf{n}'\mathbf{n}}(\mathbf{k})$ depend on the value of $E(\mathbf{k})$ itself via the dependence of the coefficients $A_{\mathbf{kn}lm}$ on $E(\mathbf{k})$ in eqn (2.4.17). It is therefore not possible to solve eqn (2.4.19) directly and one has to choose some trial value of the energy and evaluate all the matrix elements and thence the determinant; this process then has to be repeated until a particular value of the energy is obtained that makes the determinant vanish. There is a further difficulty in the evaluation of the matrix elements $\mathscr{H}^{\Phi}_{\mathbf{n}'\mathbf{n}}(\mathbf{k})$ for any given value of the energy because of the discontinuity of $\nabla\Phi_{\mathbf{kn}}(\mathbf{r})$ at r_i. When the slope of a wave function $\psi_i(\mathbf{r})$ is discontinuous, the term in $\mathscr{H}(\mathbf{r})$ that corresponds to the kinetic energy needs special treatment. Two forms of integral are often used for determining the kinetic energy, namely

$$(-\hbar^2/2m)\int \psi_i^*(\mathbf{r})\nabla^2\psi_j(\mathbf{r})\,\mathrm{d}\mathbf{r}$$

and

$$(-\hbar^2/2m)\int \nabla\psi_i^*(\mathbf{r})\,.\,\nabla\psi_j(\mathbf{r})\,\mathrm{d}\mathbf{r},$$

of which the first one is more common. However, it is the second of these expressions which is the more fundamental. It can be shown by integration by parts that, in most circumstances, these two expressions are identically equal; but if the slope of the wave function is discontinuous these two expressions differ by a surface integral which in the present application is over the surface of the sphere of radius r_i. The details of the handling of this difficulty in the evaluation of the matrix elements in the A.P.W. method are considered, for instance, in Appendix 6 of the book by Slater (1965).

Once again, as in the cellular and O.P.W. methods, it is possible to simplify group-theoretically the expansion of $\psi_{\mathbf{k}}(\mathbf{r})$ for the special points of symmetry in the Brillouin zone by symmetrizing the augmented plane waves $\Phi_{\mathbf{kn}}(\mathbf{r})$ according to the irreducible representations of $\mathbf{G}^{\mathbf{k}}$, the group of the wave vector, $\mathbf{k}$. However, modern computers are large enough to handle the expansions of $\psi_{\mathbf{k}}(\mathbf{r})$ for general wave vectors in the Brillouin zone so that the use of symmetrized augmented plane waves is perhaps less important now than it was in the past.

After its original introduction by Slater (1937) the A.P.W. method was first applied to copper by Chodorow (1939) and then the method was almost completely ignored for about two decades. However, more recently the method has been used quite extensively for a number of metals (see, for example, the book by Loucks (1967*a*)); these calculations have met with a considerable degree of success for several metals as we shall see in Chapters 4 and 5. A relativistic modification of the A.P.W. method has also been developed by Loucks (1965*d*) (see Section 2.10).

2.5. The Green's function method

In the previous sections of this book we have seen that whereas it is quite straightforward to solve the one-electron Schrödinger equation in the case when $V(\mathbf{r})$ is equal to zero it becomes very difficult to solve this equation when a realistic non-zero potential $V(\mathbf{r})$ is considered. It is, of course, also difficult to construct a realistic potential $V(\mathbf{r})$ *ab initio* but, for the moment, we assume that this has been done and that we are concerned with solving Schrödinger's equation for a given $V(\mathbf{r})$. In the O.P.W. method the wave functions of the conduction electrons in a real metal were expanded in terms of the plane waves which are solutions of the Schrödinger equation in the free-electron approximation. The Green's function method is an alternative way of making use of these known solutions of the homogeneous equation which is obtained by putting $V(\mathbf{r}) = 0$ in eqn (1.2.3), and is based on the use of the variational principle. The Green's function method was first proposed by Kohn and Rostoker (1954) and it is closely related to an earlier approach due to Korringa (1947). It is therefore often called the Korringa–Kohn–Rostoker method (or the K.K.R. method for short).

We require to solve the one-electron Schrödinger equation given in eqn (1.2.3) which we may rewrite as

$$\{(\hbar^2/2m)\nabla^2+E(\mathbf{k})\}\psi_{\mathbf{k}}(\mathbf{r}) = V(\mathbf{r})\psi_{\mathbf{k}}(\mathbf{r}). \tag{2.5.1}$$

We consider the Green's function $\mathscr{G}_{\mathbf{k}}(\mathbf{r}, \mathbf{r}')$ which is defined to be a solution of the equation

$$\{(\hbar^2/2m)\nabla^2+E(\mathbf{k})\}\mathscr{G}_{\mathbf{k}}(\mathbf{r},\mathbf{r}') = \delta(\mathbf{r}-\mathbf{r}') \tag{2.5.2}$$

which is the corresponding homogeneous equation obtained by setting $V(\mathbf{r}) = 0$ in eqn (2.5.1). $\mathscr{G}_{\mathbf{k}}(\mathbf{r},\mathbf{r}')$ is required to satisfy the boundary conditions imposed by Bloch's theorem as a result of the periodic nature of the lattice, so that

$$\mathscr{G}_{\mathbf{k}}(\mathbf{r}+\mathbf{G}_{\mathbf{n}},\mathbf{r}') = \exp(\mathrm{i}\mathbf{k}\cdot\mathbf{G}_{\mathbf{n}})\mathscr{G}_{\mathbf{k}}(\mathbf{r},\mathbf{r}'). \tag{2.5.3}$$

The problem then involves determining the Green's functions $\mathscr{G}_{\mathbf{k}}(\mathbf{r},\mathbf{r}')$ and using these Green's functions to determine solutions of eqn (2.5.1).

$\mathscr{G}_{\mathbf{k}}(\mathbf{r},\mathbf{r}')$ can be determined from eqn (2.5.2) as follows. If we suppose that $\Phi_{\mathbf{kn}}(\mathbf{r})$ is a solution of the equation

$$\{(\hbar^2/2m)\nabla^2+E(\mathbf{k})\}\Phi_{\mathbf{kn}}(\mathbf{r}) = 0, \tag{2.5.4}$$

then $\Phi_{\mathbf{kn}}(\mathbf{r})$ was determined in Section 1.5 to be of the form

$$\Phi_{\mathbf{kn}}(\mathbf{r}) = (1/\sqrt{(NV)})\exp(\mathrm{i}(\mathbf{k}+\mathbf{G}_{\mathbf{n}})\cdot\mathbf{r}), \tag{2.5.5}$$

and the functions $\Phi_{\mathbf{kn}}(\mathbf{r})$ have the property that

$$\sum_{\mathbf{n}} \Phi^*_{\mathbf{kn}}(\mathbf{r}')\Phi_{\mathbf{kn}}(\mathbf{r}) = \delta(\mathbf{r}-\mathbf{r}'). \tag{2.5.6}$$

If we write $\mathscr{P}$ for the differential operator $\{(\hbar^2/2m)\nabla^2+E(\mathbf{k})\}$ then eqn (2.5.2) can be written formally as

$$\mathscr{P}\mathscr{G}_{\mathbf{k}}(\mathbf{r},\mathbf{r}') = \delta(\mathbf{r}-\mathbf{r}'), \tag{2.5.7}$$

so that using eqns (2.5.5) and (2.5.6) we have

$$\begin{aligned}
\mathscr{G}_{\mathbf{k}}(\mathbf{r},\mathbf{r}') &= \mathscr{P}^{-1}\delta(\mathbf{r}-\mathbf{r}') \\
&= \mathscr{P}^{-1}\sum_{\mathbf{n}} \Phi^*_{\mathbf{kn}}(\mathbf{r}')\Phi_{\mathbf{kn}}(\mathbf{r}) \\
&= \sum_{\mathbf{n}} \Phi^*_{\mathbf{kn}}(\mathbf{r}')\mathscr{P}^{-1}\Phi_{\mathbf{kn}}(\mathbf{r}) \\
&= \sum_{\mathbf{n}} \frac{\Phi^*_{\mathbf{kn}}(\mathbf{r}')\Phi_{\mathbf{kn}}(\mathbf{r})}{\{-(\hbar^2/2m)(\mathbf{k}+\mathbf{G}_{\mathbf{n}})^2+E(\mathbf{k})\}} \\
&= -\frac{1}{NV}\sum_{\mathbf{n}} \frac{\exp(\mathrm{i}(\mathbf{k}+\mathbf{G}_{\mathbf{n}})\cdot(\mathbf{r}-\mathbf{r}'))}{\{-(\hbar^2/2m)(\mathbf{k}+\mathbf{G}_{\mathbf{n}})^2+E(\mathbf{k})\}}.
\end{aligned} \tag{2.5.8}$$

Equation (2.5.8) therefore gives an explicit expression for the Green's function $\mathscr{G}_{\mathbf{k}}(\mathbf{r}, \mathbf{r}')$ which satisfies eqn (2.5.2) subject to the required boundary conditions which are imposed by Bloch's theorem and given in eqn (2.5.3).

A formal expression for $\psi_{\mathbf{k}}(\mathbf{r})$ can be obtained by writing eqn (2.5.1) in the form

$$\begin{aligned}\psi_{\mathbf{k}}(\mathbf{r}) &= \mathscr{P}^{-1}V(\mathbf{r})\psi_{\mathbf{k}}(\mathbf{r}) \\ &= \int \mathscr{P}^{-1}\delta(\mathbf{r}-\mathbf{r}')V(\mathbf{r}')\psi_{\mathbf{k}}(\mathbf{r}')\,\mathrm{d}\mathbf{r}' \\ &= \int \mathscr{G}_{\mathbf{k}}(\mathbf{r}, \mathbf{r}')V(\mathbf{r}')\psi_{\mathbf{k}}(\mathbf{r}')\,\mathrm{d}\mathbf{r}'. \end{aligned} \tag{2.5.9}$$

This equation can be rewritten as

$$\psi_{\mathbf{k}}(\mathbf{r}) - \int \mathscr{G}_{\mathbf{k}}(\mathbf{r}, \mathbf{r}')V(\mathbf{r}')\psi_{\mathbf{k}}(\mathbf{r}')\,\mathrm{d}\mathbf{r}' = 0. \tag{2.5.10}$$

By constructing the quantity

$$\Lambda(\psi(\mathbf{r}), \mathbf{k}, E(\mathbf{k})) = \int \psi^*(\mathbf{r})V(\mathbf{r})\{\psi(\mathbf{r}) - \int \mathscr{G}_{\mathbf{k}}(\mathbf{r}, \mathbf{r}')V(\mathbf{r}')\psi_{\mathbf{k}}(\mathbf{r}')\mathrm{d}\mathbf{r}'\}\,\mathrm{d}\mathbf{r} \tag{2.5.11}$$

it is clear that the wave function $\psi_{\mathbf{k}}(\mathbf{r})$ satisfies the condition

$$\Lambda(\psi(\mathbf{r}), \mathbf{k}, E(\mathbf{k})) = 0. \tag{2.5.12}$$

It is also possible to show (Kohn and Rostoker 1954) that for small departures from solutions of eqn (2.5.10)

$$\delta\Lambda(\psi(\mathbf{r}), \mathbf{k}, E(\mathbf{k})) = 0. \tag{2.5.13}$$

Thus we see that a variational principle has been obtained which can be used to determine $\psi_{\mathbf{k}}(\mathbf{r})$. Using a trial function $\psi(\mathbf{r})$ which is likely to be an expansion of $\psi(\mathbf{r})$ in terms of a set of known basis functions, the quantity $\Lambda(\psi(\mathbf{r}), \mathbf{k}, E(\mathbf{k}))$ can be constructed; the coefficients in the expansion of $\psi(\mathbf{r})$ are then varied until eqns (2.5.12) and (2.5.13) are satisfied, when the required wave function $\psi_{\mathbf{k}}(\mathbf{r})$ will have been obtained. If $\psi(\mathbf{r})$ is written as

$$\psi(\mathbf{r}) = \sum_{\mathbf{n}} C_{\mathbf{kn}}\Phi_{\mathbf{kn}}(\mathbf{r}), \tag{2.5.14}$$

then

$$\Lambda(\psi(\mathbf{r}), \mathbf{k}, E(\mathbf{k})) = \sum_{\mathbf{m}}\sum_{\mathbf{n}} C^*_{\mathbf{km}}\Lambda_{\mathbf{mn}}C_{\mathbf{kn}}, \tag{2.5.15}$$

where

$$\Lambda_{\mathbf{mn}} = \int \Phi^*_{\mathbf{km}}(\mathbf{r}) V(\mathbf{r}) \Phi_{\mathbf{kn}}(\mathbf{r})\, d\mathbf{r} - \\ - \iint \Phi^*_{\mathbf{km}}(\mathbf{r}) V(\mathbf{r}) \mathscr{G}_{\mathbf{k}}(\mathbf{r}-\mathbf{r}') V(\mathbf{r}') \Phi_{\mathbf{kn}}(\mathbf{r}')\, d\mathbf{r}\, d\mathbf{r}'. \quad (2.5.16)$$

The stationary condition on $\Lambda(\psi(\mathbf{r}), \mathbf{k}, E(\mathbf{k}))$ then requires that all the partial derivatives $(\partial\Lambda(\psi(\mathbf{r}), \mathbf{k}, E(\mathbf{k}))/\partial C^*_{\mathbf{km}})$ vanish so that

$$\sum_{\mathbf{n}} \Lambda_{\mathbf{mn}} C_{\mathbf{kn}} = 0 \quad (2.5.17)$$

for all $\mathbf{m}$. The condition for the existence of a non-trivial solution to this set of equations for the coefficients $C_{\mathbf{kn}}$ is that

$$\det |\Lambda_{\mathbf{mn}}| = 0. \quad (2.5.18)$$

The energy eigenvalue $E(\mathbf{k})$ can then be obtained by solving eqn (2.5.18). If the basis functions $\Phi_{\mathbf{kn}}(\mathbf{r})$ that are chosen for the expansion of $\psi(\mathbf{r})$ are the plane wave solutions of the free-electron or empty lattice problem given in eqn (2.5.5) then $\Lambda_{\mathbf{mn}}$ reduces to the form

$$\Lambda_{\mathbf{mn}} = V(\mathbf{G}_{\mathbf{m}} - \mathbf{G}_{\mathbf{n}}) + \sum_{\mathbf{n}'} \frac{V(\mathbf{G}_{\mathbf{m}} - \mathbf{G}_{\mathbf{n}'}) V(\mathbf{G}_{\mathbf{n}'} - \mathbf{G}_{\mathbf{n}})}{\{-(\hbar^2/2m)(\mathbf{k}+\mathbf{G}_{\mathbf{n}'})^2 + E(\mathbf{k})\}}. \quad (2.5.19)$$

In practice the summation in the second term on the right-hand side of this equation is difficult to evaluate and this represents one of the main difficulties in the use of eqn (2.5.18) to determine the energy eigenvalues $E(\mathbf{k})$ and the wave functions $\psi_{\mathbf{k}}(\mathbf{r})$ of an electron with wave vector $\mathbf{k}$ in a metal.

In the above we have used the plane wave solutions of the free-electron model as the basis functions for the expansion of $\psi(\mathbf{r})$ and the subsequent construction of the secular equation, thus allowing the determination of $E(\mathbf{k})$, in eqn (2.5.18). An obvious alternative would be to use an expansion of $\psi(\mathbf{r})$ in terms of radial functions $R_l(r)$ and spherical harmonics $Y_l^m(\theta, \phi)$ as is done in the cellular method which we described in Section 2.3; the trial function $\psi(\mathbf{r})$ would then be written as

$$\psi(\mathbf{r}) = \sum_l \sum_m A_{\mathbf{k}lm} R_l(r) Y_l^m(\theta, \phi). \quad (2.5.20)$$

The coefficients $A_{\mathbf{k}lm}$ can then be varied until the conditions imposed by the variational principle,

$$\Lambda(\psi(\mathbf{r}), \mathbf{k}, E(\mathbf{k})) = 0 \tag{2.5.21}$$

and

$$\delta\Lambda(\psi(\mathbf{r}), \mathbf{k}, E(\mathbf{k})) = 0 \tag{2.5.22}$$

are satisfied. In general terms the procedure is similar to that followed in the case of the plane-wave expansion of $\psi(\mathbf{r})$; a set of simultaneous equations similar to eqn (2.5.17) can be obtained and they lead to another secular equation that replaces eqn (2.5.18). The algebraic details are somewhat tedious and are complicated by the necessity of expressing the Green's function $\mathscr{G}_{\mathbf{k}}(\mathbf{r}-\mathbf{r}')$ in terms of spherical harmonics and appropriate radial functions, which from the discussion of the A.P.W. method in Section 2.4 can be seen to be Bessel functions. We shall not reproduce all the details involved in the construction of the secular equation that is obtained when an expansion of the trial function in the form of eqn (2.5.20) is used—the interested reader is referred either to the original discussion given by Kohn and Rostoker (1954) or to the accounts given by Callaway (1964) and Segall and Ham (1968). Ziman (1965) has shown how to transform this method into a reciprocal-lattice representation which makes clearer the relationship between this method and the various plane-wave and pseudopotential methods (see also Johnson 1966; Slater 1966).

As we shall see in Chapters 4 and 5, the Green's function method has been applied to the calculation of the band structures of quite a number of metals, including, for example, the alkali metals (Ham 1960, 1962*a*), Cu (Segall 1961*b*, 1962) and Al (Segall 1961*a*, 1963). In practice the Green's function method appears to converge very rapidly and this led to the suggestion by Segall and Ham (1968) that the Green's function method should be particularly appropriate as the basis of a parametrization scheme for Fermi surfaces (see Section 2.9).

2.6. The quantum defect method and pseudopotentials

In Sections 2.2–5 we have discussed the various methods which are available for the solution of the one-electron Schrödinger equation subject to the boundary conditions imposed by the regular periodic nature of the crystal. Given a particular potential $V(\mathbf{r})$ each of these methods is capable of being used to produce a set of energy eigenvalues $E(\mathbf{k})$ throughout the Brillouin zone. In Section 2.1 we listed the main contributions to $V(\mathbf{r})$ and noted that, in practice, the calculation of these various contributions is a non-trivial operation. It is a well-known feature of *ab initio* band structure calculations that in the present state of the subject the least accurate part of the process is the construction of a realistic potential $V(\mathbf{r})$, including exchange and correlation effects, in which an electron in the individual-particle approximation is assumed to move. The accuracy of many recent computations, based on some of the methods which we have already described for determining the electronic wave functions $\psi_{\mathbf{k}}(\mathbf{r})$ and the energy eigenvalues $E(\mathbf{k})$, very often far exceeds the accuracy of the initial potential $V(\mathbf{r})$. Although detailed calculations of a large number of the contributions to $V(\mathbf{r})$ that were listed in Section 2.1 have sometimes been performed (for example by Heine (1957*a*,*b*,*c*) for Al and by Falicov (1962) for Mg) the results have generally been disappointing. The calculation of the various exchange and correlation contributions to $V(\mathbf{r})$ not only involves approximations because of the necessity of truncating the expressions involved, but also because of uncertainties about the details of the electronic wave functions. While the gross features of the band structure can often be determined fairly satisfactorily using potentials constructed in this way, it can easily happen that the uncertainties in $V(\mathbf{r})$ lead to errors in $E(\mathbf{k})$ which are quite significant near the Fermi energy E_F. It is therefore natural that attempts should be made to avoid the difficulties associated with the construction of $V(\mathbf{r})$ by devising schemes for the calculation of band structures without using any explicit analytical or numerical expression for $V(\mathbf{r})$. The two schemes that we shall discuss are the *quantum defect method* and the use of *pseudo-potential theory*.

The construction of the crystal potential can be avoided if one is prepared to make use of some experimental results at some stage in one's band structure calculation. In the quantum defect method the particular experimental results that are used consist of spectroscopic information about the energy levels of a free atom of the metal of which one requires to calculate the band structure (Kuhn and Van Vleck 1950; Brooks and Ham 1958; Ham 1962*a*, *b*). This information is used to construct wave functions for the conduction electrons in a metal that has only one conduction electron per atom in the following manner. It is assumed that the closed shells of electrons in the ion core are not significantly modified from their forms in a free atom of the metal. It is also assumed that the size of the ion core is quite small compared with the volume occupied by each atom in the metal. Under these assumptions the electrostatic field experienced by a conduction electron is nearly a pure Coulomb field over most of the unit cell and in this region, therefore, the field is very similar to that in a free atom of the metal. However, what is quite different from the case of the free atom is the form of the boundary conditions; in the free atom the wave function must tend to zero as $r \to \infty$, but in the metal the wave function has to satisfy periodic boundary conditions. Since a Coulomb potential possesses spherical symmetry it is possible to separate Schrödinger's equation by writing the wave function in the form

$$\psi(\mathbf{r}) = R_l(r) Y_l^m(\theta, \phi), \tag{2.6.1}$$

whereupon we obtain a radial equation of the form

$$\frac{\mathrm{d}^2 U_l(r)}{\mathrm{d}r^2} + \left\{\frac{2m}{\hbar^2}\left(E + \frac{|e|}{4\pi\epsilon_0 r}\right) - \frac{l(l+1)}{r^2}\right\} U_l(r) = 0, \tag{2.6.2}$$

where $U_l(r) = rR_l(r)$ and we have used a potential of the form given in eqn (2.1.2) with $Z_p(r) = 1$ since we are concerned with the outer regions of the cell.

There are two linearly independent solutions of eqn (2.6.2); these are confluent hypergeometric functions which we write as $U_{l,0}(r)$ and $U_{l,1}(r)$. $U_{l,0}(r)$ vanishes at the origin while $U_{l,1}(r)$ is singular at the origin. If a potential $-Z\,|e|/4\pi\epsilon_0 r$, where Z is a constant, applied to the whole range of r this would be the special case of a 'hydrogen-like' atom; only $U_{l,0}(r)$ would appear and the allowed values of E would be given by

$$E = -\tfrac{1}{2}m\left(\frac{Ze^2}{4\pi\epsilon_0\hbar}\right)^2\frac{1}{n^2}, \tag{2.6.3}$$

where n is an integer. For the real atom where the potential only takes the form $-|e|/4\pi\epsilon_0 r$ outside the ion core and is given by $-Z_p(r)\,|e|/4\pi\epsilon_0 r$ near to the nucleus we can still write the allowed energies in the form given in eqn (2.6.3) but n will no longer need to be an integer, that is

$$E = -\tfrac{1}{2}m\left(\frac{Ze^2}{4\pi\epsilon_0\hbar}\right)^2\frac{1}{n^2} = -\tfrac{1}{2}m\left(\frac{Ze^2}{4\pi\epsilon_0\hbar}\right)^2\frac{1}{(m-\nu)^2} \tag{2.6.4}$$

where m is an integer which is chosen so that $\nu < 1$ and ν is called the *quantum defect*. The corresponding solution of eqn (2.6.2) satisfying these atomic boundary conditions will take the form

$$U_l(r) = \alpha(n)U_{l,0}(r)+\gamma(n)U_{l,1}(r) \tag{2.6.5}$$

where the coefficients $\alpha(n)$ and $\gamma(n)$ depend on n (that is on ν or E). Because of the normalization condition it is only the ratio $\alpha(n)/\gamma(n)$, which is sometimes called the *coupling constant*, that is of interest. The coupling constant can be written explicitly as a function of ν (see Callaway (1964) p. 128). The values of ν corresponding to solutions that satisfy the boundary conditions applicable to the free atom can be determined empirically from the values of ν determined from spectroscopic data for the free atom. It is then assumed that the quantum defect ν is a smooth function of energy and its values for other energies are determined by interpolation. By using the periodic boundary conditions appropriate to the metal, the value of the coupling constant, and therefore also the value of ν, for the solution corresponding to a conduction electron with wave vector $\mathbf{k}$ in the metal can be determined. From the value of ν the corresponding value of the energy $E(\mathbf{k})$ is immediately available. The boundary conditions for the metal can be used in the form described in Section 2.3 in connection with the cellular method or in the form described in Section 2.5 in connection with the Green's function method. The usefulness of the quantum defect method has been restricted by the fact that it assumes the existence of a clear-cut separation between the conduction electrons and the ion cores and also because it neglects possible interactions among the electrons; it is therefore only likely to be suitable for use with simple monovalent metals. Moreover, except for the case of the alkali metals, the interpretation of the spectroscopic data is a difficult task.

The other approach to the determination of the band structure without using an explicit expression for the crystal potential $V(\mathbf{r})$ involves the construction of a *pseudopotential* instead. This approach has emerged as a result of the fact, which we shall encounter in detail in Chapter 4, that the free-electron model gives a very good first approximation to the actual band structures and Fermi surfaces of many of the simpler non-transition and non-rare-earth metals. For example, the Fermi surface of Na which would be a sphere in the free-electron approximation is found experimentally to depart from sphericity by rather less than 0·1 per cent—see Fig. 2.6. This is in fact the metal that comes closest of all to the free-electron model, but many other simple metals have Fermi surfaces that are remarkably close to the free-electron Fermi surfaces shown in Figs. 1.16–18; examples include the other alkali metals as well as Mg, Zn, Cd, Al, and Pb. This must mean, therefore, that as a conduction electron travels through a specimen of such a metal its scattering probability is quite small. This does not, however, imply that the potential $V(\mathbf{r})$ itself is weak, which would obviously not be true near the centres of the ion cores in the metal. It does, however, imply that the region over which $V(\mathbf{r})$ is large constitutes quite a small fraction of the volume of the unit cell of the metal, as was also assumed in the quantum defect method. It is then profitable to discuss the behaviour of the conduction electrons in a simple metal in terms of a pseudopotential. The condition of weak scattering of the conduction electrons can be considered in terms of phase shifts which are commonly used in scattering theory. The phase shift η_l can be written as $(p_l\pi+\delta_l)$ where p_l

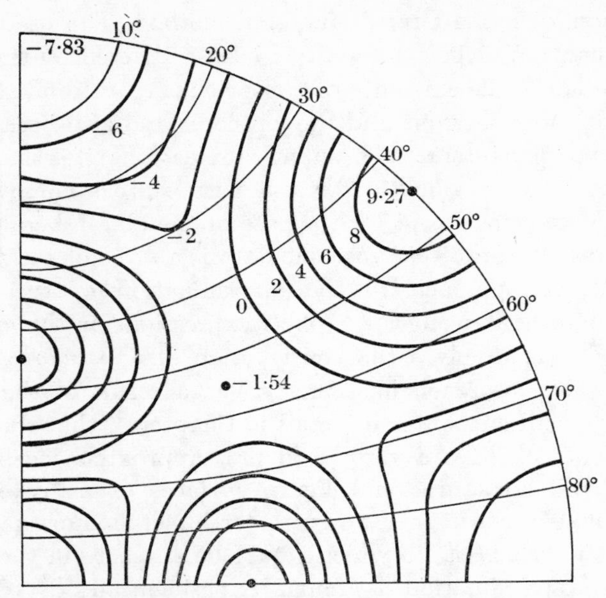

FIG. 2.6. Contours of $\Delta r/r$ of the Fermi surface of Na; units are parts in 10^4 (Lee 1966).

is an integer that is chosen so that $-\frac{1}{2}\pi < \delta_l < +\frac{1}{2}\pi$. Since the phase shift only occurs in an exponential term of the form $\exp(2i\eta_l)$ the integer p_l does not contribute to the scattering. The requirement of weak scattering means that the δ_l must be small but not that the η_l itself should necessarily be small.

The construction of a suitable pseudopotential involves an adaptation of the methods used in the quantum theory of scattering which are more conventionally applied to the scattering of beams of particles moving in free space, rather than in a solid, and colliding with scattering centres either in another beam of particles or in a solid target. It is the fact that the net scattering of the conduction electrons in a simple metal is weak that has meant that scattering theory methods have been quite simple to apply and good descriptions of the behaviour of the conduction electrons in a metal have been obtained in terms of quite a small number of parameters. These pseudopotential parameters can then be adjusted until the calculated

Fermi surface gives the best fit to a set of experimental caliper dimensions or extremal areas of cross section. For example, this has been applied to the alkali metals (Lee 1969), Mg (Kimball, Stark, and Mueller 1967), Al (Ashcroft 1963*a*, 1963*b*) and several other metals. The use of pseudopotentials in calculating band structures emerged from the O.P.W. method and owes much to the work of Phillips and Kleinman (1959); extensive discussions of the use of pseudopotentials have been given by a number of authors (Ziman 1964*b*; Lin and Phillips 1965; Harrison 1966, 1970; Heine 1969, 1970; Cohen and Heine 1970).

We mentioned in Section 2.4 that if the wave function $\psi_{\mathbf{k}}(\mathbf{r})$ is expanded in terms of plane-wave solutions of the free-electron Schrödinger equation, the convergence is very slow, but that if these basis functions are orthogonalized to the wave functions of the ion-core electrons the convergence becomes very much more rapid. Phillips and Kleinman (1959) showed that this orthogonalization has the same effect as adding an extra repulsive contribution to the true crystal potential $V(\mathbf{r})$. Because this net effective potential, which came to be called a *pseudopotential*, is weak for simple metals the net scattering of the conduction electrons in the metal is small and therefore the free-electron model gives a good first approximation to the behaviour of the electrons in the metal. We are, of course, assuming that the division of the electrons in a metal into itinerant conduction electrons and localized ion-core electrons is physically reasonable. While this is a realistic assumption for the simple metals it is unlikely to be very realistic for the more complicated transition metals and rare-earth metals. In Section 2.4 we constructed an orthogonalized plane wave

$$X_{\mathbf{k}n}(\mathbf{r}) = (NV)^{-\frac{1}{2}}\exp\{\mathrm{i}(\mathbf{k}+\mathbf{G}_n)\,.\,\mathbf{r}\} - \\ -\sum_m (NV)^{-\frac{1}{2}}\phi_{\mathbf{k},m}(\mathbf{r})\int \phi^*_{\mathbf{k},m}(\mathbf{r}')\exp\{\mathrm{i}(\mathbf{k}+\mathbf{G}_n)\,.\,\mathbf{r}'\}\,\mathrm{d}\mathbf{r}' \qquad (2.6.6)$$

(see eqns (2.4.8) and (2.4.9)), where the functions $\phi_{\mathbf{k},m}(\mathbf{r})$ are the tight-binding wave functions for the ion-core electrons.

It is convenient to use a shorthand notation for the quantity

$$\mathscr{P} = \sum_m \phi_{\mathbf{k},m}(\mathbf{r})\int d\mathbf{r}'\phi^*_{\mathbf{k},m}(\mathbf{r}')\ldots \tag{2.6.7}$$

which is a projection operator; we use the row of dots to emphasize that $\mathscr{P}$ is an operator. Equation (2.6.6) then becomes

$$X_{\mathbf{kn}}(\mathbf{r}) = (1-\mathscr{P})((NV)^{-\frac{1}{2}}\exp\{i(\mathbf{k}+\mathbf{G}_n)\cdot\mathbf{r}\}). \tag{2.6.8}$$

It will be recalled that in the O.P.W. method the wave function $\psi_{\mathbf{k}}(\mathbf{r})$ was expanded in terms of the orthogonalized plane waves $X_{\mathbf{kn}}(\mathbf{r})$,

$$\psi_{\mathbf{k}}(\mathbf{r}) = \sum_{\mathbf{n}} C^X_{\mathbf{kn}} X_{\mathbf{kn}}(\mathbf{r}), \tag{2.4.10}$$

which on substitution into the one-electron Schrödinger equation gives

$$\left(-\frac{\hbar^2}{2m}\nabla^2+V(\mathbf{r})\right)\sum_{\mathbf{n}} C^X_{\mathbf{kn}}(1-\mathscr{P})\exp\{i(\mathbf{k}+\mathbf{G}_n)\cdot\mathbf{r}\}$$
$$= E(\mathbf{k})\sum_{\mathbf{n}} C^X_{\mathbf{kn}}(1-\mathscr{P})\exp\{i(\mathbf{k}+\mathbf{G}_n)\cdot\mathbf{r}\}. \tag{2.6.9}$$

By collecting all the terms involving $\mathscr{P}$ onto the left-hand side of this equation it can be re-written in the form

$$-\frac{\hbar^2}{2m}\nabla^2\phi_{\mathbf{k}}(\mathbf{r})+W(\mathbf{r})\phi_{\mathbf{k}}(\mathbf{r}) = E(\mathbf{k})\phi_{\mathbf{k}}(\mathbf{r}), \tag{2.6.10}$$

where

$$\phi_{\mathbf{k}}(\mathbf{r}) = \sum_{\mathbf{n}} C^X_{\mathbf{kn}}\exp\{i(\mathbf{k}+\mathbf{G}_n)\cdot\mathbf{r}\} \tag{2.6.11}$$

and

$$W(\mathbf{r}) = V(\mathbf{r})+\left\{E(\mathbf{k})-\left(-\frac{\hbar^2}{2m}\nabla^2+V(\mathbf{r})\right)\right\}\mathscr{P}. \tag{2.6.12}$$

Equation (2.6.10) is of the same form as the true one-electron Schrödinger equation and in particular it has the same set of energy eigenvalues $E(\mathbf{k})$. $W(\mathbf{r})$ is called the *pseudopotential* and $\phi_{\mathbf{k}}(\mathbf{r})$ is called the *pseudo wave function*. The true wave function $\psi_{\mathbf{k}}(\mathbf{r})$ is related in a simple formal way to the pseudo wave function by

$$\psi_{\mathbf{k}}(\mathbf{r}) = (1-\mathscr{P})\phi_{\mathbf{k}}(\mathbf{r}). \tag{2.6.13}$$

There have been no new approximations involved in going from eqn (2.6.9) to eqn (2.6.10) and all that has been done is to re-arrange the terms. The importance of this re-arrangement lies in the fact that the second term in the expression for $W(\mathbf{r})$ in eqn (2.6.12) is negative and, particularly in the ion-core,

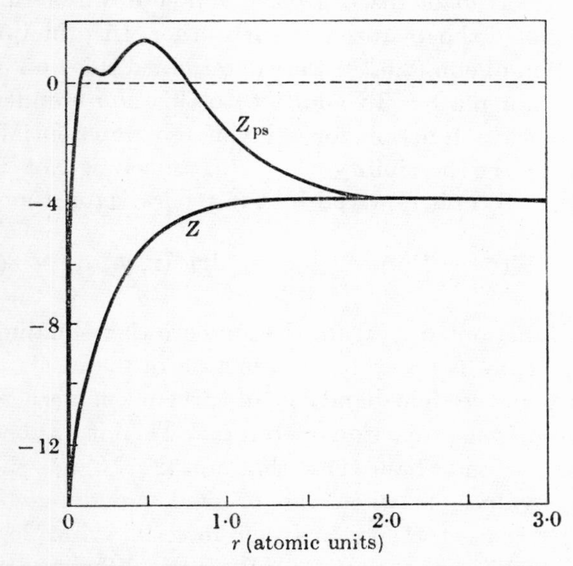

FIG. 2.7. The potential and pseudo-potential (for $l = 0$ states) of a Si^{4+} ion. The potential V is expressed in the form $V(r) = Z(r)/r$ and the pseudopotential $W(r)$ similarly in terms of $Z_{ps}(r)$. Note V and W both become equal to the Coulomb potential $-4/r$ outside the core which has a radius of about one atomic unit. (Heine 1969).

cancels out the first term $V(\mathbf{r})$ almost completely, as in Fig. 2.7. The details of this cancellation have been discussed by various authors (Antoncik 1959; Phillips and Kleinman 1959; Cohen and Heine 1961; Austin, Heine, and Sham 1962).

We have seen how to construct a pseudo Schrödinger equation of the form of eqn (2.6.10) involving the pseudopotential $W(\mathbf{r})$. However, there is no guarantee of the uniqueness of the pseudo wave function which is the solution of this equation. If $\phi_{\mathbf{k}}(\mathbf{r})$ is a solution of eqn (2.6.10) any linear combination of core wave functions can be added to $\phi_{\mathbf{k}}(\mathbf{r})$ and the result is still a solution of eqn (2.6.10) with the same eigenvalue $E(\mathbf{k})$ (Bassani and Celli 1961; Cohen and Heine 1961). Moreover, on substitution into eqn (2.6.13) it leads to the same true wave function $\psi_{\mathbf{k}}(\mathbf{r})$. Therefore the solutions of eqn (2.6.10) are not unique and this non-uniqueness arises because although the functions $X_{\mathbf{k}n}(\mathbf{r})$ were orthogonalized to the ion-core wave functions they were not orthogonalized to each other. In principle if an exact solution of eqn (2.6.10) were determined this non-uniqueness would not matter. In addition to this non-uniqueness of the pseudo wave function for a given pseudopotential $W(\mathbf{r})$, there is also the possibility that $W(\mathbf{r})$ itself is not unique. Suppose that $W(\mathbf{r})$ is replaced by the general expression

$$W(\mathbf{r}) = V(\mathbf{r})+\sum_{m} \phi_{\mathbf{k},m}(\mathbf{r})\int \mathrm{d}\mathbf{r} f^{*}(\mathbf{r}, m).... \qquad (2.6.14)$$

where the functions $\phi_{\mathbf{k},m}(\mathbf{r})$ are the ion-core tight-binding wave functions, $f(\mathbf{r}, m)$ is an arbitrary function of $\mathbf{r}$ and m, and the second term on the right-hand side of this equation is a generalized version of the projection operator $\mathscr{P}$. Then it can be shown (Austin, Heine, and Sham 1962) that eqn (2.6.10) has the same set of eigenvalues as when the original pseudopotential in eqn (2.6.12) is used. There are therefore, in principle, many pseudo Schrödinger equations all with different pseudopotentials and yet all leading to the same energy eigenvalues $E(\mathbf{k})$. Part of the skill involved in performing a pseudopotential calculation lies in obtaining the most realistic pseudopotential with the minimum expenditure of effort.

It is, perhaps, not immediately obvious that the formal manipulation that was involved in producing the expression for the pseudopotential $W(\mathbf{r})$ in eqn (2.6.12) has actually simplified the problem of solving the one-electron Schrödinger equation. In particular, whereas $V(\mathbf{r})$ was a function of position $\mathbf{r}$, the pseudopotential $W(\mathbf{r})$ is not just a simple function of $\mathbf{r}$ but involves the differential operator ∇^2 and the very complicated integral operator $\mathscr{P}$. If one were to try to construct $W(\mathbf{r})$ *ab initio* it would be even more difficult than trying to construct $V(\mathbf{r})$ *ab initio*. It is when $W(\mathbf{r})$ is small, as it appears to be for simple metals, that one can employ a quite straightforward construction of $W(\mathbf{r})$ and expect to achieve an accuracy in $E(\mathbf{k})$ that would only be obtained from the true Schrödinger equation after a much more exhaustive calculation of $V(\mathbf{r})$. It is sometimes possible to replace the complicated operator form of $W(\mathbf{r})$ in eqn (2.6.12) or eqn (2.6.14) by an ordinary function of $\mathbf{r}$ and still obtain a pseudo wave function $\phi_{\mathbf{k}}(\mathbf{r})$ and a value of $E(\mathbf{k})$ that provide good descriptions of the behaviour of the electrons in a metal; in this case the pseudopotential would be described as a *local* pseudopotential. The full operator form of the pseudopotential $W(\mathbf{r})$ without such an approximation is then sometimes described as a *non-local* pseudopotential.

In the account of pseudopotentials that we have given above it might seem that they have some necessary connection with the use of the O.P.W. method. This strong connection with the O.P.W. method is, however, to a large extent only a historical accident. It is possible to work in the formalism of any of the other methods that we have already described for calculating band structures, such as the A.P.W. method or the Green's function method, and to construct similar pseudo Schrödinger equations and pseudo wave functions. In general terms a pseudopotential $W(\mathbf{r})$ is any function, possibly including differential or integral operators, that can be used in a pseudo Schrödinger equation which has the same eigenvalues $E(\mathbf{k})$ as the true Schrödinger equation and for which the scattering of the conduction electrons is weak. Some considerable discussion of this more general approach to pseudopotentials is given by Heine (1969). It is only if the pseudopotential leads to weak scattering of the conduction electrons that it is actually likely to be of any practical use in the calculation of electronic band structures; this is much more a property of the metal in question than of one's choice of formalism in constructing $W(\mathbf{r})$.

The actual construction of a sensible pseudopotential that is not too empirical, for any given metal, is something of a skill or art in itself and we do not wish to discuss the details of this here (see, for example, the books by Harrison (1966,

1970)). If we assume for the moment that for a particular metal a pseudopotential $W(\mathbf{r})$ has somehow been obtained, then the fact that the scattering of the conduction electrons is weak means that the pseudo-Schrödinger equation can be solved to a very good approximation by a straightforward application of perturbation theory. We require to solve the pseudo-Schrödinger equation

$$\left\{-\frac{\hbar^2}{2m}\nabla^2+W(\mathbf{r})\right\}\phi(\mathbf{r}) = E(\mathbf{k})\phi(\mathbf{r}), \tag{2.6.15}$$

and we suppose that $\phi_0(\mathbf{r})$ and $E^0(\mathbf{k})$ are zero-order approximations to the exact solutions of this equation. The natural trial solutions to choose for $\phi_0(\mathbf{r})$ and $E^0(\mathbf{k})$ for an electron with wave vector $\mathbf{k}$ are the free-electron wave function $\psi_{\mathbf{k}0}(\mathbf{r}) = (NV)^{-\frac{1}{2}}\{\exp(i\mathbf{k}\,.\,\mathbf{r})\}$ and the corresponding free-electron energy $\hbar^2\mathbf{k}^2/2m$. The expression for $E(\mathbf{k})$ including first and second order corrections to $E^0(\mathbf{k})$ is therefore

$$E(\mathbf{k}) = \frac{\hbar^2\mathbf{k}^2}{2m}+\int \psi^*_{\mathbf{k}0}(\mathbf{r})W(\mathbf{r})\psi_{\mathbf{k}0}(\mathbf{r})\,d\mathbf{r}+$$

$$+\sum_{\substack{\mathbf{n}\\ \mathbf{n}\neq(0,0,0)}} \frac{\left\{\int \psi^*_{\mathbf{k}\mathbf{n}}(\mathbf{r})W(\mathbf{r})\psi_{\mathbf{k}0}(\mathbf{r})\,d\mathbf{r}\right\}\left\{\int \psi^*_{\mathbf{k}0}(\mathbf{r})W(\mathbf{r})\psi_{\mathbf{k}\mathbf{n}}(\mathbf{r})\,d\mathbf{r}\right\}}{(\hbar^2/2m)(\mathbf{k}^2-|\mathbf{k}+\mathbf{G}_\mathbf{n}|^2)} \tag{2.6.16}$$

where

$$\psi_{\mathbf{k}\mathbf{n}}(\mathbf{r}) = (NV)^{-\frac{1}{2}}\exp\{i(\mathbf{k}+\mathbf{G}_\mathbf{n})\,.\,\mathbf{r}\}. \tag{2.6.17}$$

If the pseudopotential method is to be profitable as a method for calculating band structures then one would hope to have been able to choose a pseudopotential for which it was not necessary to proceed to higher terms than the second-order terms included in eqn (2.6.16).

It is because perturbation theory is very much easier to handle than the quite complicated methods that we have described earlier in this chapter for the solution of the true Schrödinger equation, that the pseudopotential method is so very popular for the many simple metals for which it works satisfactorily. A pseudopotential calculation may appear to lack some of the elegance of a completely *ab initio* band structure calculation but it should not be despised on these grounds. One can regard the pseudopotential method as a sophisticated interpolation scheme that is based on physical rather than mathematical ideas and which enables the shape of the entire Fermi surface to be determined numerically from the necessarily rather small number of experimentally determined caliper dimensions or extremal areas of cross section. If the pseudopotential is expressed in terms of a small number of adjustable parameters, the ease with which the energy eigenvalues $E(\mathbf{k})$ can be calculated provides a very useful basis for an interpolation method to construct $E(\mathbf{k})$ all over the Brillouin zone of a metal (see Section 2.7). The term *model potential* is often used to describe a pseudopotential that is expressed in a simple functional form involving a number of adjustable parameters. If $W(\mathbf{r})$ is such a model potential then the expression $\{-(\hbar^2/2m)\nabla^2+W(\mathbf{r})\}$ is sometimes called a *model Hamiltonian.* Screened model potentials for the atoms of twenty-five elements have been given by Animalu and Heine (1965).

We shall see in Chapters 4 and 5 that the pseudopotential method for calculating the band structure of a metal and thence determining the shape of its Fermi surface has been applied successfully to quite a large number of simple metals but that for the transition and rare-earth metals it is less successful. This does not mean that it is not possible to construct a pseudopotential for the non-simple metals, but only that it is difficult, and perhaps unprofitable, to do so. One important advantage of the use of a pseudopotential is that having used the various direct experimental measurements to determine the parameters in the pseudopotential the effect of the conduction electrons on a large number of macroscopic properties of the metal can then be expressed in terms of the pseudopotential. The effect of electron–phonon interactions is often important in connection with the macroscopic properties of a metal, and one important aspect of the use of pseudopotentials is that the same pseudopotential that is used to calculate the electronic band structure of a metal, and hence the shape of the Fermi surface, can also be used to calculate the lattice vibration spectrum, or phonon dispersion relations, of the metal (see, for example, Harrison (1966, 1970)). Having determined the forms of the band structure and the phonon dispersion relations of a metal it is then important, for an understanding of many of the physical properties of a metal, to investigate the interactions among the electrons (Pines 1955) and the interactions between the electrons and the phonons (Bak 1964). We shall return to this subject in Chapter 6.

2.7. Interpolation schemes

In the previous sections of this chapter we have described various methods that can be used to calculate the band structure of a metal, that is, to calculate $E_j(\mathbf{k})$ for any given value of $\mathbf{k}$, where we use j to label the various bands at $\mathbf{k}$ (see p. 15). However, until recently, the calculation of a single eigenvalue was quite a lengthy process which was usually only possible, with the computing machinery which then existed, for wave vectors that terminate at a point of symmetry or on a line of symmetry in the Brillouin zone. The calculations are easier for these special points and lines of symmetry because it is possible to make some group-theoretical simplifications in the expansions of the wave functions in terms of sets of basis functions. The calculation of $E_j(\mathbf{k})$ is particularly lengthy in those methods in which the matrix elements or the coefficients in the boundary-condition equations are themselves functions of $E_j(\mathbf{k})$ so that various trial energies have to be investigated until an exact solution, or a best least-squares fit for $E_j(\mathbf{k})$ is obtained. It has therefore in the past been quite common only to attempt to perform direct calculations of $E_j(\mathbf{k})$ for a few important wave vectors in the Brillouin zone. These eigenvalues can then be used as the basis of some interpolation scheme for the rapid calculation of $E_j(\mathbf{k})$ over the rest of the Brillouin zone. Various interpolation schemes have been used by different workers.

It is possible to use a purely mathematical interpolation scheme in which each $E_j(\mathbf{k})$ is expanded in terms of a number of functions $f_j^n(k_x, k_y, k_z)$ with the required symmetry of the Brillouin zone

$$E_j(\mathbf{k}) = \sum_n A_j^n f_j^n(k_x, k_y, k_z). \tag{2.7.1}$$

The coefficients $A^{n}_{j_{\mathrm{l}}}$ can be found by fitting the expression in eqn (2.7.1) to the available values of $E_j(\mathbf{k})$ calculated from first principles. While this method may be satisfactory for smooth bands it is less likely to be satisfactory for bands such as those shown in Fig. 2.8 which have discontinuities in their slopes.

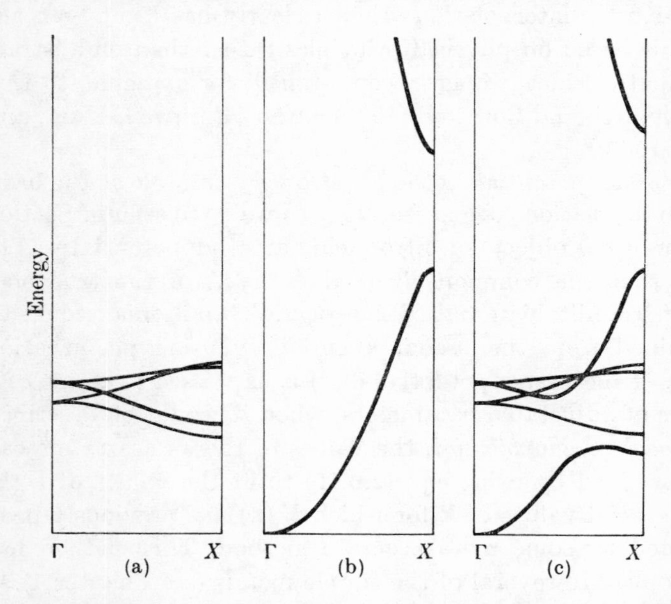

FIG. 2.8. Schematic diagram to illustrate the hybridization of (a) d-electrons and (b) s-electrons in an f.c.c. transition metal. The heavy bands all have the same symmetry (Δ_1) and therefore hybridize with each other as shown in (c). (Hodges, Ehrenreich, and Lang 1966).

Rather than using a purely mathematical interpolation scheme it is probably more rewarding to use an interpolation scheme that makes some use of the physical knowledge of the situation. A suitable interpolation scheme can then be devised using as its basis any of those methods which we have described for calculating $E_j(\mathbf{k})$ which give direct expressions for $E_j(\mathbf{k})$ and not just a set of integral equations for $E_j(\mathbf{k})$ that have to be solved by a fitting procedure. For instance an interpolation scheme based on the tight-binding method which was suggested by Slater and Koster (1954) has been used quite extensively.

Slater and Koster suggested that the various integrals in eqn (2.2.6) or eqn (2.2.7) could be regarded as adjustable parameters, which can then be determined by fitting the values of $E_j(\mathbf{k})$ at the special values of $\mathbf{k}$ for which $E_j(\mathbf{k})$ has previously been calculated by some more accurate *ab initio* method. The Slater–Koster interpolation scheme clearly has the advantage that it is based on physical principles rather than on abstract mathematical ideas; it has been applied, for example, to iron by Wohlfarth and Cornwell (1961) and by Cornwell, Hum, and Wong (1968).

The pseudopotential method is also very suitable as the basis of an interpolation scheme because, as we have seen in Section 2.6, one of the objects of introducing a pseudopotential was to ensure that the common eigenvalues $E_j(\mathbf{k})$ of the true one-electron Hamiltonian and of the pseudo Hamiltonian could be determined from the pseudopotential by using perturbation theory. If the pseudopotential $W(\mathbf{r})$ is expressed in terms of a number of adjustable parameters, when it would be described as a *model pseudopotential*, the values of these parameters can be determined by using eqn (2.6.12) to fit the values of $E_j(\mathbf{k})$ at the special values of $\mathbf{k}$ for which $E_j(\mathbf{k})$ has previously been calculated by some more accurate method. This method has been applied to several of the simple metals (see Chapter 4).

Which of these various interpolation schemes should be chosen for any given metal will depend, partly at least, on the nature of the band structure of the metal in question. If the free-electron model gives a very good description of the band structure and Fermi surface of a particular metal, then an interpolation scheme based on a local model pseudopotential is likely to be particularly successful. For transition metals, however, it is unlikely that a good understanding of the band structure will be obtained by using a local pseudopotential. An alternative method, called the *combined interpolation scheme* has been introduced for transition metals (Hodges and Ehrenreich 1965; Hodges, Ehrenreich, and Lang 1966; Mueller 1966, 1967). This method is based on the assumption that is commonly made that to a first approximation at least the conduction electrons in a transition metal can be separated into s electrons, which are highly non-localized and can be treated by plane-wave methods, and d electrons, which are very nearly localized and can therefore be treated by the tight-binding method. The wave functions for the d electrons are assumed to be given by the tight-binding wave functions

$$\phi^{d}_{\mathbf{k},m}(\mathbf{r}) = (N)^{-\frac{1}{2}} \sum_{\alpha} \exp(\mathrm{i}\mathbf{k}\,.\,\mathbf{R}_{\alpha})u_m(\mathbf{r}-\mathbf{R}_{\alpha}), \qquad (2.7.2)$$

where the $u_m(\mathbf{r})$ are atomic d functions (3d, 4d, or 5d as appropriate). The wave function for the s electrons can be expanded either in terms of simple plane waves as in eqn (2.4.1) (Hodges, Ehrenreich, and Lang 1966), or in terms of the orthogonalized plane waves (Mueller 1967):

$$X_{\mathbf{kn}}(\mathbf{r}) = (NV)^{-\frac{1}{2}} \exp\{\mathrm{i}(\mathbf{k}+\mathbf{G}_{\mathbf{n}})\,.\,\mathbf{r}\} - \sum_{m} \mu_{\mathbf{k}m}\phi_{\mathbf{k},m}(\mathbf{r}), \qquad (2.4.8)$$

where the $\phi_{\mathbf{k},m}(\mathbf{r})$ are core-electron states. In practice, in either case, the summation will be terminated at some suitable value of $\mathbf{n}$. It is then assumed that realistic wave functions for the conduction electrons in a transition metal can be obtained by allowing hybridization between the simple wave functions that we have just given for the d and s electrons but ignoring any hybridization with other atomic states. That is, we write the wave function of a conduction electron in a transition metal in the form

$$\psi_{\mathbf{k}j}(\mathbf{r}) = \sum_{m} a^{j}_{\mathbf{k}m}\phi^{d}_{\mathbf{k},m}(\mathbf{r}) + \sum_{\mathbf{n}} a^{j}_{\mathbf{kn}}X_{\mathbf{kn}}(\mathbf{r}). \qquad (2.7.3)$$

The effect of this hybridization on the band structure is illustrated in Fig. 2.8. If the potential $V(\mathbf{r})$ is known, the energy eigenvalues $E_j(\mathbf{k})$ can be found by substituting $\psi_{\mathbf{k}j}(\mathbf{r})$ given by eqn (2.7.3) into Schrödinger's equation and constructing a secular equation similar to eqns (2.4.4) and (2.4.11). However, we are not seeking to use $\psi_{\mathbf{k}j}(\mathbf{r})$ in an *ab initio* band structure calculation but in an interpolation scheme to calculate $E_j(\mathbf{k})$ all over the Brillouin zone when the values of $E_j(\mathbf{k})$ for the special points and lines of symmetry have already been found by one of the more accurate *ab initio* methods described previously. Therefore the various spatial integrals involving $V(\mathbf{r})$,

$\phi^d_{\mathbf{k},m}(\mathbf{r})$, and $X_{\mathbf{k}n}(\mathbf{r})$ are regarded as parameters that are determined by fitting $E_j(\mathbf{k})$ to the given values of $E_j(\mathbf{k})$ for the special points and lines of symmetry in the Brillouin zone. The combined interpolation scheme is therefore very similar to the interpolation scheme of Slater and Koster (1954) mentioned earlier, except that instead of using only tight-binding functions for the conduction electrons we now use both tight-binding functions and plane-wave functions. The accuracy of the combined interpolation scheme was tested by Mueller (1967)

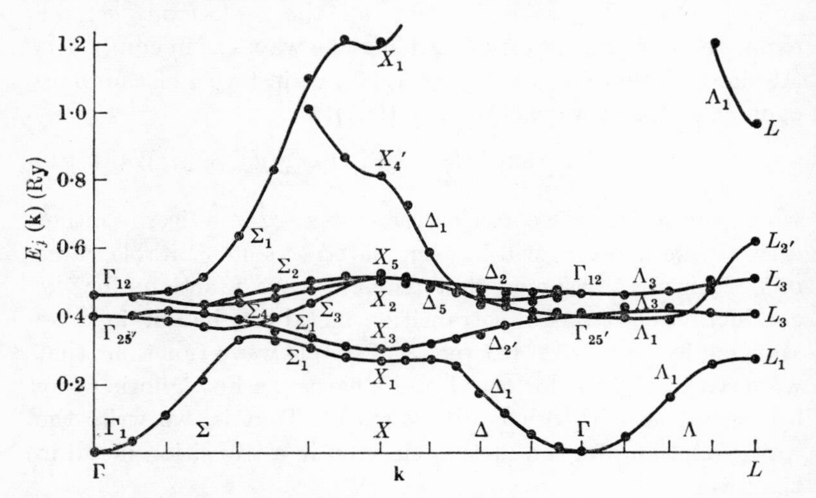

FIG. 2.9. The energy bandsof Cu. The solid circles represent the calculated values of Burdick (1963) and the solid lines represent the interpolated bands obtained using the combined interpolation scheme (Mueller 1967).

by fitting the energy bands of copper obtained by Segall (1962) and Burdick (1963), see Fig. 2.9. The root mean square error that was obtained for the interpolated energy eigenvalues was less than 0·01 Ry throughout the Brillouin zone. The problem of justifying the use of the combined interpolation scheme, and of attempting to determine from first principles the parameters that occur in the model Hamiltonians which are used, have been studied by a number of authors (Heine 1967; Hubbard 1967, 1969; Hubbard and Dalton 1968; Jacobs 1968; Phillips 1968*b*; Hum and Wong 1969; Pettifor 1969).

2.8. The density of states and the Fermi surface

In the preceding sections of this chapter we have discussed a number of methods that can be used to calculate the eigenvalues $E_j(\mathbf{k})$ of the Hamiltonian of an electron in a metal in the individual-particle approximation. In principle it would be possible to determine the potential $V(\mathbf{r})$ to any desired accuracy and then, given $V(\mathbf{r})$, also to determine the eigenvalues $E_j(\mathbf{k})$ for all wave vectors $\mathbf{k}$ in the Brillouin zone. Let us suppose that for some given metal the energy eigenvalues $E_j(\mathbf{k})$ have been calculated, either directly or by the use of some interpolation scheme, for all $\mathbf{k}$ and for all j up to some quite large number $j_{\max}$. Then it is possible by inspection to map constant energy surfaces defined by

$$E_j(\mathbf{k}) = E \tag{1.3.11}$$

where E is some constant, and therefore to identify that particular constant energy surface which is the Fermi surface, for which

$$E_j(\mathbf{k}) = E_{\mathrm{F}}. \tag{1.3.13}$$

Thus we see that the whole process of determining theoretically the shape of the Fermi surface of a metal involves calculating $E_j(\mathbf{k})$ all over the Brillouin zone and then constructing the particular constant energy surface corresponding to $E_j(\mathbf{k}) = E_{\mathrm{F}}$. However, this does assume that the actual numerical value of the Fermi energy, E_{F}, is available, which may well not be the case. Experimental values of E_{F}, determined for instance from spectroscopic measurements using soft X-rays, have not generally been found to be very satisfactory. Indeed even theoretical values of E_{F} determined from different band structure calculations on the same metal are often found to vary widely. This is because very often quite small changes in the potential $V(\mathbf{r})$ can produce surprisingly large changes in the actual numerical values of $E_j(\mathbf{k})$ without altering the general features of the shapes of the various bands very much, and therefore also without altering the shape of the calculated Fermi surface very much either. It is, therefore, highly desirable when using a particular calculated band structure for the production of a theoretical Fermi surface for some given metal, to determine E_F from the results of that calculation itself and not to use a value of E_F obtained from some other source. The calculation of E_F can be based on the fact that the number of states with $E_j(\mathbf{k}) < E_F$ must be exactly equal to the total number of conduction electrons in the metal; this calculation of E_F is usually performed by constructing the *density of states*, $n(E)$, where $n(E)\,\mathrm{d}E$ is the total number of states for which $E_j(\mathbf{k})$, irrespective of $\mathbf{k}$, lies between E and $E+\mathrm{d}E$. If a specimen of a crystal of a metal contains N unit cells the number of allowed wave vectors $\mathbf{k}$ in the Brillouin zone will be exactly N so that, allowing for the two different spin states, each band can contain up to $2N$ electrons. If the total number of conduction electrons contributed by all the atoms in a unit cell of the metal is x then the total number of bands that will be occupied is $(xN/2N)$ or $\frac{1}{2}x$ bands and the Fermi energy can be obtained from the condition that

$$\int_0^{E_F} n(E)\,\mathrm{d}E = \tfrac{1}{2}xN. \tag{2.8.1}$$

In the free-electron approximation the density of states $n(E)$ takes a particularly simple form. Since there are $2N$ allowed states per band in each Brillouin zone and since the volume of the Brillouin zone is $8\pi^3/\{\mathbf{t}_1 \cdot (\mathbf{t}_2 \wedge \mathbf{t}_3)\}$ (see Section 1.3), the number of states per unit volume of $\mathbf{k}$ space is

$$2N[8\pi^3/\{\mathbf{t}_1 \cdot (\mathbf{t}_2 \wedge \mathbf{t}_3)\}]^{-1} = NV/4\pi^3,$$

where V is the volume of each unit cell. In the extended zone scheme the constant energy surfaces corresponding to E and $E+\mathrm{d}E$ are spheres of radii k and $k+\mathrm{d}k$. The number of states between k and $k+\mathrm{d}k$ is then given by $g(k)\,\mathrm{d}k$ where

$$g(k) = 4\pi k^2(NV/4\pi^3) = NVk^2/\pi^2, \tag{2.8.2}$$

and the number of states between E and $E+\mathrm{d}E$ is $n(E)\,\mathrm{d}E$ where

$$n(E)\,\mathrm{d}E = g(k)\,\mathrm{d}k \tag{2.8.3}$$

so that

$$n(E) = \frac{NVk^2}{\pi^2}\frac{\mathrm{d}k}{\mathrm{d}E}. \tag{2.8.4}$$

However, in the free-electron approximation

$$E(\mathbf{k}) = \frac{\hbar^2}{2m} k^2 \tag{2.8.5}$$

in the extended zone scheme, so that

$$\frac{\mathrm{d}E}{\mathrm{d}k} = \frac{\hbar^2 k}{m} = \hbar\sqrt{\frac{2E}{m}} \tag{2.8.6}$$

and therefore

$$n(E) = \frac{mNVk}{\pi^2\hbar^2} = \left(\frac{NV\sqrt{(2m^3)}}{\pi^2\hbar^3}\right)E^{\frac{1}{2}}. \tag{2.8.7}$$

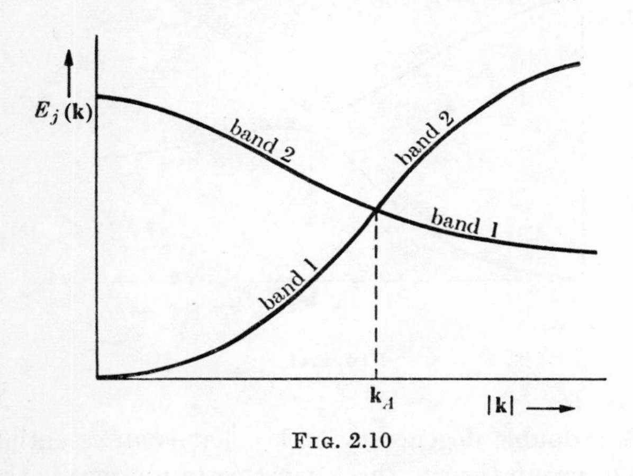

FIG. 2.10

Therefore in the free-electron approximation the density of states $n(E)$ is just a parabola. In a real metal, just as in the free-electron approximation, the eigenvalues $E_j(\mathbf{k})$ are quasi-continuous functions of $\mathbf{k}$ which we have grown accustomed to describing as bands. The subscript j is used to distinguish between the different eigenvalues at $\mathbf{k}$ and the numerical values given to j are the positive real integers, starting with 1; they are assigned so that for any given value of $\mathbf{k}$, $E_j(\mathbf{k})$ is a monotonic increasing function of j, see Section 1.3. While it is required that $E_j(\mathbf{k})$ must be continuous throughout the Brillouin zone, there is no requirement that $\nabla_{\mathbf{k}}E_j(\mathbf{k})$ should be continuous, at least in the free-electron approximation; therefore in the schematic situation shown in Fig. 2.10 the slope of band 1 and the slope of band 2 are both discontinuous at $\mathbf{k}_A$. However, $\nabla_{\mathbf{k}} E_j(\mathbf{k})/\hbar$ is the velocity of the electron in the crystal (see eqn (1.3.17)), so that in a non-free-electron-like situation the band structure of a real metal will find some way to avoid a discontinuity in $\nabla_{\mathbf{k}} E_j(\mathbf{k})$. For example, some interaction may be present that causes the two bands in Fig. 2.10 to separate slightly at $\mathbf{k}_A$ as shown in Fig. 2.11. Alternatively,

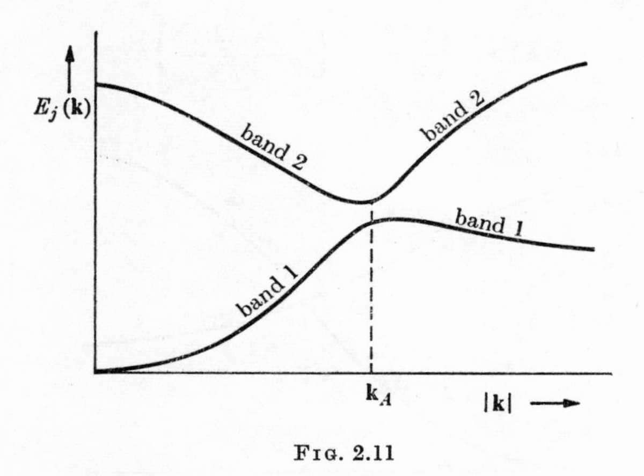

FIG. 2.11

if there is a double degeneracy at $\mathbf{k}_A$ that is an essential consequence of group theory, this almost certainly means that $\mathbf{k}_A$ is on the surface of the first Brillouin zone with a plane of symmetry or twofold rotation axis of symmetry normal to $\mathbf{k}_A$ in such a way that the bands along the direction of $\mathbf{k}_A$ flatten as they approach $\mathbf{k}_A$, their slopes vanishing at $\mathbf{k}_A$ itself, as shown in Fig. 2.12. There is therefore, again, no discontinuity in the slope, or in the velocity of an electron, at $\mathbf{k}_A$. The existence of a flat portion of an energy band in some region of the Brillouin zone means that there will be a large number of states within a very small range of energy; this may lead to a peak in the density of states curve. Such a peak is sometimes called a *Van Hove singularity* and these singularities are a topological consequence of the periodic structure of the crystal (Van Hove 1953). For a two-dimensional crystal these singularities would be

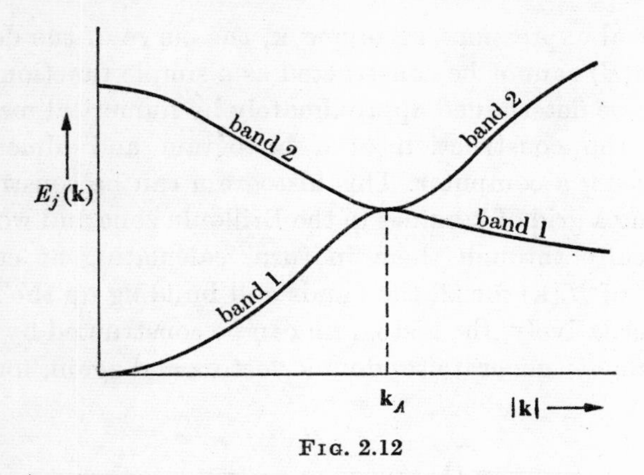

FIG. 2.12

logarithmic singularities in $n(E)$ but for a three-dimensional crystal $n(E)$ remains finite and the logarithmic singularities occur in $\mathrm{d}n(E)/\mathrm{d}E$, see Fig. 2.13. In practice, however, the cusps in $n(E)$ may escape observation as a result of low resolution of the experimental or theoretical techniques employed in determining $n(E)$.

Let us now suppose that the eigenvalues $E_j(\mathbf{k})$ for a real metal have been calculated for all $\mathbf{k}$ and for all values of j up to $j_{\max}$ where $j_{\max}$ is substantially larger than $\frac{1}{2}x$. Because the $E_j(\mathbf{k})$ have to be calculated numerically and are not obtained

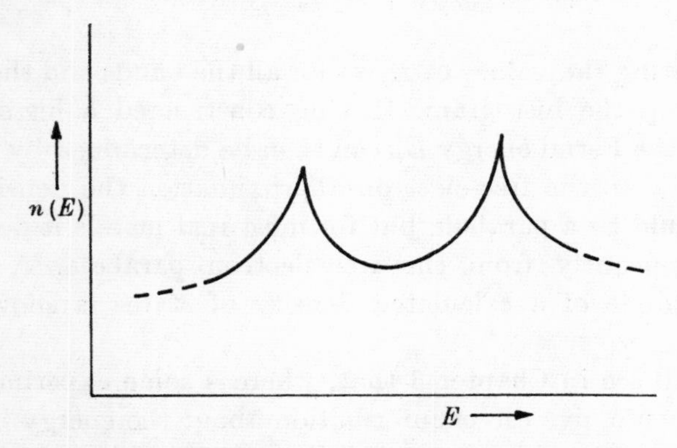

FIG. 2.13. Idealized Van Hove peaks in the density of states, $n(E)$, for a three-dimensional crystal.

as analytical expressions involving **k**, the curve of the density of states $n(E)$ cannot be constructed as a simple function of E, but has to be determined approximately by numerical methods involving the construction of a histogram and almost inevitably using a computer. This histogram can be constructed by selecting a grid of **k** values in the Brillouin zone and working systematically through them in turn, calculating at each **k** the values of $E_j(\mathbf{k})$ for all the bands and building up the histogram. Alternatively, the histogram can be constructed by using a programme to generate random **k** vectors and again, for each

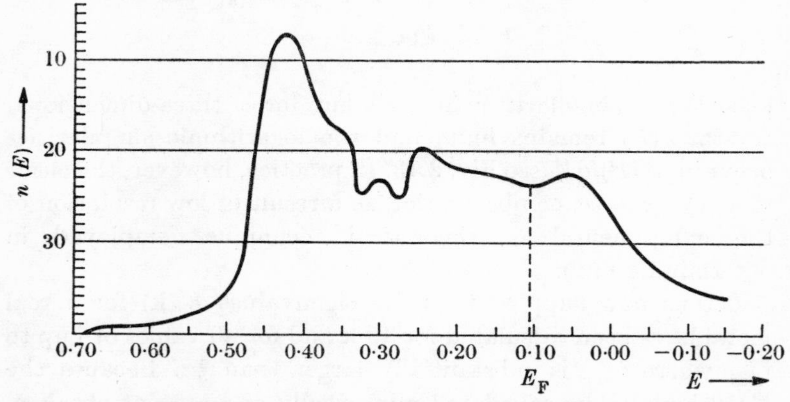

FIG. 2.14. An early example of a calculated density of states curve; $n(E)$ for Fe (Manning 1943).

k, calculating the values of $E_j(\mathbf{k})$ for all the bands and thereby building up the histogram. Having constructed a histogram for $n(E)$, the Fermi energy E_F can then be determined by using eqn (2.8.1). In the free-electron approximation the density of states would be a parabola but for most real metals it departs very significantly from the free-electron parabola. A very early example of a calculated density of states is shown in Fig. 2.14.

We shall see in Chapter 3 that, whereas some experimental measurements give direct information about the energy bands $E_j(\mathbf{k})$ in a metal and others give direct information about the dimensions of the Fermi surface of a metal, there are also some experiments that give information that can be related directly to the density of states rather than to the bands $E_j(\mathbf{k})$ or to the Fermi surface dimensions.

2.9. Parametrization of the Fermi surface

In Section 2.7 we discussed interpolation schemes in connection with the construction of the eigenvalues $E_j(\mathbf{k})$ all over the Brillouin zone using a few eigenvalues $E_j(\mathbf{k})$ that had been calculated particularly accurately for certain special wave vectors. The use of this kind of interpolation scheme is probably in decline because, with the large and fast computing machinery now available, it is quite feasible to calculate the eigenvalues $E_j(\mathbf{k})$ directly for a large number of general wave vectors in the Brillouin zone. However, there is a second kind of interpolation scheme that is probably increasing in importance as more and more accurate information becomes available about the detailed shapes of the Fermi surfaces of the various metals. As we shall see in Chapter 3, various experimental methods exist which can be used to determine either a caliper dimension of the Fermi surface in some specified direction or an extremal area of cross section of the Fermi surface normal to some specified direction. In practice it would be unrealistic to try to measure the dimensions of the Fermi surface for all possible directions of $\mathbf{k}$ in the Brillouin zone and it is usual just to make accurate measurements for the principal directions and planes in the Brillouin zone. It is then convenient to have some analytical or numerical way to parametrize the Fermi surface, that is, to be able to construct the Fermi surface, defined by $E_j(\mathbf{k}) = E_{\mathrm{F}}$, for all possible directions of $\mathbf{k}$ and for each sheet of the Fermi surface.

It is possible to use a simple mathematical interpolation scheme based on expanding $|\mathbf{k}_{\mathrm{F}}|$ in terms of a set of suitably symmetrized functions of the angles $\theta_{\mathbf{k}}$ and $\phi_{\mathbf{k}}$ which specify the direction of $\mathbf{k}_{\mathrm{F}}$. Suppose that the functions

$$X_l^{\nu}(\theta_{\mathbf{k}}, \phi_{\mathbf{k}}) = \sum_m A_l^{m\nu} Y_l^m(\theta_{\mathbf{k}}, \phi_{\mathbf{k}}) \tag{2.9.1}$$

are linear combinations of spherical harmonics where the

coefficients $A_l^{m\nu}$ have been determined group-theoretically so that the functions $X_l^\nu(\theta_{\mathbf{k}}, \phi_{\mathbf{k}})$ possess the full symmetry of the Brillouin zone of the metal in question (Von der Lage and Bethe 1947; Bell 1954; Lifshitz and Pogorelov 1954; Altmann and Bradley 1965; Altmann and Cracknell 1965; Mueller 1966). If $k_F(\theta_{\mathbf{k}}, \phi_{\mathbf{k}})$ is the magnitude of the wave vector $\mathbf{k}_F$ that terminates on the Fermi surface in the direction specified by $\theta_{\mathbf{k}}$ and $\phi_{\mathbf{k}}$, then we can expand $k_F(\theta_{\mathbf{k}}, \phi_{\mathbf{k}})$ as

$$k_F(\theta_{\mathbf{k}}, \phi_{\mathbf{k}}) = \sum_{l,\nu} B_l^\nu X_l^\nu(\theta_{\mathbf{k}}, \phi_{\mathbf{k}}) \tag{2.9.2}$$

and thereby parametrize the required Fermi surface. The $X_l^\nu(\theta_{\mathbf{k}}, \phi_{\mathbf{k}})$ can, of course, be expressed in terms of Cartesian coordinates k_x, k_y, and k_z instead of the polar coordinate angles $\theta_{\mathbf{k}}$ and $\phi_{\mathbf{k}}$. The coefficients B_l^ν can be found by fitting eqn (2.9.2) to the observed experimental data. It is not essential to use spherical harmonics and any other functions can be used, provided the functions $X_l^\nu(\theta_{\mathbf{k}}, \phi_{\mathbf{k}})$ possess the point-group symmetry of the metal. In principle any non-re-entrant sheet of the Fermi surface in any metal can be parametrized by means of a series expansion, like that in eqn (2.9.2), in terms of functions that possess the full point-group symmetry of the metal in question (Zornberg and Mueller 1966). However, if in practice the Fermi surface does not take a fairly simple form, it is likely that it will be necessary to include a very large number of terms before an accurate representation of that sheet of the Fermi surface will be obtained. For any one of the alkali metals, the Fermi surface is not very much distorted from the free-electron sphere that is centred at Γ (see Section 4.2); consequently, it is reasonable to suppose that for the alkali metals an accurate expression for $k_F(\theta_{\mathbf{k}}, \phi_{\mathbf{k}})$ will be obtained by using only the first few terms in the expansion in eqn (2.9.2). This method has been used for Cs (Okumura and Templeton 1965) and also, although it is not an alkali metal, for one sheet of the Fermi surface of Pd that happens to be roughly spherical and centered at Γ (Mueller and Priestley 1966).

If a sheet of the Fermi surface is multiply-connected it may be better to use a Fourier series expansion of $E_j(\mathbf{k})$ instead of an expansion in terms of spherical harmonics (Euwema, Stukel, Collins, De Witt, and Shankland 1969; Ketterson, Mueller, and Windmiller 1969). In this kind of expansion we write

$$E_j(\mathbf{k}) = \sum_{\mathbf{T_n}} C^j_{\mathbf{n}} \exp(i\mathbf{k} \,.\, \mathbf{T_n}), \tag{2.9.3}$$

where $\mathbf{T_n}$ is a vector of the Bravais lattice of the crystal (see eqn (1.2.11)). This expansion can be simplified by making use of the fact that $E_j(\mathbf{k})$ in the extended zone scheme possesses all the symmetry of the reciprocal lattice of the crystal. This leads to some simplification of eqn (2.9.3) because it means that if two vectors $\mathbf{T_{n_1}}$ and $\mathbf{T_{n_2}}$ are related by some rotational symmetry operation of the point group of the crystal, $C^j_{\mathbf{n}_1}$ and $C^j_{\mathbf{n}_2}$ will be equal. For the noble metals a Fourier series expansion converges very quickly. The Fermi surfaces of the noble metals are not simple closed surfaces but consist of spheres that are pulled out to reach the ⟨111⟩ faces of the Brillouin zone, as in Figs. 5.4 and 5.5. García-Moliner (1958) found that the Fermi surface of Cu could be fitted to one per cent accuracy by the very simple Fourier series expression

$$\begin{aligned} E_j(\mathbf{k}) = \alpha\{(-3 &+ \cos \tfrac{1}{2}ak_x \cos \tfrac{1}{2}ak_y + \cos \tfrac{1}{2}ak_y \cos \tfrac{1}{2}ak_z + \\ &+ \cos \tfrac{1}{2}ak_z \cos \tfrac{1}{2}ak_x) + \\ &+ r(-3 + \cos ak_x + \cos ak_y + \cos ak_z)\}, \end{aligned} \tag{2.9.4}$$

with $r = 0{\cdot}0995$ and $E_\mathrm{F}/\alpha = 3{\cdot}6301$. An extension of this series was later used to achieve a higher degree of accuracy:

$$\begin{aligned} E_j(\mathbf{k}) = -\alpha\{3 &- \cos \tfrac{1}{2}ak_x \cos \tfrac{1}{2}ak_y - \cos \tfrac{1}{2}ak_y \cos \tfrac{1}{2}ak_z + \\ &+ \cos \tfrac{1}{2}ak_z \cos \tfrac{1}{2}ak_x + \\ &+ \mathrm{C}_{200}(3 - \cos ak_x - \cos ak_y - \cos ak_z) + \\ &+ \mathrm{C}_{211}(3 - \cos ak_x \cos \tfrac{1}{2}ak_y \cos \tfrac{1}{2}ak_z - \ldots) + \\ &+ \mathrm{C}_{220}(3 - \cos ak_y \cos ak_z - \ldots) + \\ &+ \mathrm{C}_{310}(6 - \cos \tfrac{3}{2}ak_x \cos \tfrac{1}{2}ak_y + \\ &+ \cos \tfrac{3}{2}ak_x \cos \tfrac{1}{2}ak_z - \ldots) + \\ &+ \mathrm{C}_{321}(6 - \cos \tfrac{3}{2}ak_x \cos ak_y \cos \tfrac{1}{2}ak_z - \ldots)\}\ldots. \end{aligned} \tag{2.9.5}$$

This expression has now been used very widely for Cu, Ag, and Au (Roaf 1962; Halse 1969). The functions used in eqns (2.9.4) and (2.9.5) still possess the required point-group symmetry of the metal although they are constructed from plane waves rather than spherical harmonics.

If a sheet of the Fermi surface of a metal is closed but is centred at $\mathbf{k}_0$ which is some point other than Γ, it is sensible to use an expansion for $|\mathbf{k}_F - \mathbf{k}_0|$ about the point $\mathbf{k}_0$. An example of this is provided by one sheet of the Fermi surface of Ni which consists of an ellipsoid enclosing a region of holes centred at the point X. This surface was parametrized by Tsui (1967) using the expression

$$|\mathbf{k}_F - \mathbf{k}_0|^2 = k_\phi^2 k_z^2 / \{k_z^2 + (k_\phi^2 - k_z^2)\cos^2\theta\} + \\ + k_2\cos^2\theta - k_4\cos^4\theta, \quad (2.9.6)$$

where $k_\phi = k_0 + k_4\cos 4\phi + k_8\cos 8\phi$, k_2, k_4, and k_8 are constants, and $\mathbf{k}_F - \mathbf{k}_0$ is the wave vector from X to a point on the Fermi surface specified by the usual polar angles θ and ϕ referred to XW, [100], and $X\Gamma$, [001], as x and z axes.

There are many metals which have very complicated Fermi surfaces and it is often not feasible to use expansions like those given above to parametrize these Fermi surfaces because a very large number of terms would have to be included to obtain an acceptable order of accuracy. This is particularly likely to occur if the Fermi surface is re-entrant or multiply-connected. In this situation it is necessary to adopt a rather more sophisticated approach to the development of a parametrization scheme. Suppose that the potential $V(\mathbf{r})$ or the pseudopotential $W(\mathbf{r})$ is expressed in terms of a number of adjustable parameters, such as its Fourier coefficients for example. Trial values for these parameters are then assumed and one of the methods that we have described in Sections 2.2–6 is used to calculate the band structure, from which a trial Fermi surface can be constructed. This entire process can then be repeated with various different sets of values for the adjustable parameters, until the best agreement is obtained between the dimensions of the calculated Fermi surface and the experimental data. It is obviously essential to choose a method for which the calculation of the individual eigenvalues $E_j(\mathbf{k})$ is a direct and rapid process and in practice pseudopotential methods are particularly convenient to use, especially if the metal is one for which it is adequate to use a local pseudopotential. Interpolation schemes based on the use of pseudopotentials have been used for several metals (see Chapters 4 and 5 for some examples). Segall and Ham (1968) have suggested the use of the Green's function method (or Korringa–Kohn–Rostoker method) as an accurate and rapid method for calculating $E_j(\mathbf{k})$ in an interpolation scheme, but it does not appear to have been used very much in this connection (except for Cu (Cooper, Kreiger, and Segall 1969)). From the point of view of making use of the knowledge of the Fermi surface of a metal in connection with explaining various physical properties of that metal, it may be better to use a parametrization scheme that proceeds via the band structure, rather than a purely geometrical scheme for $k_F(\theta_\mathbf{k}, \phi_\mathbf{k})$. As we shall see in Chapter 6 this is because, for many properties, it is not enough just to know the shape of the Fermi surface—one needs to know the wave functions as well.

So far it is only for a relatively small number of metals that there has been any serious attempt to obtain an accurate parametrization of the Fermi surface, either by determining the coefficients in an expansion of $k_F(\theta_\mathbf{k}, \phi_\mathbf{k})$, or via a band structure calculation with a model potential or pseudopotential. However, as we shall see in Chapters 4 and 5, a large amount of data has now been collected about the Fermi surfaces of very many of the metallic elements. We can, therefore, expect that the next few years may see some considerable effort devoted to the determination of accurate parametrizations of the Fermi surfaces of the elements (Cracknell 1971*c*).

2.10. Dirac's equation and relativistic band structure calculations

So far in this chapter we have described methods for calculating band structures which have been based on the use of ordinary non-relativistic quantum mechanics, principally on the Schrödinger equation:

$$-\frac{\hbar^2}{2m}\nabla^2\psi(\mathbf{r})+V(\mathbf{r})\psi(\mathbf{r}) = E\psi(\mathbf{r}). \tag{1.2.3}$$

However, in recent years a great deal of research has been performed on the introduction of relativistic effects into band-structure calculations. In this section we shall study the more formal relativistic treatment of the motion of an electron in the periodic potential that exists in a crystalline solid and we shall identify the various additional terms which emerge and which are not present in the non-relativistic Hamiltonian used in the construction of eqn (1.2.3). Any reader who wishes to do so can, without any serious damage to his understanding of the rest of this book, omit this section and proceed to Section 2.11 in which we adopt a more empirical approach to spin–orbit coupling which is the most important of the relativistic effects.

In order to gain some insight into the problem we first consider briefly the relativistic wave equation, the Dirac equation, for a free electron before considering the case when the electron moves in the periodic potential $V(\mathbf{r})$ that exists in a crystalline solid. By the term 'free electron' in this context we do not refer to the empty lattice model described in Sections 1.5 and 1.6, but to an electron free to move in an otherwise completely empty region of space. Let us consider the time-dependent Schrödinger equation

$$\mathscr{H}\psi(\mathbf{r}, t) = i\hbar\frac{\partial\psi}{\partial t}(\mathbf{r}, t) \tag{2.10.1}$$

for a free particle. In the non-relativistic treatment the Hamiltonian $\mathscr{H}$ will just be given by

$$\mathscr{H} = \frac{1}{2m}(p_x^2+p_y^2+p_z^2). \tag{2.10.2}$$

Since the relativistic expression for the energy of an electron of momentum $\mathbf{p}$ and rest mass m is

$$E^2 = p^2c^2+m^2c^4, \tag{2.10.3}$$

it is not unreasonable to try to substitute the expression

$$\mathscr{H} = c(m^2c^2+p_x^2+p_y^2+p_z^2)^{\frac{1}{2}} \tag{2.10.4}$$

for $\mathscr{H}$ in eqn (2.10.1) in seeking to obtain a relativistic wave equation. Remembering that in quantum mechanics we use the substitution $\mathbf{p} = -i\hbar\boldsymbol{\nabla}$, we can see that while a wave equation constructed in this way is linear in $\partial/\partial t$, the components of $\mathbf{p}$ would occur in a term involving the square root of a sum of m^2c^2 and squares of the components of $\mathbf{p}$. Such an equation would not be acceptable as a relativistic wave equation and so a form for the Hamiltonian was proposed that was linear both in $\partial/\partial t$ and in the components of $\mathbf{p}$, namely

$$\mathscr{H} = c\boldsymbol{\alpha}\,.\,\mathbf{p}+\boldsymbol{\beta}mc^2 \tag{2.10.5}$$

where $\boldsymbol{\alpha}\cdot\mathbf{p}$ denotes the quantity $(\alpha_x p_x+\alpha_y p_y+\alpha_z p_z)$. For detailed arguments see, for example, Dirac (1930), Rose (1961). The time-dependent relativistic wave equation is therefore

$$\mathscr{H}\boldsymbol{\psi}(\mathbf{r}, t) = (c\boldsymbol{\alpha}\,.\,\mathbf{p}+\boldsymbol{\beta}mc^2)\boldsymbol{\psi}(\mathbf{r}, t) = i\hbar\frac{\partial\boldsymbol{\psi}}{\partial t}(\mathbf{r}, t) \tag{2.10.6}$$

and the corresponding time-independent relativistic wave equation is therefore

$$\mathscr{H}\boldsymbol{\psi}(\mathbf{r}) = (c\boldsymbol{\alpha}\,.\,\mathbf{p}+\boldsymbol{\beta}mc^2)\boldsymbol{\psi}(\mathbf{r}) = E\boldsymbol{\psi}(\mathbf{r}). \tag{2.10.7}$$

In order that the energy E in this equation should be consistent with the usual relativistic expression relating energy and momentum given in eqn (2.10.3), Dirac was able to show that

$$\boldsymbol{\alpha}_i = \begin{pmatrix} \mathbf{0} & \boldsymbol{\sigma}_i \\ \boldsymbol{\sigma}_i & \mathbf{0} \end{pmatrix}, \tag{2.10.8}$$

$$\boldsymbol{\beta} = \begin{pmatrix} 1 & 0 & 0 & 0 \\ 0 & 1 & 0 & 0 \\ 0 & 0 & -1 & 0 \\ 0 & 0 & 0 & -1 \end{pmatrix}, \tag{2.10.9}$$

where

$$\boldsymbol{\sigma}_x = \begin{pmatrix} 0 & 1 \\ 1 & 0 \end{pmatrix}, \quad \boldsymbol{\sigma}_y = \begin{pmatrix} 0 & -\mathrm{i} \\ \mathrm{i} & 0 \end{pmatrix}, \quad \boldsymbol{\sigma}_z = \begin{pmatrix} 1 & 0 \\ 0 & -1 \end{pmatrix}. \quad (2.10.10)$$

$\boldsymbol{\sigma}_x$, $\boldsymbol{\sigma}_y$, and $\boldsymbol{\sigma}_z$ are the Pauli spin matrices. The four operators $\boldsymbol{\alpha}_x$, $\boldsymbol{\alpha}_y$, $\boldsymbol{\alpha}_z$, and $\boldsymbol{\beta}$ anticommute, that is

$$\left.\begin{aligned} \boldsymbol{\alpha}_i\boldsymbol{\alpha}_j + \boldsymbol{\alpha}_j\boldsymbol{\alpha}_i &= 0 \qquad i \neq j \\ \boldsymbol{\alpha}_i\boldsymbol{\beta} + \boldsymbol{\beta}\boldsymbol{\alpha}_i &= 0 \end{aligned}\right. \quad (2.10.11)$$

where $i, j = x, y, z$, and the squares of the operators are equal to $\mathbf{1}$, that is

$$\left.\begin{aligned} (\boldsymbol{\alpha}_i)^2 &= \mathbf{1} \\ \boldsymbol{\beta}^2 &= \mathbf{1}. \end{aligned}\right\} \quad (2.10.12)$$

The fact that $\boldsymbol{\alpha}_i$ and $\boldsymbol{\beta}$ cannot be taken to be unit matrices, or multiples thereof, means that $\boldsymbol{\psi}(\mathbf{r}, t)$ must be a multi-component function of the form

$$\boldsymbol{\psi}(\mathbf{r}, t) = \begin{pmatrix} \psi_1(\mathbf{r}, t) \\ \psi_2(\mathbf{r}, t) \\ \psi_3(\mathbf{r}, t) \\ \psi_4(\mathbf{r}, t) \end{pmatrix}, \quad (2.10.13)$$

which can be written as $\boldsymbol{\psi}_{\mathbf{k}}(\mathbf{r})\exp(-\mathrm{i}Et)$ where $\boldsymbol{\psi}_{\mathbf{k}}(\mathbf{r})$ is a four-component spinor which satisfies the equation

$$\mathscr{H}\boldsymbol{\psi}_{\mathbf{k}}(\mathbf{r}) = (c\boldsymbol{\alpha} \cdot \mathbf{p} + \boldsymbol{\beta}mc^2)\boldsymbol{\psi}_{\mathbf{k}}(\mathbf{r}) = E\boldsymbol{\psi}_{\mathbf{k}}(\mathbf{r}). \quad (2.10.14)$$

$\mathbf{p}$ and $\mathbf{k}$ are related by the usual equation

$$\mathbf{p} = \hbar\mathbf{k}, \quad (2.10.15)$$

and we write p_0 for E so that

$$p_0 = E = (p^2c^2 + m^2c^4)^{\frac{1}{2}} \quad (2.10.16)$$

and eqn (2.10.14) becomes

$$\mathscr{H}\boldsymbol{\psi}_{\mathbf{k}}(\mathbf{r}) = p_0\boldsymbol{\psi}_{\mathbf{k}}(\mathbf{r}). \quad (2.10.17)$$

Although the Dirac matrices $\boldsymbol{\alpha}_i$ and $\boldsymbol{\beta}$ are four-by-four matrices it is easy to see by inspection that they can also be regarded as two-by-two supermatrices where each element of one of these supermatrices is itself a two-by-two matrix. $\psi_{\mathbf{k}}(\mathbf{r})$ can therefore be written in terms of two-component spinors, u and v:

$$\psi_{\mathbf{k}}(\mathbf{r}) = \begin{pmatrix} u \\ v \end{pmatrix} \tag{2.10.18}$$

so that, from eqn (2.10.14)

$$\left.\begin{aligned} c\boldsymbol{\sigma}\,.\,\mathbf{p}u &= (E+mc^2)v \\ c\boldsymbol{\sigma}\,.\,\mathbf{p}v &= (E-mc^2)u. \end{aligned}\right\} \tag{2.10.19}$$

u can be expressed as a linear combination of the two-component spinors $\chi(+\frac{1}{2}) = \begin{pmatrix} 1 \\ 0 \end{pmatrix}$ and $\chi(-\frac{1}{2}) = \begin{pmatrix} 0 \\ 1 \end{pmatrix}$,

$$u = a_+\chi(+\tfrac{1}{2}) + a_-\chi(-\tfrac{1}{2}) \tag{2.10.20}$$

so that $\psi_{\mathbf{k}}(\mathbf{r})$ can also be written as the same linear combination of the two functions

$$\psi_{\mathbf{k}}^{\nu}(\mathbf{r}) = \left(\frac{p_0+mc^2}{2p_0}\right)^{\frac{1}{2}} \begin{pmatrix} \chi(\nu) \\ \dfrac{c\boldsymbol{\sigma}\,.\,\mathbf{p}}{p_0+mc^2}\chi(\nu) \end{pmatrix} \exp(i\mathbf{k}\,.\,\mathbf{r}), \tag{2.10.21}$$

where the $\{(p_0+mc^2)/2p_0\}^{\frac{1}{2}}$ is a normalization factor and $\nu = \pm\frac{1}{2}$. In the non-relativistic limit $c\,|\mathbf{p}| \ll mc^2$ and eqn (2.10.21) simplifies to

$$\psi_{\mathbf{k}}^{\nu}(\mathbf{r}) = \begin{pmatrix} \chi(\nu) \\ 0 \end{pmatrix} \exp(i\mathbf{k}\,.\,\mathbf{r}), \tag{2.10.22}$$

which is the familiar non-relativistic form of the wave function with spin included. In the literature of relativistic quantum mechanics it is common to use the system of units in which $\hbar = m = c = 1$ and in this system of units eqn (2.10.21) simplifies to

$$\psi_{\mathbf{k}}^{\nu}(\mathbf{r}) = \left(\frac{1+k_0}{2k_0}\right)^{\frac{1}{2}} \begin{pmatrix} \chi(\nu) \\ \dfrac{\boldsymbol{\sigma}\,.\,\mathbf{k}}{1+k_0}\chi(\nu) \end{pmatrix} \exp(i\mathbf{k}\,.\,\mathbf{r}) \tag{2.10.23}$$

where $k_0 = E = (1+k^2)^{\frac{1}{2}}$ in these units. However, this system of units which is convenient for macroscopic relativistic problems is not the same as the system of units, sometimes called Hartree units or atomic units, which is commonly used in atomic and band-structure calculations, in which

$\hbar = m = |e| = 1$ and therefore $c \doteqdot 137$ atomic units of velocity, see p. 73). Because the expression $\hbar c/e^2$ is a dimensionless quantity it is impossible to choose a system of units in which $\hbar$, $|e|$, and c are all set equal to 1. In order to avoid possible confusion between these two systems of units when they meet in relativistic treatments of electrons in atoms, molecules, or solids, we shall retain the constants $\hbar$, m, c, and e in our formulae and equations.

So far in this section we have been concerned with the relativistic description of the behaviour of a free electron and we have seen that the Schrödinger equation for a free particle ($V(\mathbf{r}) = 0$) becomes replaced by the corresponding Dirac equation, that is by eqn (2.10.14). In order to obtain a relativistic description of the behaviour of an electron in a non-zero potential $V(\mathbf{r})$, it is necessary to determine the solutions of the Dirac equation that replaces the Schrödinger equation (1.2.3). From the form of eqn (2.10.14) it is reasonable to suppose that the corresponding Dirac equation for an electron in a potential $V(\mathbf{r})$ takes the form

$$\{c\boldsymbol{\alpha} \,.\, \mathbf{p} + \boldsymbol{\beta} mc^2 + \mathbf{E}\, V(\mathbf{r})\}\boldsymbol{\psi}(\mathbf{r}) = E\boldsymbol{\psi}(\mathbf{r}), \qquad (2.10.24)$$

where $\mathbf{E}$ is the identity matrix in four dimensions

$$\mathbf{E} = \begin{pmatrix} 1 & 0 & 0 & 0 \\ 0 & 1 & 0 & 0 \\ 0 & 0 & 1 & 0 \\ 0 & 0 & 0 & 1 \end{pmatrix}. \qquad (2.10.25)$$

If eqn (2.10.24) is used to describe the behaviour of an electron in an atom, the wave function of this electron will be a four-component spinor and will have the form

$$\boldsymbol{\psi}_{\kappa\mu}(\mathbf{r}) = \begin{pmatrix} g_\kappa(r) & \chi^\mu_\kappa(\theta, \phi) \\ \mathrm{i} f_\kappa(r) & \chi^\mu_{-\kappa}(\theta, \phi) \end{pmatrix}. \qquad (2.10.26)$$

$g_\kappa(r)$ and $f_\kappa(r)$ are solutions of the appropriate radial equation (for further details see, for example, Rose 1961). The spin-angular function $\chi^\mu_\kappa(\theta, \phi)$ is given by

$$\chi^\mu_\kappa(\theta, \phi) = \sum_{\pm} C(l\tfrac{1}{2}j;\, \mu-\nu, \nu)\, Y^{\mu-\nu}_l(\theta, \phi)\chi(\nu). \qquad (2.10.27)$$

where

$$\kappa = \begin{cases} l & \text{if } j = l-\frac{1}{2} \\ -l-1 & \text{if } j = l+\frac{1}{2} \end{cases} \tag{2.10.28}$$

and $C(l\frac{1}{2}j;\, \mu-\nu,\, \nu)$ is a Clebsch–Gordan coefficient which is given explicitly in Table 2.2. Many numerical calculations have now been performed to determine the wave functions and the energies of the electrons in free atoms using Dirac's equation rather than Schrödinger's equation.

TABLE 2.2

The Clebsch–Gordan coefficients $C(l\frac{1}{2}j;\, m-m_2,\, m_2)$

	$m_2 = +\frac{1}{2}$	$m_2 = -\frac{1}{2}$
$j = l+\frac{1}{2}$	$\left(\frac{l+m+\frac{1}{2}}{2l+1}\right)^{\frac{1}{2}}$	$\left(\frac{l-m+\frac{1}{2}}{2l+1}\right)^{\frac{1}{2}}$
$j = l-\frac{1}{2}$	$-\left(\frac{l-m+\frac{1}{2}}{2l+1}\right)^{\frac{1}{2}}$	$\left(\frac{l+m+\frac{1}{2}}{2l+1}\right)^{\frac{1}{2}}$

To calculate the band structure of a metal including relativistic effects we have a problem that is very similar to that with which we were concerned in the early sections of this chapter, except that now equation (2.10.24) replaces Schrödinger's equation. That is, it is necessary to construct the potential $V(\mathbf{r})$ and then to solve eqn (2.10.24) subject to the boundary conditions that apply in the metal. It is instructive to consider the form of the solutions of eqn (2.10.24) for a solid in the free-electron, or empty-lattice, approximation, that is, when we first require that $V(\mathbf{r})$ should possess the periodicity of the Bravais lattice of the crystal and then set $V(\mathbf{r})$ equal to zero. As we might expect, the solutions for this situation are plane waves, $\psi^{\nu}_{\mathbf{kn}}(\mathbf{r})$, which are simply related to the plane waves of eqn (2.10.21):

$$\psi^{\nu}_{\mathbf{kn}}(\mathbf{r}) = \left(\frac{p_{\mathrm{n}0}+mc^2}{2p_{\mathrm{n}0}}\right)^{\frac{1}{2}} \begin{pmatrix} \chi(\nu) \\ \dfrac{c\boldsymbol{\sigma}\cdot\mathbf{p}_{\mathrm{n}}}{p_{\mathrm{n}0}+mc^2}\,\chi(\nu) \end{pmatrix} \exp\{\mathrm{i}(\mathbf{k}+\mathbf{G}_{\mathrm{n}})\cdot\mathbf{r}\}, \tag{2.10.29}$$

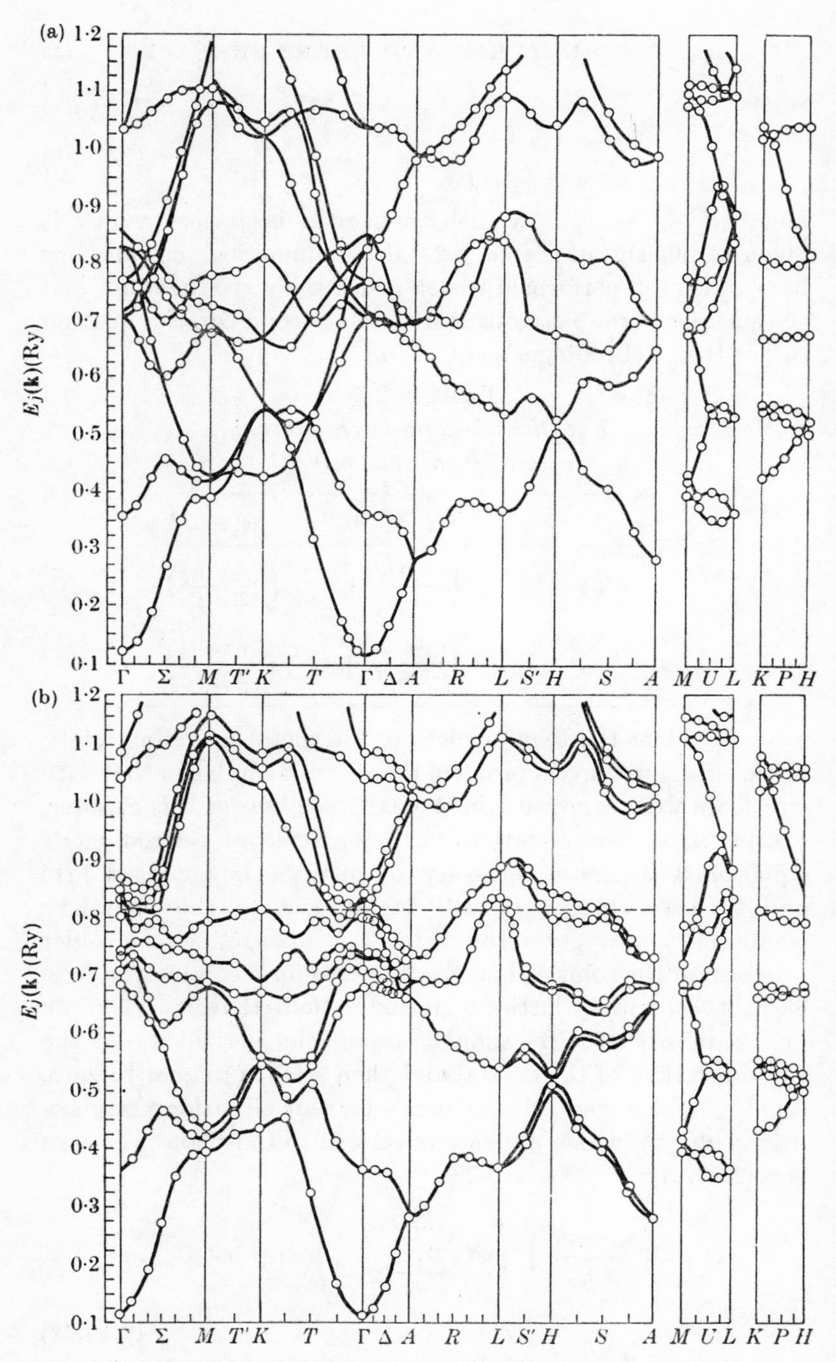

Fig. 2.15. The energy bands for Re (a) neglecting spin–orbit coupling and (b) including spin–orbit coupling (Mattheiss 1966).

where

$$\mathbf{p}_n = \hbar(\mathbf{k}+\mathbf{G}_n) \tag{2.10.30}$$

$\mathbf{G}_n$ is a reciprocal lattice vector,

$$p_{n0} = (\mathbf{p}_n^2c^2+m^2c^4)^{\frac{1}{2}} \tag{2.10.31}$$

and $\nu = \pm\frac{1}{2}$. The function $\psi^{\nu}_{\mathbf{k}n}(\mathbf{r})$ in eqn (2.10.29) is the relativistic analogue of the plane wave given in eqn (1.5.8) which is a solution of Schrödinger's equation in the free-electron approximation.

In principle, any one of the various methods described in the early sections of this chapter could be adapted to the determination of the solutions of Dirac's equation rather than Schrödinger's equation in a metal. The first complete numerical solution of the Dirac equation for a metal was determined by Soven (1965*a*, *b*) for thallium (see Section 4.4). Soven extended the orthogonalized plane wave (or O.P.W.) method (see Section 2.4) to the relativistic case by orthogonalizing plane waves of the form given in eqn (2.10.29) to the relativistic atomic wave functions for the core states. The augmented plane wave (or A.P.W.) method has been adapted by Loucks (1965*d*, 1967*a*) to relativistic calculations of the band structures of metals. As in the non-relativistic A.P.W. method, the crystal potential is approximated by a muffin-tin potential (see Section 2.1). The calculation proceeds in a manner that is similar to that outlined in Section 2.4 except that the cellular-type expansion which is used for $\psi_{\mathbf{k}}(\mathbf{r})$ in the regions near the nuclei of the atoms in the metal involves the relativistic atomic wave functions given in eqn (2.10.26), and in the region where the potential is assumed to be constant $\psi_{\mathbf{k}}(\mathbf{r})$ is given by a Dirac plane wave of the form given by eqn (2.10.29). The relativistic A.P.W. method has now been applied to the calculation of the band structures of a number of metals including Pb (Loucks 1965*c*), W (Loucks 1965*b*, 1966*a*), Pd and Pt (Andersen and Mackintosh 1968) and Re (Mattheiss 1966); further mention of the results of these calculations will be made when we discuss these metals in detail in Chapters 4 and 5. Just to illustrate the importance of relativistic effects, at least for metals of high atomic number, we illustrate in Fig. 2.15 the results of relativistic and non-relativistic band structure calculations for Re obtained by Mattheiss (1966).

2.11. Spin–orbit coupling

Instead of solving the complete Dirac equation for a crystal, it is sometimes more convenient to retain only those relativistic terms that affect the band structure most. To do this we investigate the limiting form of the Dirac equation

$$\{c\boldsymbol{\alpha}\,.\,\mathbf{p}+\boldsymbol{\beta}mc^2+\mathbf{E}V(\mathbf{r})\}\boldsymbol{\psi}(\mathbf{r}) = E\boldsymbol{\psi}(\mathbf{r}) \qquad (2.10.24)$$

when the velocity of the electron is very small compared with c, which is, of course, the case for a conduction electron in a metal. We write

$$\boldsymbol{\psi}(\mathbf{r}) = \begin{pmatrix}\psi_1(\mathbf{r})\\ \psi_2(\mathbf{r})\end{pmatrix}, \qquad (2.11.1)$$

where $\psi_1(\mathbf{r})$ and $\psi_2(\mathbf{r})$ are two-component spinor wave functions. E is the total energy of the electron which includes the rest energy mc^2 so that we write

$$E = E'+mc^2. \qquad (2.11.2)$$

By making use of the fact that both E' and $V(\mathbf{r})$ will be much smaller than mc^2 and performing some manipulation of eqn (2.10.24) (for details see, for example, Schiff 1955, p. 332–3), it is possible to show that

$$\left\{\frac{\mathbf{p}^2}{2m}+V(\mathbf{r})-\frac{\mathbf{p}^4}{8m^3c^2}-\frac{\hbar^2}{4m^2c^2}\{\nabla V(\mathbf{r})\}\,.\,\nabla+\right.$$
$$\left.+\frac{\hbar^2}{4m^2c^2}\boldsymbol{\sigma}\,.\,[\{\nabla V(\mathbf{r})\}\wedge\mathbf{p}]\right\}\psi_1(\mathbf{r}) = E'\psi_1(\mathbf{r}). \qquad (2.11.3)$$

We can study the form of the terms on the left-hand side of this equation. The first and second terms account for the non-relativistic Schrödinger equation. The third term arises from the relativistic change of mass with velocity. The fourth term has no classical analogue. The last term corresponds to the spin–orbit interaction or spin–orbit coupling.

In Section 2.10 we have mentioned that it is possible to solve Dirac's equation for an electron in a crystal by making suitable modifications to any one of the various methods described for non-relativistic calculations in the earlier sections of this chapter. However, the form of eqn (2.11.3) suggests that it might be profitable, as an alternative to solving eqn (2.10.24) directly, to start with the results of a non-relativistic band structure calculation, using any of the methods described in Sections 2.2–6, and then to introduce one or more of the extra terms in eqn (2.11.3) as a perturbation. In practice it is common in such perturbation calculations to neglect the term corresponding to the variation of mass with velocity, $(-\mathbf{p}^4/8m^3c^2)$, and the term $-(\hbar^2/4m^2c^2)\{\nabla V(\mathbf{r})\}\,.\,\nabla$ and to consider only the spin–orbit coupling term

$$\mathscr{H}_{\text{s.l}} = \frac{\hbar}{4m^2c^2}\boldsymbol{\sigma}\,.\,[\{\nabla V(\mathbf{r})\}\wedge\mathbf{p}]. \tag{2.11.4}$$

The inclusion of only the spin–orbit term gives a good approximation for metals of low atomic number, but for the metals of high atomic number the other two terms may also be significant (Johnson, Conklin, and Pratt 1963).

If we assume that $V(\mathbf{r})$ possesses spherical symmetry then

$$\nabla V(\mathbf{r}) = \frac{1}{r}\frac{\mathrm{d}V(\mathbf{r})}{\mathrm{d}r}\mathbf{r}, \tag{2.11.5}$$

so that the spin–orbit term can be written as

$$\mathscr{H}_{\text{s.l}} = \frac{1}{2m^2c^2}\frac{1}{r}\frac{\mathrm{d}V(\mathbf{r})}{\mathrm{d}r}\mathbf{s}\,.\,\mathbf{l} = \xi(r)\mathbf{s}\,.\,\mathbf{l}, \tag{2.11.6}$$

where

$$\mathbf{l} = \mathbf{r}\wedge\mathbf{p} \tag{2.11.7}$$

denotes the orbital angular momentum and

$$\mathbf{s} = \tfrac{1}{2}\hbar\boldsymbol{\sigma} \tag{2.11.8}$$

denotes the spin angular momentum. $\xi(r)$ is defined by eqn (2.11.6). Since we are considering a non-relativistic band structure calculation to which the spin–orbit coupling term in eqn (2.11.6) is added as a small perturbation, the unperturbed wave function takes the simple form given in eqn (2.10.22) for a non-relativistic wave function with spin included. Therefore, if $\psi_1(\mathbf{r})$ is written as $u_{\mathbf{k}}(\mathbf{r})\exp(i\mathbf{k}\,.\,\mathbf{r})$, where the subscript 1 has been dropped from the wave function, the function $u_{\mathbf{k}}(\mathbf{r})$ can be expanded in terms of functions which are products of spin functions, $\chi(\nu)$, and space functions with particular l and m values. That is, we may write

$$u_{\mathbf{k}}(\mathbf{r}) = \sum_{l,m} \{A_{\mathbf{k}lm}\chi(+\tfrac{1}{2})+B_{\mathbf{k}lm}\chi(-\tfrac{1}{2})\}R_l(r)Y_l^m(\theta,\phi). \quad (2.11.9)$$

The functions $R_l(r)$ are the solutions of the radial part of Schrödinger's equation and the $Y_l^m(\theta, \phi)$ are normalized spherical harmonics (see also eqn (2.3.5)). For the special points, lines, and planes of symmetry in the Brillouin zone, group theory can be used to simplify the expansion of $u_{\mathbf{k}}(\mathbf{r})$ given in eqn (2.11.9). However, since we are concerned with a system of half odd-integer spin, double groups will have to be used (see, for example, Elliott 1954; Koster, Dimmock, Wheeler, and Statz 1963; Slater 1965; Onodera and Okazaki 1966*a*, *b*; Cracknell and Wong 1967; Cracknell and Joshua 1970).

We are now in a position to describe how to calculate the changes in the energy bands $E_j(\mathbf{k})$ as a result of the introduction of $\mathscr{H}_{\mathrm{s.l}}$ as a perturbation. These will be obtained by calculating the matrix elements of $\mathscr{H}_{\mathrm{s.l}}$ using the wave functions specified by equation (2.11.9); to first-order in perturbation theory these matrix elements are $\int \psi_{\mathbf{k}}^*(\mathbf{r})\xi(r)(\mathbf{s}\,.\,\mathbf{l})\psi_{\mathbf{k}}(\mathbf{r})\,d\mathbf{r}$. To evaluate these matrix elements we need to determine the effect of the operator $\mathbf{s}\,.\,\mathbf{l}$ on the wave function $\psi_{\mathbf{k}}(\mathbf{r})$. The effect of the angular momentum operator $\mathbf{l}$ on the spherical harmonics $Y_l^m(\theta, \phi)$ is well known (see, for example, Slater 1960, p. 257)

$$\left.\begin{aligned}
\mathbf{l}_x Y_l^m(\theta,\phi) &= \tfrac{1}{2}\hbar[\{(l-m)(l+m+1)\}^{\frac{1}{2}}Y_l^{m+1}(\theta,\phi)+\\
&\qquad +\{(l-m+1)(l+m)\}^{\frac{1}{2}}Y_l^{m-1}(\theta,\phi)\\
\mathbf{l}_y Y_l^m(\theta,\phi) &= \tfrac{1}{2}i\hbar[-\{(l-m)(l+m+1)\}^{\frac{1}{2}}Y_l^{m+1}(\theta,\phi)+\\
&\qquad +\{(l-m+1)(l+m)\}^{\frac{1}{2}}Y_l^{m-1}(\theta,\phi)]\\
\mathbf{l}_z Y_l^m(\theta,\phi) &= m\hbar Y_l^m(\theta,\phi).
\end{aligned}\right\} \quad (2.11.10)$$

The corresponding expressions for the spin operator $\mathbf{s}$ are

$$\left.\begin{aligned} \mathbf{s}_x\chi(+\tfrac{1}{2}) &= \tfrac{1}{2}\hbar\chi(-\tfrac{1}{2}) & \mathbf{s}_x\chi(-\tfrac{1}{2}) &= \tfrac{1}{2}\hbar\chi(+\tfrac{1}{2}) \\ \mathbf{s}_y\chi(+\tfrac{1}{2}) &= \tfrac{1}{2}\mathrm{i}\hbar\chi(-\tfrac{1}{2}) & \mathbf{s}_y\chi(-\tfrac{1}{2}) &= -\tfrac{1}{2}\mathrm{i}\hbar\chi(+\tfrac{1}{2}) \\ \mathbf{s}_z\chi(+\tfrac{1}{2}) &= \tfrac{1}{2}\hbar\chi(+\tfrac{1}{2}) & \mathbf{s}_z\chi(-\tfrac{1}{2}) &= -\tfrac{1}{2}\hbar\chi(-\tfrac{1}{2}) \end{aligned}\right\} \tag{2.11.11}$$

which can be regarded as being derived from eqn (2.11.10) by setting $l = \frac{1}{2}$ and $m = \pm\frac{1}{2}$ and replacing $Y^m(\theta, \phi)$ by $\chi(+\frac{1}{2})$ and $\chi(-\frac{1}{2})$. Using eqns (2.11.10) and (2.11.11) it is possible to determine the effect of $\mathbf{s}\,.\,\mathbf{l} = (s_x l_x + s_y l_y + s_z l_z)$ on the functions $\chi(+\frac{1}{2})Y^m(\theta, \phi)$ and $\chi(-\frac{1}{2})Y_l^m(\theta, \phi)$; one finds that

$$\left.\begin{aligned} (\mathbf{s}\,.\,\mathbf{l})\chi(+\tfrac{1}{2})Y_l^m(\theta,\phi) &= \tfrac{1}{2}\hbar^2 m\chi(+\tfrac{1}{2})Y_l^m(\theta,\phi)+ \\ &\quad +\tfrac{1}{2}\hbar^2\{(l-m)(l+m+1)\}^{\frac{1}{2}}\chi(-\tfrac{1}{2})Y_l^{m+1}(\theta,\phi) \\ (\mathbf{s}\,.\,\mathbf{l})\chi(-\tfrac{1}{2})Y_l^m(\theta,\phi) &= -\tfrac{1}{2}\hbar^2 m\chi(-\tfrac{1}{2})Y_l^m(\theta,\phi)+ \\ &\quad +\tfrac{1}{2}\hbar^2\{(l-m+1)(l+m)\}^{\frac{1}{2}}\chi(+\tfrac{1}{2})Y_l^{m-1}(\theta,\phi). \end{aligned}\right\} \tag{2.11.12}$$

Therefore, by substituting from eqn (2.11.9) for $\psi_{\mathbf{k}}(\mathbf{r})$ and using eqn (2.11.12) it is possible to evaluate the expression $\int \psi_{\mathbf{k}}^*(\mathbf{r})\xi(r)(\mathbf{s}\,.\,\mathbf{l})\psi_{\mathbf{k}}(\mathbf{r})\,\mathrm{d}\mathbf{r}$ for the change in the energy of a band at $\mathbf{k}$, due to spin–orbit coupling. On making use of the orthonormality properties of the spherical harmonics the expression becomes

$$\int \psi_{\mathbf{k}}^*(\mathbf{r})\xi(r)(\mathbf{s}\,.\,\mathbf{l})\psi_{\mathbf{k}}(\mathbf{r})\,\mathrm{d}\mathbf{r}$$

$$= \sum_l \left(\left\{\int \xi(r)R_l^*(r)R_l(r)r^2\,\mathrm{d}r\right\} \sum_m [\tfrac{1}{2}m(A_{\mathbf{k}lm}^*A_{\mathbf{k}lm} - B_{\mathbf{k}lm}^*B_{\mathbf{k}lm})+\right.$$

$$\left. +\{(l-m)(l+m+1)\}^{\frac{1}{2}}(A_{\mathbf{k}lm}^*B_{\mathbf{k}l,m+1} + B_{\mathbf{k}l,m+1}^*A_{\mathbf{k}lm})]\right). \tag{2.11.13}$$

The evaluation of the expression in eqn (2.11.13) is particularly simple if the A.P.W. method is used (Johnson, Conklin, and Pratt 1963). Since the potential is zero outside the muffin-tin sphere, the spin–orbit term only contributes from regions within the sphere. The effect of these spin orbit matrix elements at a general point in the Brillouin zone is to add small corrections to the energy levels. At the symmetry points, many of the A_{klm} and B_{klm} will vanish because of the symmetry properties of the wave functions, but the remaining terms are sufficient to remove the degeneracies in the absence of spin–orbit coupling. These effects are illustrated for W in Fig. 2.16 (Mattheiss and Watson 1964; Mattheiss 1965). When we come to discuss systematically the band structures and Fermi surfaces of the metallic elements in Chapters 4 and 5 we shall see that

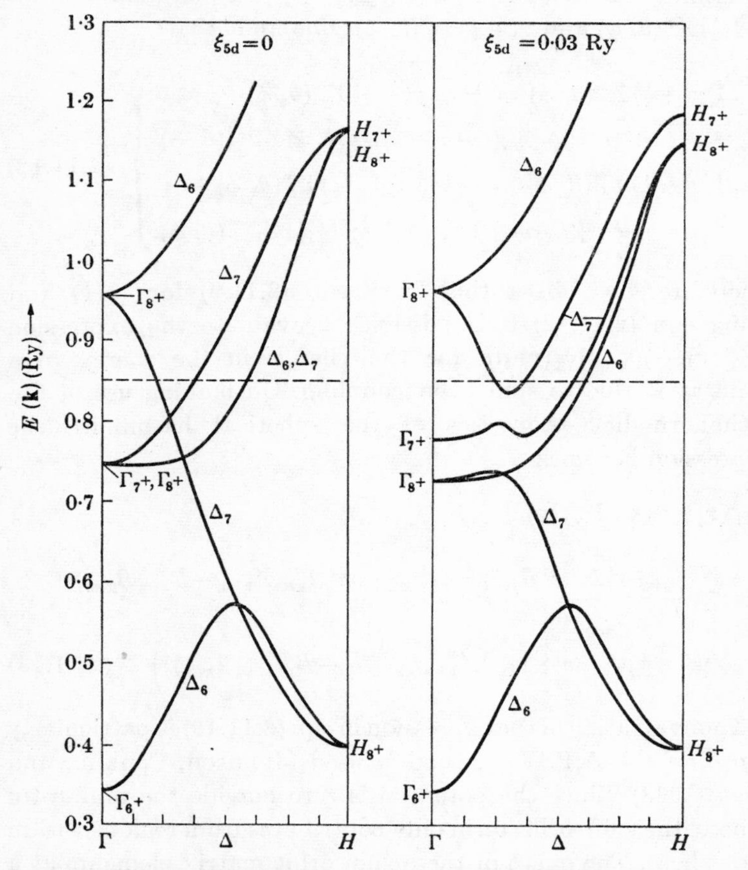

FIG. 2.16. Results of a simplified spin–orbit calculation along $\Gamma\Delta H$ for W with $\xi_{5d}(=\xi(r)) = 0$ and $\xi_{5d} = 0{\cdot}03$ Ry (Mattheiss and Watson 1965).

relativistic effects, and particularly spin–orbit coupling, have been included in recent calculations of the band structures of several of these metals. In general, for materials of high atomic number, better agreement with experiment is obtained if more relativistic corrections are made.

It can be seen from Fig. 2.16 that for a typical metal of quite high atomic number the magnitudes of the changes in the band structure introduced by spin–orbit coupling are quite small compared with E_F, the Fermi energy. For Pb, which has one of the largest atomic numbers of all metals, the neglect of spin–orbit coupling introduces an error of the order of six per cent in the area of cross section of the Fermi surface (Anderson and Gold 1965). Consequently, one might expect that the presence of spin–orbit coupling will not have any significant effect on the Fermi surface of a metal. However, this is not necessarily true because the introduction of spin–orbit coupling may lift some of the degeneracies that were essential degeneracies—that is group-theoretically necessary degeneracies—in the absence of spin–orbit coupling (Elliott 1954). Although the actual magnitudes of the splittings introduced by the inclusion of spin–orbit coupling may be quite small (for example between 10^{-4} and 10^{-7} Ry in Mg and between about 10^{-2} and 10^{-3} Ry in Tl), the fact that the splitting occurs at all will, in some cases, modify the topology of the Fermi surface quite drastically by altering the connectivities between the various sheets of the Fermi surface of a metal. This is particularly likely to be important in metals with the h.c.p. structure, although it may also arise in metals with other structures. It will be recalled that for an h.c.p. metal, in the absence of spin–orbit coupling, there is a two-fold degeneracy of the energy bands all over the large hexagonal face AHL of the Brillouin zone. For this reason a double zone is often used for h.c.p. metals, see, for example, Fig. 1.18. The introduction of spin–orbit coupling necessitates the replacement of the single-valued space-group representations by the double-valued space-group representations (Elliott 1954); this will lead, for some special $\mathbf{k}$ vectors in the Brillouin zone, to a lifting of some of the degeneracies. In particular one finds that the double-degeneracy at a general point on the top face AHL of the Brillouin zone of an h.c.p. metal is lifted; the degeneracy is also lifted at the point H (see Fig. 1.18) but it still survives along the line AL (Cohen and Falicov 1960). The calculated value of this splitting at H for Mg was found to be $\sim 10^{-4}$ Ry or less (Falicov and Cohen 1963). Although the actual splitting due to the introduction of spin–orbit coupling is quite small, it has important consequences for the general topological features of the Fermi surface of the metal. Thus, without spin–orbit coupling, the Fermi surface in band 1 and band 2 in Mg is multiply-connected with infinite extent normal to the c axis but not parallel to it. The spin–orbit splitting, however, causes the Fermi surface in band 2 to extend infinitely along the c axis. The qualitative feature of the multiple-connectivity of a sheet of the Fermi surface in a given direction can be investigated experimentally because of the striking effect that it produces in the transverse magneto-resistance (see Section 3.7).

2.12. Conclusion

At some stage it is desirable to be able to check the results of a band structure calculation against experimental evidence. In Chapter 3 we shall see that this is usually done by experimental investigations of the energies $E_j(\mathbf{k})$ or of the density of states $n(E)$ or by the measurement of some geometrical features of the Fermi surface that has been derived on the basis of the calculated bands $E_j(\mathbf{k})$. However, there are one or two experimental tests of band structure calculations that give information about the wave function $\psi_{\mathbf{k}}(\mathbf{r})$ rather than about the energies $E_j(\mathbf{k})$; these include measurements of the Knight shift and of the X-ray or neutron scattering factors. Of course, the measurement of these parameters does not give a direct detailed measurement of $\psi_{\mathbf{k}}(\mathbf{r})$ as a function of $\mathbf{r}$ for each electron in a metal, but it can be used to give some general indication of the reliability of the wave functions obtained from a given band structure calculation.

The theory of the Knight shift, that is the difference between the nuclear paramagnetic resonance frequency in a single atom of a metallic element and in a large sample of the metal, is given, for example, by Townes, Herring, and Knight (1950). The resonance frequency is shifted because the magnetic field at the nucleus is altered by a contribution due to the alignment of the spins of some of the electrons when the external field is applied. For metals possessing cubic symmetry the shift is dependent only on the s contribution (that is $l = 0$ contribution) to the wave function. The case of non-cubic metals need not detain us here (for details see, for example, Bloembergen and Rowland 1953). The difference between the applied field and the actual field at the nucleus is the contribution due to these electrons which is $(8\pi/3) \times$(mean density of spin moment at the nucleus), so that

$$\frac{\Delta B}{B} = (8\pi/3)\chi_p M \langle |\psi_{k_F}(0)|^2 \rangle_{av} \qquad (2.12.1)$$

where χ_p is the susceptibility per unit mass, M is the mass of one atom and $\langle |\psi|_{k_F}(0)|^2 \rangle_{av}$ is the average probability density, at the nucleus, of electrons at the Fermi surface. Since only s-type wave functions make non-zero contributions to $\langle |\psi_{k_F}(0)|^2 \rangle_{av}$ the measured value of $\Delta B/B$ can be used to test the average percentage s-type contribution to the wave functions $\psi_{\mathbf{k}}(\mathbf{r})$ determined by a given band structure calculation. This therefore provides some experimental check, albeit only an indirect one, on calculated wave functions $\psi_{\mathbf{k}}(\mathbf{r})$. With a little sophistication, involving the use of quantum oscillations (see Section 3.8) in the Knight shift, it is possible to measure $\langle |\psi_F(0)|^2 \rangle_{av}$ for particular cross sections of the Fermi surface, instead of just obtaining a single value representing an average over the whole Fermi surface (Goodrich, Khan, and Reynolds 1969).

The X-ray or neutron scattering factor depends on the distribution of the electronic charge density, as a function of r. The contribution to the charge density due to an electron with wave function $\psi_{\mathbf{k}}(\mathbf{r})$ is given by $e\,|\psi_{\mathbf{k}}(\mathbf{r})|^2$. For a given value of the principal quantum number n, the radial distribution of the charge density will be different for s, p, d, and f electrons. In general the radial distribution of the charge density depends on both n, the principal quantum number, and l, the angular momentum quantum number. The wave functions $\psi_{\mathbf{k}}(\mathbf{r})$ produced by a band structure calculation can be used to calculate the scattering factor which can then be compared with the results of elastic scattering measurements. This gives us a second indirect experimental method for checking the wave functions $\psi_{\mathbf{k}}(\mathbf{r})$ obtained from a band structure calculation. This method has been used for a number of metals; it is of particular interest in connection with transition metals in which it is helpful in trying to establish the magnitude of the d-type contributions to the wave functions (Weiss and De Marco 1958, 1965; Weiss and Freeman 1959; Batterman, Chipman, and De Marco 1961; Arlinghaus 1967*a*; Weiss 1970).

3

EXPERIMENTAL METHODS FOR INVESTIGATING BAND STRUCTURES AND FERMI SURFACES

3.1. Survey of methods

IN THIS chapter we shall describe the various methods that are available for the experimental investigation of the shape of the Fermi surface of a metal and that have been used, at one time or another, in that connection. We shall be concerned with describing the principles involved in the use of these methods rather than with experimental details (for further experimental details see, for example, Mackinnon 1966; Mercouroff 1967). The importance of determining the shape of the Fermi surface of a metal is to try to obtain a better understanding of all those physical properties of a metal that depend, in some way or another, on the behaviour of the conduction electrons in the metal.

Any property that involves the conduction electrons in a metal must depend on the shape of the Fermi surface and on the wave functions of the electrons at or near the Fermi surface of that metal. The detailed ways in which various physical measurements depend on the shape of the Fermi surface will vary from one phenomenon to another and, therefore, the nature and amount of information about the Fermi surface that can be extracted from different types of experimental measurements will also vary. Some of these properties of a metal are such that measurements of their behaviour give direct information about the geometry of the Fermi surface, either in the form of extremal areas of cross section or of caliper dimensions of the Fermi surface. Such experiments rely on the ability to construct a physical situation in which there is some interaction between the outside world and a selected group of electrons on the Fermi surface. These properties may be described as *microscopic*, *direct*, *topological*, or *topographical* properties (Mercouroff 1967); they include the various size effects, and the 'quantum oscillations' in a variety of phenomena as a function of applied magnetic field, of which the best known are the de Haas–van Alphen effect and the Schubnikow–de Haas effect. This kind of experiment is very informative because, if an oriented single-crystal specimen is used, it enables the dimensions of each region of the Fermi surface to be measured in particular directions or planes. If there are several incompletely filled bands in a given metal and measurements are made of a linear dimension in a particular direction or of an area of cross section normal to a particular direction, several results may be obtained corresponding to pieces of Fermi surface in different bands. Considerable care may be needed to decide which result corresponds to which band, to identify the various regions in the Brillouin zone as regions of holes or of electrons, and to ensure the correct continuity between the results as the direction of $\mathbf{k}$ is varied. There are also some properties which, although they do not give quantitative geometrical information such as caliper dimensions or cross-sectional areas, do give qualitative geometrical information about general features, such as the curvature or the connectivity of the Fermi surface of a metal. Such properties may be described as *indirect*. For example, the observation of the magnetoresistance of a metal at high magnetic fields yields information about whether open orbits are possible in the plane normal to the magnetic field and hence whether the Fermi surface is multiply-connected in one particular plane.

There is another set of properties of a metal in which it is not possible to isolate the contributions from individual groups of electrons, and the observed magnitude of any experimental number will arise as a result of contributions from all the electrons on the Fermi surface of the metal. Although the magnitudes of these bulk properties give no direct information about the geometry of the Fermi surface, they still depend on the behaviour of the electrons in the metal and in particular on the behaviour of those electrons near the Fermi surface. We may describe these as *global* (Mercouroff 1967), *bulk*, or *macroscopic* properties. Although they generally give much less direct information about the geometry of the Fermi surface, considerable use has often been made of several of them in the past because of technical difficulties associated with many of the more direct measurements. For example, measurement of the electronic contribution to the specific heat of a metal gives a measure of the density of states at the Fermi surface. These properties also include mechanical, electrical, optical, thermal, and low-field or spontaneous magnetic properties, and various transport properties. The theoretical expression for the magnitude of an experimental measurement of a parameter describing one of these properties will involve a surface integral of some function over the whole of the Fermi surface. It is, of course, possible to use the experimental measurements of one of these properties to test various proposed Fermi surface geometries by seeing which one gives the closest agreement, when substituted into the appropriate integral, with the experimental value. This is obviously rather tedious and is no longer used in Fermi surface determinations because the technical problems associated with the more direct methods can usually now be overcome; however it was used in the classic work by Pippard (1957*b*) on the determination of the shape of the Fermi surface of Cu by measuring the surface impedance in the anomalous skin effect regime.

One can view these macroscopic properties in two different ways. Formerly, when the shapes of the Fermi surfaces of many metals were not known, measurements related to these macroscopic properties of a metal were often used to deduce some information about the Fermi surface of that metal. However, now that the shapes of the Fermi surfaces of most of the metallic elements are known in considerable detail (see Chapters 4 and 5), measurements of such properties of a given metal are commonly used, together with the knowledge of the shape of the Fermi surface of that metal, to obtain information about the interactions of the conduction electrons in that metal; these interactions include the interactions of the electrons among themselves and also their interactions with some other particles, quasi-particles, or electromagnetic fields. There is, therefore, a somewhat arbitrary distinction between the discussion, in this chapter, of macroscopic properties regarded primarily in connection with determining the Fermi surface of a given metal, and the discussion, in Chapter 6, of the same, or similar, macroscopic properties in connection with studying the various interactions of the conduction electrons in a metal.

While it is relatively easy to measure the various macroscopic or bulk properties of a metal but very difficult to interpret the results in terms of the shape of the Fermi surface, the reverse is true of the microscopic (or direct or topographical) properties. As we shall see in various later sections in this chapter, the measurement of one of these microscopic properties of a metal leads to a number that is related by a quite simple formula to either a caliper dimension or an extremal area of cross section of the Fermi surface. Because measurements of these microscopic properties giving direct information about the shape of the Fermi surface correspond to selecting a particular electron or group of electrons, it is important for the success of the measurement that these electrons should not undergo a scattering event during the time in which the information is extracted. Within a metal the conduction electrons are continually undergoing collisions which cause the electron to be scattered. These collisions may be either with imperfections of the crystal lattice, such as impurity atoms, point defects, or dislocations, or with other conduction electrons (electron–electron collisions), or with the ion cores of the metal; this last kind of collision can most conveniently be regarded as collisions between the electrons and the quanta of energy of the normal modes of vibration of the metal (electron–phonon collisions). The collisions of an electron may be characterized either by a relaxation time, τ, which measures the average time between collisions, or by the average distance travelled between collisions, that is, by the mean free path, λ. If an experiment is designed to measure the energy $E_j(\mathbf{k})$, or wave vector $\mathbf{k}$, of a particular electron and the effect is not to be blurred out by the electron's collisions, then the interaction of the electron with the measuring device must occur within a time t such that

$$t \ll \tau \tag{3.1.1}$$

that is, within a distance l such that

$$l \ll \lambda. \tag{3.1.2}$$

Since the kind of influence that is applied to the metal is usually a wave motion of some kind, we can rewrite eqn (3.1.1) as a condition on the frequency of the wave

$$\frac{1}{\nu} \ll \tau$$

or

$$\nu\tau \gg 1, \tag{3.1.3}$$

since it is the frequency rather than the period of one cycle that is commonly used to characterize a wave. A slightly different way of looking at this argument is from the point of view of the uncertainty principle relating energy and time. If the energy levels $E_j(\mathbf{k})$ are to be investigated, their average lifetime τ must be sufficiently long that $E_j(\mathbf{k})\tau \gg h$, which leads directly to eqn (3.1.3). The idea behind the restrictions expressed by eqns (3.1.2) or (3.1.3) is sometimes referred to as the *Peierls–Landau condition* (Peierls 1934, 1955). The restriction in eqn (3.1.3) means that for a given specimen of a metal there is a lower limit to the frequency ν with which one can hope to observe a microscopic or topographical effect. In addition to the blurring of the energy levels as a result of collisions, there will also be the possibility of a blurring of the levels as a result of thermal agitations. In order that any quantum effects used to study the energy levels or the shape of the Fermi surface should not be blurred out as a result of thermal agitations, the quantum of energy involved must be greater than the energy kT associated with the thermal agitations. In the presence of a magnetic field the appropriate quantum of energy is $h\nu_c$ where ν_c is the cyclotron frequency of an electron in the metal. The condition for the observation of quantum effects is therefore

$$h\nu_c \gg kT. \tag{3.1.4}$$

Experimentally it is the intention to use the topographical properties to measure individual dimensions of the Fermi surface so that, except in the idealised case of a spherical Fermi surface (which never actually occurs) it is necessary to use a sample which is a single crystal. To satisfy the condition of eqns (3.1.2) and (3.1.3) it is necessary to make both the electronic mean free path λ and the corresponding relaxation time τ as long as possible. This can be achieved by using a sample of very high purity and conducting the experiments at as low a temperature as possible. Increasing the purity will reduce the number of collisions with impurities and lowering the temperature will reduce the number of collisions with phonons. To satisfy the quantum condition in eqn (3.1.4) it is also necessary to use a low temperature. By a low temperature in this context we mean at least the temperature of liquid hydrogen and preferably that of liquid helium—and often even lower temperatures such as can be obtained either by adiabatic demagnetisation or by ^{3}He/^{4}He dilution-refrigeration techniques.

The principal topological methods for the experimental investigation of the Fermi surface are listed in Table 3.1 where the relevant conditions from eqns (3.1.2), (3.1.3), and (3.1.4) that must be satisfied are given. We shall be discussing the various experimental methods given in this table in the later sections of this chapter. Although we shall not, in general, be concerned with detailed descriptions of experimental procedure we shall normally give references to writings where further information can be found.

As we shall see in the present chapter, by far the greatest amount of useful experimental information about the band structure and Fermi surface of a metal has come from detailed measurements that can be related to the geometry of the Fermi surface of that metal. However, there are some experimental measurements that can be used to give information about the energy bands, that is the curves of $E_j(\mathbf{k})$ as functions of $\mathbf{k}$, rather

TABLE 3.1

Principal topographical methods for studying the Fermi surface

	Galvano-magnetic properties	*Dimensional effects*	*Resonant absorptions*	*Propagation of waves*	*Magnetic susceptibility*
Semi-classical effects	Transverse magneto-resistance $\nu_c\tau \gg 1$	Anomalous skin effect $\delta \ll \lambda$	Cyclotron resonance $\delta \ll \lambda$ $\nu_c\tau \gg 1$	Magneto-acoustic geometric effects $\nu_c\tau \gg 1$ $k\lambda \gg 1$	
		Dimensional effects $d \ll \lambda$ $\nu_c\tau \gg 1$	Ultrasonic cyclotron resonance $k\lambda \gg 1$ $\nu_c\tau \gg 1$	Diamagnetic resonance by the Doppler effect (helicons) $k\lambda \gg 1$ $\nu_c\tau \gg 1$	
Quantum effects	Schubnikow–de Haas and similar effects $\nu_c\tau \gg 1$ $h\nu_c > kT$			Magneto-acoustic or helicon wave absorption $\nu_c\tau \gg 1$ $h\nu_c > kT$	de Haas–van Alphen effect $\nu_c\tau \gg 1$ $h\nu_c > kT$

(Mercouroff 1967)

than about the Fermi surface. The ideal experiment for investigating the band structure of a metal would be some experiment in which it was possible to measure simultaneously the energy $E_j(\mathbf{k})$ and the corresponding momentum $\hbar\mathbf{k}$ for any required conduction electron in the metal. While it might be possible, in principle, to devise such an experiment, there are formidable theoretical and experimental difficulties in the way which have so far prevented it. In practice it is only possible to study separately either the energy distribution or the momentum distribution of the conduction electrons, but not to measure $E_j(\mathbf{k})$ and $\mathbf{k}$ simultaneously. We shall discuss these two types of measurement in Sections 3.2 and 3.3 respectively before proceeding to consider, in Section 3.4 onwards, the methods that give information about the Fermi surface rather than the band structure. Where references are given in this chapter to further details of theoretical derivations of formulae or to the details of experimental techniques, we have cited convenient and generally accessible sources which are not necessarily the original sources.

3.2. Experimental determinations of the energy distribution of electrons

Almost all the standard methods for the experimental investigation of the electronic energy levels of an atom, ion, or molecule are based on some kind of optical or spectroscopic measurement. It is therefore natural to suppose that it should also be possible to study the energy levels of electrons in a solid by spectroscopic methods. However, there are complications which may arise in the case of a solid. For an atom, ion, or molecule, if one selects a particular range on the energy level diagram there will be only a finite number of energy levels within that range; although the determination of the positions of these energy levels by spectroscopic methods may sometimes be lengthy and tedious it always remains a finite problem. However, for the conduction electrons in a metal there will be a continuous distribution of energy levels. Indeed, it is the existence of continuous bands rather than of discrete lines in the spectra of solids that is the origin of the term 'band structure' for $E_j(\mathbf{k})$.

For many insulators and semiconductors it is possible to determine a considerable amount of information about the electronic band structure from optical absorption or emission measurements, see, for example, the reviews by Givens (1958) and Phillips (1966, 1968*a*). For these materials it is often possible to identify a considerable number of edges which can be ascribed to direct interband transitions, that is, transitions between different bands at the same point $\mathbf{k}$ in the Brillouin zone. Contributions to the observed spectrum of a solid arising from direct interband transitions will be expected from bands at all points $\mathbf{k}$ of the Brillouin zone but, because of the continuous, or quasi-continuous, variation of $E_j(\mathbf{k})$ with $\mathbf{k}$ in each band, sharp edges will only be expected to arise from points at which

$E_j(\mathbf{k})$ has a maximum or minimum, which usually only happens at the special points of symmetry in the Brillouin zone. For metals, however, the optical spectra often exhibit much less structure than is found for non-metals and it is usually possible only to identify relatively few direct interband transitions.

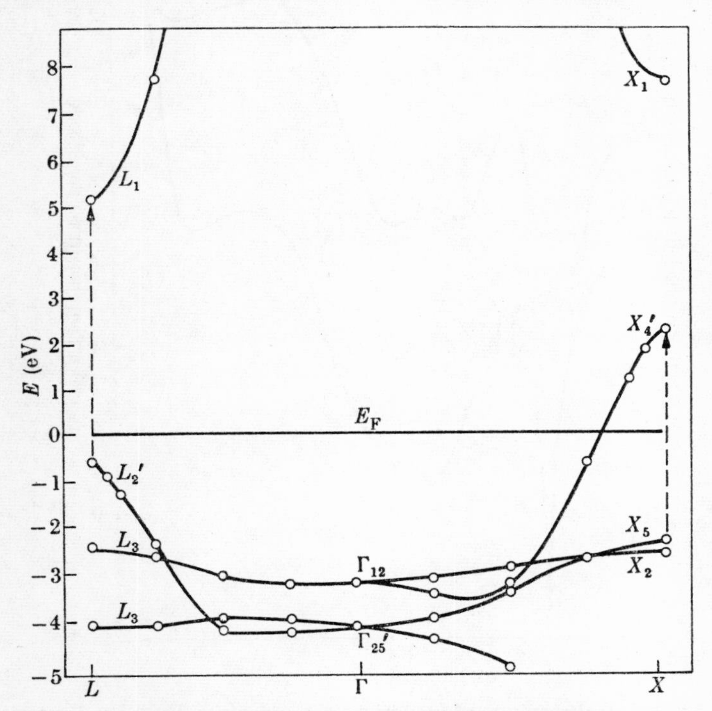

FIG. 3.1. Energy bands of copper near E_F (Phillips 1966).

Some examples of metals for which there are, at least tentative, assignments of direct interband transitions to features of the optical spectra include the noble metals, the alkaline earth metals, and Al (Phillips 1966). Features of the calculated band structure of Cu related to possible interband transitions are shown in Fig. 3.1, and the positions of these transitions relative to the measured absorption spectrum of Cu are shown in Fig. 3.2, which also shows similar data for Ag. Similar assignments of observed features in the optical spectrum have been

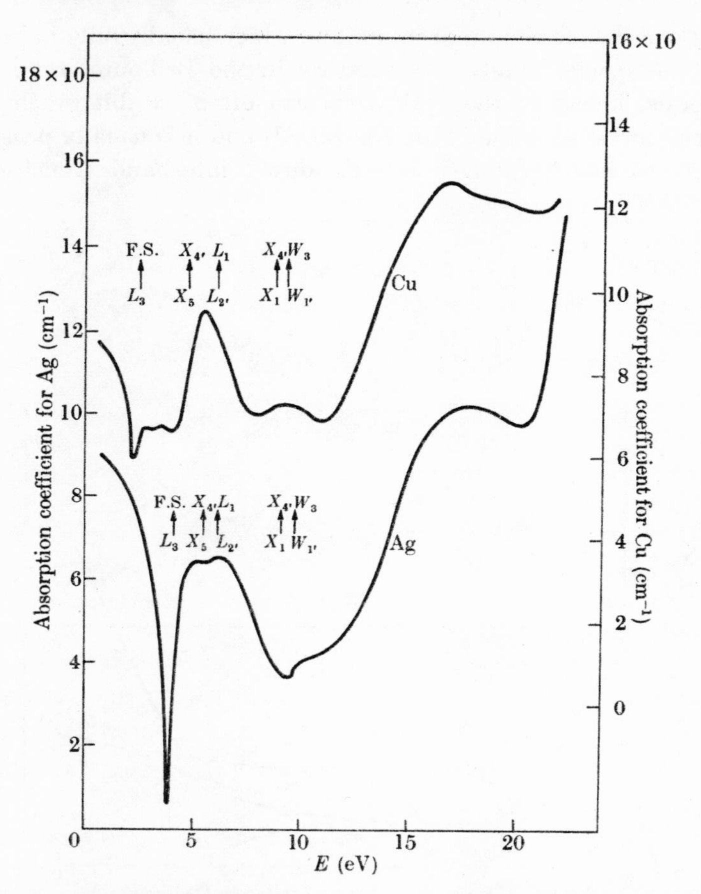

FIG. 3.2. Spectral structure of $\kappa(\omega)$ in copper and silver. Inter-band energies at symmetry points are laid above the spectra (Phillips 1966).

made for a number of other metals. However, such identifications are necessarily speculative because one does not know the value of **k** to which the transition corresponds; the value of **k** usually has to be guessed on the basis of the results of some friendly band structure calculation.

In the previous paragraph we have used the word 'optical' rather loosely. By 'optical' we do not necessarily confine ourselves to the visible region of the electromagnetic spectrum but include the infra-red and ultra-violet as well. A typical value of the Fermi energy for a metal is about 5 eV and this corresponds to a frequency, ν, of the order of 10^{15} Hz or a wavelength, λ, of the order of 2500 Å. This implies that direct emission or absorption spectroscopic measurements for the conduction electrons in a metal need to be studied in the ultra-violet region of the electromagnetic spectrum. The experimental difficulties associated with spectroscopic investigations of metals in the ultra-violet region of the spectrum are quite considerable and it is only quite recently that these difficulties have really been satisfactorily overcome. These difficulties include the problem of preparing samples of metallic surfaces sufficiently free from contamination, and also the fact that the absorption coefficient $\kappa(\omega)$ cannot be measured directly because it is so large that for ordinary sample thicknesses the amplitude of the transmitted radiation is virtually zero. Consequently it is necessary to measure the reflectivity $R(\omega)$ and to use the Kramers–Kronig relations to calculate $\theta(\omega)$, the change in phase of the wave on reflection. To determine $\theta(\omega)$ it is necessary to measure $R(\omega)$, in principle at least, for all values of ω from zero to infinity and in practice over as wide a range of ω as possible. From $R(\omega)$ and $\theta(\omega)$ it is then possible to determine $\mathrm{n}(\omega)$ and $\kappa(\omega)$, the refractive index and the absorption coefficient respectively, which are both functions of ω.

While it is possible to compare a measured absorption spectrum with a calculated band structure for a given metal and to attempt to identify spectral features with direct interband transitions, the amount of information that can be obtained in this way is actually quite restricted in practice. In addition to the fact that the measured spectra for metals exhibit relatively little structure anyway, there are other complications arising from the existence of indirect transitions, that is, transitions involving a change in $\mathbf{k}$. Additional information about features of the band structure of a metal can sometimes be obtained from measurements of the absorption or reflectivity at different temperatures (thermoreflectance measurements) and also, particularly for a magnetically ordered metal, from optical measurements performed in the presence of a magnetic field (magneto-optic measurements). As another possibility, optical measurements can be used as one of the indirect methods of Fermiology because, if one knows the wave functions and the energies of the electrons in a metal, it is possible to calculate the optical absorbing power *ab initio* and compare this, as a function of wavelength, with the results of experimental measurements (see, for example Appelbaum 1966; Jones and Lettington 1967; Overhauser 1967; Smith 1969*b*; Brust 1970).

In practice optical investigations are usually pursued in the hope of obtaining information about the density of states, $n(E)$, rather than about the bands of $E_j(\mathbf{k})$ as functions of $\mathbf{k}$. Even so, since for a solid the energy levels of the electrons are spread out into continuous, or quasi-continuous, bands it is very difficult to try to disentangle the structure of the density of states, $n(E)$, from the radiation emitted or absorbed in transitions between two levels, each of which may be anywhere in the continuous energy bands of the metal. For a long time it was necessary to study only transitions in which the lower energy level involved in the transition is a discrete energy level and the higher level is one of the various levels in the energy bands near the Fermi level. This means that, instead of ultra-violet spectroscopy, we are concerned with soft X-ray spectroscopy. Provided we ignore differences in transition probabilities, the structure of a soft X-ray emission or absorption line should show quite directly the form of the density of states curve, $n(E)$. Some considerable discussion of the experimental methods used and of the results obtained in early X-ray absorption and emission spectroscopic work on metals, which involved X-rays of energy up to a few hundred eV only, was given by Skinner and his co-workers (Skinner and Johnston 1937; Skinner 1938, 1940; Skinner, Bullen, and Johnston 1954). Some typical early results are shown in Fig. 3.3. As usual the distinction between the emission and absorption measurements is that the emission spectra give information relating to the energy bands which are occupied in the ordinary state of the metal, whereas the absorption spectra give information relating to the unoccupied levels above E_F, see

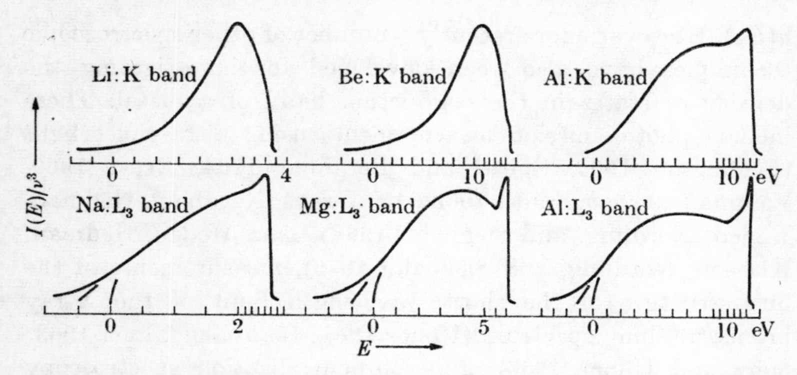

FIG. 3.3. The results of some early soft X-ray measurements on metals (Skinner 1940).

Fig. 3.4. An impression of the present state of X-ray emission and absorption techniques in connection with band structures and densities of states can be gained from the conference proceedings edited by Fabian (1968).

For very many years soft X-ray spectroscopy was the only experimental technique which was able to provide any information about the density of states, $n(E)$, except, of course, at the Fermi energy E_F itself where measurements of the specific heat and of the paramagnetic susceptibility can be related to

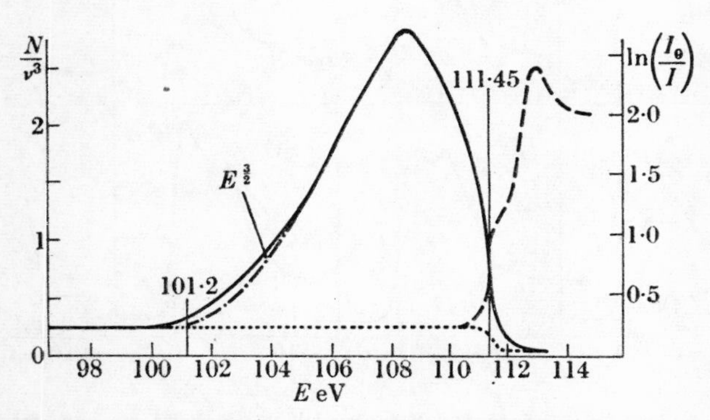

FIG. 3.4. Emission spectrum (solid curve) and absorption spectrum (broken curve) for beryllium metal (Lukirskiĭ and Brӯtov 1964).

$n(E_F)$. However, more recently a number of other spectroscopic techniques have also been developed for investigating the density of states in the conduction band of a metal. These include photo-emission measurements using ultra-violet light (Lettington 1964, Spicer and Berglund 1964, Brust 1969, Eastman 1969*a*,*b*, Smith 1969*a*) or X-rays (Nordberg, Hedman, Hedén, Nordling, and Siegbahn 1968; Baer, Hedén, Hedman, Klasson, Nordling, and Siegbahn 1970), measurements of the fine structure of the short wavelength limit of the X-ray bremsstrahlung spectrum (Ulmer 1961; Claus and Ulmer 1963; Merz and Ulmer 1966), and ion-neutralization spectroscopy (Hagstrum 1954, 1961, 1966; Hagstrum and Becker 1967). These techniques involve various different processes in which an energy transfer occurs between a conduction electron and some

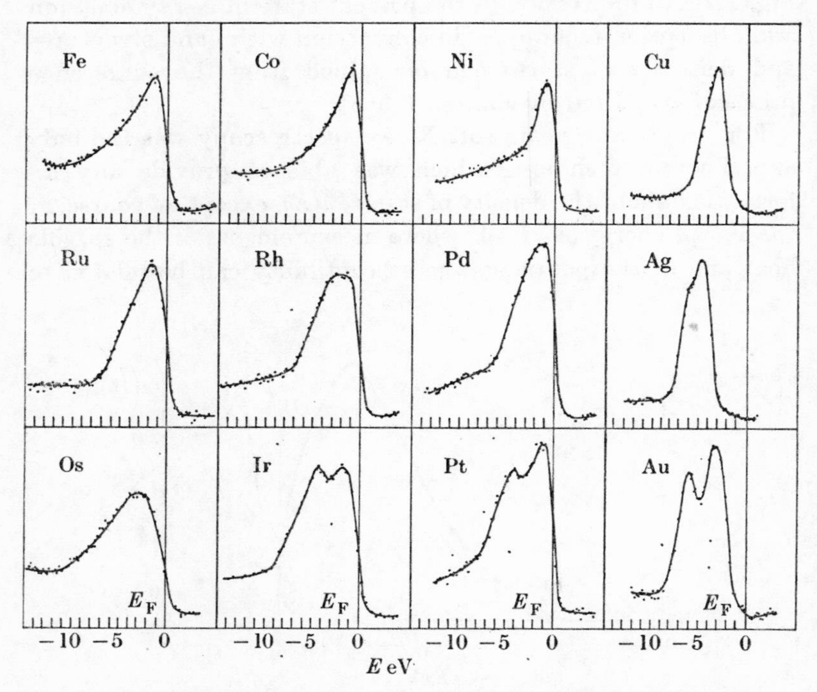

FIG. 3.5. Densities of states for several metals determined from X-ray photoemission measurements (Baer, Hedén, Hedman, Klasson, Nordling, and Siegbahn 1970).

incoming entity. For example, we consider the energy distribution of the photo-emitted electrons emerging from a sample of metal that is illuminated with ultra-violet light. If the energy of a photo-emitted electron was $E_j(\mathbf{k})$ before it was emitted from the metal, then after emission its energy is given by

$$E = h\nu - E_j(\mathbf{k}) - \phi, \tag{3.2.1}$$

where ν is the frequency of the incident ultra-violet photons and ϕ is the work function of the metal. ϕ can be removed from this equation by measuring E and $E_j(\mathbf{k})$ relative to a suitably chosen origin. If we ignore the differences between the transition probabilities for electrons with different wave vectors $\mathbf{k}$, then the number of photoelectrons emitted with energy E is proportional to the number of electrons in the metal with the particular value of $E_j(\mathbf{k})$, irrespective of $\mathbf{k}$, that fits eqn (3.2.1). Therefore the profile of the intensity distribution of the photo-emitted electrons will be the same shape as the density of states curve $n(E)$. In a similar, but obviously not identical, manner each of these other spectroscopic methods gives measurements that can be related to the density of states curve $n(E)$ throughout the conduction band. Some typical results are shown in Figs. 3.5 and 3.6. The large number of rapid fluctuations that are a feature of so many theoretical density of states calculations are too narrow to be resolved or perhaps are unreal anyway. An example of the comparison between the results of two different methods is shown in Fig. 3.7.

In any emission or absorption spectrum of a metal, or in the distribution of electrons emitted from a metal for one reason or another, there are contributions from a continuum of states in the conduction band so that the observed distribution will depend on the density of states in the conduction band. However, the intensity of any given emission or absorption process involving an electronic transition between two energies E_1 and E_2 depends not only on the density of states at E_1 and E_2 but also on the quantum-mechanical matrix element representing the transition probability between E_1 and E_2. Most of the experiments which have been performed using the methods

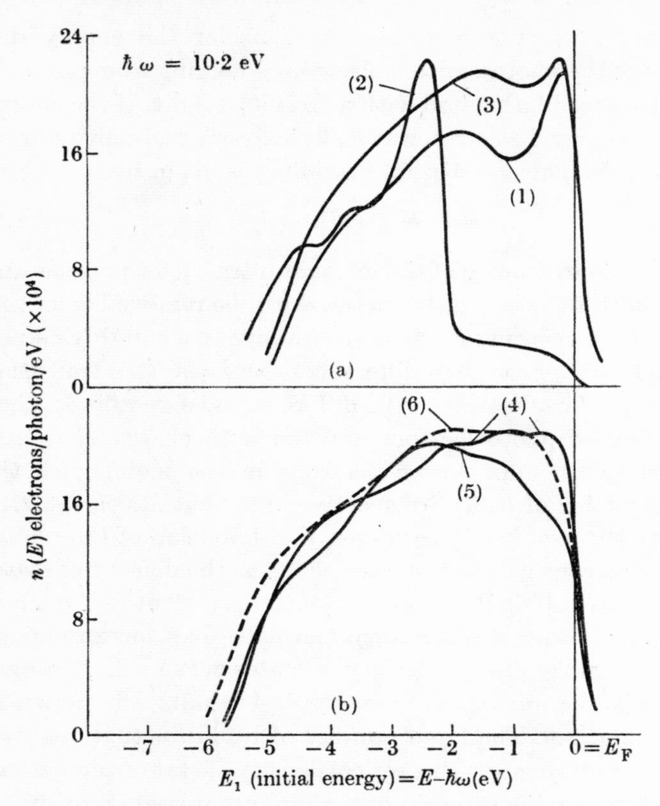

FIG. 3.6. Energy distribution of photo-emitted electrons for (1) Ni, (2) Cu ($\times$ 0·6), (3) Co, (4) Fe, (5) Cr, and (6) Mn (Eastman 1969*b*).

which we have described so far in this section have been analysed using the simplifying assumption that the transition probability is a constant. To a considerable extent it is the use of this assumption, rather than any deficiency of experimental technique, which leads to the discrepancies between the curves of the density of states produced by different methods as illustrated, for example, in Fig. 3.7. It is very difficult to calculate meaningful transition probabilities *ab initio* because the appropriate quantum-mechanical matrix elements involve the wave functions of the initial and final states of the electrons which undergo the transitions and these wave functions are generally not known very accurately.

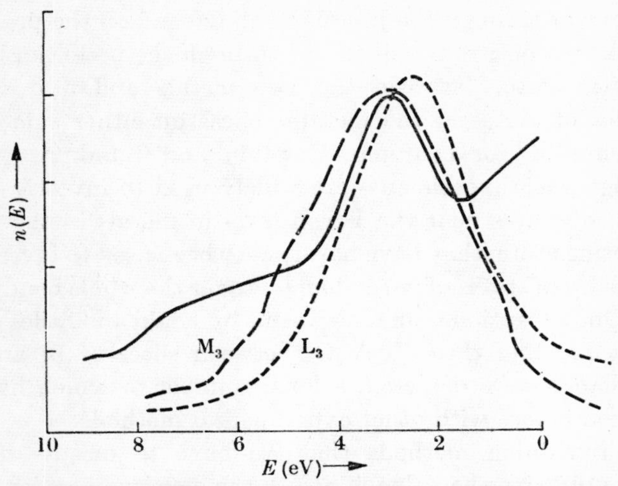

FIG. 3.7. Graph comparing the density of states for copper obtained from ion-neutralization spectroscopy (solid curve) with L_3 and M_3 soft X-ray emission spectrum results (Hagstrum and Becker 1967).

Another type of method which might seem to be profitable in connection with determining the density of states, $n(E)$, would be one based on the use of a quantum-mechanical tunneling phenomenon. That is, we might seek to determine the characteristics of a junction consisting of a sandwich of two metallic slices separated by a filling of insulating material, see Fig. 3.8.

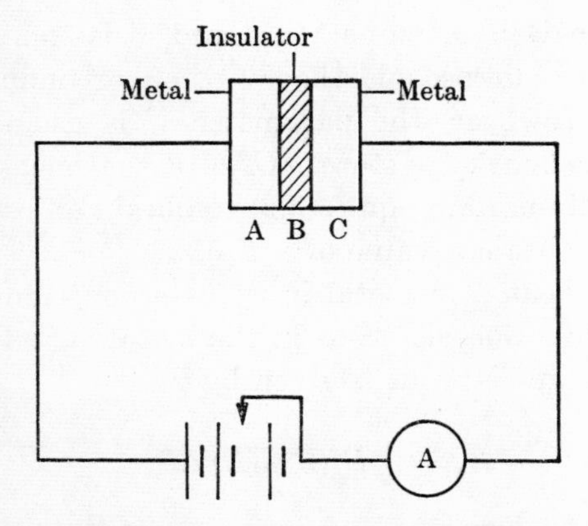

FIG. 3.8. Schematic arrangement for a tunnelling experiment.

The current through the junction will depend on the probabilities that the electrons will tunnel through the potential barrier of the insulator between the two metals and also on the densities of states in the metallic slices on either side of the junction (see, for example, Burstein and Lundqvist 1969). Although such experiments are widely used to investigate the density of states near the Fermi level in metals in the superconducting state, they have hardly ever been used to investigate the density of states of normal metals over the whole conduction band. One exception is an experiment by Esaki and Stiles (1965) using tunnelling through Al_2O_3 between slices of Bi and Al; many bands were detected, a large number of which had not been seen before with other experimental methods.

The remaining methods that we have to discuss in this section only give the value of $n(E_F)$, the density of states at the Fermi energy, and do not provide information about $n(E)$ for the other energies in the conduction band. These methods include the measurement of the superconducting transition temperature, the low temperature specific heat, and the paramagnetic susceptibility.

The superconducting transition temperature T_c can be expressed in terms of $n(E_F)$, the density of states at the Fermi energy in the normal metal (Bardeen, Cooper, and Schrieffer 1957*b*) by

$$T_c = \Theta \exp\{-1/n(E_F)V\}, \tag{3.2.2}$$

where Θ is the Debye temperature and V is some parameter involved in the microscopic theory of superconductivity (see Section 6.5). However, the parameter V is usually less well known than the density of states $n(E_F)$, so that measured values of T_c are usually used in eqn (3.2.2) to obtain an estimate of V rather than to obtain a value of $n(E_F)$.

The specific heat of a metal includes a contribution that is due to the conduction electrons in the metal. The total energy of the conduction electrons is given by

$$\bar{E}_e = \int_0^\infty Ef(E)n(E)\,dE, \tag{3.2.3}$$

where $n(E)$ is the density of states and $f(E)$ is the Fermi–Dirac distribution function. The contribution of the conduction electrons to the specific heat is therefore given by $C_e = \partial \bar{E}_e / \partial T$ and this can be reduced by a little algebraic manipulation (see, for example, Mott and Jones 1936, p. 178) to the form

$$C_e = \tfrac{1}{3}\pi^2 k^2 T n(E_F). \tag{3.2.4}$$

The electronic contribution to the specific heat of a metal is therefore proportional to T and is usually written in the form $C_e = \gamma T$ where

$$\gamma = \tfrac{1}{3}\pi^2 k^2 n(E_F). \tag{3.2.5}$$

By determining γ experimentally one therefore has a method for measuring $n(E_F)$, the density of states at the Fermi energy. The value of γ is typically of the order of a few mJ mol^{-1} K^{-2}. This means that at ordinary temperatures the electronic contribution to the specific heat is much smaller than the contribution due to the lattice vibrations, which is proportional to T^3, and it is therefore necessary to use low temperatures so that the lattice contribution becomes relatively less dominant. The total specific heat is given by

$$C = \alpha T^3 + \gamma T \tag{3.2.6}$$

so that γ is determined by plotting C/T against T^2 and determining the intercept, see Fig. 3.9. For a magnetic metal there would be a further contribution to the specific heat due to the spin waves in the metal.

It should be mentioned that the manipulation involved in obtaining eqn (3.2.4) includes an expansion of $\int^{E} E n(E)\,\mathrm{d}E$ about E_F and neglecting terms higher than second-order terms in $(E - E_F)$. Equation (3.2.5) will therefore only be valid at temperatures which are substantially lower than the temperature T_F defined by $E_F = kT_F$. Since a typical value of E_F for a metal is of the order of 5 eV, T_F is typically of the order of 58 000 K. Departures from the linear relationship $C_e = \gamma T$ can

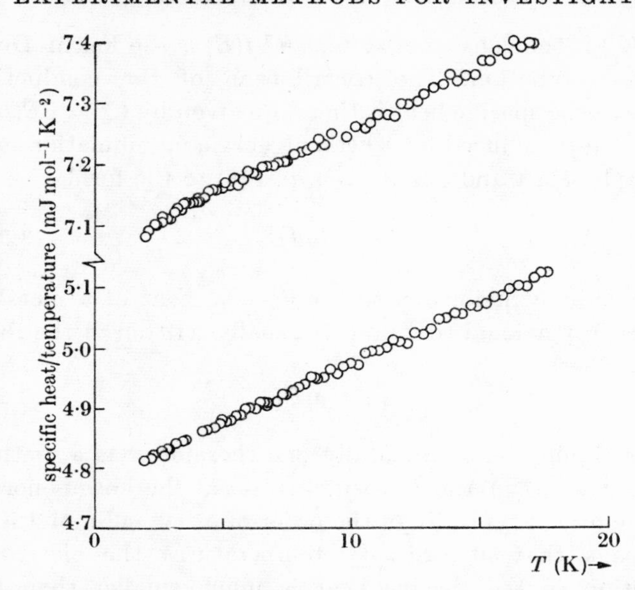

FIG. 3.9. Experimental specific heat results for a specimen of iron (lower curve) and a specimen of nickel (upper curve). The very slight downward concavity is due to the presence of a spin-wave contribution in these ferromagnetic metals (Dixon, Hoare, Holden, and Moody 1965).

be expected to occur at high temperatures (see, for example, Stoner 1936) but these are seldom important in practice.

In Table 3.2 there are some samples of experimental values of γ and a more complete set of values for all the metals is given, for example, by Gschneidner (1964). It is possible to use eqn (3.2.5) together with the expression for $n(E)$ in eqn (2.8.7) to calculate the values of γ that would be expected on the free-electron model. Values of γ calculated on the free-electron model differ from values calculated by using some more realistic band structure for the metal. Since the free-electron density of states at the Fermi level can be written as $mNVk_F/\pi^2\hbar^2$ it is common to define a *thermal effective mass* m^* such that the value of γ calculated from the more realistic band structure is $\gamma = m^*NVk_F/\pi^2\hbar^2$ where k_F is still the radius of the free-electron Fermi surface. Some typical values of m^* obtained from band structure calculations for a number of metals are

TABLE 3.2

Experimental values of γ for some metals

Metal	γ (mJ mol^{-1} K^{-2})
Li[a]	$1{\cdot}63 \pm 0{\cdot}02$
Na[a]	$1{\cdot}38 \pm 0{\cdot}02$
K[a]	$2{\cdot}08 \pm 0{\cdot}08$
Rb[a]	$2{\cdot}41 \pm^{0{\cdot}29}_{0{\cdot}16}$
Cs[a]	$3{\cdot}20 \pm 1{\cdot}05$
Al[b]	$1{\cdot}362 \pm 0{\cdot}003$
In[c]	$1{\cdot}60 \pm 0{\cdot}01$
Pb[d]	$3{\cdot}00 \pm 0{\cdot}04$

[a] Martin (1965)
[b] Dicke and Green (1967)
[c] Bryant and Keesom (1961)
[d] Van der Hoeven and Keesom (1965)

given in Table 3.3 and can be seen to include 1·64 for Li, 1·00 for Na, 0·94 for K, 1·06 for Rb, and 1·75 for Cs. It should not be assumed that there is necessarily any simple numerical relationship between the thermal effective mass defined in this way and any of the other various effective masses which can be defined in a variety of other different ways. In the early days of band structure calculations the calculated values of γ were viewed with considerable suspicion and the agreement with experimental results was not very impressive. However, more recently it has been realized that there are systematic reasons for the discrepancies. As well as the values of γ obtained using the value of $n(E_F)$ calculated from the band structure, there are additional contributions to γ as a result of electron–phonon interactions and electron–electron interactions (see Chapter 6). Again it is convenient to think of these additional contributions in terms of a thermal effective mass and some typical values of these contributions to the thermal effective mass are given in Table 3.3.

The magnetic susceptibility of a metal contains a paramagnetic contribution arising from the re-orientation of the spins of the conduction electrons in the presence of an applied

TABLE 3.3

Electron–electron and electron–phonon contributions to the thermal effective mass for various metals

	$(m^*/m)_{BS}$	$(\delta m/m)_{e-e}$	$(\delta m/m)_{e-p}$	$(m^*/m)_{calc}$			$(m^*/m)_{exp}$
Li	1·64	0·03	0·34	2·25	1·96	1·78	2·19
Na	1·00	0·06	0·18	1·24	1·15	1·13	1·25
K	0·94	0·11	0·15	1·18	1·35	1·24	1·23
Rb	1·06	0·13	0·17	1·38	1·56	1·38	1·36
Cs	1·75	0·15	0·18	2·33	2·86	1·97	1·63
Al	1·06	−0·01	0·49	1·57	—	—	1·60
In	0·91	∼0·0	0·60	1·45	—	—	1·46
Pb	0·90	0·00	1·05	1·85	—	—	2·20

(After Cracknell 1969*c*)

Note

$$\left(\frac{m^*}{m}\right)_{\text{total}} = \left(\frac{m^*}{m}\right)_{\text{BS}}\left\{1+\left(\frac{\delta m}{m}\right)_{\text{e-e}}+\left(\frac{\delta m}{m}\right)_{\text{e-p}}\right\}$$

where $(m^*/m)_{\text{BS}}$ is the band-structure effective mass, and $(\delta m/m)_{\text{e-e}}$ and $(\delta m/m)_{\text{e-p}}$ are the contributions to the thermal effective mass arising from electron–electron interactions and electron–phonon interactions, respectively.

magnetic field. This paramagnetic susceptibility of the spins can, like the electronic term γT in the specific heat, be shown to be proportional to $n(E_{\text{F}})$, the density of states at the Fermi level. In the absence of an applied magnetic field and, for the moment, omitting any consideration of magnetically-ordered metals, the density of states $n(E)$ will be the same for the two spin states $m_s = \pm\frac{1}{2}$, see Fig. 3.10, so that there will be equal numbers of electrons with spin up and spin down and the sample of metal has no net magnetic moment associated with the conduction electrons. When considering the theory of the paramagnetism of the conduction electrons in a metal it is customary to assume that the application of a (not too large) magnetic field does not distort the shapes of the energy bands $E_j(\mathbf{k})$, but simply introduces a separation between the spin–up bands and the spin–down bands. In an external magnetic field $\mathbf{H}_0$ the energy of an electron will be changed from $E_j(\mathbf{k})$ to $\{E_j(\mathbf{k})-g_s m_s \mu_{\text{B}} B_0\}$ where g_s is the magnetogyric ratio of an electron $(=2)$ and μ_{B} is the Bohr magneton. All the states with spin up are therefore lowered ($m_s = +\frac{1}{2}$; up $=$ parallel to $\mathbf{B}_0$)

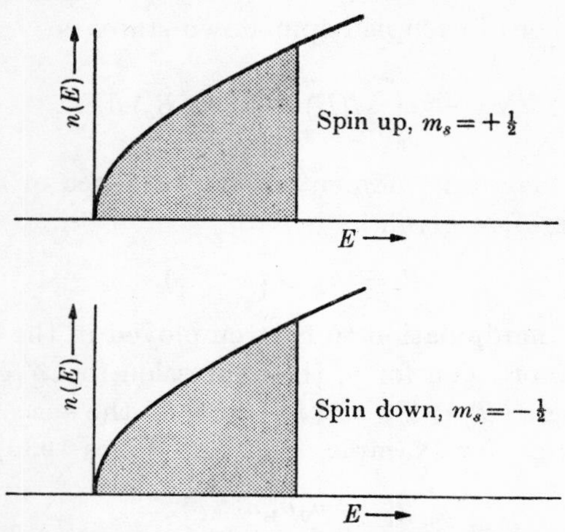

Fig. 3.10. Schematic density of states, $n(E)$, for spin-up and spin-down states with no magnetic field.

and the states with spin down ($m_s = -\frac{1}{2}$) are raised; some of the electrons that were in spin-down states therefore spill over into spin-up states, see Fig. 3.11. The number of occupied spin-up states is therefore

$$N_\uparrow = \int_{-\mu_B B_0}^{\infty} f(E) n(E + \mu_B B_0) \, dE, \tag{3.2.7}$$

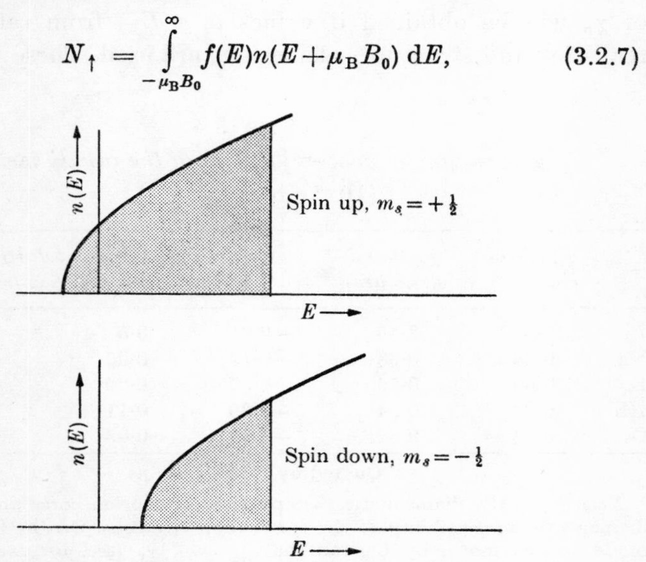

Fig 3.11. Schematic density of states, $n(E)$, for spin-up and spin-down states with an external magnetic field applied.

and the number of occupied spin–down states is

$$N_{\downarrow} = \int_{\mu_B B_0}^{\infty} f(E)n(E-\mu_B B_0)\,\mathrm{d}E, \tag{3.2.8}$$

and the net magnetic moment in the presence of the applied field $\mathbf{B}_0$ is therefore given by

$$\mathscr{M} = \mu_B(N_{\uparrow} - N_{\downarrow}). \tag{3.2.9}$$

By a similar manipulation to that employed in the determination of the expression for γ, the expression for $\mathscr{M}$ can be simplified so long as $T \ll T_F$ and in terms of the susceptibility χ_p one obtains (see, for example, Mott and Jones 1936)

$$\chi_p = \mu_0 \mu_B^2 n(E_F) \tag{3.2.10}$$

per atom. It is therefore possible by measuring the paramagnetic susceptibility of a metal to obtain another experimental determination of the density of states at the Fermi level. Values of χ_p for the alkali metals calculated by Mott and Jones (1936) using the values of $n(E_F)$ given by the free-electron model (see eqn (2.8.7)) are given in Table 3.4. Substantially different values of χ_p will be obtained if values of $n(E_F)$ from rather more realistic band structure calculations are used. These values are

TABLE 3.4

The magnetic susceptibilities of the alkali metals

(10^{-6} e.m.u. gm^{-1})

	χ_p (*free-electron*)	χ_p (*band structure*)	χ_d^1	χ_L (*free-electron*)	*total* χ, *measured experimentally*[1]
Li	1·5	2·46	−0·01	−0·5	3·54
Na	0·68	0·68	−0·18	−0·23	0·70
K	0·60	0·56	−0·37	−0·20	0·54
Rb	0·32	0·34	−0·30	−0·11	0·22
Cs	0·24	0·42	−0·30	−0·08	0·20

[1] Quoted by Kittel (1956).

Note. χ_d is the diamagnetic susceptibility of the ion cores and χ_L is the diamagnetic susceptibility of the conduction electrons. On the free electron model it was shown by Landau that $\chi_L = -\frac{1}{3}\chi_p$ (see, for example, Mott and Jones 1936, p. 204–6).

also given in Table 3.4 and are, of course, related to the free-electron values by ratios equal to the band-structure thermal effective masses given in Table 3.3. The comparison of calculated values of χ_p with experiment, or the use of measured values of the susceptibility of a metal to determine the density of states at the Fermi energy is less straightforward than it might appear to be at first sight. This is because, in addition to the magnetic

TABLE 3.5

Measured values of γ and χ_p.

(γ is in units of 10^{-4} cal mol^{-1} K^{-2} and χ_p is in units of 10^{-6} e.m.u.; according to eqn (3.2.11) the ratio χ_p/γ should then be equal to 5·76)

	Sc	Ti	V	Cr	Mn	Fe	Co	Ni	Cu
γ	24·7	8·5	22	3·7	33	12	17·7	12	1·64
χ_p	315	158	295	165	527	—	—	—	−5
	Y	*Zr*	*Nb*	*Mo*	*Tc*	*Ru*	*Rh*	*Pd*	*Ag*
γ	24·5	7·25	17·5	5·05		8·0	11·7	25·6	1·58
χ_p	191	122	206	90	268	43	102	567	−19
	La	*Hg*	*Ta*	*W*	*Re*	*Os*	*Ir*	*Pt*	*Au*
γ	16·1	6·3	14·0	2·9	5·85	5·6	7·5	16·3	1·77
χ_p	118	75	154	59	67	10	26	189	−28

(After Lomer 1962*b*)

moment associated with the spins of the conduction electrons there are a number of other contributions to the measured magnetic susceptibility of a metal.

In principle from eqns (3.2.5) and (3.2.10) the ratio of χ_p/γ should be a constant,

$$\frac{\chi_p}{\gamma} = \frac{3\mu_B^2}{\pi^2 k^2}, \tag{3.2.11}$$

which is independent of $n(E_F)$. This ratio should therefore be a constant, not only in the free-electron approximation, but also when the true band structure of the metal is used. By inspection of sets of values of χ_p and γ for real metals (Table 3.5), one can see that the observed experimental values of the ratio χ_p/γ depart quite substantially from the ideal value of $5{\cdot}76 \times 10^{-2}$

where χ_p is in e.m.u. mol^{-1} and γ is in cal. mol^{-1} K^{-2}. While some of the discrepancies are probably connected with errors occurring in the estimated values of the diamagnetic susceptibility and also with the possibility of the presence of paramagnetic contributions associated with the orbital motions of the electrons, this is not the whole story. The reasons for the discrepancies also include the fact that we have regarded the conduction electrons in a metal as a system of independent non-interacting fermions. We have already mentioned the effects of electron–phonon interactions and electron–electron interactions on γ. The various interactions of the conduction electrons should, therefore, also be included in the calculation of χ_p. Direct *ab initio* calculations of these contributions to χ_p have not, so far, received much attention; indeed the measured departures of the value of χ_p/γ from its ideal value of $3\mu_B^2/\pi^2 k^2$ have been used in connection with the determination of the parameters in Landau's phenomenological theory of electron–electron interactions (see Chapter 6).

3.3. Experimental determinations of the momentum distribution of electrons

Whereas in the previous section we were concerned with methods aimed at investigating the probability distribution of the energies of the conduction electrons in a metal, we now turn our attention to methods which are primarily concerned with studying the distribution of the electrons as a function of their momentum $\hbar\mathbf{k}$ rather than of their energy E. Of course, if all the details of the complete band structure of a metal were available this would mean that $E(\mathbf{k})$ was completely known as a function of $\mathbf{k}$ so that the density of states with respect to energy, $n(E)\,\mathrm{d}E$, could readily be converted into the density of states with respect to $\mathbf{k}$, $g(|\mathbf{k}|)\,\mathrm{d}|\mathbf{k}|$, and *vice versa*. However, for a large number of metals it has very often been the case that the details of the band structure were not very well known, so that it has often been useful to have separate measurements aimed at identifying features of the distribution of the electrons with respect to E rather than with respect to $\mathbf{k}$.

3.3.1. Compton scattering line shapes

The reader is assumed to be familiar already with the existence of the Compton effect in which an incident X-ray photon is scattered inelastically by collision with an electron. In the very simplest treatment of the process the electron is assumed to have been at rest initially and the collision is analyzed by using the conditions imposed by the laws of the conservation of momentum and the conservation of energy, see Fig. 3.12. If the angular

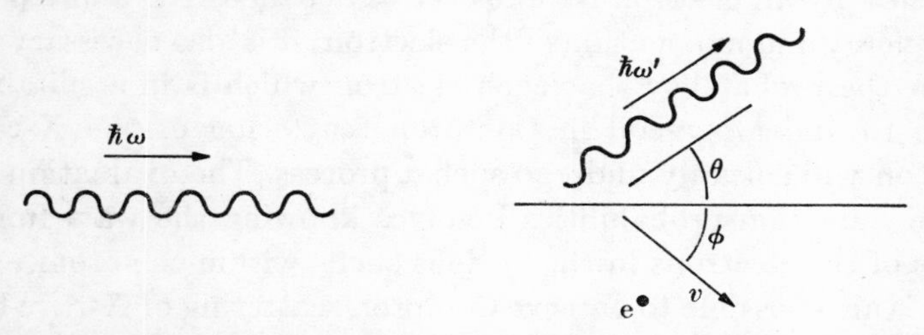

FIG. 3.12. Diagram to illustrate Compton scattering of an X-ray photon by an atom with the ejection of an electron.

frequencies of the incident and scattered X-ray photons are ω and ω' respectively the law of the conservation of energy requires that

$$\hbar\omega + mc^2 = \hbar\omega' + \frac{mc^2}{\sqrt{1-v^2/c^2}}, \tag{3.3.1}$$

where m is the rest mass of the electron, while the law of conservation of momentum requires that

$$\frac{\hbar\omega}{c} = \frac{\hbar\omega'}{c}\cos\theta + \frac{mv\cos\phi}{\sqrt{1-v^2/c^2}} \tag{3.3.2}$$

and

$$0 = \frac{\hbar\omega'}{c}\sin\theta - \frac{mv\sin\phi}{\sqrt{1-v^2/c^2}}. \tag{3.3.3}$$

If the scattered radiation is observed in the direction θ, then these conservation laws are enough to determine the unknowns ω' (the angular frequency of the scattered X-rays) and v and ϕ (the speed and direction of the recoiling electron). This very simple analysis is based on the assumption that the electron was originally free. However, Compton scattering usually involves electrons which are not actually free but which are bound in the various shells within an atom. Even for an individual atom which is assumed to be in free space one will not obtain a complete description of Compton scattering simply by using the equations obtained from the laws of conservation of energy and conservation of momentum. It is not enough to know that the scattering (through an angle θ) of an X-ray photon of wavelength λ by an electron requires certain changes ΔE and $\Delta \mathbf{p}$ in the energy and momentum of the electron; it is also necessary to know the probability that each electron which is in a suitable state to undergo recoil in Compton scattering of the X-ray photon will actually undergo such a process. The evaluation of these transition probabilities involves knowing the wave functions of the electrons in the various shells within the atom.

It is also possible to observe Compton scattering of X-rays by electrons in solids as well as by free electrons, or by the electrons in individual free atoms. In Section 3.2 we saw that spectroscopic measurements of the line shapes of X-ray emission or absorption lines could be used to obtain information about the distribution of the energy levels of the conduction electrons in a metal. In Compton scattering, instead of there being a complete transfer of all the energy, $\hbar\omega$, of the photon, only part of the energy of the incoming photon is transmitted to the electron; the remainder emerges as a photon of lower energy. It should therefore be possible to use measurements of the line shape in Compton scattering of X-rays by a metal to obtain information about the distribution of the electrons in its conduction band. It has been realized for several decades that Compton scattering measurements could be used to obtain information about the conduction electrons in a metal in this way and a review of the early work on this subject is given, for example, by Compton and Allison (1967) and Cooper and Leake (1967). However, it is only quite recently that it has been possible to perform the line-shape measurements with sufficient accuracy to enable useful information about the density of states to be obtained (Cooper and Leake 1967; Phillips and Weiss 1968).

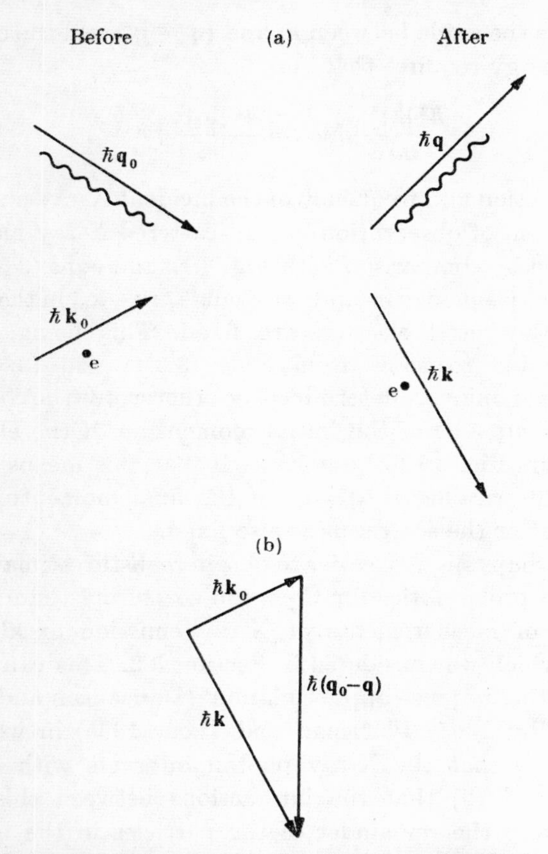

FIG. 3.13. Momentum diagrams for Compton scattering by a conduction electron in a metal. The initial and final momenta of the X-ray photon are $\hbar\mathbf{q}_0$ and $\hbar\mathbf{q}$, and for the electron the initial and final momenta are $\hbar\mathbf{k}_0$ and $\hbar\mathbf{k}$.

The theory of the line-shape of the scattered X-rays in Compton scattering by the electrons in a solid has been considered by a number of authors. The diagram shown in Fig. 3.12 was for a scattering process by an electron that was assumed to be initially at rest. If the initial momentum of the electron is $\hbar\mathbf{k}_0$ rather than zero, the scattering diagram will be altered slightly, see Fig. 3.13(a). The conservation of momentum requires (see Fig. 3.13(b)) that

$$\hbar^2\,|\mathbf{k}|^2 = \hbar^2\,|\mathbf{k}_0|^2+\hbar^2\,|\mathbf{q}_0-\mathbf{q}|^2-2\hbar^2\,|\mathbf{k}_0|\,|\mathbf{q}_0-\mathbf{q}|\cos\phi, \quad (3.3.4)$$

where ϕ is the angle between $\mathbf{k}_0$ and $(\mathbf{q} - \mathbf{q}_0)$ and the conservation of energy requires that

$$\frac{\hbar^2 |\mathbf{k}|^2}{2m} + \hbar\omega' = \frac{\hbar^2 |\mathbf{k}_0|^2}{2m} + \hbar\omega. \tag{3.3.5}$$

If the direction and frequency of the incident X-ray photon and the direction of observation of the scattered X-ray photon are fixed, then by comparison with Fig. 3.12 and eqns (3.3.1–3.3.3) the angular frequency ω' and the change $\hbar(\mathbf{k}-\mathbf{k}_0)$ in the momentum of the recoil electron are fixed. This being so, it is then possible to show from eqns (3.3.4) and (3.3.5) that $\hbar |\mathbf{k}_0| \cos \phi$ is uniquely determined; in other words the component parallel to $(\mathbf{q}_0-\mathbf{q})$ of the initial momentum of the electron is fixed. From Fig. 3.13(b) one can see that this means that the component parallel to $(\mathbf{q}_0-\mathbf{q})$ of the final momentum of the electron after the scattering is also fixed.

It is perhaps slightly easier to obtain realistic estimates of the transition probabilities for Compton scattering than it was in the case of measurements of X-ray emission or absorption spectra which we considered in Section 3.2. This can be done by using the 'impulse approximation' (Duncanson and Coulson 1945; Kilby 1965; Platzman and Tzoar 1965) in which it is assumed (i) that the X-ray photon interacts with only one electron and (ii) that the interactions between this ejected electron and the remainder of the particles in the metal are constant during the time of the collision. The first of these assumptions is reasonable so long as the energy of the incoming photon is large compared with the energies of the conduction electrons. The second assumption is valid so long as the energy of the recoiling electron is also very much greater than its energy was before the scattering occurred; this ensures that the collision time will be much shorter than the time required for the re-arrangement of the other electrons which remain in the conduction band and are not directly involved in the scattering. If the initial wave function of the conduction electron was $\psi_{\mathbf{k}_0}(\mathbf{r})$ and if the final state of the ejected electron is described by a plane wave with wave vector $\mathbf{k}$, then it is possible to show that in the impulse approximation, the cross section for Compton scattering is given (see, for example, Phillips and Weiss 1968) by

$$\frac{\mathrm{d}\sigma}{\mathrm{d}q} \propto \sum_{\mathbf{k}} \left| \int \mathrm{e}^{\mathrm{i}\mathbf{k}\cdot\mathbf{r}} \mathrm{e}^{\mathrm{i}(\mathbf{q}-\mathbf{q}_0)\cdot\mathbf{r}} \psi_{\mathbf{k}_0}(\mathbf{r})\, \mathrm{d}\mathbf{r} \right|^2 \delta\{E(\mathbf{k}) - E(\mathbf{k}_0) + \hbar\omega' - \hbar\omega\}. \tag{3.3.6}$$

We have already seen that for a given choice of $\mathbf{q}_0$ and $\mathbf{q}$, the components of $\mathbf{k}_0$ and $\mathbf{k}$ parallel to $(\mathbf{q}_0-\mathbf{q})$ are fixed. The expression for the intensity of the scattered X-rays in the direction under observation therefore only contains contributions from those electrons with wave vectors that terminate on one particular plane normal to $(\mathbf{q}_0-\mathbf{q})$.

If one makes reasonable assumptions about the form of the conduction-electron wave function $\psi_{\mathbf{k}_0}(\mathbf{r})$, eqn (3.3.6) can be used to determine the number of occupied states in a particular plane normal to $(\mathbf{q}_0-\mathbf{q})$ in the Brillouin zone. Then by varying the size of the angle 2θ through which the X-rays are scattered and observing the variation of the X-ray intensity with angle, that is by recording the line shape of the scattered X-ray beam, it is possible to study the variation in the density of states as a function of momentum. It is, of course, necessary to subtract the core electrons' contribution to the Compton scattering from the total measured scattered intensity in order to obtain the contribution that is due to the conduction electrons. So far experiments have been performed on a number of light metals with some considerable success (for example, Li, Be, Na, Mg, and Al by Phillips and Weiss 1968) and Hartree–Fock atomic wave functions were used as the basis for the construction of the conduction electron wave functions $\psi_{\mathbf{k}_0}(\mathbf{r})$ as well as for the calculation of the core electrons' contributions to the scattering (Weiss, Harvey, and Phillips 1968), see Fig. 3.14. Because of the heavy absorption of X-rays by elements of high atomic number it would be much more difficult to obtain useful information about the density of states from Compton line-shape measurements on metals of high atomic number.

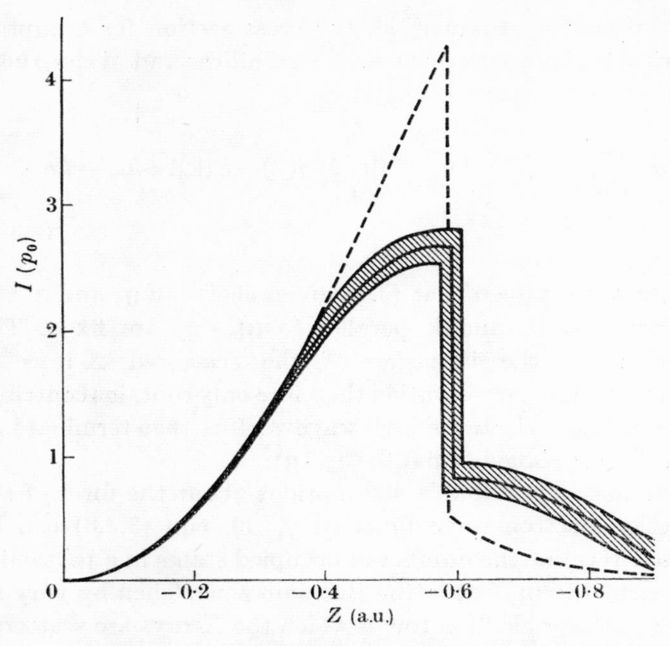

FIG. 3.14. Momentum density for the conduction electrons in lithium metal obtained from Compton scattering line-shape measurements after subtracting the background and the effect of the core electrons determined on the basis of a Hartree–Fock calculation (Phillips and Weiss 1968). The broken curve is after Geldart, Houghton, and Vosko (1964).

3.3.2. The Kohn effect

If there are n atoms in each fundamental unit cell of a crystal there will be $3n$ normal modes with frequencies $\omega_j(\mathbf{k})$ $(j = 1, 2, \ldots, 3n)$ for any given wave vector $\mathbf{k}$.‡ The properties of the wave vectors of the phonons, which are the quanta of energy in the normal modes of vibration of a crystal, are determined by the underlying symmetry of the crystal in exactly the same way as the properties of the wave vectors of the electrons. Bloch's theorem (see Section 1.2) applies to the wave functions that describe the vibrational motions of the positive ions in a metal just as much as it applies to the translational motions of the

‡ Some authors use $\mathbf{q}$ rather than $\mathbf{k}$ for the wave vector of a normal mode of vibration of a crystal.

conduction electrons in the metal. The unit cell of the vector space of **k** is the same Brillouin zone that we have used so far almost exclusively in connection with the motions of the electrons. The frequencies $\omega_j(\mathbf{k})$ of the normal modes of vibration of a crystalline solid will, generally, be smooth and continuous functions of **k** which are called the phonon dispersion relations of the crystal, see Section 6.4.

In a metal the vibrations of the positive ions which make up the lattice are screened by the conduction electrons. It was suggested by Kohn that whenever the wave vector **k** of a phonon is equal to, or differs by a reciprocal lattice vector from, an extremal chord of the Fermi surface parallel to **k**, there should be a kink in the phonon dispersion relations (Kohn 1959*a*; Woll and Kohn 1962). It is the appearance of such anomalies in the phonon dispersion curves which is described as the *Kohn effect* and the anomalies themselves are sometimes called *Kohn anomalies*.

It is possible to demonstrate the plausibility of the existence of the Kohn effect in real metals by considering the conduction electrons as a free-electron gas. The presence of a phonon with wave vector **k** will mean that a Fourier analysis of the charge distribution arising from the ion cores in the metal will contain a term of the form $\rho_0\exp(i\mathbf{k}\,.\,\mathbf{r})$. The effect of this on a gas of free electrons with a spherical Fermi surface of radius k_F is to induce a corresponding periodic charge density in the electron gas of the form $-F(k)\rho_0\exp(i\mathbf{k}\,.\,\mathbf{r})$ where $k = |\mathbf{k}|$ and

$$F(k) = \frac{1}{\pi a_0 k^2}\left[1+\frac{k_F}{k}\left(1-\frac{k^2}{4k_F^2}\right)\ln\left|\frac{k+2k_F}{k-2k_F}\right|\right]. \qquad (3.3.7)$$

This function $F(k)$ is continuous and finite at $k = 2k_F$, but its slope $\mathrm{d}F(k)/\mathrm{d}k$ possesses a logarithmic singularity at $k = 2k_F$. This indicates an abrupt decrease in the ability of the electrons to screen the component of the ionic charge distribution of wave vector **k** once k exceeds $2k_F$. This is because the charge density $\rho_0\exp(i\mathbf{k}\,.\,\mathbf{r})$ associated with a phonon at **k** causes virtual excitations between two electronic states $\mathbf{k}_1$ and $\mathbf{k}_1+\mathbf{k}$ on the Fermi surface without any change of energy so long as $k < 2k_F$.

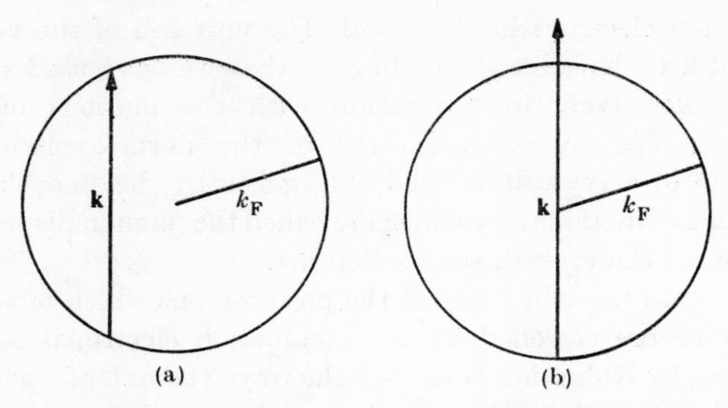

FIG. 3.15. Virtual excitations for (a) $|\mathbf{k}| < 2k_F$ and (b) $|\mathbf{k}| > 2k_F$ (Kohn 1959*a*)

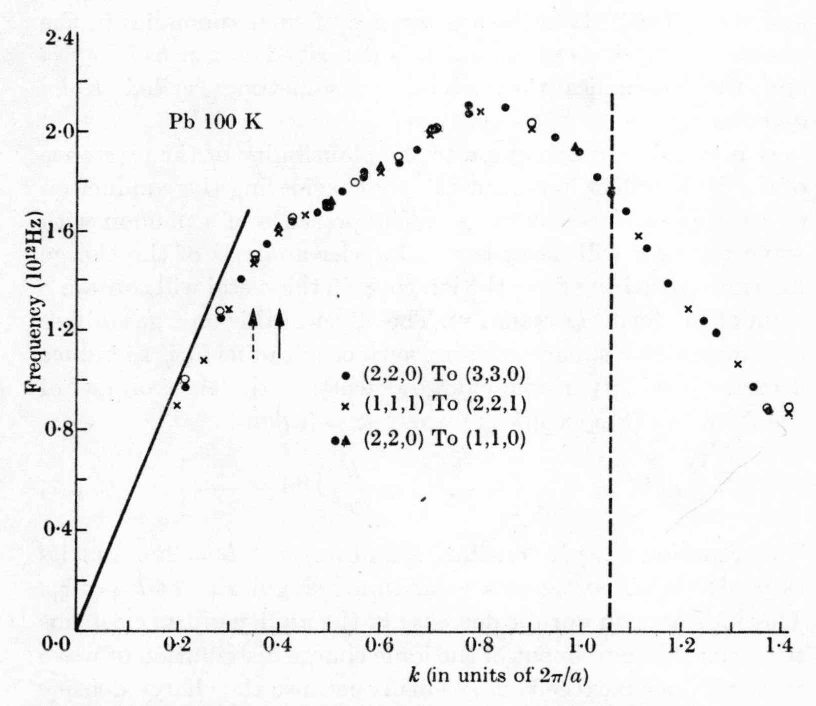

FIG. 3.16. Measured phonon dispersion curve for longitudinal polarization for the [110] direction in lead. A Kohn anomaly is indicated by the solid vertical arrow at $k \simeq 0{\cdot}4(2\pi/a)$. For free electrons it would appear at $0{\cdot}35(2\pi/a)$ as indicated by the dashed arrow (Brockhouse, Rao, and Woods 1961).

However, once k exceeds $2k_F$, or for a more general shape of Fermi surface once $|\mathbf{k}|$ exceeds the extremal chord length of the Fermi surface parallel to $\mathbf{k}$, such excitations are no longer possible, see Fig. 3.15.

The Kohn effect was first observed experimentally in the phonon dispersion curves of Pb determined by inelastic neutron scattering (Brockhouse, Rao, and Woods 1961), see Fig. 3.16, and was subsequently observed in X-ray scattering measurements of the phonon dispersion relations of the same metal (Paskin and Weiss 1962). Kohn anomalies have now also been observed in the phonon dispersion relations of a number of other metals. However, the anomalies in the phonon dispersion relations are usually quite small and, therefore, rather difficult to detect; indeed the anomalies are sometimes so slight that they cannot be seen at all by the naked eye and only become visible in curves of $d\omega_j(\mathbf{k})/d\mathbf{k}$. It should also be possible to observe Kohn anomalies in the spin-wave dispersion relations of a magnetic metal.

3.3.3. *Positron annihilation*

It is well known that a positron is an unstable particle which will sooner or later annihilate with an electron with the production of radiation that will be in the γ-ray part of the electromagnetic spectrum. If the electron that is annihilated was strongly bound to the nucleus of an atom then it is possible for a single γ-ray to be produced when momentum will be conserved by the recoil of the nucleus. However, if the positron is annihilated by a free electron, the emission of only one photon would violate the conservation of momentum and it is only possible to conserve momentum if the annihilation radiation consists of two or more γ-rays. The probability of annihilation with the emission of n photons decreases rapidly with increasing n and nearly all the work which has been performed has been concerned with the two–photon case. The two–photon annihilation is the most probable process and for this case it is only possible to conserve momentum if the two γ-rays are nearly antiparallel. The deviation of the directions of the

γ-rays from exact antiparallelism enables the initial momentum of the annihilating particles to be determined. In a similar way the annihilation of positrons by the conduction electrons in a metal can lead to the production of two nearly antiparallel γ-rays and, if one assumes that the positron was initially thermalized, then the departure from collinearity of the two

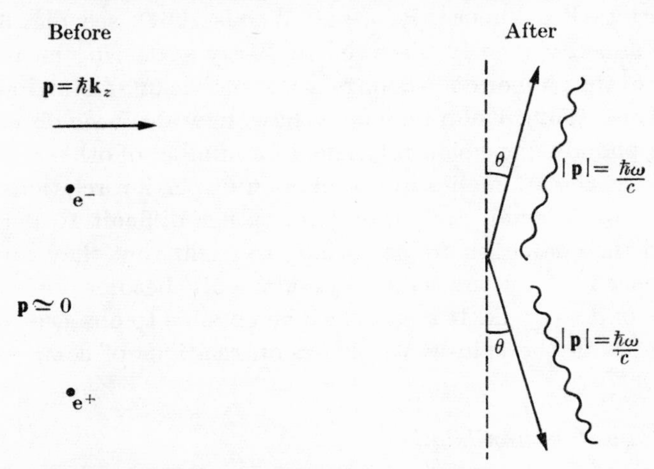

FIG. 3.17. Momentum diagram to illustrate the annihilation of a positron with a conduction electron to produce two γ-rays.

γ-rays enables one to determine the momentum of the conduction electron that was annihilated. The energy of the two γ-rays comes almost entirely from the rest masses of the two particles which are annihilated and therefore $\hbar\omega \doteqdot mc^2$. If the angle 2θ measures the departure of the two γ-rays from collinearity (see Fig. 3.17) then the conservation of momentum requires that

$$\hbar k_0 = 2\,\frac{\hbar\omega}{c}\sin\theta = 2\,\frac{mc^2}{c}\sin\theta \tag{3.3.8}$$

so that

$$k_0 = \frac{2mc\sin\theta}{\hbar}. \tag{3.3.9}$$

Therefore, by observing the coincident arrival of γ-rays in two counters pointing in the directions suggested by Fig. 3.17, one knows that these γ-rays arose from the annihilation of an electron with a component k_0 along the direction bisecting the directions of the γ-rays given by eqn (3.3.9). The count-rate will, therefore, depend on the number of states in a slice of occupied states at $k_0 = 2mc \sin \theta/\hbar$, and also on the transition probabilities of the various electrons in this slice. As we have already noted in several different connections, the determination of the transition probability for some process involving a conduction electron in a metal involves the wave functions of the electrons, which are generally not very well known. In the absence of any more accurate information it has been common to assume that the transition probability for positron annihilation does not vary too drastically with changes in **k**, so that the count-rate as a function of angle can readily be related to the density of states, $g(k)$, as a function of the magnitude of the component of **k** in a particular direction.

The first angular correlation measurements on the two-γ annihilation of positrons by the conduction electrons in a metal were performed by Beringer and Montgomery (1942) on Cu and Pb and since then very many similar experiments have been performed on a large number of different metals. Some results for several metals are shown in Fig. 3.18 and it can be seen that the cut-off is not complete and there is a tail to the curve of count-rate against θ. Although one often obtains values of caliper radii of the Fermi surface which appear to be quite reasonable, the margin of error is quite high. By repeating the experiment for various orientations it should in principle be possible to use the cut-off to determine the caliper radius of the Fermi surface as a function of orientation. Indeed it has been possible to observe anisotropy in some cases, as for example in Fig. 3.19. However, the accuracy of any quantitative deductions is always fairly low and so the observation of cut-offs in the angular correlation curves for two-γ annihilation of positrons is not usually used in experiments on the detailed point by point mapping of the Fermi surface of a metal with a complicated Fermi surface. For such detailed mapping there are other more accurate methods available, see Sections 3.8 and 3.9, and it is

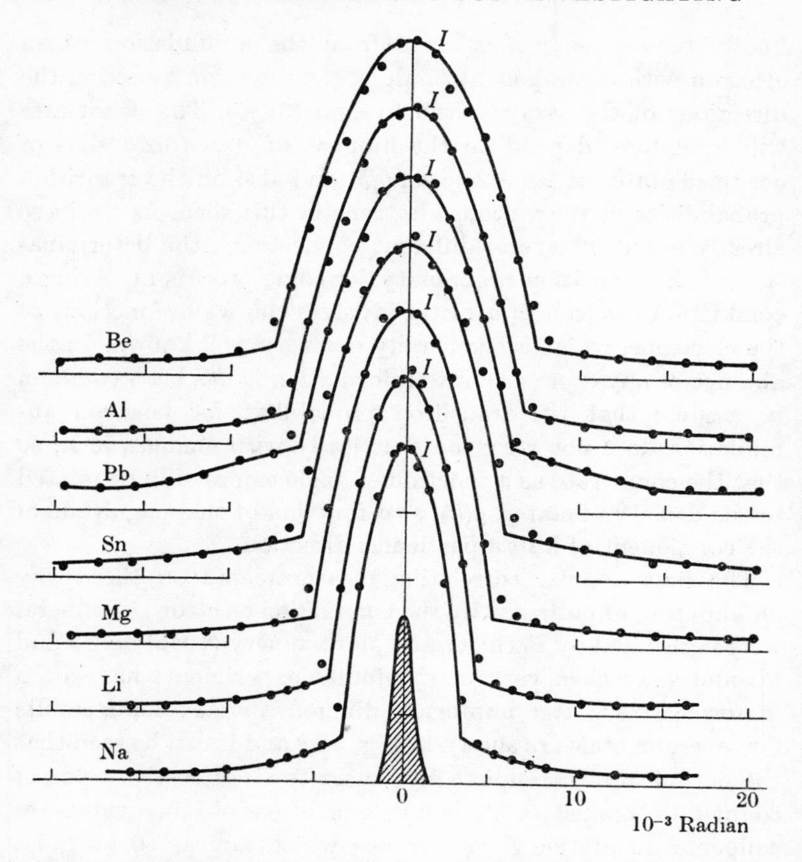

Fig. 3.18. Measured curves of the angular distribution of two-photon positron annihilation radiation from a number of metals. The shaded curve illustrates the resolution (Lang, De Benedetti, and Smoluchowski 1955).

probably more common now to regard the use of positron annihilation as one of the *indirect* methods of Fermiology, in which case a postulated band structure and Fermi surface would be used with the corresponding wave functions $\psi_{\mathbf{k}}(\mathbf{r})$ to calculate the complete shape of the angular correlation curve of the two-γ positron annihilation radiation and compare this with experiment. In general good agreement between theoretical calculations and experimental measurements of the angular correlation curve will only be obtained if allowance is made for

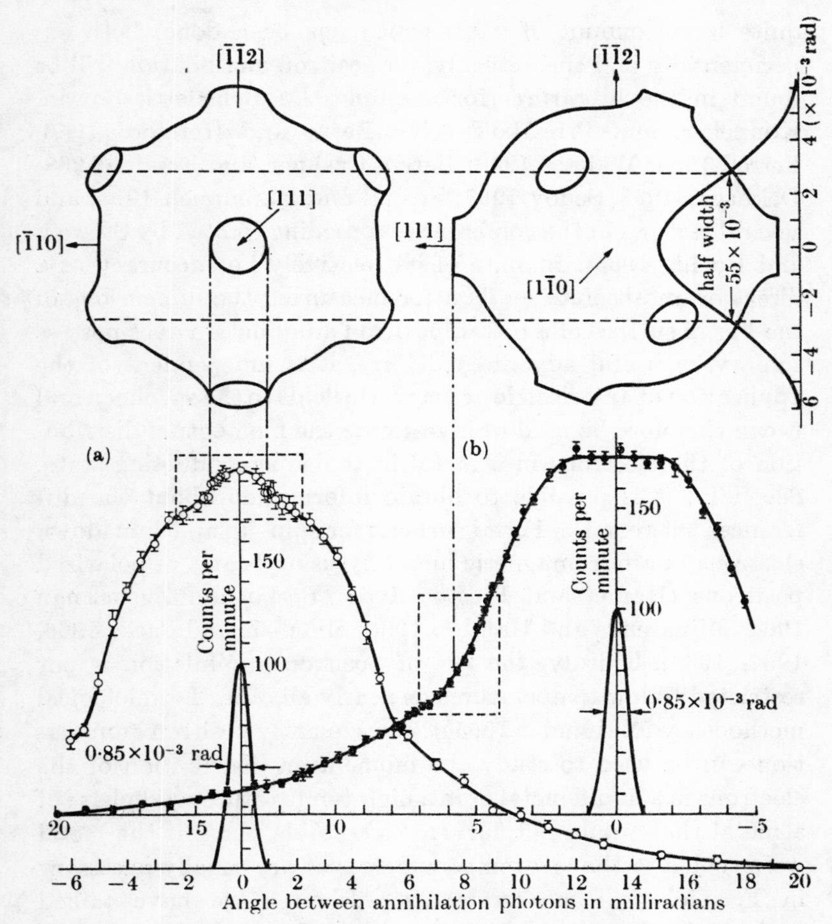

FIG. 3.19. Angular correlation curves for positron annihilation radiation from copper. Schematic diagrams of the side views of the Fermi surface and the vertical and the horizontal resolution curves are shown (Fujiwara and Sueoka 1966).

annihilation by electrons in the ion cores and for the effect of Coulomb interactions between the electrons and positrons (Kahana 1963; Carbotte and Kahana 1965; Majumdar 1965*a*, Carbotte 1966, 1967; Carbotte and Salvadori 1967; Garg and Saraf 1969). The complete theory of the annihilation of positrons by the conduction electrons in a metal contains many refinements that we have not mentioned; several discussions of the quite large amount of work which has been done, both experimentally and theoretically, on positron annihilation will be found in the literature (for example, De Benedetti, Cowan, Konneker, and Primakoff 1950; Berko and Hereford 1956; Ferrell 1956; Wallace 1960; Berko, Cushner, and Erskine 1968; Dekhtyar 1968; Sedov 1968; Stroud and Ehrenreich 1968; and several reviews in the conference proceedings edited by Stewart and Roellig 1965). In spite of its relatively low accuracy as a direct or an absolute method for measuring the dimensions of the Fermi surface of a metal, positron annihilation does possess some very useful advantages. First, it is independent of the application of any electric or magnetic fields to the specimen and it can therefore be used to investigate the momentum distribution of the electrons in a metal in the superconducting state. Secondly, it is possible to obtain information about the differences between the Fermi surfaces for spin–up and spin–down electrons in a ferromagnetic metal by using beams of polarized positrons (Hanna and Preston 1958; Berko and Zuckerman 1964; Mijnarends and Hambro 1964; Mihalisin and Parks 1966, 1967, 1969). Thirdly, the use of positron annihilation is not restricted to low temperatures as nearly all other Fermiological methods are (Majumdar 1965*b*); consequently positron annihilation can be used to study the momentum distribution of the electrons in a liquid metal or in a high-temperature solid phase of a metal that would not survive as a stable phase if the metal were cooled to the low temperatures that are usually necessary in Fermiological measurements. Although we have talked exclusively about two-photon angular correlation experiments to study the momentum distribution of the conduction electrons in a metal using positron annihilation, the same information could be obtained from a study of the line width of the positron annihilation radiation line described in terms of Doppler broadening due to the motions of the electrons (see Hotz, Mathiesen, and Hurley 1968).

3.4. The anomalous skin effect

To observe some property of a conduction electron in a metal it is necessary to do so in a time that is shorter than the relaxation time, τ, and this can be done by using an electromagnetic wave with a frequency that satisfies the condition

$$\nu\tau \gg 1 \tag{3.1.3}$$

which was given in Section 3.1. The longest relaxation time that can reasonably be expected is of the order of 10^{-9} s which places a lower limit of 10^9 Hz on the frequencies that can be used. This places the experimentalist in the microwave region of the electromagnetic spectrum. At frequencies of this order of magnitude the skin effect is already important and this means that, unless thin-film samples are used, a high-frequency alternating electromagnetic field is not able to penetrate into the body of the sample, but is confined to a layer of thickness δ at the surface of the metal. δ is known as the *skin depth* of the metal. As far as high-frequency electromagnetic radiation is concerned this means that the samples commonly used are, to all intents and purposes, semi-infinite. Moreover, misleading results can very easily be obtained if the irregularities of the surface are of dimensions comparable with the skin depth, or if the thermal or mechanical history of the surface differs from that of the bulk of the sample.

The value of δ can be obtained from classical electromagnetic theory by the use of Maxwell's equations. It can be shown (see Bleaney and Bleaney 1965, p. 267) that, if ω is the angular frequency of the oscillating electric field, the electric field at a depth z below the plane surface of a metal decays exponentially according to

$$E_x = (E_x)_0 e^{-z/\delta} e^{j(\omega t - z/\delta)}, \tag{3.4.1}$$

where

$$\delta = (\tfrac{1}{2}\sigma\omega\mu\mu_0)^{-\frac{1}{2}}. \tag{3.4.2}$$

This means that the amplitude of the alternating electric field of the electromagnetic wave decays exponentially in the z direction, and at a depth δ below the surface has fallen to 1/e of its original value. For example, for Cu at 1200 MHz (wavelength = 0·25 m) δ is approximately equal to $1{\cdot}9 \times 10^{-6}$ m. Equation (3.4.2) is based entirely on macroscopic arguments and assumes implicitly that the mean free path, λ, is very small in comparison with δ. The quantity that is physically observable in microwave work is the surface impedance Z, which is defined as the tangential electric field, $(E_x)_0$, at the surface of the metal divided by the current density in the direction of $(E_x)_0$ integrated over z, the depth in the metal. That is

$$Z = (E_x)_0 \Big/ \int_0^\infty j_x \mathrm{d}z. \tag{3.4.3}$$

In optical work the observable quantity is the reflectivity of the metal.

From eqn (3.4.2) we can see that as the frequency ν is increased, in order to satisfy eqn (3.1.3) better, so δ, the classical skin depth, will become smaller. At room temperatures and for most frequencies, δ is much larger than λ; for example, for Cu at room temperature and at a frequency of 1200 MHz (i.e. wavelength $= 0{\cdot}25$ m) a rough estimate gives $\lambda/\delta \simeq 2\times 20^{-2}$ (Pippard 1954). If the frequency is increased to optical frequencies δ becomes very small indeed. Alternatively, if the temperature is lowered, λ will increase. In either case it will eventually happen that the value of δ will become very much smaller than the mean free path λ. In this situation the simple theory ceases to be appropriate and we find what is called the *anomalous skin effect*. This appears to have been first observed by London (1940) who, in his observations of the resistance of Sn at a wavelength of 0·2 m, noticed that the surface resistance at 4 K was several times higher than would have been expected from the insertion of the measured d.c. conductivity into the classical skin effect formula in eqn (3.4.2). London correctly attributed the discrepancy to the long mean free path of the conduction electrons in comparison with δ, although he did not analyze the effect in detail. The behaviour of the surface impedance of Cu as a function of temperature at a frequency of 1200 MHz is shown in Fig. 3.20 (Chambers 1952). The dotted line indicates the behaviour expected using δ from eqn (3.4.2); the departure from this straight line marks the anomalous skin effect regime, and the vertical broken line marks the point at which $\lambda/\delta = 1$. In the case of Fig. 3.21 the value of λ/δ rises to

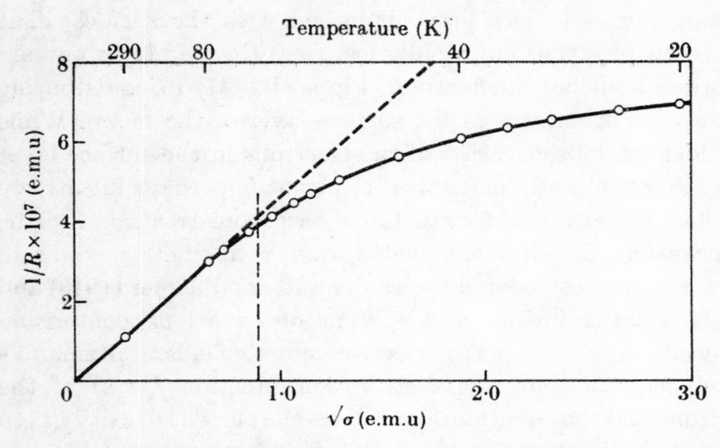

FIG. 3.20. The surface resistance of copper (after Chambers 1952).

about 40. For samples of other metals whose residual resistance is lower, the value of λ/δ can be made much larger.

The main contribution to the current in the skin layer in the anomalous skin effect regime ($\lambda \gg \delta$) comes from those electrons which have velocities nearly parallel to the surface (that is

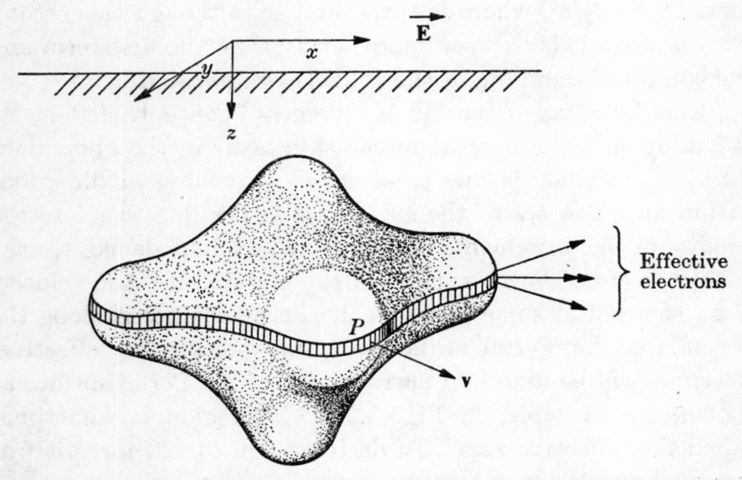

FIG. 3.21. A belt of 'effective' electrons on a Fermi surface (Ziman 1964*a*).

making angles of less than about δ/λ with the surface). The remaining electrons are rapidly lost from the skin layer and are regarded as being 'ineffective' (Pippard 1947) in contributing to the flow of current in the surface layer of the metal. While this idea of the separation of the electrons in the surface layer into 'effective' and 'ineffective' electrons is perhaps intuitively sensible it is also well founded in the rigorous treatment of the propagation of electromagnetic waves in metals. Such a treatment was considered by Reuter and Sondheimer (1948) and involves the solution of a system of equations comprising Maxwell's equations for the electromagnetic fields, Boltzmann's kinetic equation for the distribution function $f_{\mathbf{k}}(\mathbf{r}, t)$ of the electrons and the equation that relates the current density $j(\mathbf{r})$ to this distribution function (for details see Reuter and Sondheimer 1948; Dingle 1953*a*, *b*, *c*; Chambers 1969*a*). As a result of such an analysis it was shown that under anomalous skin effect conditions it is impossible to describe the electromagnetic fields in a metal in terms of a single exponentially damped wave (Pippard, Reuter, and Sondheimer 1948; Pippard 1954) and the precise expression for the field contains two terms. One of these is connected with the effective electrons and decays rapidly as the distance, z, from the surface increases. The second term has the form $(\exp -\xi)/\xi^2$, where $\xi = z/\lambda$, and so, although this second term is small it is damped more slowly than the first term and the combined damping of the fields is complicated.

By making use of the "ineffectiveness" concept, that is, by assuming that the current in the skin layer in the anomalous skin effect regime is due to electrons travelling in directions within an angle δ/λ of the surface, it is possible to relate the results of measurements of the surface impedance to the geometry of the Fermi surface of the metal. Since the velocity of an electron at some point on the Fermi surface is along the normal to the Fermi surface at that point, the 'effective' electrons will be found on narrow belts of the Fermi surface as shown, for example, in Fig. 3.21. This region is sometimes called the 'effective zone'. In deriving eqn (3.4.2) for δ it was assumed that the metal was an isotropic or a cubic medium and that the electrical conductivity, σ, could be represented by a single number. In the anomalous skin effect regime this is no longer the case and the conductivity has to be allowed the freedom to have the number of components allowed to a tensor of rank two, that is, Ohm's law takes the form

$$j_m = \sum_n \sigma_{mn} E_n. \tag{3.4.4}$$

It can then be shown (Pippard 1965, Chapter 5) that the components of the conductivity tensor σ_{mn} can be expressed in terms of integrals of the radius of curvature ρ of the Fermi surface around the belt of effective electrons, or the effective zone; thus, for example,

$$\sigma^{\mathbf{q}}_{xx} \approx \frac{e^2}{4\pi^2\hbar\,|\mathbf{q}|} \oint |\rho_y|\, \mathrm{d}k_y, \tag{3.4.5}$$

where $\mathbf{q}$ is the wave vector of the electromagnetic wave. Then, in principle at least, by measuring the surface impedance Z in the anomalous skin effect regime, and using various trial shapes of the Fermi surface in equations of the form (3.4.5), it is possible to determine the shape of the Fermi surface. The method has one or two disadvantages: to obtain complete information about the shape of the Fermi surface it is necessary to repeat the measurements of the surface impedance using specimens with several different relative orientations of the crystallographic axes and the surface of the specimen, because it is only possible to rotate the $\mathbf{E}$ vector within the plane of the surface of the metal. Moreover, while the anomalous skin effect can be used for metals with a fairly simple Fermi surface, such as the noble metals Cu, Ag, and Au, there is the additional complication that if there are several sheets of the Fermi surface, the measured quantity (the surface impedance) represents the result of a summation over all these pieces of Fermi surface and it is not possible to isolate experimentally the separate contributions. Historically some of the most important early experimental investigations of the Fermi surface of a metal were performed on Cu using the anomalous skin effect in the microwave region of the electromagnetic spectrum (Pippard 1957*b*). However, it is clear that even for a metal with a relatively simple Fermi surface, the analysis of the results is inclined to be tedious. The use of the anomalous skin effect would not have been very suitable for determining the shapes of the complicated Fermi surfaces that are possessed by a large number of the metallic elements (see Chapters 4 and 5). Other techniques are now available which produce results that are easier to analyse and in which it is possible to observe separately the contributions from different sheets of the Fermi surface. The anomalous skin effect is, of course, not confined to the microwave region of the spectrum, but is relevant to waves of higher frequencies as well and in particular in the optical region (see, for example, Dingle 1953*a*,*b*,*c*; Shkliarevskiĭ and Padalka 1959; Lenham and Treherne 1966). However, the task of the extraction of useful Fermi surface dimensions from optical reflectivity measurements is even more formidable than for the case of microwave surface impedance measurements.

In this section we have assumed that the metallic specimen under investigation was a single crystal. If the surface impedance of a polycrystalline specimen of a metal is measured, then it is possible to determine the total area A of the Fermi surface, (Chambers 1952; Fawcett 1961*b*; Pippard 1965).

3.5. The energies of electrons in metals in applied magnetic fields

The path of an electron moving in free space in a magnetic induction **B** is a helix, and the projection of this trajectory onto a plane normal to **B** is a circle. For convenience we can neglect the component of the motion parallel to **B** and loosely describe the orbit as a circle. The force that causes the circular motion is the Lorentz force $e\mathbf{v} \wedge \mathbf{B}$ on the moving electron so that, neglecting the component of the motion parallel to **B**,

$$evB = \frac{mv^2}{r}, \tag{3.5.1}$$

or

$$\omega_c = \frac{eB}{m} \tag{3.5.2}$$

where $B = |\mathbf{B}|$. The frequency ω_c, which is independent of the radius of the orbit, is called the *cyclotron frequency*. In a metal a conduction electron is by no means free since it is acted on by all the ion cores and all the other conduction electrons in the metal; consequently we should expect the orbit of a conduction electron to be quite distorted from the circular shape of the orbit of a free electron. However, it is quite easy, in principle at least, to determine the shapes of these orbits because they can be related to the shape of the Fermi surface, and it is possible to use an equation of the form (3.5.2) where the free-electron mass m is replaced by an effective mass m_c^* which is called the *cyclotron mass*.

In a magnetic field an electron satisfies Newton's second law of motion in the form

$$\hbar\dot{\mathbf{k}} = e\mathbf{v} \wedge \mathbf{B}, \tag{3.5.3}$$

where $\hbar\dot{\mathbf{k}}$ is the rate of change of momentum of the particle. We know that in real space this would mean that the electron would travel in an orbit with shape determined by $\mathbf{v}$ and $\mathbf{B}$. Equation (3.5.3) means that, in $\mathbf{k}$ space, the rate of change of $\mathbf{k}$ is a vector that is normal to $\mathbf{B}$. Therefore, in $\mathbf{k}$ space the electron's wave vector also moves in an orbit that is in a plane normal to $\mathbf{B}$. By differentiating eqn (3.5.3) we find

$$\hbar\ddot{\mathbf{k}} = e\dot{\mathbf{v}} \wedge \mathbf{B}, \tag{3.5.4}$$

so that the acceleration $\ddot{\mathbf{k}}$ which describes the $\mathbf{k}$-space orbit is normal to the acceleration $\dot{\mathbf{v}}$ of the orbit in real space. On the other hand by integrating eqn (3.5.3) we find

$$\mathbf{k} = \frac{e}{\hbar}\,\mathbf{r} \wedge \mathbf{B}. \tag{3.5.5}$$

This means that the orbits in real space and reciprocal space are geometrically similar but the orbit in $\mathbf{k}$ space is rotated through $\pi/2$ with respect to the real-space orbit (see Fig. 3.22) and its radius is $eB/\hbar$ times the radius of the real-space orbit. This means that an electron which had a velocity $\mathbf{v}$ before the magnetic field was applied will move in a spiral of which the

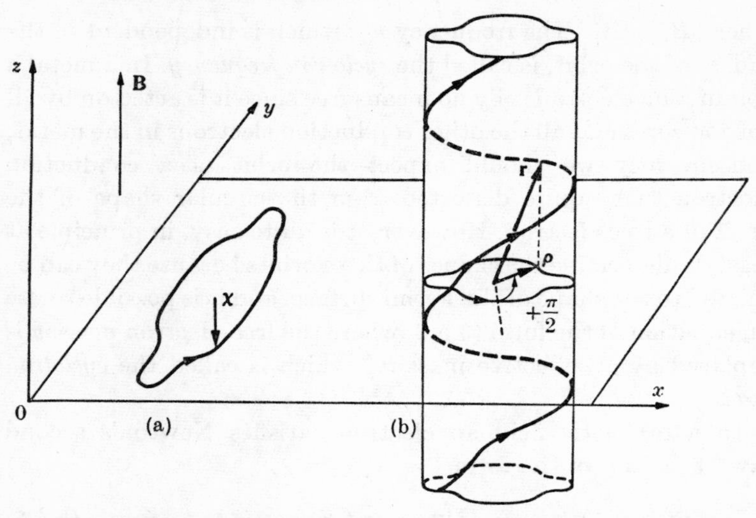

Fig. 3.22 Relation between the orbit of an electron in a magnetic field in (a) **k**-space and (b) real space (After Mercouroff 1967).

projection on a plane normal to **B** is a closed orbit. In **k** space the wave vector of the electron was stationary (if we ignore collisions) but it starts to precess in a similar orbit to the real-space orbit when **B** is applied. For the case of a spherical Fermi surface which arises in the empty lattice (or free-electron) model, see Section 1.5, the cross sections of the constant-energy surfaces are always circles as, indeed, we would expect them to

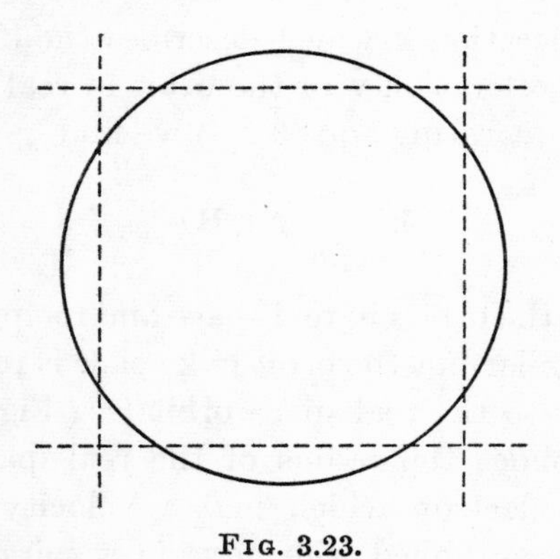

Fig. 3.23.

be from the consideration of electrons in free space in a magnetic field. However, even in the empty lattice model a complication arises if the Fermi surface intersects the boundary of the Brillouin zone. This is illustrated schematically in Fig. 3.23. Since all unit cells of the reciprocal lattice are equivalent, that is

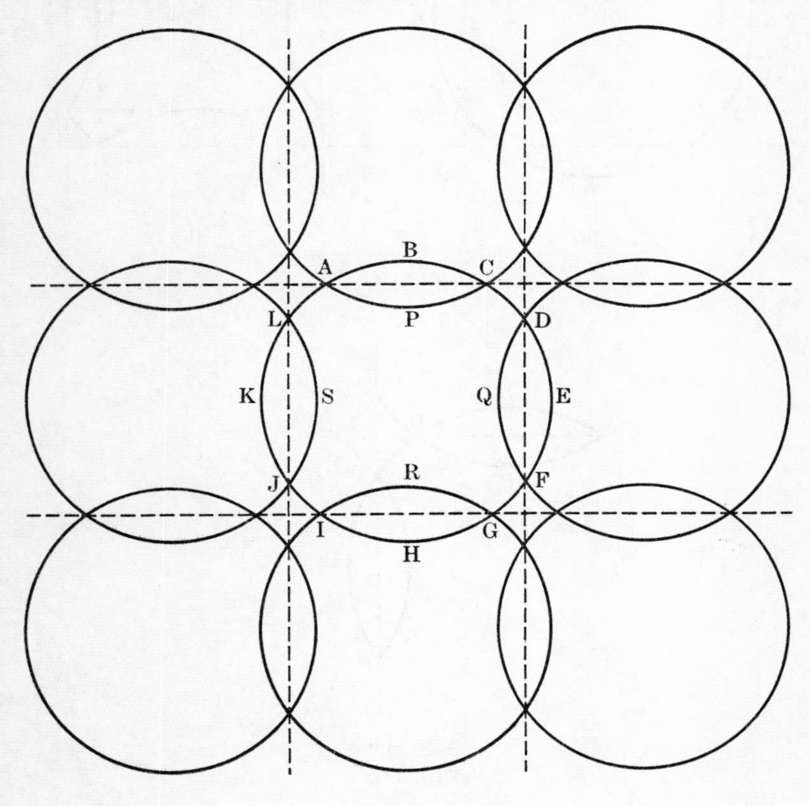

FIG. 3.24

all Brillouin zones are equivalent, the circular cross section of the constant-energy surface can be repeated through the plane of the diagram in the so-called 'repeated zone scheme', see Fig. 3.24. Any one Brillouin zone in this repeated zone scheme is physically indistinguishable from any of the other Brillouin zones. Therefore, while we certainly can regard an electron as

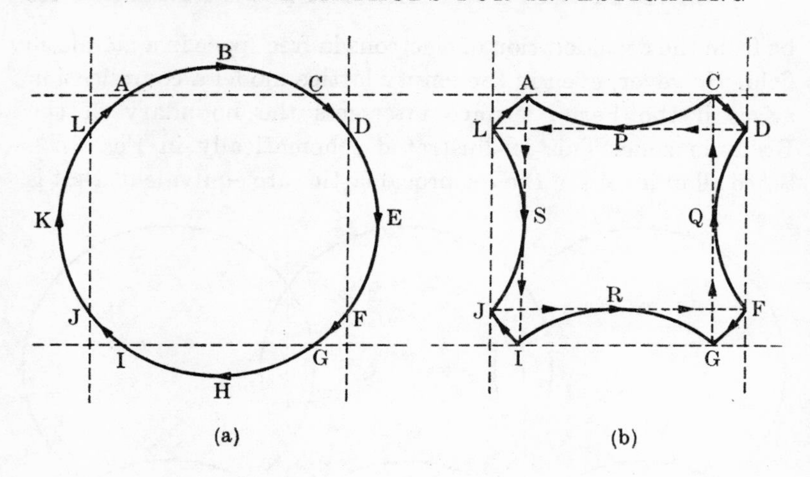

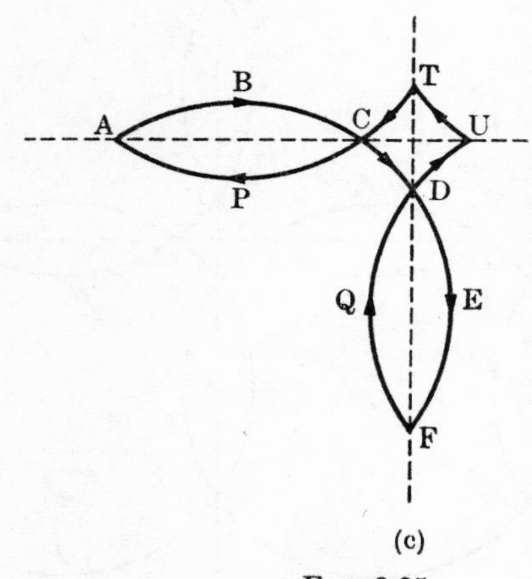

FIG. 3.25

traversing the circular path ABCDEFGHIJKL as in Fig. 3.25(a), there are a number of other ways of describing this orbit. For example, we may not wish to use the extended zone scheme and may prefer to confine ourselves to a single Brillouin zone; in this case the four arcs ABC, DEF, GHI, and JKL are equivalent to IRG, LSJ, CPA, and FGD, respectively, and the electron now follows the path IRG/CD/LSJ/FG/CPA/IJ/FQD/LA/I which is illustrated in Fig. 3.25(b). The parts of the path indicated by broken lines join equivalent points in reciprocal space and therefore are traversed instantaneously. There are many other ways of drawing the orbit of which the orbit ABCDEF/TCPA/UT/FQDU/A shown in Fig. 3.25(c) is only one example.

In the development of Brillouin zone theory in Sections 1.2–4 we said that the values of $\hbar\mathbf{k}$, the momentum of an electron in a crystal, were quantised and that the allowed values of $\mathbf{k}$ were specified by

$$\mathbf{k} = m_1\mathbf{g}_1/N_1 + m_2\mathbf{g}_2/N_2 + m_3\mathbf{g}_3/N_3. \tag{1.2.12}$$

These allowed wave vectors terminate at a set of points which form a regular lattice. This was a result of Bloch's theorem which, in turn, was relevant because the potential $V(\mathbf{r})$ was a periodic function in three directions. When a magnetic field is applied, these quantization conditions cannot necessarily be expected to be still valid. If we take the example of a simple cubic lattice and $\mathbf{B}$ is applied in the z direction, then $V(\mathbf{r})$ is still periodic in z and Bloch's theorem will still apply to the k_z component of $\mathbf{k}$. However, $\mathbf{k}$ now describes an orbit which, on the free-electron model, is a circle of radius $eBr/\hbar$, where r is the radius of the real-space orbit. The angular velocity of the electron in its real-space orbit is governed by

$$Bev = \frac{mv^2}{r}, \tag{3.5.6}$$

that is,

$$\omega_c = \frac{eB}{m}, \tag{3.5.7}$$

where ω_c is just the cyclotron frequency. If we make use of the correspondence principle then these orbits must satisfy

$$\oint \mathbf{p}\,.\,\mathrm{d}\mathbf{r} = (n+\tfrac{1}{2})2\pi\hbar \tag{3.5.8}$$

(Onsager 1952). The areas of the orbits in $\mathbf{k}$ space are then given by

$$\oint \mathbf{k}\,.\,\mathrm{d}\mathbf{k} = \frac{1}{\hbar}\,.\,\frac{eB}{\hbar}\,(n+\tfrac{1}{2})2\pi\hbar, \tag{3.5.9}$$

that is,

$$\mathscr{A}_n = \frac{2\pi eB}{\hbar}\,(n+\tfrac{1}{2}). \tag{3.5.10}$$

So the areas of the orbits in **k** space, and therefore also in real space, are quantised. Therefore instead of the allowed components k_x and k_y of the wave vector **k** in a simple cubic lattice taking the stationary allowed values specified by the square lattice of points in Fig. 3.26(a), they move round on one or other of a set of quantised circular orbits shown in Fig. 3.26(b). Each of these circular orbits has a considerable degree of degeneracy and indeed it is possible to show (Ziman 1964*a*, p. 273) that the

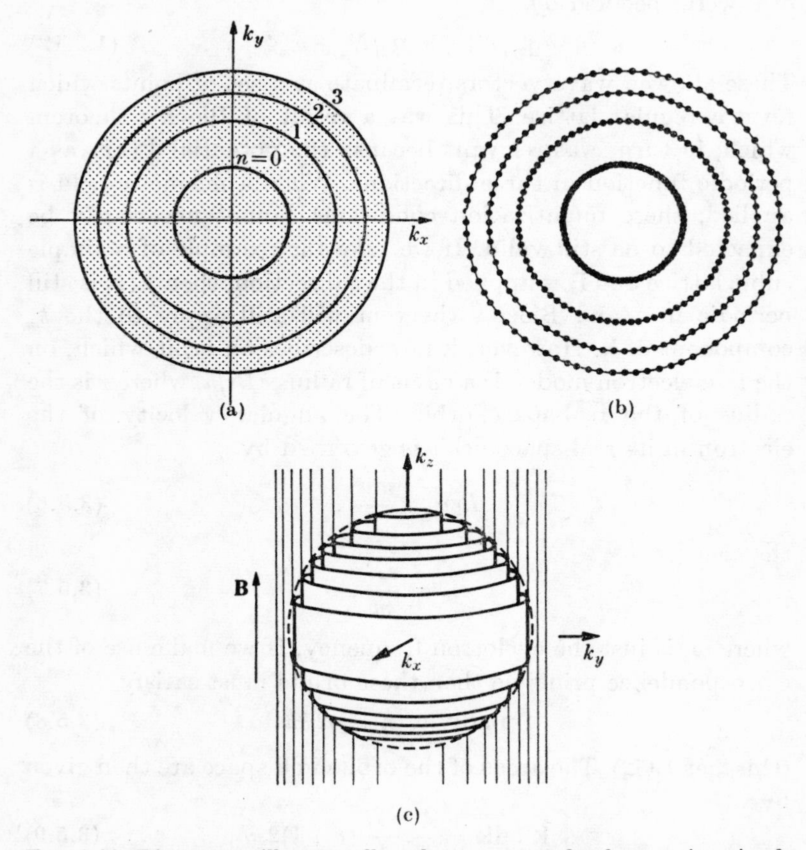

FIG. 3.26. Diagrams to illustrate allowed wave vectors for electrons in a simple cubic metal: (a) allowed values of k_x and k_y with no magnetic field, (b) allowed values of k_x and k_y with applied magnetic field parallel to k_z, and (c) coaxial cylinders of allowed states with magnetic field parallel to k_z (Chambers 1956*b*, Ziman 1964*a*).

degeneracy of the circular orbits is just the number of states that, in Fig. 3.26(a), were previously associated with the area between two successive orbits. In other words, the average number of states per unit area of the $k_x k_y$ plane is unchanged by the addition of the magnetic field. When **B** is applied, each of the states in Fig. 3.26(a) moves to the nearest available circular orbit states and then moves round in this orbit at an angular frequency of ω_c. If we include the third direction k_z, which is quantised in units of $\mathbf{g}_3/N_3$ we see that the allowed states then consist of circular orbits in a plane parallel to the $k_x k_y$ plane, but at different heights $m_3\mathbf{g}_3/N_3$ above or below it. That is, these orbits lie on a set of coaxial cylinders whose common axis is the k_z axis, see Fig. 3.26(c); adjacent orbits on any one of these cylinders are separated by $(\mathbf{g}_3/N_3)$. The energy levels of a conduction electron in a metal in the presence of a magnetic field which have 'condensed' in this way are often called the *Landau levels* (Landau 1930).

There is a considerable lack of theoretical rigour in the approach that we have adopted in this section. Strictly it is necessary to construct the proper Hamiltonian operator for the electrons in a metal in the presence of a magnetic field and then to attempt to determine the eigenfunctions and eigenvalues of this Hamiltonian, which is quite a difficult task (Luttinger 1951; Adams 1952*a*,*b*; Kohn 1959*b*; Blount 1962*a*; Roth 1962; Wannier 1962; Chambers 1966; Butler and Brown 1968; Zak 1968; Langbein 1969). In constructing such a Hamiltonian the momentum $\mathbf{p}$ would be replaced by $(\mathbf{p}+e\mathbf{A}(\mathbf{r}))$ where $\mathbf{A}(\mathbf{r})$ is the magnetic vector potential at $\mathbf{r}$. However, for the purposes of this chapter the pictorial approach that we have adopted in this section is adequate.

There is another point which is of considerable importance in connection with the behaviour of the conduction electrons in a metal in a magnetic field; this is the phenomenon of *magnetic breakdown* (Cohen and Falicov 1961; Stark and Falicov 1967). For the consideration of magnetic breakdown it is convenient to use the extended zone scheme and we suppose, for example, that we have a hypothetical Fermi surface with the cross section shown in Fig. 3.27; this is similar to the cross section shown in Fig. 3.24 except that we have removed the points of contact and produced a set of isolated surfaces. The small cross sections at X represent holes in band 1 and the elliptical cross sections represent electrons in band 2. In a magnetic field we should therefore expect to find two kinds of closed orbits possible for the conduction electrons, namely elliptical orbits of the type

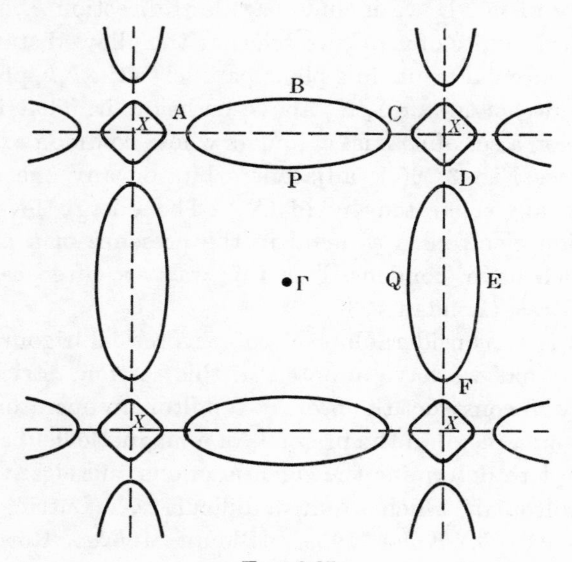

FIG. 3.27.

ABCP or DEFQ, and roughly circular orbits around the other type of cross section centred at the corners X of the Brillouin zone. This is quite different from the situation illustrated in Fig. 3.24 where an extremely varied selection of extended or open orbits is possible. However, it may be possible, if the magnetic field is sufficiently large, for the electrons to jump across the spaces separating adjacent elliptical and circular orbits. If this occurs the connectivity shown in Fig. 3.24 will be obtained and a variety of open and extended orbits, of which some examples are shown in Fig. 3.25, would be possible. Thus we have a significant qualitative difference between the behaviour of the conduction electrons in a low magnetic field, when only a small finite number of distinct closed orbits arise, and in a high magnetic field, when a variety of complicated open and extended orbits are possible. Several examples of possible modifications of electron and hole orbits as a result of magnetic breakdown are illustrated in Fig. 3.28.

The question arises as to the magnitude of $\mathbf{B}_0$, the critical field separating these two types of behaviour for a given metal, and whether this value of $\mathbf{B}_0$ is likely to be obtained in practice for that metal. The two types of orbit shown in Fig. 3.27 are connected with different bands and we suppose that E_g is a typical value of the gap between band 1 and band 2 in the region of $\mathbf{k}$ space near to A or C. Then if an electron has a cyclotron frequency $\omega_c\,(=eB/m_c^*)$ the critical field for the occurrence of magnetic breakdown will be given (Blount 1962*a*) by

$$E_F\hbar\omega_c \sim E_g^2. \qquad (3.5.11)$$

The value of E_g that would satisfy this condition is of the order of $10^{-8}\,|\mathbf{B}_0|$ eV, where $\mathbf{B}_0$ is in gauss, whereas typical gaps in ordinary band structures are seldom as small as 0·1 eV. The observation of magnetic breakdown would therefore usually require the use of impossibly large magnetic fields. However, there are certain metals in which very small gaps exist, usually as a result of spin–orbit coupling but also possibly as a result of the existence of points of accidental degeneracy near to E_F. One of the first examples of such a metal to be discussed extensively was Mg, but there are several other metals in which magnetic breakdown is also of importance. The phenomenon of magnetic breakdown is particularly important in h.c.p. metals because the introduction of spin–orbit coupling lifts the double-degeneracy that would otherwise be present all over the large hexagonal face at the top of the Brillouin zone (see p. 142). The possible existence of magnetic breakdown in a metal is important if one is seeking to determine cross sections of the Fermi surface of a metal by some method involving the use of a magnetic field. One has to be able to determine whether any given orbit is directly related to the Fermi surface or whether it arises as a

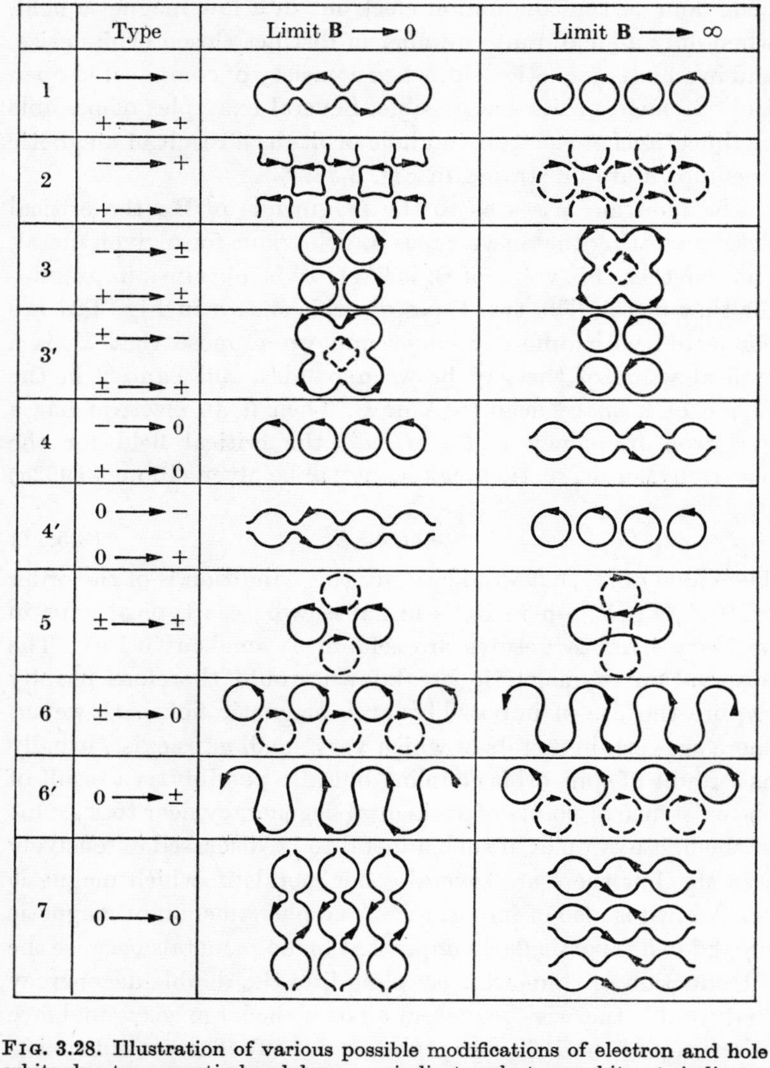

	Type	Limit $\mathbf{B} \to 0$	Limit $\mathbf{B} \to \infty$
1	$- \to -$ $+ \to +$		
2	$- \to +$ $+ \to -$		
3	$- \to \pm$ $+ \to \pm$		
3′	$\pm \to -$ $\pm \to +$		
4	$- \to 0$ $+ \to 0$		
4′	$0 \to -$ $0 \to +$		
5	$\pm \to \pm$		
6	$\pm \to 0$		
6′	$0 \to \pm$		
7	$0 \to 0$		

FIG. 3.28. Illustration of various possible modifications of electron and hole orbits due to magnetic breakdown. − indicates electron orbits, + indicates hole orbits, and 0 indicates open orbits (Mercouroff 1967).

result of magnetic breakdown. Also, if one is using the known Fermi surface of a given metal in order to explain some property involving the behaviour of the conduction electrons in a magnetic field, one has to remember to take into account the possibility of magnetic breakdown if different sheets of the Fermi surface are close together in **k** space.

While on the subject of the behaviour of the conduction electrons in a metal in the presence of a magnetic field, we would just mention the observation of electron spin resonance in metals, although such experiments give only very indirect information about the band structure and Fermi surface of a metal. Not unnaturally the results of conduction-electron spin resonance experiments are interpreted in terms of a g-factor and for simple metals we should expect the g-factor to be close to that of a free electron. Thus it is the case for the alkali metals that the 'g shift' δg, the departure from the free-electron g-factor, has such small values that for a long time even the sign of δg was in doubt for these metals. The smallest value appears to be that of Li for which $\delta g = (-6{\cdot}1 \pm 0{\cdot}2) \times 10^{-5}$ (Van der Ven 1968), while the value for Cs is $\delta g = (1{\cdot}1 \pm 0{\cdot}2) \times 10^{-2}$ (Schultz and Shanabarger 1966). Similarly small values of δg have been found for Be and Mg (Cousins and Dupree 1965; Orchard-Webb and Cousins 1968). However, the small values of the g-shift for these examples should not be taken to imply that δg is small for all metals. Indeed there are some cases in which the g-factor for conduction electrons is enormous. For example in Bi g-factors as large as 200 were predicted for certain orientations (Cohen and Blount 1960) and have now been confirmed both directly in spin resonance experiments and also from the spin-splitting of magnetothermal oscillations (for references see Cracknell 1969c p. 784). Such very large values of the g-value occur in metals in which there are sheets of the Fermi surface enclosing very small pockets of carriers.

3.6. Cyclotron resonance

In eqn (3.5.2) we gave the well-known expression for the cyclotron frequency, ω_c, for a free electron in an applied magnetic field. We have seen in Section 3.5 that in a real metal the shapes of the orbits of the electrons can be quite substantially distorted from circles. It is possible to determine an expression for the angular frequency at which the electron travels round this distorted orbit and we can still regard this as being the cyclotron frequency, ω_c, for this particular orbit. From eqn (3.5.3)

$$\hbar \frac{\mathrm{d}k}{\mathrm{d}t} = ev_{\perp}B, \tag{3.6.1}$$

where $v_{\perp}$ is the component of $\mathbf{v}$ normal to $\mathbf{B}$ and $B = |\mathbf{B}|$. The period T $(= 2\pi/\omega_c)$ for this orbit can then be found by integrating eqn (3.6.1) so that

$$T = \frac{2\pi}{\omega_c} = \int_0^T \mathrm{d}t = \frac{\hbar}{eB} \oint \frac{\mathrm{d}k}{v_{\perp}}, \tag{3.6.2}$$

or

$$\omega_c = \frac{2\pi eB}{\hbar} \bigg/ \oint \frac{\mathrm{d}k}{v_{\perp}}. \tag{3.6.3}$$

It would then seem not unreasonable to expect a resonant absorption of energy to occur if an oscillating electromagnetic field with a frequency of ω_c is applied to the specimen of the metal. This phenomenon would be called *cyclotron resonance*. For free electrons the integral on the right-hand side of eqn (3.6.3) can easily be evaluated analytically and must lead to the value of ω_c given in eqn (3.5.2). For a real metal, however, it is not easy to determine $v_{\perp}$ and therefore the integral in the expression for ω_c in eqn (3.6.3) is not easy to evaluate. It is then customary, by analogy with eqn (3.5.2), to write

$$\omega_c = \frac{eB}{m_c^*} \tag{3.6.4}$$

where m_c is then defined to be the *cyclotron mass*. It is not unreasonable to expect that m_c^* will be quite close to m for a metal such as Na for which the Fermi surface is very close to the free-electron Fermi surface and does not intersect the Brillouin zone boundary. However, where these conditions are not satisfied, it is possible for the value of m_c^* to depart quite considerably from m. The cyclotron mass m_c^* is one example of an effective mass but it should be clear from eqns (3.6.3) and (3.6.4) that it is a property of the complete orbit and not of a particular point in $\mathbf{k}$ space. It is not the same as the thermal effective mass or the dynamical effective mass of an electron in a solid which was mentioned in Section 1.3 and which is a tensor of rank two defined by

$$\frac{1}{m_{ij}(\mathbf{k})} = \frac{1}{\hbar^2} \frac{\partial^2 E_p(\mathbf{k})}{\partial k_i \, \partial k_j} . \tag{1.3.14}$$

It is possible to obtain an analytical expression for m_c, in terms of the geometrical details of the orbit, by manipulation of eqns (3.6.3) and (3.6.4),

$$m_c^* = \frac{\hbar^2}{2\pi} \frac{\partial \mathscr{A}}{\partial E} \tag{3.6.5}$$

(see Ziman 1964*a*, p. 252) where $\mathscr{A}$ is the area of the cross section of the Fermi surface corresponding to the orbit of the electron in real space.

We have not, so far, mentioned any of the experimental difficulties involved in observing cyclotron resonance in metals. The first condition, which is not peculiar to metals, is that the electron must be able to travel a reasonable distance along the orbit before it is scattered; this means that the scattering frequency $1/\tau$ must be much lower than the cyclotron frequency, so that

$$\omega_c \tau \gg 2\pi. \tag{3.6.6}$$

τ is fixed by the temperature for the intrinsic scattering, and by the purity of the specimen for the impurity scattering. For the usual values of τ that occur, the frequencies that must be used to satisfy the condition $\omega_c \tau \gg 2\pi$ are very high so that the skin depth is very small. Indeed the frequencies required are so high that the skin depth is much smaller than the mean free path, λ, of the electrons in the metal. We are therefore in the extreme anomalous regime mentioned already in Section 3.4. Also the skin depth is very much smaller than the radius of the orbit of an electron so that, for an arbitrary orientation of $\mathbf{B}$ relative to the surface of the metal, the oscillating electromagnetic field will be ineffective over most of the orbit of an electron. One might hope to be able to avoid this difficulty by arranging to have the magnetic field **B** normal to the surface of the metal; however, in this orientation the resonance is very broad and difficult to detect (see Ziman 1964*a*).

Although the direct observation of straightforward cyclotron resonance presents considerable experimental difficulties there is a slightly different kind of cyclotron resonance which is much easier to observe. Let us consider the case when **B** is tangential to the surface of a specimen of a metal. If the mean free path of

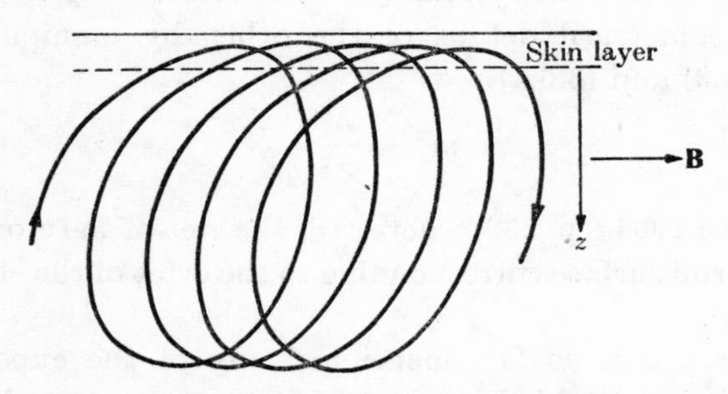

FIG. 3.29. Electron orbit giving rise to cyclotron resonance (Pippard 1960*a*).

the electron is so large that the electron travels round the orbit many times before being scattered, it can absorb a little energy each time that it passes through the region near the surface of the metal where the oscillating electromagnetic field is effective, as shown in Fig. 3.29. It is not necessary that the frequency of the oscillating electromagnetic field should be equal to the cyclotron frequency, but only that the phase of the oscillating field should be the same each time that the electron returns to the layer near the surface of the metal. This condition will be satisfied by any angular frequency ω for which

$$\omega = n\omega_c. \tag{3.6.7}$$

This phenomenon, which then occurs not only at the cyclotron frequency but at multiples thereof, is called *Azbel'–Kaner cyclotron resonance* (Fawcett 1956; Azbel' and Kaner 1957, 1958). Although the condition given in eqn (3.6.7) for the observation of Azbel'–Kaner cyclotron resonance can be obtained by simple arguments, the full mathematical treatment is very complicated. As in the case of the anomalous skin effect described in Section 3.4, it is also necessary here to solve a complicated set of equations including Maxwell's equations for the electromagnetic fields, Boltzmann's equation, and the equation relating the current density $\mathbf{j}(\mathbf{r})$ to the distribution function obtained from the solution of Boltzmann's equation. In this case, however, the situation is slightly more complicated because of the presence of the steady magnetic field. The details are discussed, for example, by Azbel' and Kaner (1956, 1958), Pippard (1960*a*, 1965), and Chambers (1969*a*). It is possible to adapt the 'ineffectiveness' concept to this situation to show that only relatively few of the electrons actually contribute to the resonance (see, for example, Heine 1957*d*; Chambers 1965, 1969*a*).

Equation (3.6.7) can be written as

$$\omega = \frac{neB}{m_c^*} \tag{3.6.8}$$

or

$$\frac{1}{B} = \frac{ne}{\omega m_c^*}. \tag{3.6.9}$$

This means that the surface impedance of the metal is periodic in $1/B$ and from the period it is possible to determine m_c^*, the cyclotron mass, and therefore $(\partial\mathscr{A}/\partial E)$. It is only the orbits that correspond to extremal values of $(\partial\mathscr{A}/\partial E)$ which contribute appreciably to the resonance. An example of these oscillations is illustrated in Fig. 3.30 for metallic Cu. Experimentally the situation is quite similar to that used in anomalous skin effect measurements (see Section 3.4). That is, one has to measure the surface impedance of a specimen of metal at microwave frequencies, but there is now the difference that there is also a steady magnetic field present parallel to the surface. The information that is extracted about the shape of the Fermi surface is quite different; from the measurements of the surface impedance Z in the absence of a magnetic field it was possible to obtain a number that was an integral involving the curvature of certain regions of the Fermi surface, whereas from the periodicity of Z as a function of $1/B$ in Azbel'–Kaner cyclotron resonance measurements it is possible to determine m_c^* and therefore $\partial\mathscr{A}/\partial E$. For a metal with a complicated Fermi

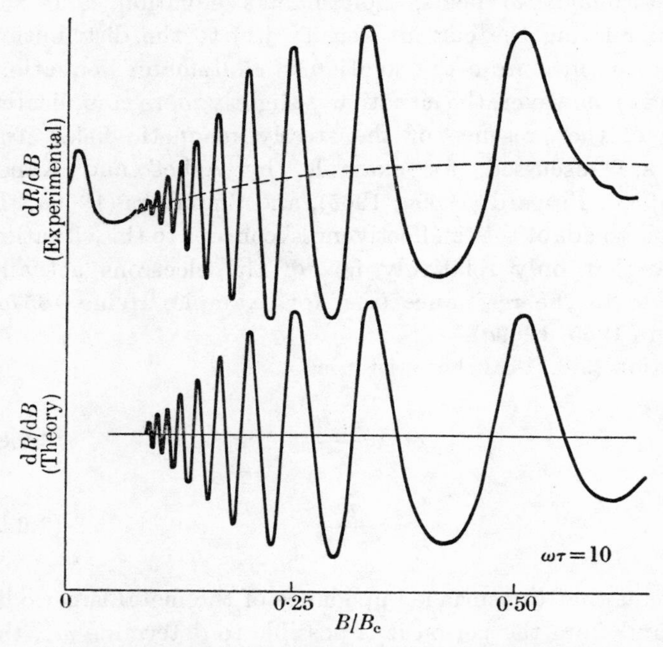

FIG. 3.30. Comparison between a typical cyclotron resonance absorption derivative trace (for Cu) and the theoretical prediction of Azbel' and Kaner for $\omega\tau = 10$ (Kip, Langenberg, and Moore 1961).

surface there may be several different extremal orbits that can lead to Azbel'–Kaner cyclotron resonance. In this situation it may be difficult to disentangle the various values of ω_c from the experimental results; this is especially difficult because, for the values of τ in samples that are usually available, n is restricted to quite small numbers (see below). Because different extremal orbits may behave in different ways if the field **B** makes some small angle, ψ, with the surface of the metal, it is sometimes possible to use tipped fields to distinguish between resonances from different orbits. A number of experimental investigations of Azbel'–Kaner cyclotron resonance in tipped fields have been performed (Grimes and Kip 1963; Grimes, Kip, Spong, Stradling, and Pincus 1963; Koch, Stradling, and Kip 1964; Spong and Kip 1965; Smith 1967).

We have mentioned the existence of the skin depth as a difficulty in the way of observing 'ordinary' cyclotron resonance at $\omega = \omega_c$ in metals. One way to overcome this difficulty is with *acoustic cyclotron resonance*, using ultrasonic rather than electromagnetic waves. The experimental difficulties associated with observing cyclotron resonance using ultrasonic waves have been concerned primarily with satisfying the condition $\omega_c\tau \gg 2\pi$ (see Table 3.1). τ is, of course, governed by the purity of the sample so that, as usual, one requires crystals of the highest possible purity. For the best values of τ commonly obtainable this means that frequencies must be used that, by ultrasonic standards at least, are very high indeed. The phenomenon of ultrasonic cyclotron resonance was observed in Ga by Roberts (1961) at a frequency of 115 MHz. However, what we have said so far in this section should not be taken to imply that it is impossible for any electromagnetic wave to penetrate into a metal beyond the skin depth. While this would, of course, be the case at ordinary frequencies in the absence of an external magnetic field, it is possible to show that when there is an external magnetic field present, the propagation of certain kinds of electromagnetic waves is possible within the metal. Such waves may be either circularly polarized waves, called *helicons*, or plane polarized waves, called *Alfvén waves* (for further discussion see Section 6.2).

Let us suppose that a circularly polarized wave, with angular frequency ω and wave vector $\mathbf{k}$ propagates in a metal parallel to the direction of a stationary applied magnetic field. This wave may be either a helicon or a circularly polarized ultrasonic wave. Then, provided that B $(= |\mathbf{B}|)$ is not too large, there will be some electrons with component v_B of velocity parallel to $\mathbf{B}$ such that their cyclotron frequency is equal to the frequency of the alternating field which they experience as they move through the circularly polarized wave and therefore these electrons absorb energy from the wave. Because of the Doppler effect, the angular frequency of the wave will appear to the electrons to be

$$\omega' = \omega \pm v_{\mathrm{B}} \, |\, \mathbf{k} \,| \tag{3.6.10}$$

rather than ω. This absorption is referred to as *Doppler-shifted cyclotron resonance*, although this term is sometimes used only for the case of helicons rather than for the case of ultrasonic waves as well. The electrons that are involved in the absorption satisfy

$$\omega_{\mathrm{c}} = v_{\mathrm{B}} \, |\, \mathbf{k} \,| \,, \tag{3.6.11}$$

so that if B is varied it will be different groups of electrons that will be responsible for causing the absorption for different values of B. For large values of B there will be no electrons that have high enough velocities along $\mathbf{B}$ to cause the absorption so that, at some critical value, B_{c}, of the field there will be an absorption edge which is called the *Kjeldaas edge* (Kjeldaas 1959; Stern 1963). At this absorption edge the values of v_{B} and ω_{c} that satisfy eqn (3.6.11) will refer to some limiting point on the Fermi surface. By using the measured values of $\mathbf{k}$ and the critical field B_{c} and by performing some manipulation of eqn (3.6.11), it is possible to determine the Gaussian curvature of the Fermi surface at this limiting point (Stern 1963). For fields lower than the critical field B_{c}, the attenuation, in the ultrasonic case, will exhibit magnetoacoustic geometric oscillations (Eckstein 1966)—see Section 3.9.3. In the ultrasonic case the practical use of the Kjeldaas edge is somewhat restricted by the fact that it is only possible to produce circularly polarized ultrasonic waves for propagation along a few high-symmetry directions. This difficulty does not apply to helicons which propagate along the direction of $\mathbf{B}$ which can be arranged to lie along any given direction in the metal. In addition to the ability to propagate in any given direction rather than only along certain high-symmetry directions, helicons also have the advantage that in practice their absorption edge is often more clearly defined than in the ultrasonic case.

The condition for the observation of cyclotron resonance is

$\omega_c\tau \gg 2\pi$ (see Table 3.1). The number of oscillations that can be observed in practice is limited, as one might expect, by the mean free path λ. The resonances are usually observed by varying **B** and keeping ω, the microwave frequency, fixed. The resonances are all on the low-field side of the field corresponding to the conventional cyclotron resonance ($n = 1$). As the magnitude of **B** is reduced, so the cross sectional area of the helix shown in Fig. 3.29 increases and, since λ is fixed, this means that the electron performs fewer revolutions before it is scattered. The amplitude of the resonance signal therefore decreases as n increases (corresponding to $|\mathbf{B}|$ decreasing), until it disappears altogether when the electron does not even complete one revolution, that is, when $\omega_c\tau \sim 2\pi$, or $\omega\tau/n \sim 2\pi$ or

$$2\pi n \sim \omega\tau. \qquad (3.6.12)$$

The number of oscillations that can be observed is therefore of the order of $\omega\tau$. For typical microwave frequencies that are used and typical sample purities that are available, the number of oscillations that can be observed is usually of the order of 20 or thereabouts. Therefore, in addition to using the period in $1/B$ of the oscillations of Z to determine the cyclotron frequency, it is also possible to use the envelope of the decay of the oscillations with decreasing B to determine the relaxation time τ and hence the mean free path λ. If $\omega_c\tau$ is quite large, so that the decay of the oscillations is not appreciable, it is still possible to determine τ from the line-width of the resonance. The expressions describing the decay and the line-width in terms of τ must, of course, be obtained from the proper mathematical analysis of Azbel'–Kaner cyclotron resonance which we have not attempted to describe. The value of τ or λ measured in this way is an average value over the 'orbit' in **k**-space corresponding to the orbit in real space that is responsible for causing the resonance. Even though it is not possible to determine $\tau_{\mathbf{k}}$ the relaxation time, or $\lambda_{\mathbf{k}}$ the mean free path for an individual state with wave vector **k**, this is substantially better than some other methods which only give a value of τ or λ that is an average over the whole of the Fermi surface (see the beginning of Section 3.9.1).

3.7. Galvanomagnetic and related properties

In the present section we shall consider the ways in which the effect of a magnetic field on some of the transport properties, in particular the electrical resistivity, of a metal can be used to obtain some information about the shape of the Fermi surface of that metal. For the moment we shall only be concerned with the behaviour of the electrons in a metal in relatively small magnetic fields, by which we mean that the cyclotron frequency ω_c $(= eB/m_c^*)$ is sufficiently small that an electron is only able to complete a small fraction of one revolution before it is scattered, that is $\omega_c \ll 2\pi/\tau$ or $\omega_c\tau \ll 2\pi$. Looked at another way this means that the paths of the electrons between collisions will deviate only slightly from straight lines. What happens in large magnetic fields, such that $\omega_c\tau \gg 2\pi$, will be discussed later in this section.

We are accustomed to regard the electrical resistivity, ρ, as a simple scalar number, but of course this is only true for crystalline materials that are cubic, and for isotropic non-crystalline materials. That is, we write

$$\mathbf{E} = \rho\mathbf{j}, \tag{3.7.1}$$

or, in terms of the conductivity σ rather than the resistivity ρ,

$$\mathbf{j} = \sigma\mathbf{E}, \tag{3.7.2}$$

where $\sigma = 1/\rho$. The resistivity, ρ, or the conductivity, σ, of a cubic crystal or an isotropic non-crystalline material can then be determined by applying an electric field $\mathbf{E}$ to a specimen of the material and measuring $\mathbf{j}$, which will be in the direction of $\mathbf{E}$. However, for other materials which are neither cubic nor isotropic it has to be recalled that, strictly speaking, ρ and σ are not scalars but tensors of rank two, so that

$$E_p = \sum_q \rho_{pq} j_q \tag{3.7.3}$$

or

$$j_p = \sum_q \sigma_{pq} E_q. \tag{3.7.4}$$

For our present purpose we prefer to think in terms of the resistivity, ρ_{pq}, rather than the conductivity, σ_{pq}. For a crystalline non-cubic material the number of independent components of ρ_{pq} will depend on the point-group symmetry of the crystal (for details see Nye 1957). For a cubic or isotropic material then

$$\rho_{pq} = \rho\delta_{pq}, \qquad (3.7.5)$$

where ρ is the ordinary scalar resistivity. However if a non-zero magnetic field $\mathbf{B}$ is applied to such a crystal then its resistivity can no longer be represented by a single scalar number ρ and one has to use eqns (3.7.3) and (3.7.4) rather than eqns (3.7.1) and (3.7.2) even for cubic or isotropic materials. The components of the resistivity tensor will be functions of the field $\mathbf{B}$ and can be written as $\rho_{pq}(\mathbf{B})$ and the simple relation in eqn (3.7.5) no longer holds. The changes $\{\rho_{pp}(\mathbf{B})-\rho_{pp}(0)\}$ in the diagonal components of the resistivity in the presence of a magnetic field are normally referred to as the *magnetoresistance;* if $\mathbf{B}$ is in the z direction then changes in $\rho_{xx}(\mathbf{B})$ and $\rho_{yy}(\mathbf{B})$, which specify the resistivity for a current flowing normal to $\mathbf{B}$, are called the *transverse magnetoresistance;* the change in $\rho_{zz}(\mathbf{B})$, which specifies the resistivity for a current flowing in the direction of $\mathbf{B}$, is called the *longitudinal magnetoresistance.* If $\mathbf{B}$ is in the z direction and a current flows in the y direction then we should expect an electric field $\mathbf{E}$ to appear in the x direction normal to $\mathbf{j}$ and $\mathbf{B}$; this is the *Hall effect* and its magnitude is related to the off-diagonal tensor component $\rho_{xy}(\mathbf{B})$ which we should not normally expect to be zero.

In the case of the free-electron model the expressions for $\rho_{pq}(\mathbf{B})$ are particularly simple. Suppose that a current is flowing in the y direction in a metal. An electron which is moving with a velocity $\mathbf{v}$ in the y direction in a magnetic field applied in the z direction, experiences a Lorentz force of $e\mathbf{v} \wedge \mathbf{B}$ in the direction of the x-axis, as shown in Fig. 3.31. Consequently all the electrons drift sideways in the x direction causing an electric field $\mathbf{E}$ to be set up in the specimen. A steady state is then set up when the force $e\mathbf{E}$ on each electron due to this electric field just balances the Lorentz force $e\mathbf{v} \wedge \mathbf{B}$ on that electron. In the particular configuration of $\mathbf{j}$ and $\mathbf{B}$ that we have chosen this means that

$$eE_x = ev_y B_z. \tag{3.7.6}$$

We can eliminate v_y by making use of the relationship $\mathbf{j} = ne\mathbf{v}$ so that

$$E_x = \frac{1}{ne} B_z j. \tag{3.7.7}$$

Since the Hall coefficient R is defined as $E_x/B_z j$ we have, for the free-electron model,

$$R = \frac{1}{ne}. \tag{3.7.8}$$

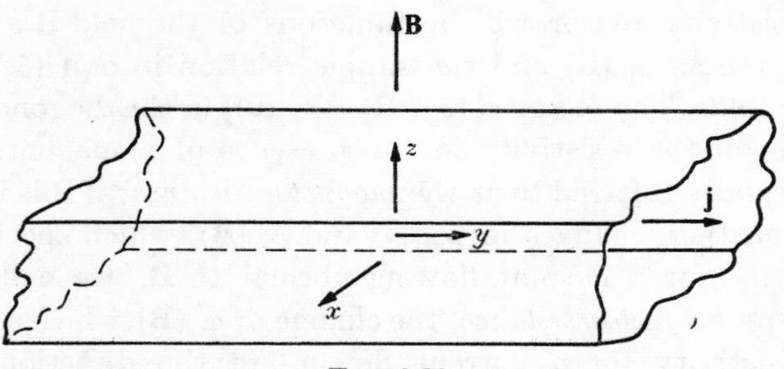

FIG. 3.31

The y components of the velocities of the electrons are not altered by the presence of the field B_z, so that in the free-electron approximation there is no transverse magnetoresistance. Also, if a magnetic field is applied parallel to $\mathbf{j}$, the Lorentz force $e\mathbf{v} \wedge \mathbf{B}$ vanishes so that, in the free-electron approximation, there is no longitudinal magnetoresistance either. The galvanomagnetic properties of a metal in the free-electron approximation are therefore very simple: the Hall coefficient is a constant ($= 1/ne$) and the magnetoresistance is zero. The galvanomagnetic properties of a real metal are more complicated than this but, of course, also more useful from the point of view of Fermiological investigations. The magnetoresistance is no longer zero and the Hall coefficient, R, no longer takes the simple value of $R = 1/ne$ (although for liquid metals the value of R is often extremely close to the free-electron value). To describe properly a transport property such as the resistivity requires the use of Boltzmann's equation (see Chapter 6), but we shall adopt a simplified approach in this section based on a hypothetical metal with just two partially full bands.

Once the Fermi surface of a real metal departs more or less from the free-electron model it is no longer particularly profitable to use the extended zone scheme. For most real metals in the reduced-zone scheme there will be sheets of the Fermi surface in several different bands; once departures from the free-electron model are introduced there is no reason to suppose that the specially simple free-electron results for the Hall effect and the magnetoresistance will still apply. It is convenient to describe bands which are less than half full as containing carriers which are electrons, while bands which are more than half full are described as containing carriers which are holes. Suppose that there are two types of carriers in a metal with conductivity σ_1 and σ_2 respectively. One can use this simple two-band model to demonstrate that for a real metal the magnetoresistance is no longer zero and the Hall coefficient departs from the ideal free-electron value of $1/ne$. By using a metallic specimen in the form of a wire, one constrains the current to flow in a direction parallel to the length of the wire. However, because of the presence of the external magnetic field and the consequent transverse Hall voltage that is produced, the total electric field $\mathbf{E}$ in the wire will not, in general, be parallel to $\mathbf{j}$. $\mathbf{E}$ can be written as the sum of two contributions, $\mathbf{E}_1$ $(= \mathbf{j}/\sigma)$ parallel to the length of the wire and the transverse Hall field $\mathbf{E}_2$ $(= \mathbf{v} \wedge \mathbf{B} = R\mathbf{j} \wedge \mathbf{B})$, so that

$$\mathbf{E} = \mathbf{j}/\sigma + R\mathbf{j} \wedge \mathbf{B}. \tag{3.7.9}$$

If there are two groups of carriers then we can write equations similar to eqn (3.7.9) for the two components $\mathbf{j}_1$ and $\mathbf{j}_2$ of the current associated with the two groups of carriers, so that

$$\mathbf{E} = \mathbf{j}_1/\sigma_1 + R_1\mathbf{j}_1 \wedge \mathbf{B} \tag{3.7.10}$$

and

$$\mathbf{E} = \mathbf{j}_2/\sigma_2 + R_2\mathbf{j}_2 \wedge \mathbf{B}, \tag{3.7.11}$$

and the total current $\mathbf{j}$ can be written as

$$\mathbf{j} = \mathbf{j}_1 + \mathbf{j}_2. \tag{3.7.12}$$

For a specimen in the form of a wire the total current density $\mathbf{j}$ must be parallel to the length of the wire, whereas in the presence of a magnetic field, the total electric field $\mathbf{E}$ will not, in general, also be in this direction, see Fig. 3.32. Equations

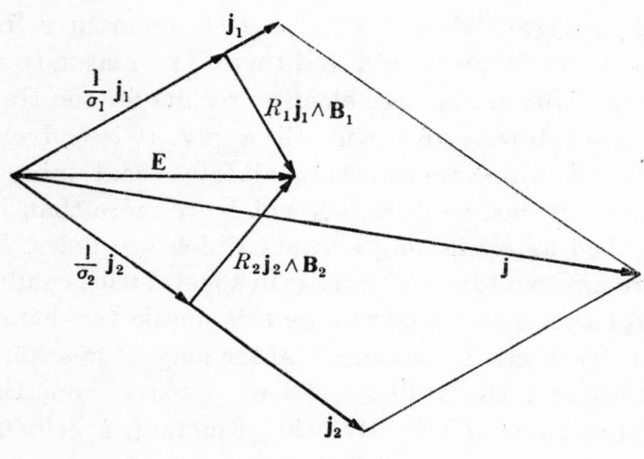

FIG. 3.32.

(3.7.10) and (3.7.11) can be inverted to give expressions for $\mathbf{j}_1$ and $\mathbf{j}_2$ in terms of $\mathbf{E}$ which can then be substituted into eqn (3.7.12) and, after some further manipulation, it can then be shown (Ziman 1964*a*, pp. 215–6) that the Hall constant R is given, for low magnetic fields, by

$$R = \frac{\sigma_1^2 R_1 + \sigma_2^2 R_2}{(\sigma_1 + \sigma_2)^2}, \tag{3.7.13}$$

where R_1 and R_2 are the Hall constants that the sets of carriers would have if they were present in the metal on their own. Therefore, if one type of carriers is electrons and the other type is holes, R_1 and R_2 will be of opposite sign and the contributions from the two types of carriers will tend to cancel out. The sign of the Hall constant will therefore indicate, for a given metal, whether the majority of the current is carried by electrons or by holes, while the magnitude of the Hall constant will give an indication of the difference between the numbers of majority and minority carriers. Although the two-band model that we have used in obtaining eqn (3.7.13) is a grossly over-simplified model of the behaviour of the electrons in a real metal carrying a current in the presence of a magnetic field, it does illustrate how the observed Hall effect in most metals arises as a sum of a number of contributions.

The simple two-band model also leads to an expression for the magnetoresistance; in the configuration used in Fig. 3.31 this is, of course, the transverse magnetoresistance, which we denote by $\Delta\rho_t/\rho_0$. The conductivity of a wire is defined as the ratio of $|\mathbf{j}|$ to the component of $\mathbf{E}$ parallel to $\mathbf{j}$ and it is possible to show (Ziman 1964*a*, p. 216) that

$$\frac{\Delta\rho_t}{\rho_0} = \frac{\sigma_1\sigma_2(\beta_1-\beta_2)^2B^2}{(\sigma_1+\sigma_2)^2+(\beta_1\sigma_1+\beta_2\sigma_2)^2B^2}, \tag{3.7.14}$$

where $\beta_1 = R_1\sigma_1$ and $\beta_2 = R_2\sigma_2$. Although the simple two-band model which leads to this result is only a crude approximation to the description of the properties of the electrons in a real metal, eqn (3.7.14) does manifest the usual behaviour of the field-dependence of the transverse magnetoresistance. For low magnetic fields the second term in the denominator on the right-hand side of eqn (3.7.14) will be negligible, while for large values of $\mathbf{B}$ this term will dominate; consequently for low magnetic fields

$$\frac{\Delta\rho_t}{\rho_0} \propto B^2, \tag{3.7.15}$$

while for large fields $\Delta\rho_t/\rho_0$ will saturate to a constant value, see Fig. 3.33. In spite of the oversimplifications used in this model this behaviour is exhibited by quite a large number of metals.

The above discussion of the Hall effect and the magneto-resistance is only valid for a relatively low magnetic field because it involves assuming that there will only be a small bending of the path of an electron between two successive collisions; that is, the field is sufficiently small that the cyclotron frequency is very much less than the collision frequency, $1/\tau$, or $\omega_c\tau \ll 2\pi$.

The arguments that we have given so far in this section were first of all based on treating the conduction electrons as free electrons and then introducing a single collision time τ and a single effective mass for the carriers in each band. In such a treatment, the detailed shapes of the pockets of carriers in each band, that is the detailed shapes of the various sheets of the Fermi surface, do not enter the argument at all. However, as with a number of other 'effects' considered in this chapter, once

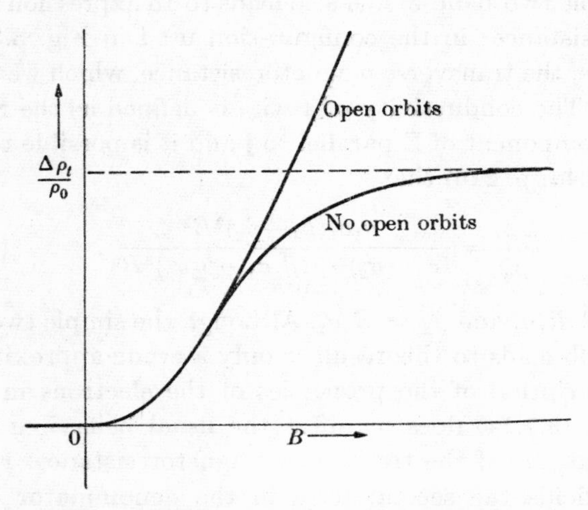

FIG. 3.33. Schematic diagram to illustrate the behaviour of the transverse magnetoresistance, $\Delta\rho_t/\rho_0$, as a function of B for a given direction of $\mathbf{B}$.

the electrons are able to traverse a distance of the order of one complete passage of an orbit between successive collisions (that is $\omega_c\tau \sim 2\pi$) the actual shape of that orbit becomes important in the theory. When $\omega_c\tau \sim 2\pi$ the theory of the galvanomagnetic properties of a metal should really be described by using Boltzmann's equation which is essential to the proper discussion of the transport properties of a solid (see Section 6.6). For the moment we simply quote the results that emerge from the use of Boltzmann's equation for the galvanomagnetic properties of a metal in a large magnetic field (Lifshitz, Azbel', and Kaganov 1956).

Provided no open orbits are possible in the cross sections of the Fermi surface normal to **B**, the general features exhibited by $\Delta\rho_t/\rho_0$ in the expression (3.7.14) still appear. Thus at low fields $\Delta\rho_t/\rho_0 \propto B^2$ while at high fields $\Delta\rho_t/\rho_0$ tends to a constant value. However, if open orbits are possible in the cross-section of the Fermi surface normal to **B**, then $\Delta\rho_t/\rho_0$ may increase as B^2 without saturation for indefinitely large magnetic fields. It is therefore possible to use an experimental study of the behaviour of the transverse magnetoresistance, $\Delta\rho_t/\rho_0$, as a function of field **B** and of orientation to determine whether or not open orbits can exist at all in the metal under consideration. If open orbits are found for any orientation of **B** relative to the crystallographic axes this shows that at least one sheet of the Fermi surface of that metal must be multiply-connected. By studying carefully the relative orientations of **B** and the crystallographic axes for which open orbits occur, it may be possible to distinguish between proposed Fermi surfaces exhibiting different forms of multiple-connectivity. For a compensated metal, that is, a metal which has equal numbers of positive and negative carriers, it is likely that the values of β_1 and β_2 will be quite similar so that we can see from eqn (3.7.14) that the transverse magnetoresistance will only be very small. Therefore, the conventional transverse magnetoresistance measurements of $\Delta\rho_t/\rho_0$ as a function of B for each of the various orientations of **B** provides little information about open orbits in a compensated metal. However, one can use a slightly different method which involves slowly rotating an applied magnetic field, and measuring the induced torque about an axis perpendicular to the plane of rotation (Moss and Datars 1967). When the current is due to closed orbits only, the induced torque is small, but when the magnetic field is perpendicular to an open orbit, the current of the open orbit is in a plane perpendicular to the open orbit direction and is independent of magnetic field strength. Induced currents in this plane are determined by the open orbit conductivity and cause the torque. Measurements of this effect in Hg have been made by Moss and Datars (1967).

There is, of course, a complication that occurs in connection with the use of the non-saturation of $\Delta\rho_t/\rho_0$ to detect open orbits. This is the fact that whereas in the absence of a magnetic field a particular cross section of the Fermi surface may not possess any open cross sections, the application of a magnetic field may lead to open orbits as a result of magnetic breakdown. If the field $\mathbf{B}_0$ at which magnetic breakdown occurs is sufficiently small, then $\Delta\rho_t/\rho_0$ will exhibit the non-saturation behaviour that is typical for the case of open orbits. However, if $\mathbf{B}_0$, the field at which magnetic breakdown occurs is large, then $\Delta\rho_t/\rho_0$ may first of all appear to saturate and then at about $\mathbf{B} = \mathbf{B}_0$ start to rise again in a parabolic fashion as shown in Fig. 3.34.

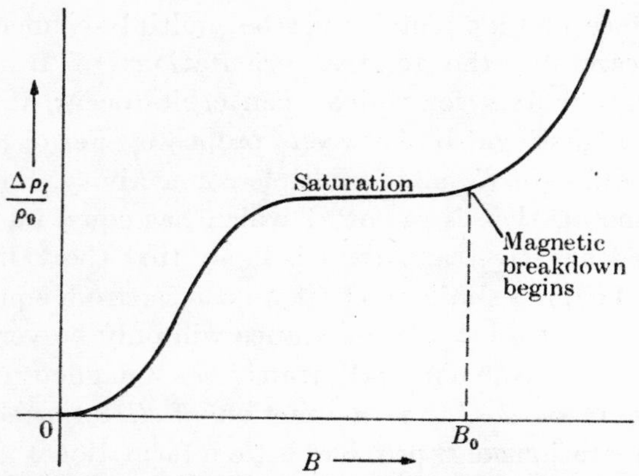

FIG. 3.34. Diagram to illustrate the behaviour of the transverse magneto-resistance under conditions of magnetic breakdown for a given direction of **B**.

So far we have been concerned with the transverse magneto-resistance and its use in connection with the determination of whether or not the Fermi surface of a given metal is multiply-connected. The information that can be obtained from the longitudinal magnetoresistance does not give such direct information about the topology of the Fermi surface of a metal, although it does give information about the relaxation times, $\tau_{\mathbf{k}}$, of the conduction electrons in a metal (see Section 6.6).

There is a close similarity between the transport of heat by the conduction electrons in a metal under the influence of a temperature gradient, and the transport of electric charge, also by the conduction electrons in a metal, under the influence of an electric field. According to the free-electron model (see, for example, Kittel 1956), it can be shown, on the basis of simple kinetic theory arguments and the use of Fermi–Dirac statistics, that the ratio of the thermal conductivity, κ, to the electrical conductivity, σ, is a constant at any given temperature, or that

$$\frac{\kappa}{\sigma T} = \frac{1}{3}\frac{\pi^2 k^2}{e^2}, \tag{3.7.16}$$

where it is assumed that the relaxation time, τ, is the same in the two cases. The assertion that $\kappa/\sigma T$ is a constant is known as the *Wiedemann–Franz law*, although for a real metal the actual value of $\kappa/\sigma T$ may depart from the free-electron value given by eqn (3.7.16). This is because the thermal conductivity will include a contribution due to the lattice vibrations ('phonon drag' see p. 482) which is not present in the electrical conductivity. It is customary to write

$$\frac{\kappa}{\sigma T} = L, \tag{3.7.18}$$

where L is called the *Lorentz number* or the *Lorentz ratio*. It is therefore not surprising that in the presence of a magnetic field the thermal conductivity behaves in a very similar way to the electrical conductivity. If a magnetic field is applied normal to the direction of heat flow in a metal, a temperature gradient will appear in the direction normal to both **B** and the heat flow; this is the thermal analogue of the Hall effect and is called the *Righi–Leduc effect* or the *thermal Hall effect* (see Lipson 1964; Amundsen 1968, 1969). The thermal Hall effect corresponds to the appearance of non-diagonal elements in the second-rank tensor that describes the thermal resistivity of a metal in the presence of a magnetic field. The presence of the magnetic field will also cause changes in the diagonal elements of the thermal resistivity tensor, which means that there will be changes in the transverse and longitudinal thermal resistance in the presence of

B. The existence of such *thermal magnetoresistance* or *magnetothermal resistance* has now been detected in a number of metals (some examples are quoted by Jan 1957) and leads to the same kind of information about the topology of the Fermi surface of a metal as can be obtained, probably more easily and without complications arising from phonon drag, from electrical magnetoresistance experiments.

In this section we have considered the behaviour of the electrical and thermal conductivity tensors of a metal in a

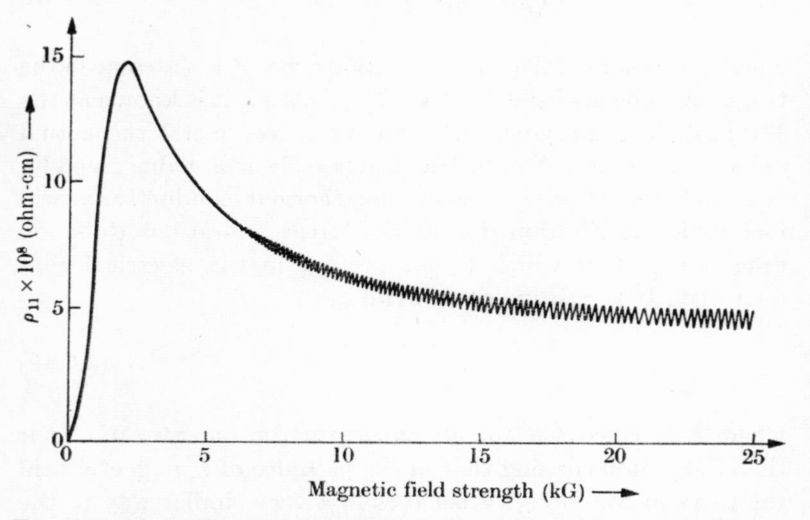

FIG. 3.35. The transverse magnetoresistance of Mg showing the B^2 behaviour and saturation at (relatively) low fields followed by the onset of Schubnikow–de Haas oscillations at higher fields (Falicov and Stark 1967).

magnetic field both for very low fields and also at higher fields such that $\omega_c \tau > 2\pi$. However, if the temperature is sufficiently low or **B** is sufficiently large that the condition

$$\hbar\omega_c \gg kT \tag{3.7.18}$$

is also satisfied; then super-imposed on the behaviour mentioned above there will also be quantum oscillations in the galvanomagnetic properties of a metal, see Fig. 3.35. These quantum oscillations will be discussed in more detail in the next section.

3.8. The de Haas–van Alphen effect and related effects

In Section 3.5 we described the formation of the Landau levels in a crystalline solid in the presence of a magnetic field. The existence of these levels leads to a striking behaviour in which many of the physical properties of a metal become oscillatory functions of the magnetic field, with a constant period in $1/B$. In Section 3.5 we were mostly concerned with the formation of the Landau levels in the free-electron approximation. If we now consider an electron in a real metal in which $V(\mathbf{r})$ is non-zero, the electron will still travel, between collisions, in a helix of which the projection on a plane normal to $\mathbf{B}$ is a closed orbit. In the absence of any magnetic field the wave vector $\mathbf{k}$ of an electron would remain constant until the electron undergoes a collision. However, in a field $\mathbf{B}$ the point $\mathbf{k}$ will precess in the plane normal to $\mathbf{B}$, in an orbit that is geometrically similar to the projection normal to $\mathbf{B}$ of the helical path of the electron in the metal; as in the free-electron case there will be an angle of $\frac{1}{2}\pi$ between the orientations of these orbits (see eqn (3.5.5)) although these orbits are no longer circular as they were for free electrons, see Fig. 3.22. The frequency of precession of $\mathbf{k}$ in one of these orbits is just the cyclotron frequency ω_c for that orbit (see eqn (3.6.3)). For a given plane in $\mathbf{k}$-space normal to $\mathbf{B}$, the area enclosed by one of these orbits cannot vary continuously but has to satisfy the quantization condition

$$\mathscr{A}_n = \frac{2\pi e B}{\hbar}(n+\gamma), \tag{3.8.1}$$

which is similar to eqn (3.5.10) for free electrons. This means that in a similar way to the free-electron case illustrated in Fig. 3.26, the allowed wave vectors in a plane normal to $\mathbf{B}$ condense onto a set of orbits which correspond to intersections of this plane with the surfaces corresponding to constant energy surfaces in the absence of a magnetic field, and which satisfy eqn (3.8.1). When the third direction k_z is included it is clear that the allowed wave vectors of an electron in a metal will lie in orbits on the surfaces of a set of coaxial tubes whose axis is the

k_z axis (see Fig. 3.36). Once again the degeneracies of these orbits are such that the average number of states per unit volume of **k** space is exactly the same as it was before the magnetic field was applied. Although we know that **k** is only quasi-continuous, we still commonly regard the Fermi surface in the

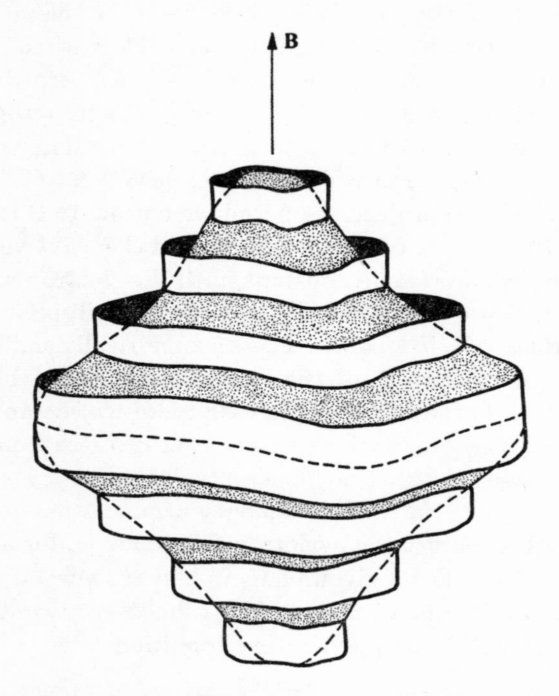

FIG. 3.36. Tubes of quantized allowed states in a magnetic field (Ziman 1960).

absence of a magnetic field as a smooth continuous surface in a three-dimensional space. We can assume that the energy separation between successive Landau levels in a metal in a typical magnetic field will be very much less than E_F. If we use a typical value of **B** and set $\mathscr{A}_n$ in eqn (3.8.1) to be of the order of a typical area of cross section of the Fermi surface of a metal, we shall find that n must be of the order of 10^4. This means that, although the distribution of the allowed wave vectors in **k** space is altered, **k** is still quasicontinuous, at least on the scale of typical pieces of Fermi surface. Since the wave vectors have only been redistributed, while maintaining the original number of states per unit volume of **k** space, the shape of the Fermi surface will be the same as it was in the absence of the magnetic field. In a semi-metal the separation between the Landau levels may be of the same order of magnitude as the overlap between the conduction and valence bands, but we shall not consider this 'quantum limit' regime here (for further details see McClure and Spry 1968; Abrikosov 1969).

If B, the magnitude of **B**, is increased steadily the orbits in **k** space will expand since, from eqn (3.8.1), $\mathscr{A}_n \propto B$; every so often there will arise the special situation in which one of the allowed orbits coincides exactly with the extremal cross section of the Fermi surface normal to **B**. Comparing Figs. 3.26(c) and 3.36 this means that one of the tubes fits exactly onto the Fermi surface. Strictly, as the magnetic field is varied, the Fermi level will oscillate about the value that it takes in the absence of an applied magnetic field; these oscillations in the Fermi level have been observed as oscillations in the contact potential for several metals (Kaganov, Lifshitz, and Sinel'nikov 1957; Verkin, Pelikh, and Eremenko 1965; Caplin and Shoenberg 1965; Whitten and Piccini 1966). There are several other properties of a metal which exhibit an oscillatory behaviour as B increases; each oscillation corresponds to one of these tubes passing through the Fermi surface and out into the unoccupied region of **k** space. There is no violation of the 'conservation of energy levels' because although tubes of allowed orbits are passing out through the Fermi surface every so often as B increases, so also the degeneracy of each of the remaining orbits increases. These oscillations were first observed in the electrical resistivity (Schubnikow and de Haas 1930*a–d*) and in the diamagnetic susceptibility (de Haas and van Alphen 1930*a*, *b*, 1932), but they have now been observed in a large number of other properties as well. The occurrence of these oscillations in the electrical resistivity is usually called the *Schubnikow–de Haas effect* and their occurrence in the diamagnetic susceptibility is usually called the *de Haas–van Alphen* effect. Although it is the electrical resistivity and the diamagnetic susceptibility in which these oscillations have most commonly been observed, they have also been observed in many other properties, in which case they are usually just referred to as *quantum oscillations*. Quantum oscillations have now been observed in the velocity of ultrasonic waves (Mavroides, Lax, Button, and Shapira 1962; Alers and Swim 1963), the attenuation of ultrasonic waves (Gibbons 1961, Gibbons and Falicov 1963, see below), the Hall constant (Gerritsen and de Haas 1940), the Knight shift (Jones and Williams 1964), the thermoelectric power (Steele and Babiskin 1955), the thermal conductivity (Steele and Babiskin 1955), the surface impedance (Khaĭkin, Mina, and Edel'man 1962), the contact potential (see above), the optical reflection and transmission coefficients (Azbel' 1958; Mattis and Dresselhaus 1958; Boyle and Rodgers 1959; Dresselhaus and Mavroides 1964*a*), and the specific heat (Kunzler and Hsu 1960; Kunzler, Hsu, and Boyle 1962). The quantum oscillations in a number of these properties are discussed in some detail by Kahn and Frederikse (1959). The quantum oscillations in the specific heat of a metal, which are usually called *magnetothermal oscillations*, can be observed by following the temperature variations of a thermally isolated sample in a varying magnetic field and recently this has become quite an important technique in Fermiological measurements.

As is the case in several of the other phenomena discussed in this chapter the derivation of the complete expression for the behaviour of any one of these oscillatory properties as a function of **B** is very complicated. However, the complete formulae are fortunately not very important for our present purposes since the importance of these quantum oscillations for Fermiological studies lies in the fact that the oscillations are periodic in $1/B$, and from the periods it is possible to determine extremal areas of cross section of the Fermi surface normal to **B**. If $\mathscr{A}_0$ is the extremal cross-sectional area of the Fermi surface there will be a peak, or a trough, in the observed property every time one of the areas $\mathscr{A}_n$ of equation (3.8.1) is exactly equal to $\mathscr{A}_0$, that is, whenever

$$\mathscr{A}_0 = \frac{2\pi e B}{\hbar}(n+\gamma). \tag{3.5.10}$$

The observed quantity is therefore periodic in $(1/B)$ with period given by

$$\Delta\left(\frac{1}{B}\right) = \frac{2\pi e}{\hbar \mathscr{A}_0}. \tag{3.8.2}$$

Therefore, by observing the period of the oscillations in a quantity such as the diamagnetic susceptibility, as a function of $1/B$, it is possible to determine the extremal area of cross section of the Fermi surface normal to **B**. An example of de Haas–van Alphen oscillations in Zn is illustrated in Fig. 3.37.

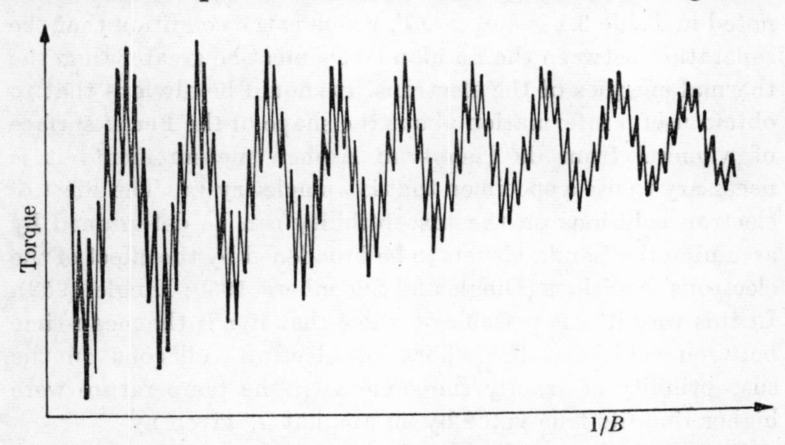

FIG. 3.37. An example of de Haas–van Alphen oscillations in Zn recorded by the torque method. Two frequencies are present and they are associated with different sections of the arms of the monster shown in Fig. 5.7. (Joseph and Gordon 1962).

Of all the various properties in which quantum oscillations have been observed it is probably the oscillations in the diamagnetic susceptibility which have been most extensively used in Fermiological investigations. Reviews of work on the de Haas–van Alphen effect at various stages in its development have been given by Shoenberg (1939, 1952*a*, 1957) and Kahn and Frederikse (1959). The derivation of the complete expression for the diamagnetic susceptibility χ_L of the conduction electrons

(Peierls 1933*a*, *b*; Blackman 1938; Landau 1939) is a very long and complicated mathematical procedure which it would hardly be profitable for us to study here; the outlines of the argument are given, for example, by Ziman (1964*a*). The amplitude of the oscillations is known to depend on several factors; these include (i) the curvature of the Fermi surface at the extremum at which $\mathscr{A}_0$ is being determined, (ii) the temperature, and (iii) the purity of the sample. For the observation of the de Haas–van Alphen effect the electrons must have a mean free path between collisions with impurities or collisions with phonons that is not appreciably shorter than the length of an orbit; hence the condition $\omega_c\tau \gg 2\pi$ given in Table 3.1. The second condition noted in Table 3.1 is $\hbar\omega_c > kT$, which is the condition that the separation between the Landau levels must be greater than the thermal energies of the electrons. It should be obvious that to obtain useful information about the shape of the Fermi surface of a metal from de Haas–van Alphen measurements it is necessary to use a specimen that is a single crystal. The effect of electron collisions on the susceptibility can be determined by assuming the Landau levels to be broadened by the effect of the electrons' collisions (Dingle and Shoenberg 1950; Dingle 1952). In this way it was possible to show that if τ is the mean time between collisions, the effect of electron collisions on the susceptibility is exactly the same as if the temperature were higher than its true value by an amount T' given by

$$T' = \frac{\hbar}{2\pi k\tau}\,. \tag{3.8.3}$$

Indeed, it had been pointed out as an empirical fact much earlier (Shoenberg 1939) that the field and temperature variation of the susceptibility of Bi could be made to agree much more closely with the theoretical expression if the temperature had been for some reason about 1 K higher than its measured value. T' is sometimes called the *Dingle temperature*. Since τ depends on the purity of the metal as well as on the intrinsic properties of the pure single-crystal metal, the Dingle temperature will vary between different specimens of a given metal; its magnitude is an indication of the concentration of impurities in the specimen. In addition to the broadening of the Landau levels as a result of electron collisions it is also possible to observe spin-splitting of the Landau levels. This spin-splitting manifests itself as a splitting of the peaks in the quantum oscillations in various properties and has been observed in magnetothermal oscillations, (Boyle, Hsu, and Kunzler 1960) and in the magneto-acoustic 'giant quantum oscillations' which are described below, as well as in the de Haas–van Alphen effect (Saito 1963). Under conditions of very strong de Haas–van Alphen magnetization, a crystal may be able to achieve a state of lower free energy by forming domains having equal and opposite magnetization (Condon 1966*a*, *b*; Condon and Walstedt 1968).

The use of the de Haas–van Alphen effect, or of the quantum oscillations in any other property, has the advantage that because the value of n is so large ($\sim 10^4$) the extremal areas of cross section are very sharply defined. Also, if the Fermi surface has several sheets in different bands the oscillations will contain several frequencies, with one frequency corresponding to each value of $\mathscr{A}_0$. In such a situation it should be possible to observe enough oscillations to enable the individual periods to be unravelled by a Fourier analysis because the value of n is so large. This separation of the contributions from different bands is not so easy for some of the other phenomena for which similar quantities to n are restricted to small values of the order of 50 or less (for example, Azbel'–Kaner cyclotron resonance (see Section 3.6) and magnetoacoustic geometric oscillations (see Section 3.9.3)). On the other hand, the use of the quantum oscillations has the disadvantage that it leads to values of cross-sectional areas of the various pieces of the Fermi surface, so that it is necessary to employ some inversion or parametrization scheme to determine the caliper dimensions of the Fermi surface (Lifshitz and Pogorelov 1954; Roaf 1962; Ashcroft 1965; Mueller 1966; Mueller and Priestley 1966; Foldy 1968), whereas some other methods lead to values of the caliper linear dimensions in the form of radii of the Fermi surface (see Section 3.9). While measurements of cross-sectional areas enable any postulated Fermi surface to be tested and, if necessary, modified, the interpretation of direct measurements of caliper dimensions is less tedious and less liable to ambiguity of interpretation. Moreover the de Haas–van Alphen effect only enables the orientation and the area—but not the actual location in the Brillouin zone—of each cross section of the Fermi surface to be determined. Another difficulty is that if two, or more, different periods are present for some particular magnetic field direction, they may give rise to significant oscillations with combination frequencies and, in particular, the difference frequencies (Shoenberg 1968; Shoenberg and Templeton 1968). Of course, as with any other technique that involves the use of a magnetic field there is the danger that the shape of the Fermi surface, and in particular its connectivity, may be altered by the presence of the magnetic field. In the case of the de Haas–van Alphen effect this means that one may obtain areas of cross section that correspond to orbits which arise as a result of magnetic breakdown (see p. 197) between different sheets of the Fermi surface that are close together. In addition to using the period of the de Haas–van Alphen oscillations for measuring $\mathscr{A}_0$, the extremal area of cross section, it is also possible to determine $\partial\mathscr{A}/\partial E$ (at $\mathscr{A}_0$) from measurements of the decay of the amplitude of these oscillations with increasing temperature and at a fixed field $\mathbf{B}$: this information is usually expressed in terms of an effective mass. As with cyclotron resonance, as $\mathbf{B}$ is reduced so also ω_c decreases, and the condition $\omega_c\tau \gg 2\pi$ will eventually be fulfilled less satisfactorily and the amplitude of the quantum oscillations will, again, decrease. It should then also be possible to determine the relaxation time, τ, from measurements of the decay of the amplitude of the de Haas–van Alphen oscillations with decreasing field $\mathbf{B}$ (see Section 6.6).

In addition to the de Haas–van Alphen effect, the quantum oscillations in the electrical resistivity (the Schubnikow–de Haas effect), and those in the attenuation coefficient for ultrasonic waves (magnetoacoustic quantum oscillations) have also been used quite extensively in Fermiological measurements.

Since the original observation of the Schubnikow–de Haas oscillations in Bi they have now been observed in a large number of other metals. The expression for the resistivity ρ of a metal in the presence of a magnetic field is different for the longitudinal resistivity (with parallel electric and magnetic fields) and the transverse resistivity (with crossed electric and magnetic fields). For details of the derivation of expressions for the resistivity we refer the reader, for example, to the article of Kahn and Frederikse (1959). Quantum oscillations in the attenuation of ultrasonic waves in a metal were first observed by Reneker (1958) in Bi. Since then quantum oscillations in the ultrasonic attenuation coefficient have been observed in quite a large number of other metals as well (for references see, for example, Roberts 1968). The observed periods show no significant differences from those obtained by using the de Haas–van Alphen effect. Ultrasonic attenuation measurements could therefore, in principle, be used instead of de Haas–van Alphen measurements in Fermiological investigations. However, in practice, it is usually better to perform both types of measurement because, for a given metal, some cross sections of the Fermi surface may be more prominent in de Haas–van Alphen measurements than in ultrasonic measurements, while the converse may be true for other cross sections.

The quantum oscillations in the diamagnetic susceptibility or the electrical resistivity of a metal are functions which vary quite slowly with $1/B$ and their general appearance is similar to that of a sine curve. When the sound frequency is low and the mean free path is short, the quantum oscillations in the ultrasonic attenuation coefficient also behave in this way. But if the frequency is increased, and if the mean free path is very long, the sinusoidal nature of the oscillations changes; there is an intermediate region when the oscillations take on a saw-tooth appearance and, finally, when $\omega\tau \gg 2\pi$ the oscillations become much more "spike-like", that is, larger and sharper, see Fig. 3.38, and they are then called *giant quantum oscillations*. The existence of these giant quantum oscillations was predicted by Gurevich, Skobov, and Firsov (1961); they were first observed

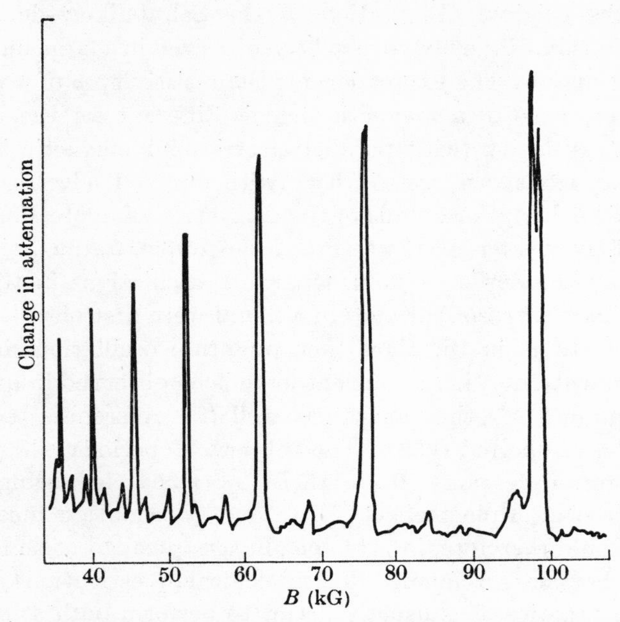

Fig. 3.38. The attenuation of 50 MHz longitudinal ultrasonic waves in Ga showing the giant quantum oscillations at 1·40 K with **k** ||**B**|| **b** (Shapira and Lax 1965).

experimentally in Zn by Korolyuk and Prushchak (1961) and they have now also been observed in several other metals. The conditions for the observation of these oscillations include the use of very high frequencies and very pure specimens. In a very pure metal the scattering of an ultrasonic wave by impurities will be negligible and the principal mechanism for the attenuation of the wave will be scattering by the electrons. The only electrons that contribute significantly to the attenuation are those for which the component of their velocity parallel to **k** is very nearly equal to v, where **k** and v are the wave vector and the velocity of the ultrasonic wave. For these electrons k_z must be very close to k_0 where

$$\frac{\hbar k_0 \cos\theta}{m^*} = v \tag{3.8.4}$$

and θ is the angle between $\mathbf{k}$ and $\mathbf{B}$. It is then possible to show (see, for example, Shapira 1968) that the usual expression for the period of the de Haas–van Alphen type of quantum oscillations

$$\Delta\left(\frac{1}{B}\right) = \frac{2\pi e}{\hbar \mathscr{A}_0} \tag{3.8.2}$$

still applies to these giant quantum oscillations except that while $\mathscr{A}_0$ is still an area of cross section of the Fermi surface normal to $\mathbf{B}$ it is no longer an extremal area of cross section, but the cross section at $k_z = k_0$. However, since the ultrasonic velocity is much smaller than a typical value for the Fermi velocity (a factor of about 10^{-3} would be typical) it can be seen from eqn (3.8.4) that unless θ is very close to 90°, the cross-sectional area $\mathscr{A}_0$ at $k_z = k_0$ will be very close to the cross-sectional area at $k_z = 0$. It would not be appropriate for us to become involved in an extensive discussion of the theory of the line shape and the amplitude of the giant quantum oscillations, (see, e.g., Shapira 1968). Because the peaks are so sharp it is particularly easy to observe their spin-splitting as a result of the splitting of the Landau levels, which occurs when the spins of the conduction electrons are taken into consideration (Shapira 1964), see Fig. 3.39. It is possible to use this spin-splitting to determine the g-factor for the conduction electrons (see Section 3.5). Giant quantum oscillations in the ultrasonic velocity, rather than the attenuation, have also been observed in Bi (Beletski, Korolyuk, Obolenski, and Khotkevich 1970). We shall see in Section 3.9.3 that in addition to the quantum oscillations and the giant quantum oscillations in the ultrasonic attenuation, there are some further magnetoacoustic 'effects' which can be used in connection with the determination of the shape of the Fermi surface of a metal.

3.9. Size effects

In all the different experimental methods that we have discussed so far in this chapter, the size and shape of the sample of the metal under investigation was assumed not to be important, although it was usually necessary for the sample to be

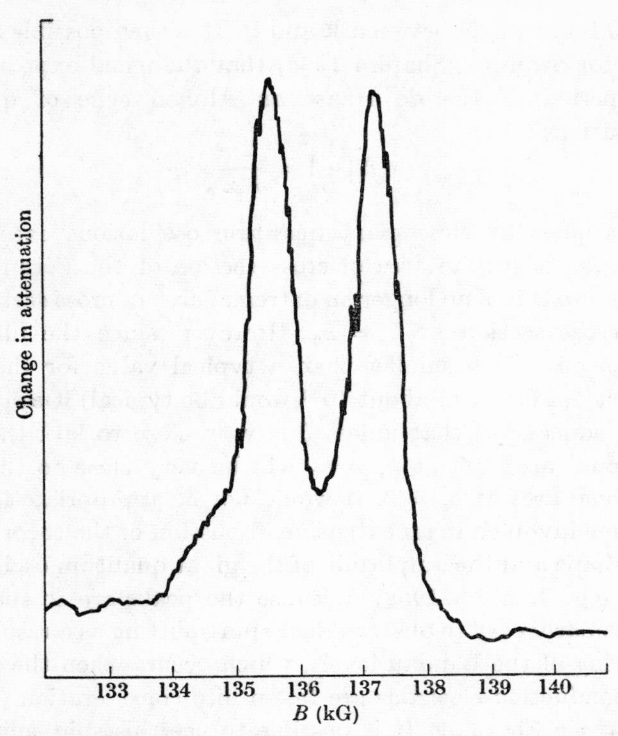

Fig. 3.39. The attenuation of 50 MHz longitudinal ultrasonic waves in Ga at 1·40 K with **k** ∥**B**∥ **b** showing the spin-splitting of the peak at 136 kG (Shapira and Lax 1965).

an oriented single crystal. Most of the phenomena involved were assumed to involve conduction electrons which, at least for the time during which they were being observed, were not close to any of the surfaces of the specimen. That is, for the purpose of our description of the phenomena, the specimen of the metal under investigation was assumed to be of infinite extent. In one or two phenomena, such as the anomalous skin effect and cyclotron resonance, the explanation in terms of the behaviour of the conduction electrons in the metal does not involve a supposedly infinite specimen, but is concerned with the behaviour of the conduction electrons within a region near the surface of a semi-infinite specimen. However, in practice no real specimen of a metal is either infinite or semi-infinite and there are situations in which the actual size and shape of the specimen may become important. This is particularly likely to occur if the specimen is in the form of a thin film, plate, or wire. We also include magnetoacoustic geometric resonances in our discussion of size effects, although in this case it is not the physical boundary of the specimen, but rather the wavelength of the ultrasonic wave which gives rise to special features in the absorption coefficient of a piece of metal in the presence of a magnetic field.

3.9.1. Direct current size effects

Size effects in the ordinary d.c. electrical resistivity of a metal were discovered a long time ago; a review of the early work is given by Sondheimer (1952) and a more recent review is given by Chambers (1969*a*). The electrical resistivity of a large specimen of a metal is made up of two components, one due to the intrinsic scattering mechanisms that are present even in a perfect single-crystal specimen of the metal, and the other due to scattering by impurities in the crystal. If the collisions experienced by an electron with velocity $v_{\mathbf{k}}$ are represented by a relaxation time $\tau_{\mathbf{k}}$ (the mean time between collisions), or as a mean free path $\lambda_{\mathbf{k}}$ where $\lambda_{\mathbf{k}} = v_{\mathbf{k}}\tau_{\mathbf{k}}$, it can readily be shown that for a cubic metal the d.c. electrical conductivity in the absence of a magnetic field is given by the following integral over the Fermi surface (Pippard 1960*a*; Ziman 1964*a*; Lifshitz and Kaganov 1965):

$$\sigma = \frac{e^2}{12\pi^3\hbar}\int_A \lambda_{\mathbf{k}}\,\mathrm{d}A_{\mathrm{F}}, \tag{3.9.1}$$

or

$$\sigma = \frac{e^2}{12\pi^3\hbar}\lambda A \tag{3.9.2}$$

where λ is the average value of the mean free path $\lambda_{\mathbf{k}}$ over the whole Fermi surface and A is the total area of the Fermi surface. It is only possible to determine the product λA from measurements of the d.c. conductivity. λ can obviously be estimated by using the value of A given by the free-electron model and values of λ of between about 100 Å and 600 Å at 0°C are obtained for some typical metals such as Li, Na, Cu, Ag, and Au (see, for example, Table 6.1 of Cusack 1958). It is also possible to determine the ratio σ/λ from eddy-current heating experiments (Cotti 1963, 1964; Cotti, Fryer, and Olsen 1964).

When using the idea of the mean free path λ in connection with the description of some property of a solid, one usually assumes that the specimen of the material is very large and that collisions of the electrons with the surfaces are not important. That is, the linear dimensions of the specimen are very much larger than λ which therefore denotes the average distance travelled by electrons within the material between collisions. However, if an electron is near to the surface of a specimen of a metal it will suffer additional scattering as a result of collisions with the surface. Consequently λ is decreased and the electrical resistivity is increased. This increase in the resistivity of a metal will only be significant if a large fraction of the electrons in the specimen are involved in collisions with the surface, and this situation can be brought about by reducing the size of the specimen relative to λ, by using a thin wire, plate, or film, and by reducing the temperature so as to increase the value of λ. There were some very early observations of an increase in the electrical resistivity of a metal with the sample in the form of a thin film instead of a bulk specimen (Stone 1898; Patterson 1901; Thomson 1901; Lovell 1936, 1938; Fuchs 1938; Andrew 1949). An approximate theory of the size effect in the d.c. electrical resistivity of a specimen in the form of a thin film was given by Thomson (1901) and the currently accepted treatment was originally given by Fuchs (1938) (see also the reviews of Sondheimer 1952; Chambers 1969*a*). The enhancement of the resistivity of a thin film over the bulk value as a function of d/λ, where d represents the thickness, was calculated by Fuchs (1938) and is illustrated in Fig. 3.40. The corresponding analysis for a wire of square cross section was performed by MacDonald and Sarginson (1950) and for a wire of circular cross-section by Dingle (1950) and Graham (1958); the results for circular cross sections are illustrated in Fig. 3.40. These theoretical analyses

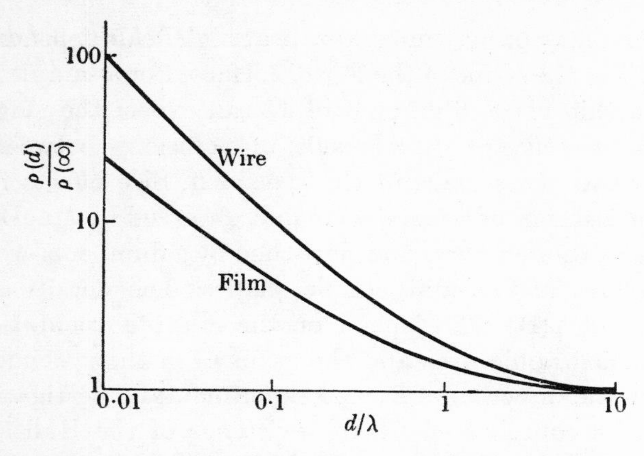

FIG. 3.40. The enhancement of apparent resistivity of a thin film and of a thin wire, as a function of d/λ, for the free-electron model (Chambers 1969*a*).

for thin films and wires do not include allowance for the anisotropy in the mean free path which can be expected to be associated with the high degree of anisotropy which exists in the Fermi surfaces of most real metals. Moreover, with only a few exceptions (for example, Holwech and Risnes (1968) on Al), most experiments have been performed on polycrystalline specimens. It is therefore not always easy to make meaningful comparisons between experimental results and the theoretical curves in Fig. 3.40, but discussions of these experimental results have been given by Sondheimer (1952) and by Chambers (1969*a*).

So far we have been concerned with size effects for direct currents in the absence of any magnetic field. Further examples of size effects can be seen in the electrical transport properties of a metal in the form of a thin plate, film, or wire when the specimen is in a magnetic field, **B**. If a large specimen of a metal is placed in a magnetic field **B**, the paths of the electrons, between collisions, will be changed from straight lines into helices and the electrical resistance of the metal in the presence of **B**, that is the *magnetoresistance* of the metal, will be some function of **B**. We have seen in Section 3.7 that for a bulk metal $\rho(\mathbf{B})$, the resistivity in a magnetic field **B**, increases as B^2 at low fields **B** and may or may not saturate at high fields depending on the details of the shape of the Fermi surface. For a sample in the form of a thin plate, film, or wire we can expect the magneto-resistance to change as a result of collisions between the electrons and the surface of the specimen. Size effects in the magnetoresistance of a metal were first observed by MacDonald (1949) using thin Na wires and since then by a number of workers using various metals and sample shapes. The details of the behaviour of $\rho(\mathbf{B})$ will depend on the relative orientation of the crystallographic axes and the surfaces of the specimen, as well as on the direction of **B**. The experimental investigation of $\rho(\mathbf{B})$ will be complicated by the existence of the Hall effect; indeed, unless the Hall electric field is either zero or uniform across the sample, the detailed theoretical treatment is also very complicated, although the situation of **B** and **j** both parallel to the surface of a thin film but at right angles to each other has been investigated (MacDonald and Sarginson 1950; Ditlefson and Lothe 1966; Blatt, Burmester, and La Roy 1967). As the field **B** is increased one would expect $\rho(\mathbf{B})$ to revert to the bulk value when **B** becomes large enough for the diameters of the orbits of the electrons to be smaller than d, the thickness of the specimen. If this happens at a critical field $\mathbf{B}_c$, then the extremal Fermi surface diameter will be given by $eB_c d/\hbar$ (see eqn (3.5.5)). The zero Hall electric field can be achieved, for a free-electron metal at least, by having the magnetic field parallel to the electric current density, and size effects in this situation have been studied for both thin wires and thin films (Chambers 1950, Koenigsberg 1953, Azbel' 1954, MacDonald and Barron 1958, Kao 1965). Once again $\rho(\mathbf{B})$ reverts to the bulk value when the diameters of the orbits of the electrons become less than d.

An interesting phenomenon occurs if **B** is normal to a thin film of metal and $d \ll \lambda$ (Sondheimer 1950, 1952; Gurevich 1958). For this orientation of **B** the resistivity $\rho(\mathbf{B})$ exhibits oscillations which are periodic in **B** and which are called *Sondheimer oscillations*. If $\lambda \gg d$ most of an electron's collisions will occur at the surfaces of the film. The helical trajectory of an electron moving across the film from one surface to the other is cut off at some fraction of a revolution of the helix depending on the value of **B**. The magnitude of this fraction determines the oscillating component of $\rho(\mathbf{B})$. The period ΔB can be shown to be given by

$$\Delta B = \pm\frac{2\pi m_c^* \bar{v}_z}{ed} = \mp\frac{\hbar}{ed}\left(\frac{\partial \mathscr{A}}{\partial k_z}\right)_E, \tag{3.9.3}$$

where **B**, which is normal to the film, is in the z direction (for details of the derivation see, for example, Sondheimer 1950, Gurevich 1958, Chambers 1969*a*). The Sondheimer oscillations

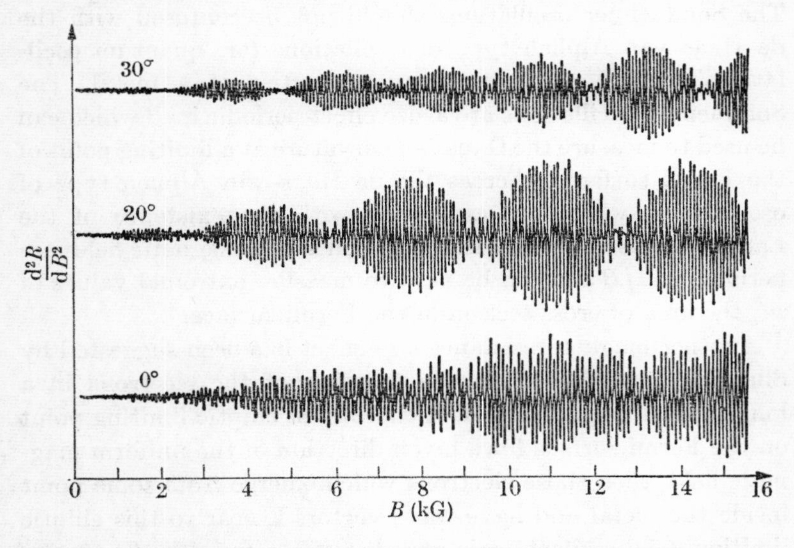

FIG. 3.41. Sondheimer oscillations in Ga for various angles of tilt between **B** and the normal to the sample surface (Munarin and Marcus 1964).

have been observed in a number of metals (Babiskin and Siebenmann 1957; Førsvoll and Holwech 1962; Zebouni, Hamburg, and Mackey 1963) and some results of measurements of these oscillations in Ga are shown in Fig. 3.41 (Munarin and Marcus 1964; Pippard 1966; Munarin, Marcus, and Bloomfield 1968). As one might perhaps expect from the other phenomena considered in this chapter, the amplitude of these oscillations will only be appreciable when $(\partial\mathscr{A}/\partial k_z)_E$ is extremal, or when there is an abrupt cut-off in the periods, as will occur if there is a point on the Fermi surface that is tangential to a plane k_z = constant. Such a point is sometimes called an *elliptic limiting point*. The period ΔB can be expressed in terms of the Gaussian curvature $K = (\rho_1\rho_2)^{-1}$, where ρ_1 and ρ_2 are the principal radii of curvature of the Fermi surface at this point; since it can be shown that $|\mathrm{d}\mathscr{A}/\mathrm{d}k_z| = 2\pi/\sqrt{K}$ the period can be written as

$$\Delta B = \frac{2\pi\hbar}{edK^{\frac{1}{2}}}. \tag{3.9.4}$$

The Sondheimer oscillations should not be confused with the de Haas–van Alphen type of oscillations (or 'quantum oscillations') which occur in many properties of a metal. The Sondheimer oscillations are a size effect periodic in B which can be used to measure the Gaussian curvature at a limiting point of the Fermi surface, whereas the de Haas–van Alphen type of oscillations, which are associated with the existence of the Landau levels in a metal in the presence of a magnetic field, are periodic in $1/B$ and can be used to measure extremal values of $\mathscr{A}$, the area of cross section of the Fermi surface.

Another magnetoresistance size-effect has been suggested by Sharvin (1965) based on the focusing of the electrons in a longitudinal magnetic field. If there is an elliptic limiting point on the Fermi surface for a given direction of the uniform magnetic field, then those electrons which emerge from some point inside the metal and have wave vectors $\mathbf{k}$ near to this elliptic limiting point, collect again at a distance L from the first point where

$$\frac{2\pi n}{\sqrt{K}} = eBL \tag{3.9.5}$$

where n is an integer. To observe this focusing effect it would be necessary to measure the resistance of the sample between two very small contacts, for example using thin wires touching the surface of the sample at two points on opposite faces of the sample. If the conditions for focusing are satisfied for these points there will be a periodic variation in the resistance of the sample with B. Investigations of the effect could give information on the Gaussian curvature of the Fermi surface, and on the value of the mean free path λ and its temperature dependence. This effect was first observed by Sharvin and Fisher (1965) in Sn.

There is a complication that we have ignored so far in this section and which is relevant to situations both with and without a magnetic field. This is the fact that in a thin film of a metal with thickness d less than the mean free path, λ, an electron only experiences a periodic potential in the direction normal to the film for a restricted distance, so that Bloch's theorem does not apply for this direction. This leads to a quantization of the component of $\mathbf{k}$ normal to the film, corresponding to standing waves fitted into d, the thickness of the film. This quantization of the energy levels is analogous to the formation of the Landau levels in a metal in a magnetic field and leads to oscillations in a number of properties of the metallic film as a function of thickness (Lutskii, Korneev, and Elinson 1966; Ogrin, Lutskii, and Elinson 1966; Lutskii and Fesenko 1968; Garcia, Kao, and Strongin 1969). If a magnetic field is also applied to the film it may be possible to determine non-extremal areas of cross section of the Fermi surface of a metal (Lutskii and Fesenko 1968, Fesenko 1969).

Although quantitative descriptions of the size effects that we have mentioned so far must depend on the details of the shapes of the band structure and Fermi surface of the metal in question, many of these size effects do not lead to any very direct information concerning the geometry of the Fermi surface of the metal. However, there are certain other size effects which do lead to rather direct information about the geometry of the Fermi surface of a metal, and it is these size effects, which include the radio-frequency size effect (or Gantmakher effect) and magnetoacoustic geometric oscillations, that we shall consider in the next two sections.

3.9.2. The radio-frequency size effect (Gantmakher effect)

This technique, which is now widely used in connection with the experimental investigations of the Fermi surfaces of metals, was a much later arrival on the scene than most of the other techniques that are discussed in this chapter.

It will be recalled that in Section 3.4 we discussed the anomalous skin effect in the absence of an applied magnetic field, while in Section 3.6 we discussed the observation of Azbel'–Kaner cyclotron resonance when a magnetic field is applied parallel to the plane surface of a semi-infinite block of metal. The condition for the existence of Azbel'–Kaner cyclotron resonance is that

$$\omega = n\omega_c. \tag{3.6.7}$$

However, if the specimen is in the form of a thin parallel-sided plate instead of a semi-infinite block, and the size of the magnetic field is steadily decreased, then the radii of the orbits of the electrons will also increase steadily and there will suddenly be a cut-off when the diameter of the orbits exceeds the sample thickness and no more cyclotron resonance anomalies will be observed in $Z(\mathbf{B})$, the surface impedance as a function of $\mathbf{B}$. If the field is decreased still further the electrons will collide with the surface of the sample before completing one orbit and Azbel'–Kaner cyclotron resonance will no longer be possible. If the thickness of the sample is d and the critical field at which the cut-off occurs is $\mathbf{B}_c$ then from eqn (3.5.5) the diameter of the orbit in $\mathbf{k}$ space will be $eB_c d/\hbar$. This Azbel'–Kaner cyclotron resonance cut-off was observed in Sn by Khaĭkin (1961), see Fig. 3.42.

A little later Gantmakher (1962*a*, *b*) showed that anomalies in $Z(\mathbf{B})$ could be observed at much lower frequencies than those required for cyclotron resonance, namely at about 1 MHz in the radio-frequency part of the electromagnetic spectrum instead of in the microwave region. There are in fact a variety of somewhat different radio-frequency size effects which tend to be known collectively as the *Gantmakher effect*. The radio-frequency size effects have now become well established as a common and useful method for the experimental determination of caliper dimensions of the Fermi surface of a metal. In the simplest of these effects a magnetic field is applied parallel to the surfaces of a thin plate of the metal under investigation and the surface impedance $Z(\mathbf{B})$ is measured as a function of $\mathbf{B}$. The difference from ordinary Azbel'–Kaner cyclotron resonance is that the

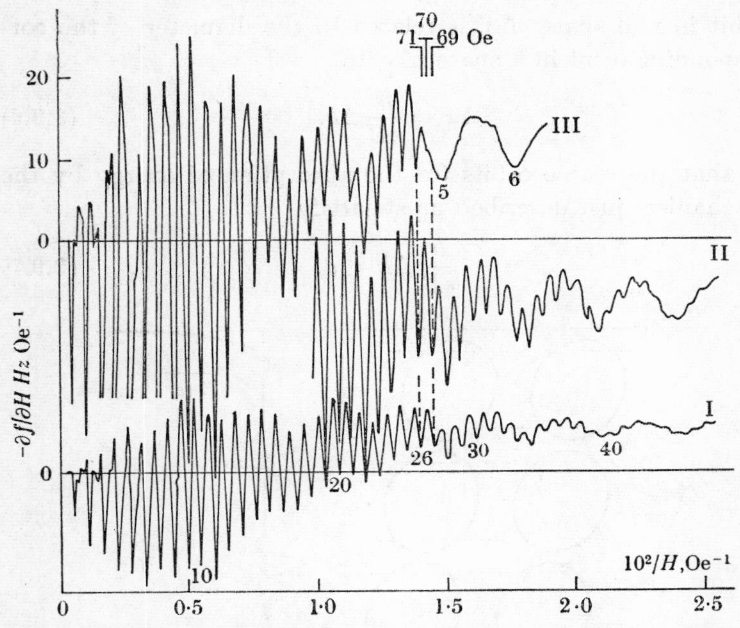

FIG. 3.42. Trace of cyclotron resonances on Sn single crystals of thickness 2 mm (curves I and II) and of thickness 0·982 mm (curve III). It can be seen that in curve III resonances of order 27 and higher are missing. The orders of the resonances are shown under curves I and III; the traces of some deep resonances on curve III are cut off (Khaikin 1961).

frequency of the electromagnetic wave is now much lower so that the thickness of the specimen assumes an essential role in the experiment. So long as the metal is sufficiently pure the usual anomalous skin effect condition $\delta \ll \lambda$ still holds and, as illustrated in Fig. 3.29, an electron travels many times round its orbit before it is scattered. Since the applied frequency ω is very much lower than ω_c, the cyclotron frequency, the phase of the electromagnetic wave seen by the electron on two successive passages through the skin layer hardly changes at all and the electron absorbs energy coherently from the wave, which leads to a corresponding reduction in $Z(\mathbf{B})$ below its zero-field value. Because the specimen is in the form of a thin plate this process only occurs for those orbits for which the diameter, in real space, is less than d, the thickness of the sample. The diameter of an orbit in real space, Δx, is related to the diameter of the corresponding orbit in **k** space, Δk, by

$$\Delta x = \frac{\hbar}{eB} \Delta k, \tag{3.9.6}$$

so that the viable orbits for the absorption of energy by the mechanism just described must satisfy

$$\frac{\hbar}{eB} \Delta k < d. \tag{3.9.7}$$

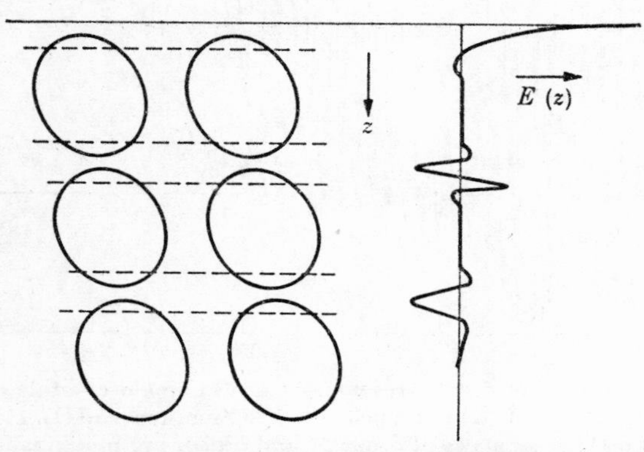

FIG. 3.43. Diagram illustrating schematically the field spikes set up below the surface of a metal by chains of interacting orbits (Chambers 1969*a*).

If the field **B** is decreased slowly the diameters of the orbits of the electrons will increase and fewer orbits will now satisfy the condition in eqn (3.9.7). In particular there will be an anomaly in $Z(\mathbf{B})$ whenever **B** passes through a value for which some orbit corresponding to an extremal value of Δk on the Fermi surface suddenly fails to satisfy eqn (3.9.7). If an anomaly occurs at $\mathbf{B}_c$ then the corresponding extremal diameter of the Fermi surface will be given (Kaner 1958) by

$$\Delta k_e = \frac{eB_c}{\hbar} d. \tag{3.9.8}$$

The situation is actually a little more complicated than we have just suggested because, for orbits in real space that correspond to extremal orbits on the Fermi surface in **k** space, it is possible to set up chains of coupled orbits, see Fig. 3.43, and there will therefore be an anomaly in $Z(\mathbf{B})$ whenever an integral number of these orbits fits exactly into the thickness d of the plate, that is when

$$n\ \Delta k_e = \frac{eB}{\hbar} d, \tag{3.9.9}$$

so that $B = nB_c$. An example of some experimental results is shown in Fig. 3.44. In addition to the experimental difficulties

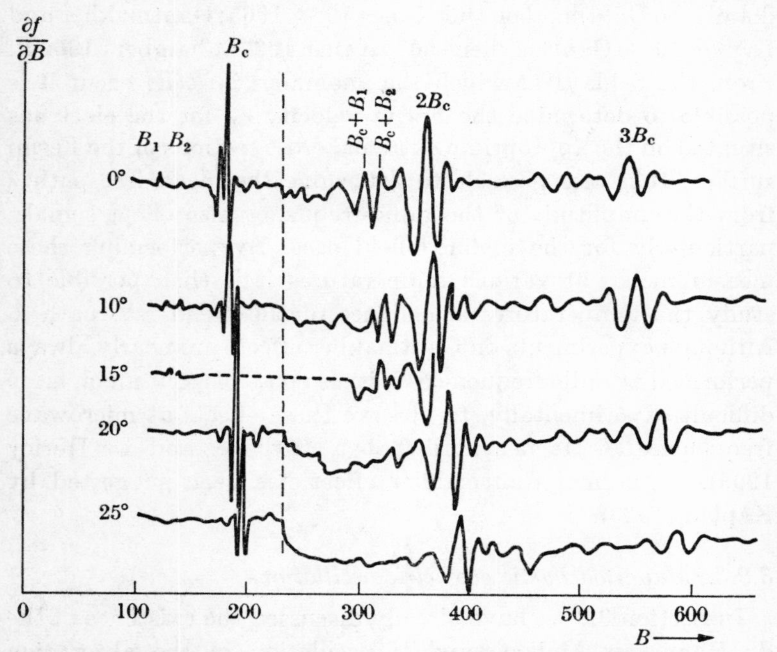

FIG. 3.44. Illustration of anomalies in the surface impedance of Sn at B_c, $2B_c$, and $3B_c$ using a frequency of 3·06 MHz. The angle between the constant field and the high-frequency field is indicated beside each curve. The ordinate scale on the right of the broken vertical line is magnified by a factor of 9 compared with that on the left-hand side and the units of B are not specified. (Gantmakher 1962*b*).

involved in producing thin samples of sufficient purity to ensure that the condition $\delta \ll \lambda$ is satisfied, there is also the difficulty that, for a given specimen, measurements of the type just described only give the caliper dimensions of the Fermi surface for certain directions in $\mathbf{k}$ space determined by the orientation of the crystallographic axes relative to the surfaces of the specimen. To obtain complete measurements of the dimensions of the Fermi surface of a given metal it will therefore be necessary to use a number of different specimens with different orientations of the crystallographic axes.

If the steady field **B** is no longer applied exactly parallel to the surface of the sample but is tilted at some angle ψ to the surface, it is also possible to observe size effect signals (for details see Gantmakher and Kaner 1963, 1965; Gantmakher and Krylov 1964; Gantmakher 1967; Kaner 1967; Chambers 1969*a*). From the fields **B** at which the anomalies in $Z(\mathbf{B})$ occur it is possible to determine the Fermi velocity v_F for the electrons situated on the appropriate extremal cross sections of the Fermi surface. It is also possible to determine the mean free path λ from the amplitude of the radio-frequency size effect signals, particularly for the inclined field case. By performing these measurements at various temperatures it is then possible to study the temperature dependence of the mean free path λ. Although experiments on Gantmakher effects are nearly always performed at radio frequencies it is also possible, but much more difficult experimentally, to observe these effects at microwave frequencies (D'Haenens, Libchaber, Laroche, and Le Hericy 1968). A thermal Gantmakher effect has been suggested by Kaplan (1969).

3.9.3. *Magnetoacoustic geometric oscillations*

In Section 3.8 we have already discussed the existence of the de Haas–van Alphen type of oscillation in the absorption coefficient for ultrasonic waves in a metal (magnetoacoustic quantum oscillations) as a result of the existence of the Landau levels. The experimental techniques for the observation of these quantum oscillations and of the geometric oscillations are very similar, and it is perhaps a little artificial to separate them from one another in our discussion. Nevertheless the physical explanations of the two kinds of oscillation are rather different, as also are the kinds of information that can be obtained about the shape of the Fermi surface.

Historically it was the geometric oscillations that were

observed first by Bömmel (1955) in experiments on Sn at 4·2 K using longitudinal ultrasonic waves of frequency 10·3 MHz, see Fig. 3.45. Bömmel's initial results were quickly explained by Pippard (1957*a*). If an ultrasonic wave is propagating through a metal it causes an oscillating electric field to be set up in the metal as a result of the displacements of the ions caused by the wave. This electric field will possess similar wave-like properties to those of the ultrasonic wave itself. Thus we have what is, in effect, an electric wave which is able to pass through the metal and is not hampered by skin-depth problems as an ordinary

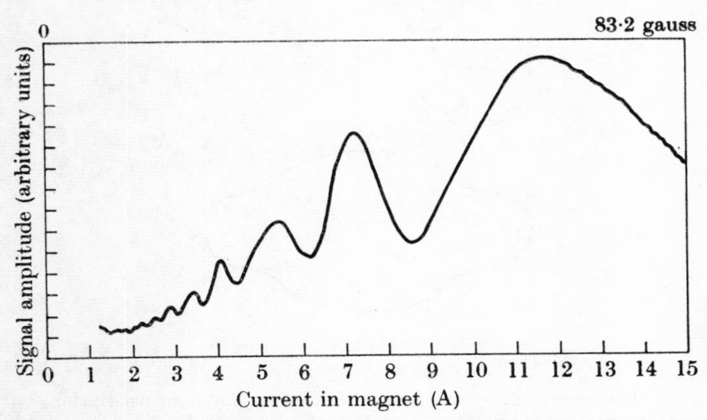

FIG. 3.45. An example of magnetoacoustic geometric oscillations in Sb for **k** || **B** with longitudinal ultrasonic waves of frequency 138 MHz (Eckstein, Ketterson, and Eckstein 1964).

electromagnetic wave would be. What is particularly important is that the velocity of this wave is very much slower (typically of the order of 10^3 m s^{-1}) than the velocities of the conduction electrons (for which a typical value of the Fermi velocity is of the order of 10^6 m s^{-1}). As far as these electrons are concerned the ultrasonic wave appears to produce an electric field which varies sinusoidally and is almost stationary. If a magnetic field is applied to the metal, the conduction electrons follow paths in the metal which are scale reproductions of orbits given by intersections of the Fermi surface with planes normal to **B** (see Section 3.5).

We suppose that an ultrasonic wave propagates in the x direction and is a transverse wave with the vibrations in the y direction, where both these directions are normal to the field **B** which is in the z direction. The electric field **E** associated with the ultrasonic wave is supposed normal to **B** and to the direction of propagation of the wave. The spatial variation of **E** is shown at the top of Fig. 3.46(a). We consider two points A and B on

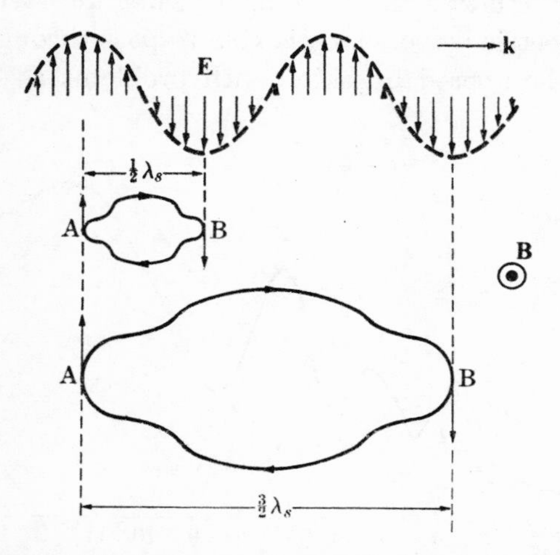

FIG. 3.46. Diagram to illustrate magnetoacoustic geometric oscillations with a transverse ultrasonic wave. **B** is normal to the plane of the figure and orbits in real space for two different values of B are shown (Mercouroff 1967).

opposite sides of the orbit such that the velocity of an electron at A is exactly antiparallel to its velocity at B. If the distance AB is equal to an odd number of half wavelengths of the ultrasonic wave, the electric fields experienced by an electron at A and at B will also be exactly antiparallel, see Fig. 3.46. In this situation there will therefore be a maximum absorption of energy from the ultrasonic wave, somewhat similar to the acceleration of a charged particle in a cyclotron. The condition for the maximum absorption of energy is therefore

$$\text{AB} = (n+\tfrac{1}{2})\lambda_s = (n+\tfrac{1}{2})\frac{2\pi}{|\mathbf{k}|}, \tag{3.9.10}$$

where λ_s is the wavelength and $\mathbf{k}$ is the wave vector of the ultrasonic wave. AB is simply related by a scale factor and a rotation of $\pi/2$ to a corresponding length A′B′ on the orbit of the electron in $\mathbf{k}$ space. If we write A′B′ as $|\mathbf{k}_{A'}-\mathbf{k}_{B'}|$ or $2\,|\mathbf{k}_{A'}|$ then eqn (3.9.10) becomes

$$2\,|\mathbf{k}_{A'}| = \frac{eB}{\hbar}\,\text{AB} = \frac{eB}{\hbar}\,(n+\tfrac{1}{2})\lambda_s. \tag{3.9.11}$$

In terms of the period $\Delta(1/B)$ in the absorption coefficient $\alpha(\mathbf{B})$ as a function of B, this condition becomes

$$\Delta\left(\frac{1}{B}\right) = \frac{e\lambda_s}{\hbar}\,\frac{1}{2\,|\mathbf{k}_{A'}|}, \tag{3.9.12}$$

so that from observations of the periodicity in $1/B$ of the ultrasonic attenuation coefficient, the caliper dimension, A′B′, normal to $\mathbf{B}$ and $\mathbf{k}$, of the Fermi surface can be determined. As usual it is only the extremal values of the caliper dimensions that contribute significantly to the periodic behaviour of the attenuation coefficient. If there are several sheets of the Fermi surface in different bands there will, in general, be one set of oscillations corresponding to each sheet.

In eqn (3.9.12) we have seen that the attenuation coefficient is periodic in $(1/B)$ and this oscillatory behaviour is often described as magnetoacoustic 'geometric resonance'. However, the use of this term may be misleading since the peaks in the attenuation coefficient $\alpha(\mathbf{B})$ are not the sharp peaks that would be characteristic of a true resonance phenomenon; the variation of the attenuation coefficient $\alpha(\mathbf{B})$ with $1/B$ is much closer to a sinusoidal variation and it is therefore recommended (for example, by Roberts 1968) that the term 'geometric resonances' should not be used. We shall use the term *magnetoacoustic geometric oscillations* to distinguish these oscillations from the magnetoacoustic quantum oscillations that were discussed in Section 3.8. The geometrical condition given in eqn (3.9.10) is in fact unnecessarily naïve; it would be more correct to write

$$\text{AB} = (n+\gamma)\lambda_s = (n+\gamma)\,\frac{2\pi}{|\mathbf{k}|}\,. \tag{3.9.13}$$

For a circular orbit it can be shown (see Ziman 1964*a*, p. 269) that the maxima in $\alpha(\mathbf{B})$ occur at the maxima of the Bessel function $J_1(qx)$ where $x = \frac{1}{2}$AB so that, in this case, $(n+\gamma)$ takes the values 1·22, 2·23, 3·24, with γ having the limiting value of $\frac{1}{4}$. For a perfectly rectangular orbit, on the other hand, $\gamma = 0$ (Pippard 1965, p. 124). For a given **B** and n, the mean free path must be at least of the same order of magnitude as AB, that is,

$$\lambda \sim n\lambda_s \tag{3.9.14}$$

in order to observe the geometric oscillations. λ_s is determined by the ultrasonic frequency and it is seldom possible to increase λ/λ_s to any very large value and therefore the value of n is quite small in practice. In the early days of the experimental work it was considered quite an achievement if oscillations were observed which pushed n into double figures. More recently oscillations with values of n up to about 50 in Ga (Roberts 1961) and about 60 in Mg (Ketterson and Eckstein 1966) have been recorded, see Fig. 3.47. It should be noted that while increasing the mean free path λ increases the number of oscillations that can be observed, it does not sharpen the peaks in $\alpha(\mathbf{B})$.

We have only described the magnetoacoustic geometric oscillations for the case of transverse waves with wave vector **k** normal to **B** and where the displacements of the medium are normal to both **B** and **k**, but similar oscillations can be expected to occur if longitudinal waves are used instead of transverse waves (Pippard 1957*a*). However, following the earliest observations and the suggestion by Pippard (1957*a*) of the mechanism causing the oscillations, a number of other related magnetoacoustic 'effects' were also predicted and observed. These include the use of other rectilinear or orthogonal configurations of **B**, **k**, and the ultrasonic velocity. They also include the observation of oscillations in the velocity (Beattie and Uehling 1966), rather than the attenuation, of the ultrasonic waves. If tilted magnetic fields are used 'surf-riding resonance' can occur, the observation of which enables the Fermi velocity for the group of electrons to be determined. We do not propose to describe these various refinements here (the interested reader

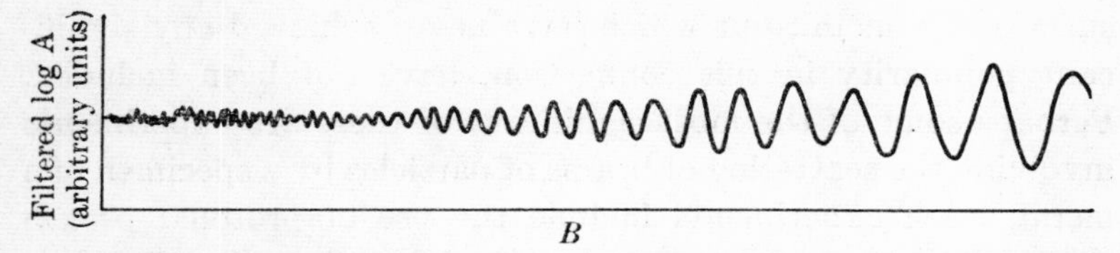

FIG. 3.47. Magnetoacoustic geometric oscillations in Mg showing 59 maxima obtained with electronic filtering techniques, used for eliminating wide background variations and superimposed oscillation series (adapted from Ketterson and Stark 1967).

should consult the works of Gurevich 1959*a*, *b*; Cohen, Harrison, and Harrison 1960; Pippard 1960*b*, 1965; Spector 1960, 1962, 1966; Mackinnon, Taylor, and Daniel 1962; Quinn 1963; Stolz 1963*a*, *b*; Kaner and Fal'ko 1964; Roberts 1968). We have already mentioned ultrasonic cyclotron resonance and the existence of an absorption edge (the *Kjeldaas edge*) for large magnetic fields in the propagation of ultrasonic waves along the direction of **B**, see Section 3.6.

An extensive survey of the results of magnetoacoustic investigations of the Fermi surfaces of a large number of metals is given by Roberts (1968). In addition to the determination of the shape of the Fermi surface, the study of ultrasonic attenuation in a metal, not necessarily in the presence of a magnetic field, can be used to help investigate electron–phonon interactions in that metal, see Chapter 6.

3.10. Conclusion

In this chapter we have discussed various experimental techniques that can be used to obtain information about the Fermi surfaces of metals. As mentioned in Section 3.1 there is inevitably a somewhat arbitrary distinction between the phenomena discussed in this chapter and those that will be discussed in Chapter 6. In the present chapter we have been primarily concerned with methods that are commonly used now, or have been successfully used in the past, to obtain qualitative or quantitative information about the Fermi surfaces of various metals. Thus certain experiments which could in principle be used to obtain information about the band structure and Fermi surface of a metal, but which have never achieved any significant popularity in this connection, have not been included. Perhaps some of the most significant of these are experiments involving the scattering of beams of particles by a specimen of a metal. Such experiments include the use of protons: proton channelling (see, for example, Machlin, Petralia, Desalvo, Missiroli, and Zignani 1968), or of electrons: low-energy electron diffraction, commonly referred to as LEED (see, for example, Stern and Taub 1970; Dederichs 1972).

The experiments described in the present chapter are usually performed on pure single-crystal specimens under normal, that is non-superconducting, conditions. The whole question of the origins of superconductivity and of the significance of the Fermi surface of a metal in the superconducting state will be discussed later, see Chapter 6. We simply note here that many of the methods described in the present chapter could not be used to investigate the Fermi surface of a metal in the superconducting state, although there are one or two exceptions such as positron annihilation.

It is also usual to determine the Fermi surface of a metal with a sample of the highest degree of purity obtainable and in the absence of any kind of mechanical strain. However, of late quite a considerable number of Fermi surface measurements have been made both with samples of a metal containing known impurities and with pure samples subjected to some mechanical strain. It might appear that workers have turned to the study of such systems as a result of the reduction in the number of pure metals that remain to be investigated. However, such a viewpoint is probably unduly cynical. For example, measurements on a sample containing small controlled amounts of a known impurity sometimes provide useful information about the shape of the Fermi surface of a pure metal, see Section 7.2. If the valence of the impurity is different from the valence of the host, one can assume, to a first approximation at least, that the energy bands are unaltered and that the only effect of the impurity is to change the number of conduction electrons per atom in the metal. This can be particularly useful for a metal in which some of the sheets of the Fermi surface enclose small pockets of carriers, such as Bi (see p. 303) and Zn and Cd (see p. 324). If the concentration of impurities is large, that is, one has a non-dilute alloy, the situation may become more complicated and we leave the discussion of this problem until later (see Chapter 7).

If a metal is subjected to a homogeneous strain this alters one or more of the basic vectors $\mathbf{t}_1$, $\mathbf{t}_2$, and $\mathbf{t}_3$ of the Bravais lattice. There will be consequential changes in the reciprocal lattice vectors $\mathbf{g}_1$, $\mathbf{g}_2$, and $\mathbf{g}_3$ and therefore in the dimensions of the Brillouin zone. By changing the relative positions of the atoms in a metal the various contributions to the potential $V(\mathbf{r})$ will be altered and therefore the details of the band structure will be altered too. In general we should therefore expect to find changes in both the dimensions and the connectivities of the Fermi surface of a metal subjected to a homogeneous strain. For a semi-metal, which is compensated and has only small isolated pockets of carriers, it is interesting to look for evidence of the metal-insulator transition which may occur for certain semi-metals when they are subjected to strain. However, if the strain is isotropic and if one makes use of the free-electron model (see eqn (1.5.7)) one would expect to find a simple scaling of the curves of $E_j(\mathbf{k})$ against $|\mathbf{k}|$ and therefore also a simple scaling of the dimensions of the Fermi surface by the same factor as that by which the dimensions of the Brillouin zone are scaled. In such circumstances the connectivity of the Fermi surface would not be altered. Therefore by measuring changes in the dimensions of the Fermi surface as a function of pressure, and comparing the results with calculations based on the scaling of the bands in the free-electron model, one can obtain a qualitative picture of the extent of the validity of the free-electron model for the metal in question. Even if the strain is not isotropic, it is still possible to calculate the expected changes in the dimensions of the Fermi surface based simply on scaling of the energy bands as a result of the changes in the lattice constants of the crystal under a homogeneous, but not necessarily isotropic, strain; however it has to be remembered that the connectivities as well as the dimensions of the Fermi surface may now be altered. The results of these calculations can then also be compared with the results of experimental measurements to determine how much of the experimental change is due to scaling, and how much is due to other changes in the potential $V(\mathbf{r})$. Many of the experimental methods that we have discussed in this chapter have now been used to investigate changes in the Fermi surface of a metal under tension or subjected to hydrostatic pressure. These include the use of the de Haas–van Alphen effect, the Schubnikow–de Haas effect, the magnetoresistance, and optical measurements; a general review of the effects of pressure on the band structures of various metals has been given by Drickamer (1965) and several references to work on particular metals are cited by Cracknell (1971*b*). Part of the reason for the interest in the effect of strain on the Fermi surface lies in the connection which can be shown to exist between the deformation tensor and the propagation characteristics for ultrasonic waves in a metal (Pippard 1960*b*; Perz, Hum, and Coleridge 1969; Testardi and Condon 1970).

4

THE FERMI SURFACES OF THE METALLIC ELEMENTS. I. SIMPLE METALS.

4.1. Introduction

IN THIS chapter and the next chapter we shall give a systematic account of the Fermi surfaces of all the metallic elements. We shall not attempt to give a complete set of references to work on these metals; references to work performed up to about the end of the 1960's are cited elsewhere (see Cracknell 1969*c*, 1971*a*, *b*). In general the present state of knowledge of the topology and the detailed dimensions of the Fermi surface of each metallic element was not discovered all at once but was accumulated as a result of various band structure calculations based on the methods described in Chapter 2 and of numerous experiments based on the techniques discussed in Chapter 3. Early reviews of results for various metals have been given by a number of authors (Chambers 1956*b*; Haering and Mrozowski 1960; Boyle and Smith 1963; Shoenberg 1964, 1969*a*; Lomer and Gardner 1969; Cracknell 1969*c*, 1971*a*, *b*).

The periodic table provides the most sensible classification for the study of the Fermi surfaces of the various metallic elements, because this classification is itself based on the electronic structures of the atoms of these elements. In Table 4.1 a convenient version of the periodic table is given with the division into s-block, p-block, d-block, and f-block elements. It is most useful for our present purposes to consider together those metals which are in the same group of the periodic table because such elements have very similar properties, whereas neighbours in horizontal periods exhibit gradual changes in their chemical, physical, and electronic properties. This chapter will be concerned with 'simple' metals which, roughly speaking, include those metals for which the free-electron model provides at least a first approximation to the shape of the Fermi surface.

TABLE 4.1

The periodic table

	s-block		*d-block*										*p-block*					
Group	IA	IIA											IIIB	IVB	VB	VIB	VIIB	O
	Li	Be											B	C	N	O	F	Ne
	Na	Mg	IIIA	IVA	VA	VIA	VIIA	—	VIII	—	IB	IIB	Al	Si	P	S	Cl	A
	K	Ca	Sc	Ti	V	Cr	Mn	Fe	Co	Ni	Cu	Zn	Ga	Ge	As	Se	Br	Kr
	Rb	Sr	Y	Zr	Nb	Mo	Tc	Ru	Rh	Pd	Ag	Cd	In	Sn	Sb	Te	I	Xe
	Cs	Ba	La[1]	Hf	Ta	W	Re	Os	Ir	Pt	Au	Hg	Tl	Pb	Bi	Po	At	Rn
	Fr	Ra	Ac[2]															

f-block

[1]Ce	Pr	Nd	Pm	Sm	Eu	Gd	Tb	Dy	Ho	Er	Tm	Yb	Lu
[2]Th	Pa	U	Np	Pu	Am	Cm	Bk	Cf	Es	Fm	Md	No	Lw

(Cooper 1968).

Chapter 5 will be concerned with 'non-simple' metals for which it is unlikely that the free-electron model gives even a first approximation to the shape of the Fermi surface; Chapter 5 therefore includes the transition metals, the lanthanide (or rare earth) metals, and the actinide metals. The distinction between 'simple' and 'non-simple' metals is somewhat arbitrary and, for convenience, we take a 'simple' metal to be one that belongs to the s-block or p-block of the periodic table and a 'non-simple' metal to be one that belongs to the d-block or f-block of the periodic table. This division is not entirely satisfactory because some authors regard the metals of group IB (Cu, Ag, Au) and group IIB (Zn, Cd, Hg) as simple metals on the grounds that, as we shall see in Chapter 5, it is still possible to see some relationship between the Fermi surfaces of these metals and the appropriate free-electron Fermi surfaces, whereas we include these metals with the transition metals in Chapter 5 because the d-bands, although full, are quite close to the Fermi level. In the present chapter we also have to be somewhat arbitrary about drawing the line between metallic and non-metallic elements. Thus Ge, Se, Si, grey Sn, and Te are regarded as semiconductors and are excluded although their band structures and constant energy surfaces have been studied in considerable detail, whereas graphite, though perhaps not usually regarded as a metal, does have a small overlap between the conduction and valence bands and therefore is included in our discussion. Fr and Ra are excluded on the grounds that, for fairly obvious reasons, hardly anything is known about their electronic structures.

At the start of each section we give the electronic structure of a free atom of each element in that group. This forms a convenient starting point for the discussion of the metallic states of these elements although it must be remembered that the concept of orbits and shells that has been developed for describing the electronic structures of atoms will generally only be useful for those electrons in a metal that are localized on one atom or ion core. The electrons that are deep inside the ion core are localized and can be regarded as belonging to well-defined and approximately **k**-independent levels. However, it is possible that electrons which one might, at first sight, have thought to be localized are in fact spread out into bands with a definite band structure. This is illustrated nicely by the 3s and 3p electrons in the band structure calculated by Mattheiss (1964*a*) for the non-metal solid A, see Fig. 2.1. So long as such bands remain separated from the conduction band by a large energy gap E_0 ($E_0 \gg kT$) this broadening is not very important. However, it is important to realize that there is not necessarily always a clear distinction in a metal between the electrons in the ion cores and the conduction electrons. At the start of each section we also indicate the crystal structure of each of the metallic elements discussed in that section.

4.2. Group IA

Li, Na, K, Rb, Cs

Li	Lithium	2.1	b.c.c.
Na	Sodium	2.8.1	b.c.c.
K	Potassium	2.8.8.1	b.c.c.
Rb	Rubidium	2.8.18.8.1	f.c.c.
Cs	Caesium	2.8.18.18.8.1	f.c.c.

The crystal structures indicated above apply at room temperature. Li and Na partially transform to the h.c.p. structure below 78 K and 35 K respectively (Barrett 1956) but similar transformations were not found for K, Rb, and Cs down to 1·2 K (Barrett 1955). Since Fermi surface determinations are usually performed above 1·2 K the Fermi surfaces that have been found for K do apply to the high-temperature form. The possibility of a direct determination of the Fermi surface of b.c.c. Li above 78 K is unlikely (see Chapter 3) except possibly by positron annihilation experiments, and there would seem to be little chance of freezing in the b.c.c. structure at a sufficiently low temperature either. Even if one felt that a determination of the shape of the Fermi surface of the h.c.p. phase of Li would be useful for understanding the low-temperature properties of Li, this would still be a difficult experimental problem. With a transition temperature of 78 K it would not only be difficult to grow a single crystal of h.c.p. Li but also, if one did succeed in growing such a crystal, it would always have to be kept at a temperature below 78 K. However, for Na the transition temperature is only 35 K, so that there is a good chance that as a single crystal b.c.c. sample of Na is cooled below 35 K the b.c.c. structure will be frozen in. The Fermi surface determination would then automatically be a determination of the Fermi surface of the high-temperature form in a metastable state below the transition temperature (Lee 1966). It is, of course, important in such experiments to check that the sample still has the b.c.c. structure, either by X-ray work at the low temperature or by checking the rotational symmetry of the effect that is being used to determine the shape of the Fermi surface. For the other alkali metals any possible transition to the h.c.p. structure is unimportant because it must be at a temperature below that normally used in Fermi surface studies. In the alkali metals the Fermi surface is only very slightly distorted from the free-electron sphere which was described in Section 1.6 and which is completely contained within the first Brillouin zone. This means that the effect of the periodic lattice on the conduction electrons is very small, so that even when a phase transformation occurs in these metals there will be very little change in the contours of the energy surfaces or in the shape of the Fermi surface.

Some of the earliest calculations of the energies of electrons in metals were performed on the alkali metals (Wigner and Seitz 1933, 1934; Herring and Hill 1940; von der Lage and Bethe 1947; Howarth and Jones 1952). More recently these metals have continued to receive a large proportion of the time and energy of those theoreticians engaged in band structure calculations; this is doubtless because of the relative simplicity of the electronic structures of these metals. The general conclusion of all these calculations was that the band structures of the alkali metals are quite similar to the free-electron energy bands for a monovalent metal with the appropriate crystal structure. This can be seen in the values of a number of parameters such as the density of states, the Fermi energy, and the cyclotron, thermal, and optical effective masses, which were calculated for the alkali metals by Ham (1962*b*), see Table 4.2. The Fermi surfaces of Li and Cs were found to be the most distorted from the spherical free-electron Fermi surface and exhibited bulges in the [110] directions, while that of Na was

TABLE 4.2

Calculated parameters of the Fermi surfaces of the alkali metals

	Li	Na	K	Rb	Cs
a (A.U.)	6·651	8·109	10·049	10·742	11·458
m_t/m_0	1·64	1·00	1·07	1·18	1·75
$(m_t/m_0)_s$	1·32	1·00	1·01	0·99	1·06
m_a/m_0	1·45	1·00	1·02	1·06	1·29
m_t/m_a	1·13	1·00	1·06	1·11	1·36
$m_{c[110]}/m_0$	1·48	1·00	1·035	1·07	1·46
$m_{c[100]}/m_0$	1·65	1·00	1·063	1·16	1·92
$m_{c[111]}/m_0$	1·82	1·00	1·092	1·25	2·38
$\mathscr{A}_{[110]}/\mathscr{A}_0$	0·976	1·00	0·995	0·979	0·94
$\mathscr{A}_{[100]}/\mathscr{A}_0$	0·993	1·00	1·001	0·996	0·99
$\mathscr{A}_{[111]}/\mathscr{A}_0$	1·011	1·00	1·007	1·013	1·04
A/A_0	1·06	1·00	1·03	1·06	1·12
$(A/A_0)^2$	1·11	1·00	1·06	1·13	1·25
$k_{[110]}/k_F$	1·023	1·00	1·007	1·018	1·08
$k_{[100]}/k_F$	0·973	1·00	0·994	0·980	0·94
$k_{[111]}/k_F$	0·983	1·00	0·994	0·980	0·94

a is the lattice constant, m_t thermal mass, m_a optical mass, m_c cyclotron mass, $\mathscr{A}$ extremal area of cross section of the Fermi surface, A surface area of the Fermi surface, k radius of Fermi surface. A_0 and $\mathscr{A}_0$ denote the corresponding areas of a spherical Fermi surface and k_F its radius. (Adapted from Ham 1962*b*.)

found (Hughes and Callaway 1964) to be spherical to within about one-quarter per cent. According to a band structure calculation by Callaway (1961), the Fermi surface for b.c.c. Li is distorted from the spherical free-electron Fermi surface by having bulges of the order of five per cent along the [110] directions. By performing band structure calculations for various values of the lattice constant, corresponding to known possible changes in the external pressure, Ham (1962*b*) predicted that the Fermi surface of Li should not touch the

Brillouin zone boundary even under substantial pressure, while contact should occur for Cs at a slight compression.

A review of the information that could be obtained about the electronic band structures and the topology of the Fermi surfaces of the alkali metals by studying their macroscopic properties was given by Cohen and Heine (1958). Their general conclusions, which were in agreement with the results of band structure calculations and which have largely been borne out by later experiments, were that the Fermi surface of Na is nearly spherical, the Fermi surface of K is somewhat distorted, the Fermi surfaces of Rb and Cs are rather more distorted, while the Fermi surface of Li is also considerably distorted and might even touch the Brillouin zone boundary. This means that there is a steady trend of increasing distortion from sphericity as the atomic number is increased, with the exception that the behaviour of Li is somewhat anomalous. This anomalous behaviour of Li arises because the Li atom has a very small ion core, $Li^{+}(1s^{2})$, so that the valence electron is, in general, closer to the nucleus than in the other alkali metals. Its wave function is therefore more distorted from a free-electron wave function and, consequently, the energy bands and the Fermi surface itself can be expected to be more distorted from the free-electron model than they are in the other alkali metals.

The most direct experimental determinations of the shapes of the Fermi surfaces of the alkali metals have been based on the use of the de Haas–van Alphen effect. A special field-modulated technique was developed by Shoenberg and Stiles (1963) which is capable of detecting and measuring very small departures from a spherical Fermi surface. This technique has been applied to Na (Shoenberg and Stiles 1964; Lee 1966), K (Shoenberg and Stiles 1964; Lee and Falicov 1968) and Rb (Shoenberg and Stiles 1964). Contours of the Fermi surfaces of Na, K, and Rb obtained from de Haas–van Alphen measurements are shown in Figs. 4.1–3. Results for Cs were obtained by Okumura and Templeton (1965) using the pulsed field method. The pulsed field method yields results of considerably greater accuracy in Cs than it does in K and Rb for which the distortions from

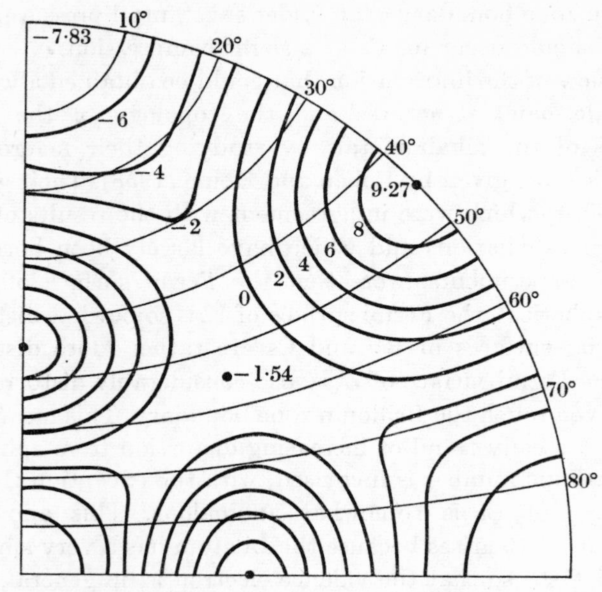

Fig. 4.1. Contours of $\Delta r/r$ of the Fermi surface of Na; units are parts in 10^4 (Lee 1966).

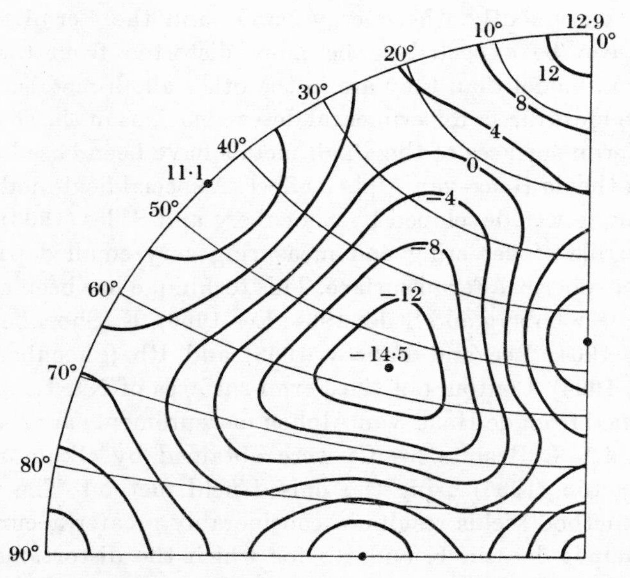

Fig. 4.2. Contours of $\Delta r/r$ of the Fermi surface of K; units are parts in 10^4 (Shoenberg and Stiles 1964).

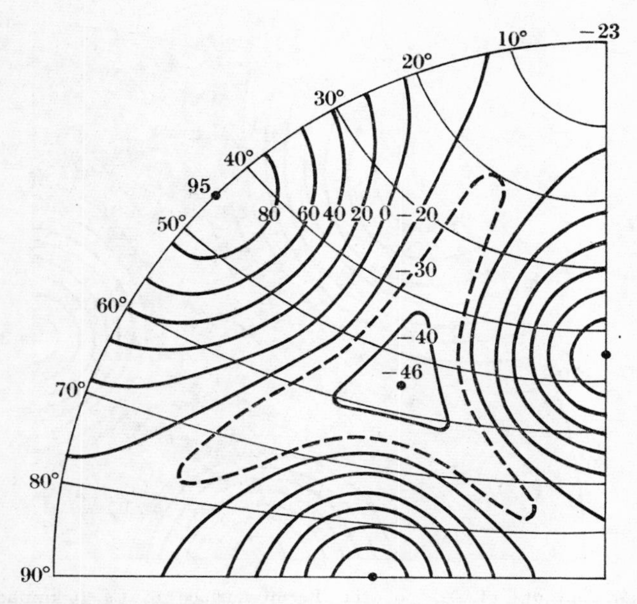

FIG. 4.3. Contours of $\Delta r/r$ of the Fermi surface of Rb; units are parts in 10^4 (Shoenberg and Stiles 1964).

sphericity are so small. In the [110] direction Okumura and Templeton observed bumps in the Fermi surface and contours of $\Delta r/r$ in Cs are plotted on a stereogram in Fig. 4.4 from which it will be seen that the distortion is considerably larger than in Rb.

A large number of other experimental investigations of the electronic structures of the alkali metals has been performed using a variety of techniques, including positron annihilation, magnetoacoustic effects, cyclotron resonance and radio-frequency size effects. However, in most cases they do not provide any extra information about the shapes of the Fermi surfaces of these metals beyond that already provided by the de Haas–van Alphen measurements. We have already mentioned the experimental difficulties involved in studying the Fermi surface of Li because of the inconvenient temperature at which the phase change occurs. However, experiments on positron

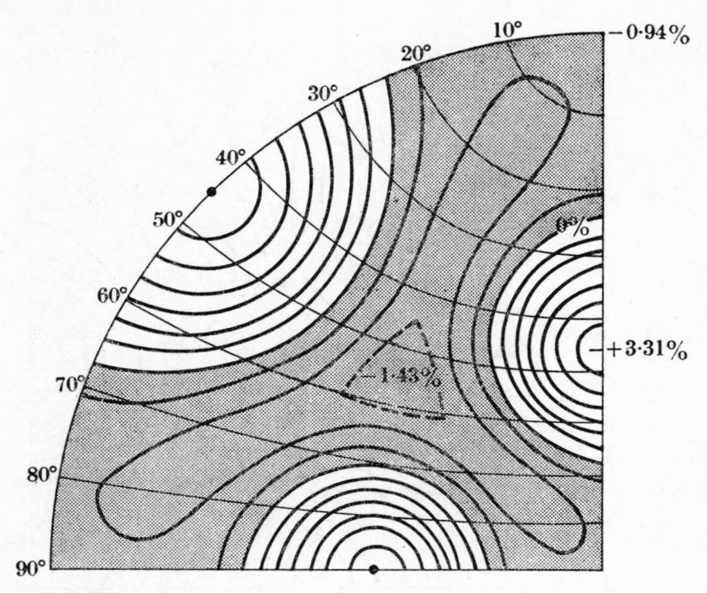

FIG. 4.4. Contours of $\Delta r/r$ of the Fermi surface of Cs (Okumura and Templeton 1965).

annihilation showed that the Fermi surface of Li was quite distorted from sphericity with $|\mathbf{k}_{110}|$ being about five or six per cent greater than $|\mathbf{k}_{100}|$ (Donaghy, Stewart, Rockmore, and Kusmiss 1964; Donaghy and Stewart 1967*a*,*b*). Detailed theoretical predictions of the positron annihilation angular correlation curves for Li were made by Melngailis and De Benedetti (1966) using the O.P.W. method to determine the band structure and the conduction electron wave functions. By including both the Fermi surface topology and the effect of the wave functions, but not just the Fermi surface topology on its own, good agreement with the experimentally observed anisotropies was obtained. Quantitative agreement with experiment was obtained when, in addition to all other effects, the enhancement factor due to electron–positron attraction was included.

It was suggested by Overhauser (1962) that several phenomena in the behaviour of certain alkali metals, in particular K, could be explained by assuming that the electronic ground state of an alkali metal consists of a spin-density wave (or sinusoidal antiferromagnetic ordering of the electron spins). More recently, the suggestion was modified to that of a charge-density wave instead. However, the evidence for the existence of these spin-density waves, or charge-density waves, is rather indirect and they have proved rather too elusive to be observed in any direct and convincing way (see Cracknell 1971*b* p. 25).

4.3. Group IIA

Be, Mg, Ca, Sr, Ba

Be	Beryllium	2.2	h.c.p.
Mg	Magnesium	2.8.2	h.c.p.
Ca	Calcium	2.8.8.2	f.c.c.
Sr	Strontium	2.8.18.8.2	f.c.c.
Ba	Barium	2.8.18.18.8.2	b.c.c.

The crystal structures indicated above apply at room temperature, although Ca and Sr undergo phase transitions to h.c.p. and b.c.c. structures, respectively, at high temperatures (Jayaraman, Klement, and Kennedy 1963). The electrical resistivities of Ca and Sr exhibit large changes at very high pressures (Stager and Drickamer 1963; Drickamer 1965); attempts have been made to explain this behaviour in terms of structural phase changes but it now seems more likely to be due to changes in the electronic band structure with pressure, without any change from the f.c.c. crystallographic structure.

Early experimental work on the galvanomagnetic properties of Be indicated that Be has a closed Fermi surface although later investigations of the magnetoresistance of Be at higher fields which were performed by Alekseevskiĭ and Egorov (1963*a*, *b*; 1964) indicated that in fields larger than about 50 kG magnetic breakdown occurs and open orbits appear parallel to the [0001] direction in Be. The shape of the Fermi surface of Be was established primarily as a result of band structure calculations (Loucks 1964; Loucks and Cutler 1964; Terrell 1964, 1966) and de Haas–van Alphen measurements (Watts 1963,

1964*b*). Loucks and Cutler (1964) determined the band structure of Be at a large number of general points in the Brillouin zone using a self-consistent O.P.W. calculation. The calculated density of states was found to be in agreement with soft X-ray emission and absorption data and with the experimental value

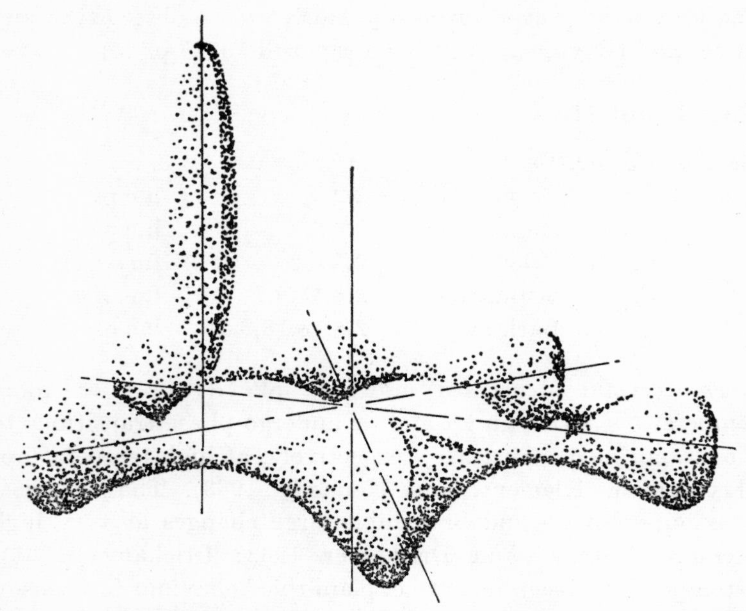

FIG. 4.5. The Fermi surface of Be; cigar and coronet (Loucks and Cutler 1964).

of the electronic specific heat. The Fermi surface was found to consist of

(i) a region of holes in band 2 which resembles a coronet and

(ii) two identical pockets of electrons in band 3 similar in shape to a cigar with a triangular cross section.

This Fermi surface can be produced by introducing a certain amount of distortion to the free-electron Fermi surface of a divalent h.c.p. metal shown in Fig. 1.18. The Fermi surface of

Be is illustrated in Fig. 4.5 and some principal cross-sections are shown in Fig. 4.6.

The free-electron Fermi surface for a divalent h.c.p. metal has sheets in the first four bands in the single-zone scheme and they are illustrated in Fig. 1.18. This Fermi surface consists of

(i) a 'monster' of holes in bands 1 and 2 and
(ii) pockets of electrons in bands 3 and 4 with the shapes of a lens at Γ, and of butterflies, cigars, and caps at various positions around the edges of the Brillouin zone.

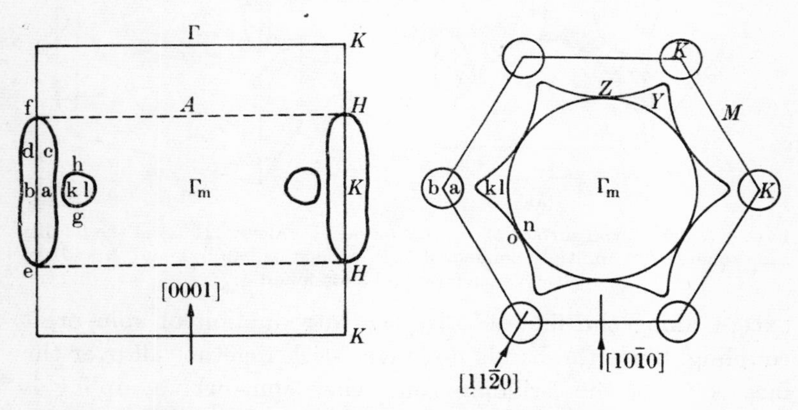

FIG. 4.6. Two sections through the Fermi surface of Be (Watts 1963).

Although some of these sheets of the Fermi surface are missing in Be, early experimental work verified the presence of all of them in the Fermi surface of Mg. A band structure calculation for Mg was performed by Falicov (1962) using the O.P.W. method and the shape of the Fermi surface deduced on the basis of this calculation is illustrated in Fig. 4.7. This model is topologically equivalent to the free-electron model for a divalent h.c.p. metal although the sizes of the various sheets of the Fermi surface are rather different. For an h.c.p. metal the effects of spin–orbit coupling are particularly important on the hexagonal top face *AHL* of the Brillouin zone; the double degeneracy of the energy bands that exists all over this face in the absence of spin–orbit coupling will be lifted everywhere,

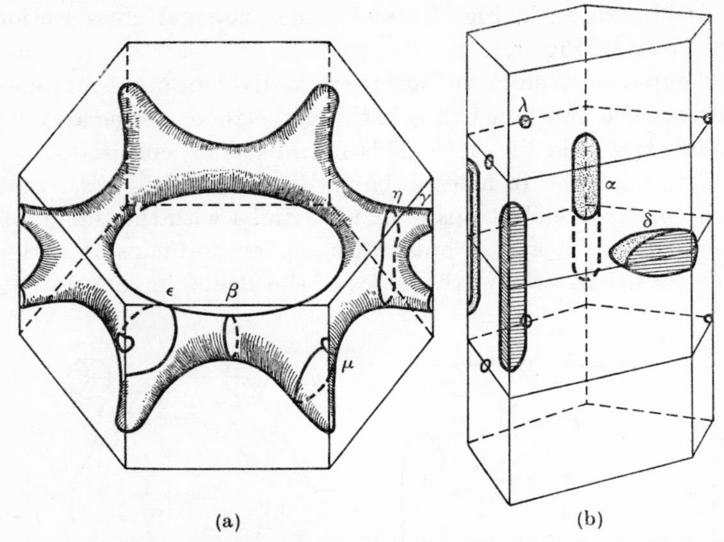

FIG. 4.7. The Fermi surface of Mg calculated by Falicov (1962) in the double zone scheme: (a) multiply-connected hole surface in bands 1 and 2, and (b) electron surfaces in bands 3 and 4.

except along the line AL, by the introduction of spin–orbit coupling. Since the bands no longer stick together all over the face AHL of the Brillouin zone when spin–orbit coupling is included, the double-zone scheme that is commonly used for h.c.p. metals is no longer valid. The magnitude of the spin–orbit splitting of the energy bands at the corners H of the Brillouin zone of Mg was calculated by Falicov and Cohen (1963) to be $4{\cdot}16\times10^{-4}$ Ry, $5{\cdot}52\times10^{-7}$ Ry, and $4{\cdot}20\times10^{-4}$ Ry for three different bands. Because of the relative smallness of these spin–orbit splittings, magnetic breakdown effects must be expected to occur at relatively low magnetic fields, in which case the electrons will ignore the gaps and re-establish the connectivity that would have been obtained if one had chosen to neglect spin–orbit coupling; in practice magnetic breakdown does indeed occur in Mg at quite low fields, of the order of 0·2 kG. As a result of numerous de Haas–van Alphen measurements on Mg, as well as some galvanomagnetic and magneto-acoustic measurements, agreement was obtained with the general features of the Fermi surface calculated by Falicov (1962) for Mg which is topologically the same as the free-electron Fermi surface. However, a new and much more detailed study of the Fermi surface of Mg by using magneto-acoustic geometric resonances (Ketterson and Stark 1967) and the de Haas–van Alphen effect (Stark 1967) showed that the Fermi surface dimensions of Mg are much closer to those of the free-electron model than in Falicov's predictions. This can be illustrated with the values obtained for the dimensions of the lens in band 3 (see Table 4.3). Dimensions of the other sheets

TABLE 4.3

Dimensions of the lens in band 3 in Mg *(atomic units)*

Direction	*Free-electron model*	*Calculated (Falicov 1962)*	*Experimental (Ketterson and Stark 1967)*
ΓA	0·085	0·058	0·080
ΓK	0·341	0·253	0·312
ΓM	0·341	0·255	0·312

of the Fermi surface of Mg are given by Ketterson and Stark (1967). As we shall see in Section 5.3 there is, not unnaturally, a close similarity between the Fermi surface of Mg and the Fermi surfaces of the divalent h.c.p. metals Zn and Cd of group IIB.

The Fermi surfaces of the divalent f.c.c. metals Ca and Sr in the free-electron approximation are illustrated in Fig. 1.16. Each of them consists of a multiply-connected hole surface in band 1 and a set of lens-shaped pockets at L on the surface of the Brillouin zone in band 2.

The de Haas–van Alphen effect in Ca was studied by Condon and Marcus (1964) and periods corresponding to three distinct extremal areas of cross section of the Fermi surface were observed. The results were interpreted in terms of a Fermi surface suggested by Harrison (1963*b*) on a nearly free-electron model, see Fig. 4.8, but there appear to be more orbits predicted on Harrison's model than the three actually observed by Condon and Marcus. Band structure calculations have been performed for Ca (Altmann and Cracknell 1964; Vasvari, Animalu, and Heine 1967; Vasvari and Heine 1967; Vasvari 1968; Chatterjee and Chakraborti 1970; Altmann, Harford, and Blake 1971) and Sr (Cracknell 1967*b*; Vasvari, Animalu, and Heine 1967). In each of these calculations a Fermi surface

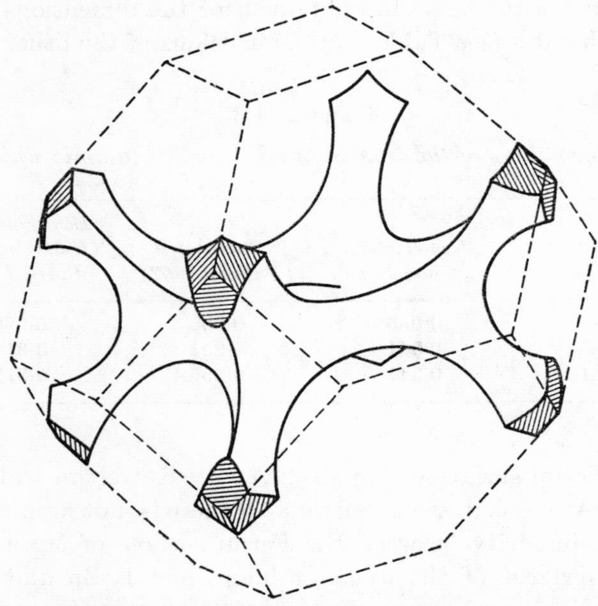

Fig. 4.8. The multiply-connected hole Fermi surface for Ca in band 1 proposed by Harrison (1963*b*).

was produced that departed significantly from the free-electron model for a divalent f.c.c. metal. Although the lens-shaped pockets of electrons in band 2 were still similar to these pockets in the free-electron model, it was originally thought that the multiply-connected regions of holes in band 1 becomes separated into a number of isolated small pockets of holes. There is some discrepancy between the suggested positions of these pockets of holes; according to Altmann and Cracknell (1964) and Chatterjee and Chakraborti (1970) they are at K whereas according to Vasvari (1968) they are at W, see

Fig. 4.9. However, from more recent calculations it appears that the Fermi surface in band 1 is multiply—connected after all, see Fig. 4.9(c) (Altmann, Harford, and Blake 1971). The volume associated with holes in the Brillouin zone in band 1 must be equal to the volume of the region occupied by electrons in band 2, that is the metal is compensated. However, the actual volume of holes, or of electrons, is quite small and this feature is reflected in the behaviour of the electrical resistance of Ca and Sr under pressure. The resistance of Ca rises steeply at a pressure of about 150 kbar followed by a fall at about 300 kbar, see Fig. 4.10(a); the high value

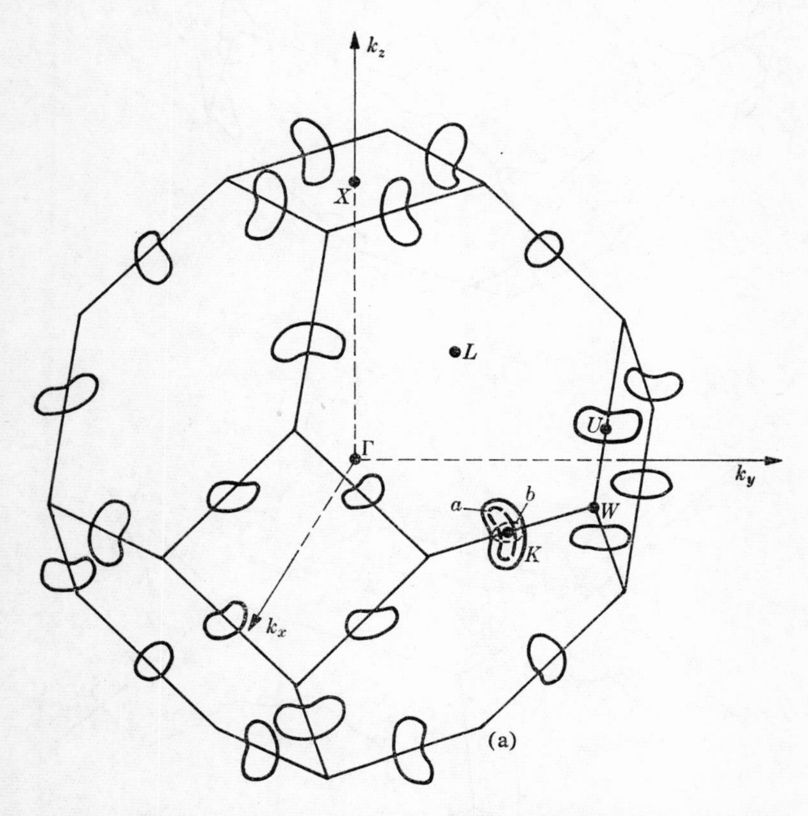

FIG. 4.9. The hole Fermi surface of Ca in band 1 calculated by (a) Altmann and Cracknell (1964), (b) Vasvari (1968), and (c) Altmann, Harford, and Blake (1971).

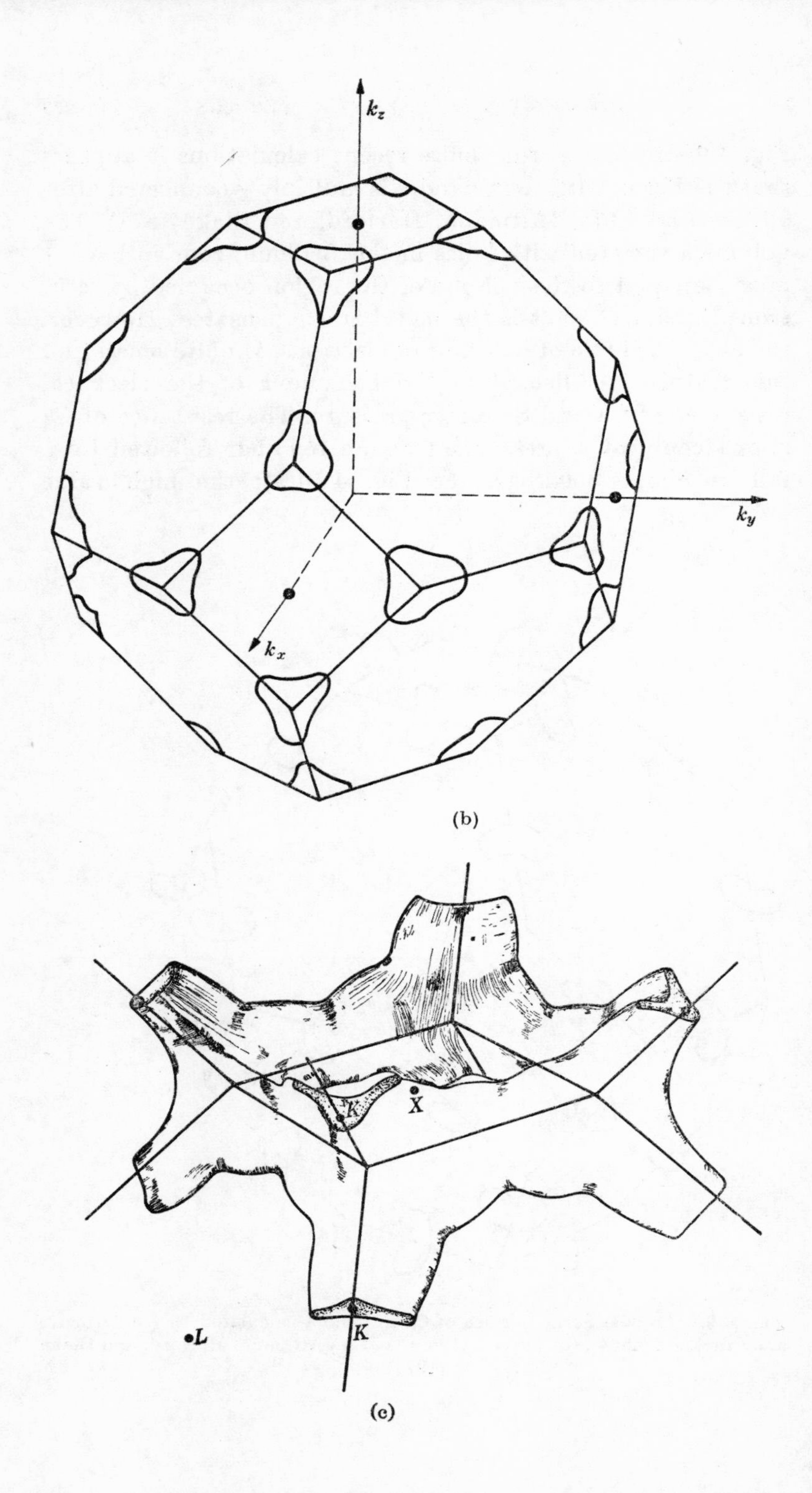

(b)

(c)

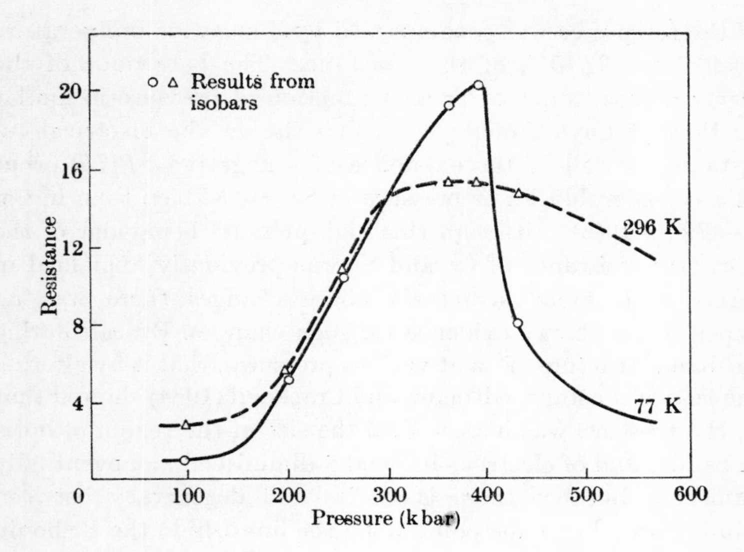

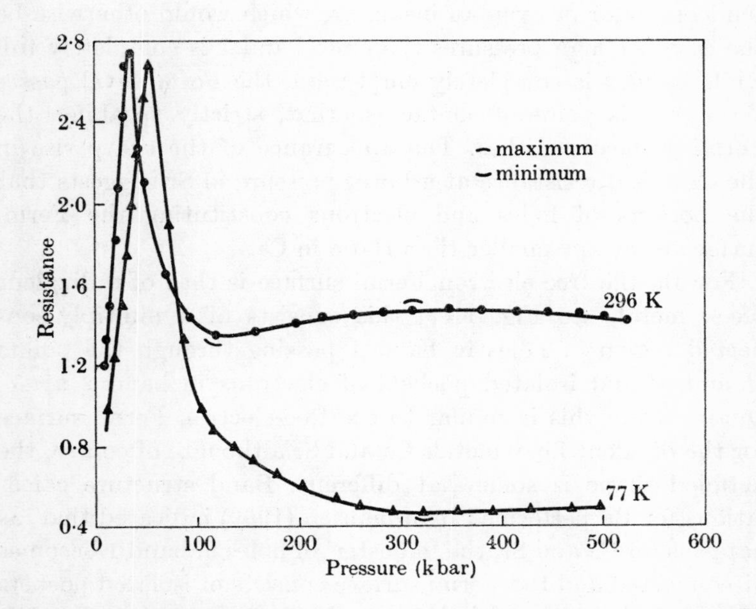

FIG. 4.10. The resistance (arbitrary units) versus pressure of (a) Ca, and (b) Sr (Stager and Drickamer 1963).

of the resistance was accompanied by a negative temperature coefficient, $\partial R/\partial T$, of the resistance. The behaviour of the electrical resistance of Sr as a function of pressure is similar to that of Ca although the sharp rise in the electrical resistance as well as the existence of a negative $\partial R/\partial T$ occur at a considerably lower pressure in Sr ($\sim$35 kbar) than in Ca, see Fig. 4.10(b). Although this high-pressure behaviour of the electrical resistance of Ca and Sr was previously explained in terms of assumed structural phase changes there was no experimental X-ray evidence for such changes. By calculating the band structure of Ca at various pressures, that is by altering the lattice constant, Altmann and Cracknell (1964) showed that as the pressure was increased so the size of the region of holes in band 1 and of electrons in band 2 diminished and eventually vanished. Because there is an isolated degeneracy between bands 1 and 2 at some point along the line LW in the Brillouin zone, Ca remains a semi-metal rather than becoming an intrinsic semiconductor or even an insulator, which would otherwise be the case. At high pressures therefore band 1 is completely full while band 2 is completely empty and the Fermi level passes through this point of contact so that, strictly speaking, the Fermi surface vanishes. The appearance of the sharp rise in the electrical resistance at a lower pressure in Sr suggests that the pockets of holes and electrons constituting the Fermi surface of Sr are smaller than those in Ca.

For Ba the free-electron Fermi surface is that of a divalent b.c.c. metal, see Fig. 1.17; this consists of a multiply-connected region of holes in band 1 passing through the points H and P and isolated pockets of electrons in band 2 at N. Qualitatively this is similar to the free-electron Fermi surface for the divalent f.c.c. metals Ca and Sr although, of course, the detailed shape is somewhat different. Band structure calculations for Ba performed by Johansen (1969) indicated that, as happens for Ca and Sr, the 'monster' of holes in band 1 becomes disconnected and the Fermi surface consists of isolated pockets of holes in band 1 and isolated pockets of electrons in band 2. The pockets of holes in band 1 were found to be at P and each of them was described as a 'tetracube', while the pockets of electrons in band 2 were found to be at H and each of them was described as a 'superegg'. While the tetracube of holes at P is an obvious survivor of the free-electron monster in band 1, it is a little surprising that the pocket of electrons in band 2 should appear at H rather than at N as one might have expected

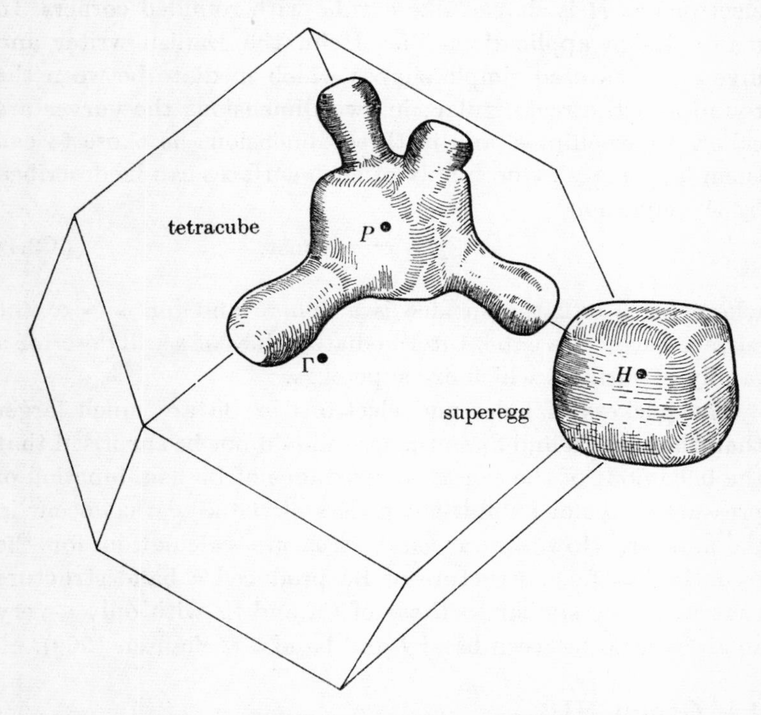

FIG. 4.11. The Fermi surface of Ba calculated by Johansen (1969) (see also Fig. 5.45) (McEwen 1971).

from the free-electron model. These two sheets of the Fermi surface of Ba are illustrated in Fig. 4.11 and three sets of oscillations attributable to the tetracube have been observed in the de Haas–van Alphen measurements of McEwen (1969, 1971).

The terms 'tetracube' and 'superegg' were introduced originally by Andersen and Loucks (1968) in connection with the rare-earth metal Eu which, being a divalent b.c.c. metal with atomic number quite close to that of Ba, has a Fermi surface very similar to that of Ba (see Section 5.10). The region of holes at P is shaped like a rounded cube and has ellipsoidal projections arranged tetrahedrally on four of the eight corners, hence the name 'tetracube'. The region of electrons at H is shaped like a cube with rounded corners. In many design applications Piet Hein, the Danish writer and inventor, has used simple shapes which mediate between the round and the rectangular; in two dimensions the curves are called 'super-ellipses' and in three dimensions he chose to call them 'supereggs'. One family of such surfaces can be described by the equation

$$x^n+y^n+z^n = \text{const.} \tag{4.3.1}$$

where for $n = 2$ the surface is a sphere and for $n \to \infty$ the surface becomes a cube. Intermediate values of n will describe a family of surfaces which are supereggs.

The pockets of holes and electrons in Ba are much larger than those in Ca and Sr so that we should not be surprised that the behaviour of the electrical resistance of Ba as a function of pressure does not exhibit steep rises of the kind that occur in Ca and Sr. However, a band structure calculation for the hypothetical f.c.c. structure of Ba produced a band structure that was very similar to those of Ca and Sr with only a very small overlap between band 1 and band 2 (Johansen 1969).

4.4. Group IIIB

Al, Ga, In, Tl

[B	Boron	2.(2, 1)]	—
Al	Aluminium	2.8.(2, 1)	f.c.c.
Ga	Gallium	2.8.18.(2, 1)	orthorhombic, see below
In	Indium	2.8.18.18.(2, 1)	tetragonal, see below
Tl	Thallium	2.8.18.32.18.(2, 1)	h.c.p.

The face-centered tetragonal structure of In, which belongs to the space group $F4/mmm$ (D_{4h}^{17}), is simply related to the face-centered cubic structure by a distortion along the c-axis ($a = 4{\cdot}58$ Å, $c = 4{\cdot}94$ Å). Ga has a complicated orthorhombic structure ($Abma$, D_{2h}^{18}) with 8 Ga atoms in the conventional unit cell at

$$[000;\ 0\tfrac{1}{2}\tfrac{1}{2}]\pm[x0z;\ \tfrac{1}{2}+x,\ \tfrac{1}{2},\ \bar{z}],$$

where $x = 0{\cdot}079$ and $z = 0{\cdot}153$ ($a = 4{\cdot}52$ Å, $b = 4{\cdot}51$ Å, $c = 7{\cdot}65$ Å). A discussion of the reasons for the existence of this complicated structure in terms of the behaviour of the electrons is given by Heine (1968). Tl transforms to the b.c.c. structure above 230°C.

Al contains three conduction electrons per atom and therefore, in the free-electron approximation, band 1 is full, band 2 is nearly all full but the Fermi surface does not touch the Brillouin zone boundary, in band 3 the Fermi surface contains sets of arms that intersect the Brillouin zone boundary and form a connected 'monster', and band 4 contains a few very small isolated pockets of electrons (see Fig. 1.16). The initial investigations of the Fermi surface of Al were made by means of the de Haas–van Alphen effect (Shoenberg 1952*a*; Gunnerson 1957). This work was followed by an O.P.W. band structure calculation performed by Heine (1957*a*,*b*,*c*) in which many different contributions to the potential were included. The energy eigenvalues obtained, which were quite close to the free-electron values, were used by Harrison (1959) as the starting point of a pseudopotential interpolation scheme to deduce the shape of the Fermi surface of Al. For the Fermi surface computed in this way the band 4 pockets of electrons of the free-electron model disappeared and the dimensions of the band 2 Fermi surface were altered slightly although that surface was not altered topologically. The Fermi surface in band 3 was also slightly distorted from the free-electron model. This sheet of the Fermi surface proposed by Harrison showed the same connectivities as the free-electron Fermi surface, see Fig. 4.12, but there remained some doubt as to whether all the

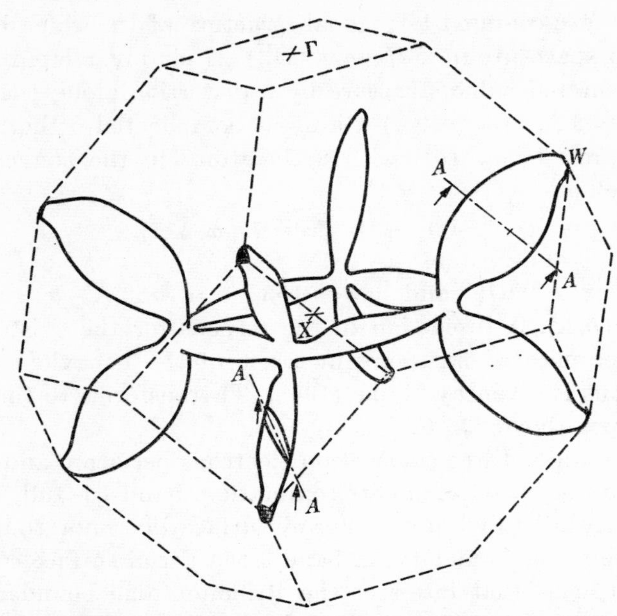

FIG. 4.12. The Fermi surface of electrons of Al in band 3 proposed by Harrison (1959); note that X and not Γ is at the centre of the zone.

arms were connected together or not. Magnetoresistance measurements by Alekseevskiĭ and Gaĭdukov (1959*a*) and Lüthi (1960) suggested that there are no open orbits possible on the Fermi surface of Al. Although several further band structure calculations were performed for Al, the accuracy of these calculations in band 3 in the crucial regions near the corner of the Brillouin zone at W was inadequate to fix the connectivities of the Fermi surface in band 3. Moreover, while further experimental measurements on Al were made using both the de Haas–van Alphen effect and magnetoacoustic effects, it proved difficult to establish the connectivity at W of the sheet of the Fermi surface in band 3 because of the possibility of magnetic breakdown. However, at last it was established, principally as a result of a pseudopotential analysis by Ashcroft (1963*b*) and subsequent experimental measurements by Vol'skiĭ (1964) and Larson and Gordon (1967). The important difference between the conclusions of Ashcroft and the model proposed by Harrison (see Fig. 4.13) is that according to Ashcroft, band 3 at W is very slightly above the Fermi level. This alters the connectivities and the monster in band 3 becomes dismembered by becoming disconnected at W, see Fig. 4.13. When the connectivity at W is broken, the dis-

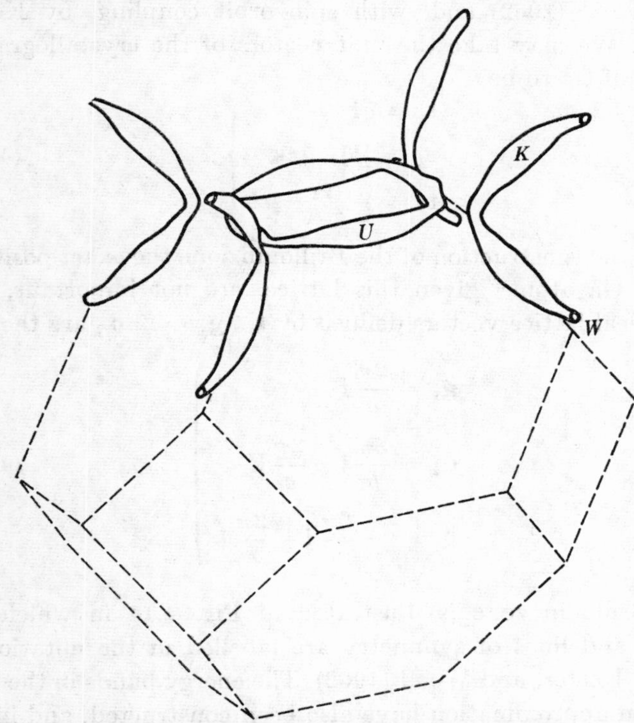

Fig. 4.13. The Fermi surface of electrons in band 3 of Al proposed by Ashcroft (Larson and Gordon 1967).

membered monster in band 3 becomes reduced to a set of square rings in various orientations where each ring consists of four sausage-like arms joined together. Subsequent low-field de Haas–van Alphen measurements (Larson and Gordon 1967) and measurements of quantum oscillations of the surface resistance of Al (Vol'skiĭ 1964) demonstrated fairly convincingly the dismembering of the monster.

As mentioned at the beginning of this section the conventional orthorhombic unit cell of Ga contains 8 Ga atoms. The fundamental unit cell therefore contains 4 Ga atoms so that, with a valence of 3, an equivalent of 6 energy bands must be occupied throughout the Brillouin zone. The free-electron Fermi surface for Ga has been investigated by Slater, Koster, and Wood (1962) and, with spin–orbit coupling, by Koster (1962). We may take the unit vectors of the crystallographic lattice of Ga to be

$$\left.\begin{aligned} \mathbf{t}_1 &= a\mathbf{i} \\ \mathbf{t}_2 &= \tfrac{1}{2}b\mathbf{j}+\tfrac{1}{2}c\mathbf{k} \\ \mathbf{t}_3 &= -\tfrac{1}{2}b\mathbf{j}+\tfrac{1}{2}c\mathbf{k} \end{aligned}\right\}, \tag{4.4.1}$$

and in the construction of the Brillouin zone the exact positions of the Ga atoms, given this lattice, are not important. The reciprocal lattice vectors defined by $\mathbf{t}_i \cdot \mathbf{g}_j = 2\pi\delta_{ij}$ are then

$$\left.\begin{aligned} \mathbf{g}_1 &= \frac{2\pi}{a}\,\mathbf{i} \\ \mathbf{g}_2 &= \frac{2\pi}{b}\,\mathbf{j}+\frac{2\pi}{c}\,\mathbf{k} \\ \mathbf{g}_3 &= -\frac{2\pi}{b}\,\mathbf{j}+\frac{2\pi}{c}\,\mathbf{k} \end{aligned}\right\}. \tag{4.4.2}$$

The Brillouin zone is illustrated in Fig. 4.14 in which the points and lines of symmetry are labelled in the notation of Slater, Koster, and Wood (1962). The energy bands in the free-electron approximation have also been constructed, and in the neighbourhood of the Fermi level they are exceedingly complicated (see Fig. 5 of Slater, Koster, and Wood 1962) with the result that there are pieces of Fermi surface in several bands. For example, at Γ nine bands are occupied but at X only two bands are occupied. Not only is this free-electron Fermi surface exceedingly complicated, involving bands 3, 4, 5, 6, 7, 8, and 9, but also it is clear that any quite small changes to the free-electron bands as the result of using a non-zero potential, will make drastic changes to the topology of the Fermi surface.

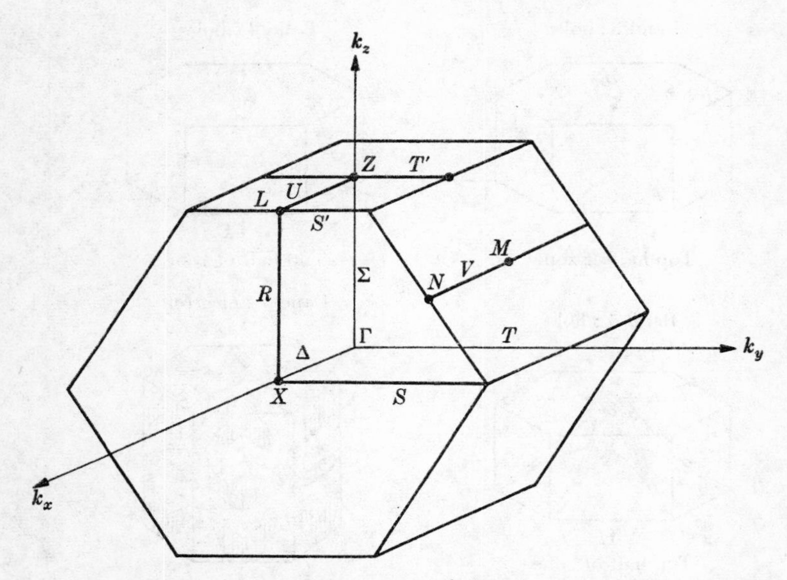

FIG. 4.14. The Brillouin zone for Ga (Slater, Koster, and Wood 1962).

Also the presence of pieces of Fermi surface in so many bands makes it difficult to interpret experimental data on the de Haas–van Alphen, magnetoacoustic, and other 'effects'. The various pieces of the free-electron Fermi surface for Ga are illustrated in Fig. 4.15. It will be noticed that this free-electron Fermi surface appears to exhibit hexagonal symmetry. This hexagonal symmetry is accidental and not quite exact; it arises as a result of the special values of b and c which lead to angles on the hexagonal face of about 59° and 61°.

A band structure calculation for Ga has been performed by Wood (1966) using the A.P.W. method and Fig. 4.16 shows the Fermi surface constructed by Goldstein and Foner (1966) from the sections given by Wood. There are many differences from the free-electron Fermi surface shown in Fig. 4.15. The small pieces of Fermi surface in bands 3, 4, and 9 have disappeared and several pieces of Fermi surface in bands 7 and 8 have also disappeared. The pseudohexagonal symmetry which was such a marked feature of the free-electron model is not present in the

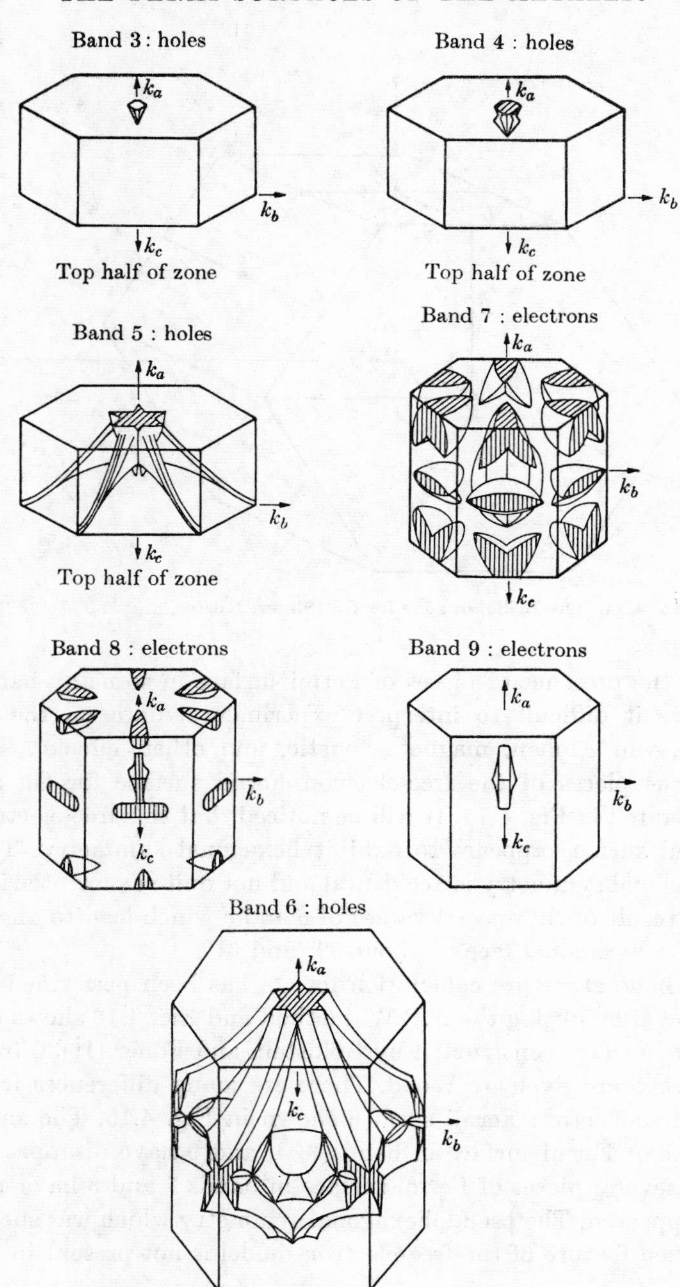

FIG. 4.15. The Fermi surface of Ga in the free-electron approximation (Reed and Marcus 1962).

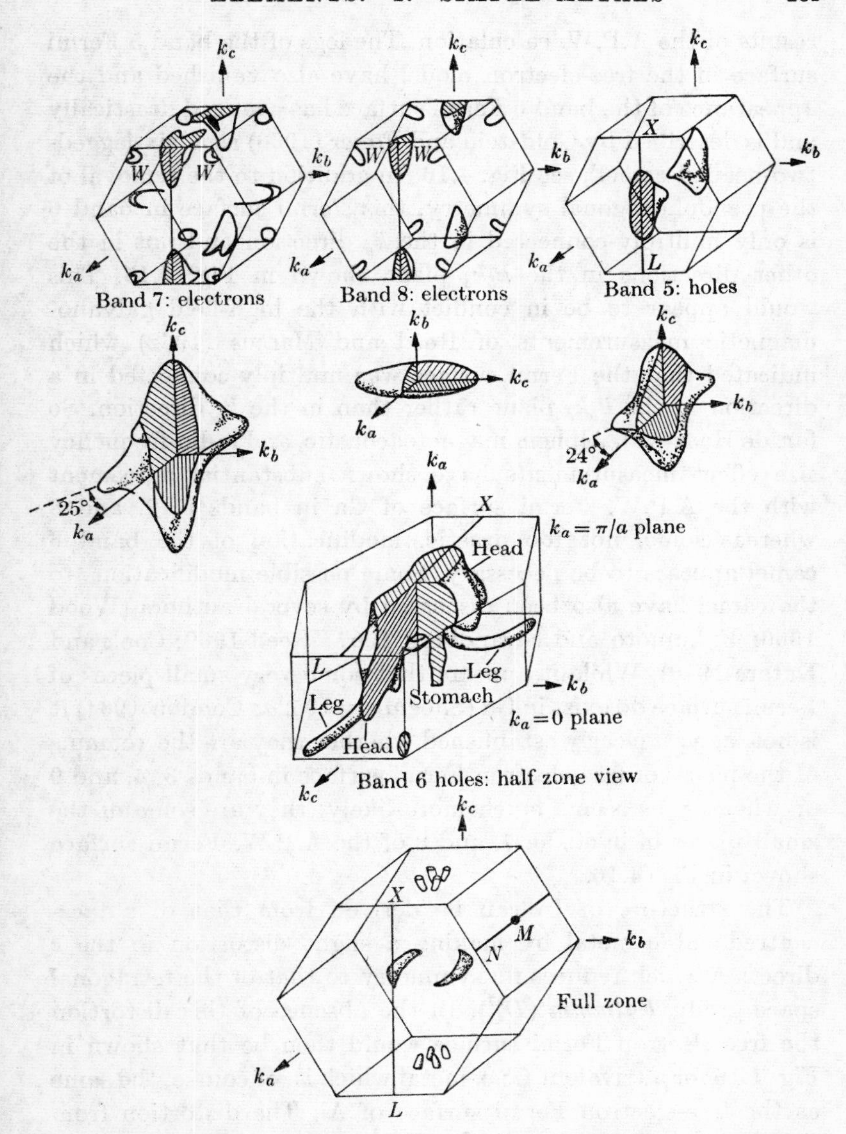

FIG. 4.16. The Fermi surface of Ga constructed by Goldstein and Foner (1966) from the results of the A.P.W. calculation of Wood (1966).

results of the A.P.W. calculation. The legs of the band 5 Fermi surface in the free-electron model have also vanished and the appearance of the band 6 Fermi surface has changed drastically and is described by Goldstein and Foner (1966) as a 'six-legged-two-headed-camel', see Fig. 4.16. In addition to the removal of the pseudohexagonal symmetry, this Fermi surface in band 6 is only multiply-connected in the k_a direction and not in the other directions in the $k_b k_c$ plane shown in Fig. 4.15. This would appear to be in conflict with the high-field galvanomagnetic measurements of Reed and Marcus (1962) which indicated that the Fermi surface was multiply-connected in a direction in the $k_b k_c$ plane rather than in the k_a direction. So far, de Haas–van Alphen, magnetoacoustic, and radio-frequency size effect measurements have shown substantial agreement with the A.P.W. Fermi surface of Ga in bands 5, 7, and 8 whereas some, not too drastic, modification of the band 6 camel appears to be necessary. Some possible modifications to the camel have also been discussed by several authors (Wood 1966; Fukumoto and Strandberg 1967; Reed 1969; Cook and Datars 1970). While it appears that some very small pieces of Fermi surface do exist in Ga (Shoenberg 1952*a*; Condon 1964) it is not, as yet, clearly established whether they are the remains of the pieces of free-electron Fermi surface in bands 3, 4, and 9 or whether, as seems much more likely, they are some of the small pieces in bands 6, 7, and 8 of the A.P.W. Fermi surface shown in Fig. 4.16.

The structure of In can be derived from that of a face-centred cubic metal by making a slight distortion in the c direction which reduces its symmetry to that of the tetragonal space group $F4/mmm$ (D_{4h}^{17}). In the absence of this distortion the free-electron Fermi surface would then be that shown in Fig. 1.16 for a trivalent f.c.c. metal which is, of course, the same as the free-electron Fermi surface of Al. The distortion from the f.c.c. structure consists of a stretching by about eight per cent in the c axis direction, and consequently in reciprocal space the Brillouin zone is slightly contracted in the k_z direction. This has an important effect on the connectivity of the Fermi surface in band 2 in the free-electron model which now just touches the Brillouin zone boundary at those W points in the planes normal to k_x and k_y, see Fig. 4.17, and therefore becomes multiply-connected in the k_z direction. In the free-electron model, the Fermi surface in band 3 in In has the same connectivity as in an undistorted face-centred cubic metal. However, various galvanomagnetic experiments failed to reveal any

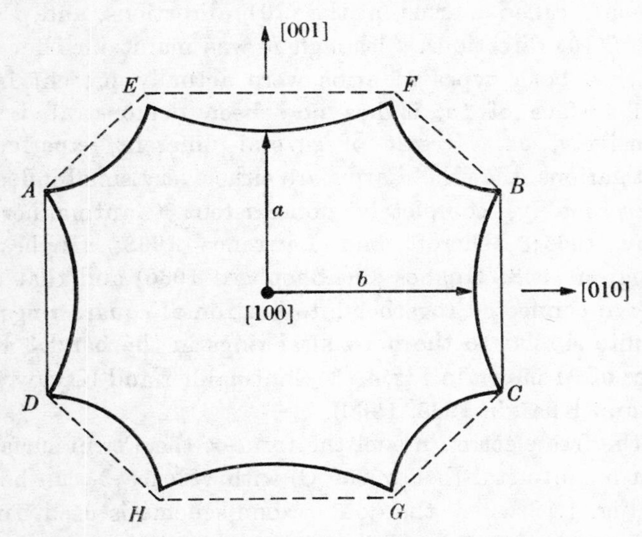

FIG. 4.17. Cross section of the free-electron Fermi surface of In in band 2 (Rayne 1963).

evidence of open orbits in In and this indicates that the Fermi surface in band 2 does not touch the Brillouin zone boundary after all and that the Fermi surface in band 3 becomes disconnected in some way. Experimental work on the de Haas–van Alphen, magnetoacoustic, and radio-frequency size effects showed that the free-electron model is a fairly good first approximation to the Fermi surface of In although the changes are sufficiently large to alter the connectivities and to remove all possibilities of open orbits (in the absence of magnetic breakdown). The Fermi surface of In in band 2 is therefore not multiply connected and its general appearance is very similar to that of the free-electron Fermi surface in band 2 of a trivalent f.c.c. metal, see Fig. 1.16. The linear dimensions of this closed Fermi surface in band 2 of In are about ten per cent smaller than in the free-electron model (for details see Rayne 1963; Mina and Khaĭkin 1966). Because of the tetragonal distortion from the f.c.c. structure, the band 3 Fermi surface of In in the free-electron model consists of two different kinds of arms, commonly called α arms, in the $\langle 101 \rangle$ directions, and β arms, in the $\langle 110 \rangle$ directions. Although it was maintained for some time that both types of arms were actually present in the Fermi surface of In, it has now been demonstrated fairly conclusively, as a result of several different experimental investigations, that the α arms are either very small indeed or, more probably, completely non-existent (Gantmakher and Krylov 1965; Ashcroft and Lawrence 1968; Hughes and Lettington 1968; Hughes and Shepherd 1969) and that the β arms are connected together into horizontal square rings that are quite similar to the horizontal rings in the band 3 Fermi surface of Al shown in Fig. 4.13 (Gantmakher and Krylov 1965; Mina and Khaĭkin 1965, 1966).

In the free-electron model the form of the Fermi surface of Tl, an h.c.p. metal (below 230°C) with valence 3, can be seen from Fig. 1.18 where the double zone scheme is used. In this double-zone scheme band 1 and band 2 are completely full, there is a complicated multiply-connected Fermi surface in band 3 and band 4, and the Fermi surface in band 5 and band 6 consists of isolated pockets of electrons. In Section 4.3 we have already discussed the particular importance of spin–orbit coupling for h.c.p. metals because its introduction alters the connectivity of the Fermi surface across the hexagonal face AHL and destroys the usefulness of the double-zone scheme. Because the atomic number of Mg is quite small the gaps introduced on AHL by spin–orbit coupling are small and can easily be overcome by magnetic breakdown if quite a small external magnetic field is present. However, the calculated splittings for Tl given in Table 4.4 can be seen to be considerably larger than those quoted previously for Mg (see

TABLE 4.4

Spin-splitting at K and H in Tl

Symmetry without spin	*Symmetry with spin*	*Energy* (Ry)	*Spin-splitting* (Ry)
K_5	K_9 K_8	0·477 0·483	0·006
K_1	K_7	0·603	—
H_2	H_4+H_6 H_8	0·548 0·548	$<$ 0·001
H_3	H_5+H_7 H_9	0·655 0·673	0·018
H_1	H_9 H_8	0·713 0·735	0·022

(Soven 1965*a*)

page 266), which is not surprising because the atomic number of Tl (81) is very much larger than the atomic number of Mg (12). To perform a meaningful band structure calculation for Tl it is therefore necessary to include relativistic effects *ab initio* by solving the Dirac equation for an electron in a crystal (see Section 2.9), rather than just including spin–orbit coupling by perturbation theory as an afterthought at the end of a band structure calculation using one of the common methods described in Sections 2.2–6 based on the use of the non-relativistic Schrödinger equation.

A relativistic band structure calculation for Tl based on the O.P.W. method was performed by Soven (1965*a*, *b*). The Fermi surface of Tl calculated by Soven on the basis of this band structure consisted of six non-equivalent sheets distributed among band 3, band 4, band 5, and band 6 in the single-zone scheme. It is convenient to describe this Fermi surface in terms of a Brillouin zone centred at A instead of at Γ, and it consists of

(i) in band 3, a large closed surface, sometimes called a 'cookie', centred at A, enclosing a region of holes, see Fig. 4.18;

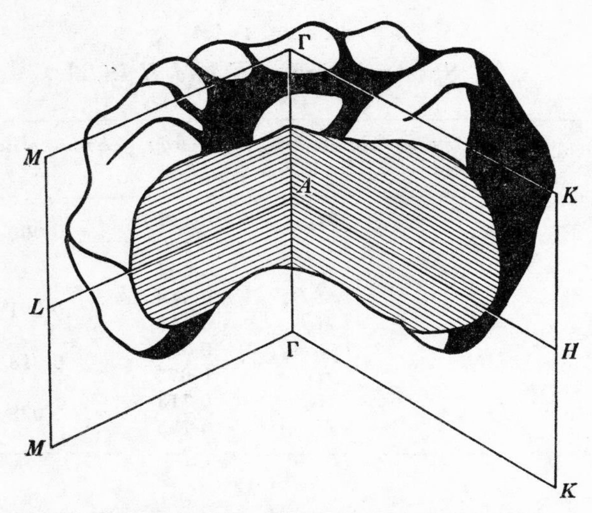

FIG. 4.18. The Fermi surface of Tl in band 3 (Soven 1965*a*).

(ii) in band 3, a smaller closed sheet centred at *M* also enclosing a region of holes;

(iii) in band 4, a honeycomb-like network, bounding occupied states, spanning the rectangular faces of the Brillouin zone, see Fig. 4.19. There are 12 posts protruding from this network which are roughly parallel to [0001], but

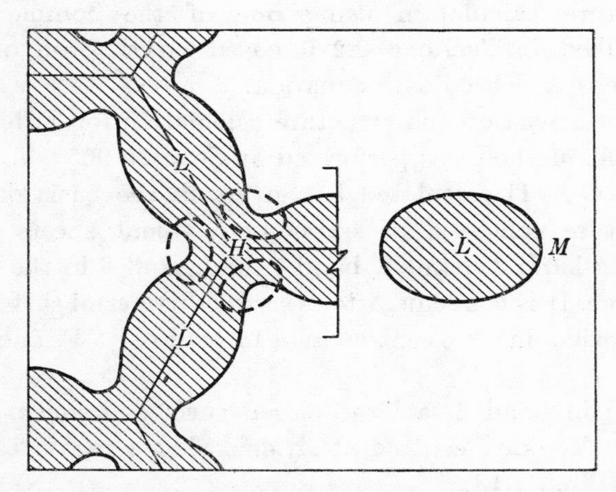

FIG. 4.19. Cross section *AHL* of the Fermi surface of Tl in band 4 (Soven 1965*a*).

some doubt remained as to whether or not these posts were long enough to join up with each other to form a Fermi surface that is multiply-connected in the [0001] direction as well as in the horizontal plane. The resolution of this difficulty experimentally is complicated by the fact that even if the posts are not joined together the size of the gap would be sufficiently small that there would be the possibility of magnetic breakdown in any experiments in which a magnetic field is applied,

(iv) in band 4, if the posts in (iii) are not multiply-connected in the [0001] direction, small occupied pockets remain on the ΓMK planes between adjacent posts,

(v) in band 5, small pockets of electrons at H, and

(vi) in band 6, another set of smaller pockets of electrons at H.

The relationship to the free-electron Fermi surface for a trivalent h.c.p. metal is not very close.

This calculated Fermi surface was compared by Soven (1965*b*) with the experimental results which existed at that time; these results consisted of galvanomagnetic, de Haas–van Alphen, and magnetoacoustic work. The dimensions of the cookie in band 3 are well established and are tabulated by Eckstein, Ketterson, and Priestley (1966). In connection with the Fermi surface in band 4 the de Haas–van Alphen results of Priestley (1966) were not completely decisive, but the absence of any large observed area corresponding to extended orbits parallel to [0001] together with the form of the angular dependence of one of the periods which was assigned to the posts seemed to indicate that the posts were not multiply-connected in the [0001] direction. This conclusion is also supported by the failure to observe open orbits in the [0001] direction either in magnetoresistance measurements (Alekseevskiĭ and Gaĭdukov 1962; Mackintosh, Spanel, and Young 1963; Milliken and Young 1966) or in magnetoacoustic measurements (Eckstein, Ketterson, and Priestley 1966; Coon, Grenier, and Reynolds 1967). It therefore seems that the general features of the predictions of Soven about the Fermi surface in band 4 have been moderately well confirmed and it seems fairly certain that the posts are not multiply-connected in the [0001] direction, although in sufficiently large fields magnetic breakdown can be expected to occur between the posts (Young 1967). The remaining sheets of the Fermi surface of Tl predicted by Soven consisted of a number of small isolated pockets in various bands. In addition to the cookie in band 3, and the multiply-connected region of electrons in band 4, it was shown as a result of de Haas–van Alphen measurements (Priestley 1966; Ishizawa and Datars 1969; Capocci, Holtham, Parsons, and Priestley 1970) that there are two further small closed pieces of Fermi surface in Tl. One set of small pockets with the symmetry of $mmm(D_{2h})$ could be either holes in band 3 at M or electrons in band 4 along ΓM, but it was not feasible to distinguish between these two possibilities. The second set of pockets were dumb-bell shaped and were thought to be centred at H and in band 5; this result differs from the predictions of Soven in which there are two roughly ellipsoidal pockets of holes at H, one in band 5 and the other in band 6.

4.5. Group IVB

C, Sn, Pb

C	Carbon	2.(2, 2)	see below
[Si	Silicon	2.8.(2, 2)]	
[Ge	Germanium	2.8.18.(2, 2)]	
Sn	Tin	2.8.18.18.(2, 2)	tetragonal, see below
Pb	Lead	2.8.18.32.18.(2, 2)	f.c.c.

In the form of diamond C is an insulator while Si, Ge, and grey Sn, which have the same structure, are semiconductors; we shall not be concerned with any of these materials. C, in the form of graphite, is a semi-metal with its well-known hexagonal structure described by the space group $P6_3mc$ (C_{6v}^4) with atoms at the positions $(0, 0, 0)$, $(0, 0, \frac{1}{2})$, $(\frac{1}{3}, \frac{2}{3}, z)$, and $(\frac{2}{3}, \frac{1}{3}, z+\frac{1}{2})$ where z is very small, see Fig. 4.20. If z is exactly equal to zero the space-group symmetry is increased to that of $P6_3/mmc$ (D_{6h}^4). Metallic Sn has a tetragonal structure which belongs to

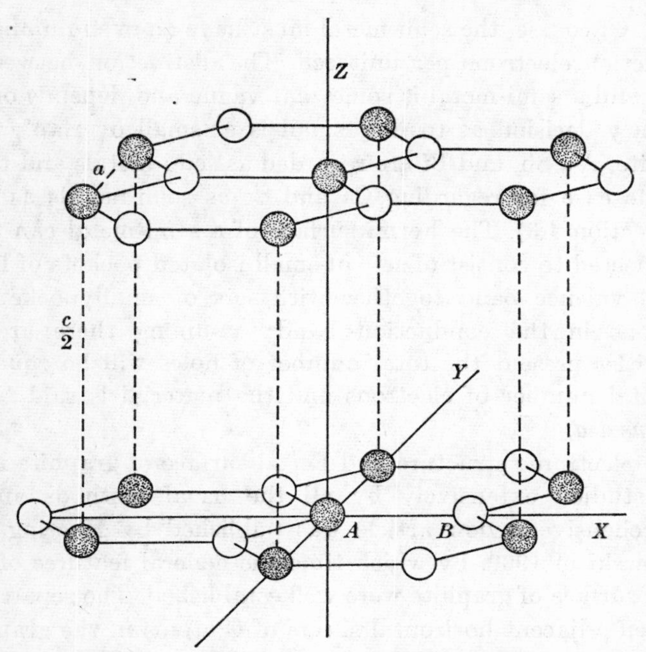

FIG. 4.20. The structure of graphite (McClure 1964).

the space group $I4_1/amd$ (D_{4h}^{19}) with atoms situated at (0, 0, 0), ($\frac{1}{2}$, $\frac{1}{2}$, $\frac{1}{2}$), ($\frac{1}{2}$, 0, $\frac{1}{4}$), and (0, $\frac{1}{2}$, $\frac{3}{4}$). The transition temperature between the low-temperature form (grey Sn) and the high-temperature form (metallic Sn) is 13·2°C but it is a matter of everyday experience that metallic Sn can exist for long periods in a metastable state at temperatures below the transition point. In this section where we refer to Sn we mean metallic (white) Sn.

The electrical resistivity of graphite is considerably higher than that of most metals so that graphite can be regarded as a semi-metal, along with As, Sb, and Bi (see Section 4.6). The distinction between a semi-metal and a semiconductor is that in a semiconductor there is a finite gap between the valence band, which is full at absolute zero, and the conduction band, which is empty at absolute zero, whereas in a semi-metal there is a small but finite overlap of the valence and conduction bands. Of course, the semi-metal must have an even number of conduction electrons per unit cell. The distinction between a metal and a semi-metal is somewhat vague and depends on an arbitrary decision as to how small is a 'small overlap'; thus graphite, As, Sb, and Bi are regarded as semi-metals and there is some case for regarding Ca and Sr as semi-metals as well (see Section 4.3). The Fermi surface of a semi-metal can then be expected to consist of sets of small isolated pockets of holes in the valence band together with sets of small pockets of electrons in the conduction band; assuming there are no impurities present the total number of holes will be equal to the total number of electrons and the material is said to be *compensated.*

The electronic structure and Fermi surface of graphite have been studied extensively by all the usual methods and a comprehensive review article was published by Haering and Mrozowski in 1960, by which time the general features of the Fermi surface of graphite were well established. The separation between adjacent horizontal sheets of C atoms in the graphite structure is nearly $2\frac{1}{2}$ times as large as the separation between neighbouring C atoms in a given horizontal sheet. For this reason it has been very common to consider the behaviour of the electrons in graphite as a two-dimensional problem involving the motion of electrons in a single layer; interactions between the layers can then be introduced afterwards as a perturbation (Coulson 1947; Wallace 1947). Of the four valence electrons of each C atom, three form covalent bonds with the three nearest-neighbour C atoms in the same horizontal sheet. The remaining electron is much less tightly bound and occupies the valence band. Since there are two C atoms per unit cell there are just enough electrons to fill the valence band if there is no overlap between the valence band and the conduction band. This two-dimensional treatment suggested that there was a point of contact between the conduction band and the valence band at the corners of the Brillouin zone. When the proper three-dimensional structure is considered, the Brillouin zone has the same general appearance as that of the h.c.p. structure except that the axial ratio c/a is much larger than that of the h.c.p. structure; consequently the Brillouin zone for graphite is very much more squat than the h.c.p. Brillouin zone.

Slonczewski and Weiss (1958) performed a perturbation calculation to determine the band structure of three-dimensional graphite using the four tight-binding wave functions with wave vector **k** corresponding to one of the corners of the Brillouin zone for the two-dimensional problem of a single layer. The region of particular interest is the vertical line HKH which joins two corners of the Brillouin zone. The expressions obtained by Slonczewski and Weiss for the energies in the region of HKH represented a set of four hyperboloids of revolution:

$$\begin{aligned} E(\mathbf{k}) &= \tfrac{1}{2}(E_1+E_3)\pm[\tfrac{1}{4}(E_1-E_3)^2+(\gamma_0\sigma)^2]^{\frac{1}{2}} \\ E(\mathbf{k}) &= \tfrac{1}{2}(E_2+E_3)\pm[\tfrac{1}{4}(E_2-E_3)^2+(\gamma_0\sigma)^2]^{\frac{1}{2}}; \end{aligned} \tag{4.5.1}$$

where

$$\begin{aligned} E_1 &= \Delta+2\gamma_1\cos(\tfrac{1}{2}\xi) \\ E_2 &= \Delta-2\gamma_1\cos(\tfrac{1}{2}\xi) \\ E_3 &= 2\gamma_2\cos^2(\tfrac{1}{2}\xi) \\ \sigma &= \tfrac{1}{2}(\sqrt{3})a\kappa \\ \xi &= k_z c. \end{aligned}$$

$\boldsymbol{\kappa}$ is the wave vector measured from the line HKH in a horizontal plane and Δ, γ_0, γ_1, and γ_2 are adjustable parameters. The bands given by Slonczewski and Weiss (1958) along HKH are shown for one choice of these parameters in Fig. 4.21. On this set of energy bands there will be expected to be a thin vertical cigar-shaped pocket of holes in the valence band centred at K together with a pocket of electrons at each end of the cigar; these regions of electrons and holes will just touch because of the crossing of the bands in Fig. 4.21.

The problem of the determination of the best set of values for the parameters Δ, γ_0, γ_1, and γ_2 (as well as γ_3 and γ_4 which were omitted from eqn (4.5.1)) to use in the Slonczewski–Weiss model to give the best fit to all the various pieces of experimental data is discussed at length by Haering and Mrozowski.

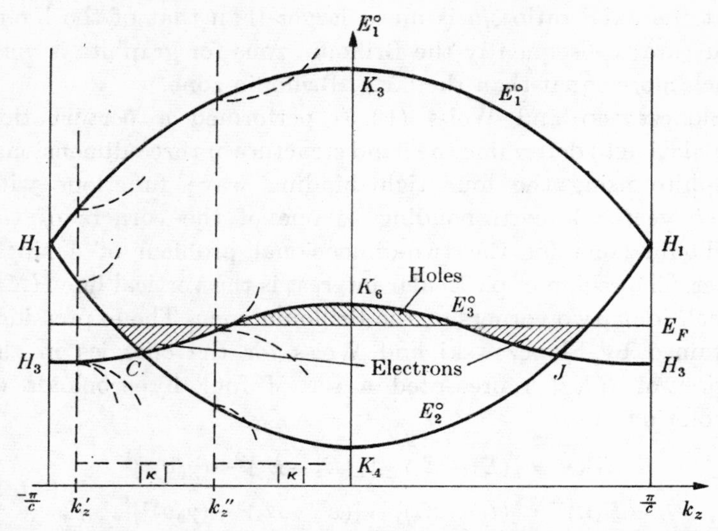

Fig. 4.21. The band structure of graphite calculated by Slonczewski and Weiss (1958).

The experimental data involved included principally the de Haas–van Alphen effect, the Hall effect, and cyclotron resonance (for references see Haering and Mrozowski 1960). Using the de Haas–van Alphen data of Shoenberg (1952*a*) and Berlincourt and Steele (1955), values of γ_1, γ_2, and Δ were calculated by McClure (1957) for a range of values of γ_0. The Fermi surface deduced by McClure is shown in Fig. 4.22, where it should be noted that the electron pockets do not quite reach the point H. More recent determinations of the parameters in the Slonczewski–Weiss model involving magnetoreflection measurements (Dresselhaus and Mavroides 1964*b*, *c*, 1966) have led to a substantially modified band structure in which the two fold degenerate level at H is below the Fermi level so that the pockets of electrons now reach H and there will be a small pocket of electrons in band 3 at H. In the extended zone scheme this is represented by extending the main pocket of electrons beyond H. A new set of measurements of the Schubnikow–de Haas effect (Soule, McClure, and Smith 1964) and of cyclotron resonance (Williamson, Surma, Praddaude,

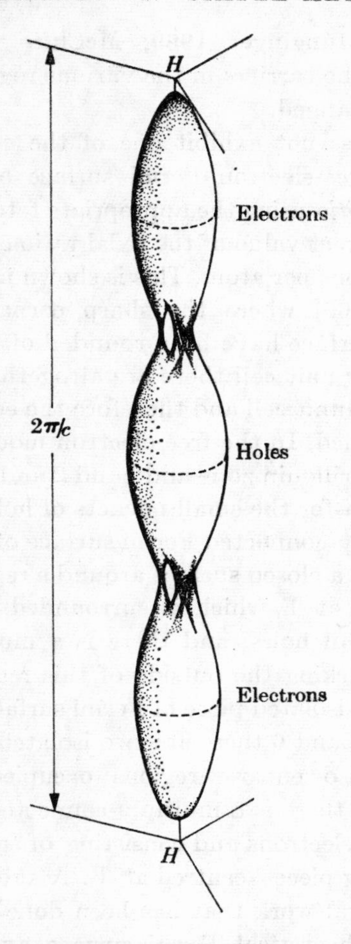

FIG. 4.22. The Fermi surface of graphite proposed by McClure (1957); it now seems likely that the electrons and holes should be interchanged (McClure 1971).

Patten, and Furdyna 1966) provided direct confirmation of the principal features of the Fermi surface of graphite as described and illustrated above, although the details of the region in which the electron and hole Fermi surfaces meet are difficult to verify directly. However, it now seems likely (Schroeder, Dresselhaus, and Javan 1968; Kechin 1969; van

Haeringen and Junginger 1969; McClure 1971) that the identification of the carriers in the various regions in Fig. 4.22 should be interchanged.

Because Sn does not exhibit one of the common metallic structures, the free-electron Fermi surface of Sn has to be constructed *ab initio* using the appropriate tetragonal Brillouin zone with the correct value of the axial ratio c/a and with four conduction electrons per atom. This is shown in Fig. 4.23 (Gold and Priestley 1960) where the sharp corners on the free-electron Fermi surface have been rounded off. Since there are two Sn atoms per unit cell there are altogether eight conduction electrons per unit cell and therefore the equivalent of four bands must be filled. In the free-electron model band 1 is full throughout the Brillouin zone and band 2 and band 3 are very nearly full, except for the small pockets of holes at W in band 2 and the multiply-connected Fermi surface of holes in band 3. In band 4 there is a closed surface around a region occupied by electrons centred at Γ which is surrounded by a multiply-connected region of holes, and there is a multiply-connected Fermi surface marking the outside of this region of holes. In band 5 there is an isolated piece of Fermi surface at Γ enclosing electrons, and in band 6 there are two isolated pieces of Fermi surface which also enclose regions occupied by electrons. Finally in band 5 there is a multiply-connected piece of Fermi surface enclosing electrons and consisting of 'pears' centred at H and 'connecting pieces' centred at V. A substantial amount of the experimental work that has been done on Sn was done when the only theoretical Fermi surface available was this free-electron Fermi surface: naturally the various workers attempted to explain their experimental results in terms of the free-electron Fermi surface.

The band structure and Fermi surface of Sn were calculated by Weisz (1966) using a local-pseudopotential approximation, including spin–orbit coupling, and fitting the parameters with the results of the radio-frequency size effect measurements of Gantmakher (1962*b*, 1963, 1964). Because the atomic number of Sn is relatively high (50) one would expect spin–orbit coupling effects to be important. Although the actual spin–orbit interaction in Sn is weaker than in the very heavy elements, its importance is enhanced by the fact that the crystal structure has relatively low symmetry. A group-theoretical analysis of the electronic energy bands in Sn shows that there are substantial reductions in the degeneracies of the bands when spin–orbit coupling is introduced (Mase 1959*b*; Miąsek and Suffczyński 1961*a*, *b*; Suffczyński 1961). From the calculated band structure Weisz obtained a Fermi surface that, although distorted, was still recognizably related to the free-electron Fermi surface of Sn. The pocket of holes at W in band 2 vanishes, and in band 3 the regions of holes around W disappear leaving isolated cylinders along the lines XP. In band 4 some electrons are introduced along ΓH near W so that there is a neck connecting the previously isolated electron pocket at Γ in band 4 (b in Fig. 4.23) to the other, already multiply-connected, region occupied by electrons in band 4 (a in Fig. 4.23). In band 5 the cigar of electrons around Γ disappears but the network involving the pears is not substantially changed. In band 6 the small pockets of electrons near K (labelled ν in Fig. 4.23) disappear while the cigars near V (labelled ρ in Fig. 4.23) become amalgamated into a vertical shape with a square cross section with slight prongs in the directions of the original cigars. Several of the de Haas–van Alphen results of Gold and Priestley (1960), although generally less accurate than Gantmakher's radio-frequency size effect results, showed satisfactory agreement with the extremal areas of cross section of the theoretical Fermi surface calculated by Weisz. However, some other observed de Haas–van Alphen periods did not agree with the theoretical model of Weisz and also the theoretical model predicted more de Haas–van Alphen periods than had actually been observed experimentally. The existence of a multiply-connected Fermi surface, both in the free-electron model and in the calculations of Weisz, is in agreement with the results of magnetoresistance measurements on Sn (Alekseevskiĭ and Gaĭdukov 1962). After the calculation by Weisz (1966) of the Fermi surface of Sn an extensive series of further de Haas–van Alphen measurements were made and

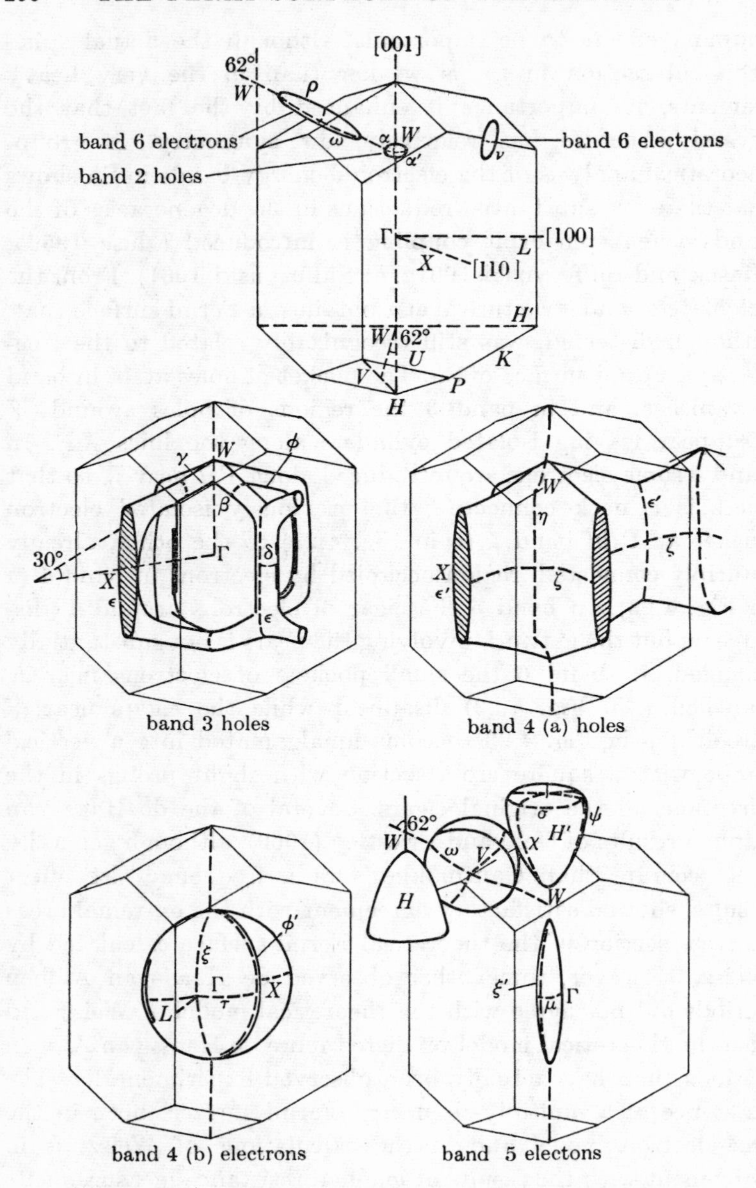

Fig. 4.23. The Fermi surface of Sn on the free-electron model (with the sharp corners rounded off) (Gold and Priestley 1960).

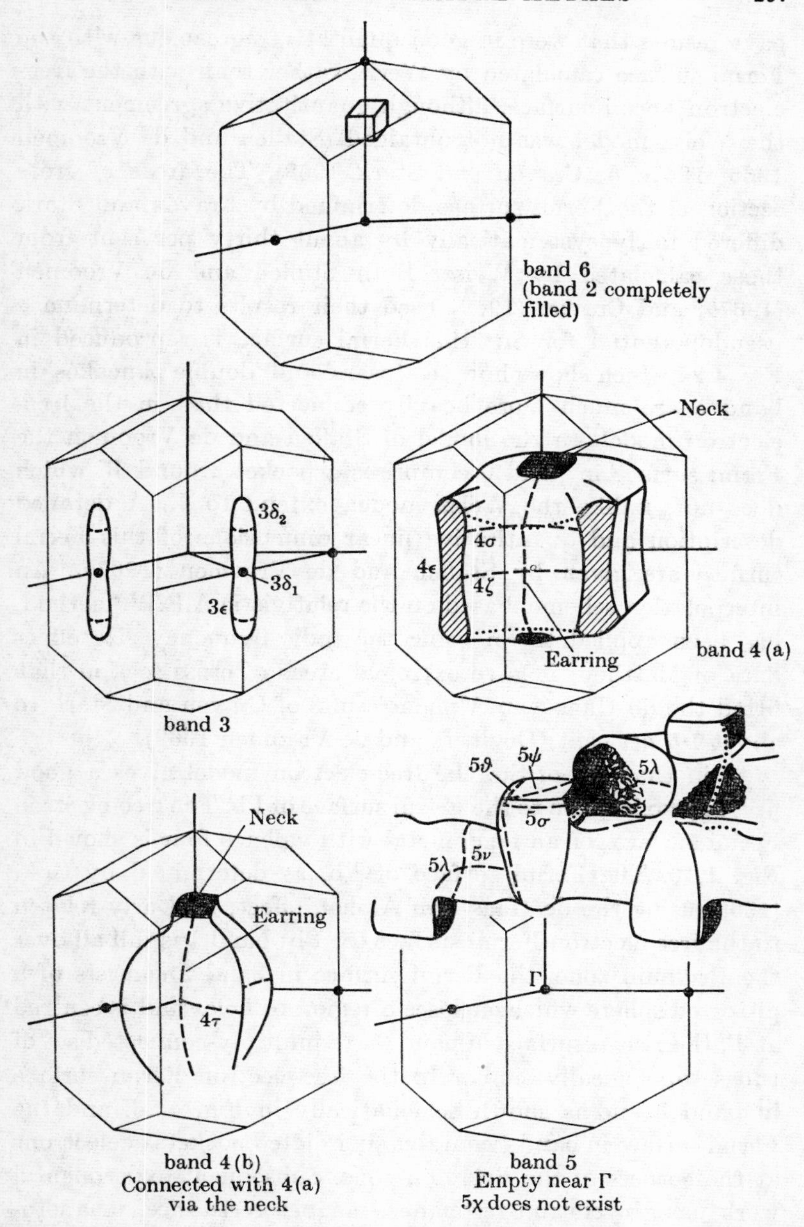

Fig. 4.24. The Fermi surface of Sn according to Stafleu and de Vroomen (1967*b*).

gave results that were in good qualitative agreement with the Fermi surface calculated by Weisz, rather than with the free-electron Fermi surface, although quantitative agreement with the Weisz model was not obtained (Stafleu and de Vroomen 1966, 1967*a*, *b*; Craven and Stark 1968). The areas of cross section of the Fermi surface determined by Craven and Stark differed fairly systematically by about thirty per cent from those calculated by Weisz. Both Stafleu and de Vroomen (1967*b*) and Craven (1969) used their results to determine a pseudopotential for Sn; this Fermi surface is reproduced in Fig. 4.24 which shows how the 'pears' and 'double pancakes' in band 5 are much more heavily connected than in the free-electron model. In the model of Stafleu and de Vroomen the Fermi surface in band 6 comprises a pocket around W which does not, unlike the Weisz model, extend to V. A detailed description and quantitative linear dimensions of this Fermi surface are given by Stafleu and de Vroomen (1967*b*). An interpolation scheme, based on the relativistic A.P.W. method, has been applied to Sn using the radio-frequency size effect data of Matthey; it gave extremal areas of cross section that fitted the de Haas–van Alphen results of Craven and Stark to about 1·5 per cent (Devillers and de Vroomen 1969).

As in the case of Sn, the free-electron model gives a good first approximation to the Fermi surface of Pb. The free-electron Fermi surface for an f.c.c. metal with valence four is shown in Fig. 1.16. The Fermi surface of Pb, as determined by Gold (1958) using the de Haas–van Alphen effect, is clearly related to the free-electron Fermi surface for Pb; band 1 is full all over the Brillouin zone, the Fermi surface in band 2 consists of a distorted sphere which encloses a region of holes and is centred at Γ, the Fermi surface in band 3 is a multiply-connected set of tubes topologically similar to the free-electron Fermi surface in band 3 and is shown schematically in Fig. 4.25, and the Fermi surface in band 4 contains six isolated pockets of electrons at the corners of the Brillouin zone. Subsequent experimental work using cyclotron resonance, magnetoresistance, magneto-acoustic, and de Haas–van Alphen measurements showed that the conclusions of Gold (1958) about the sheets of the Fermi

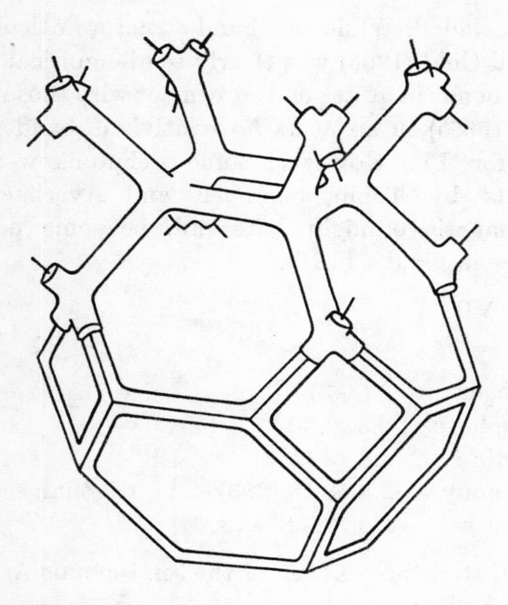

FIG. 4.25. Schematic drawing of the Fermi surface of Pb in band 3 (Gold 1958).

surface in bands 2 and 3 were substantially correct but led to the conclusion that band 4 is completely empty. Dimensions of the sheets of the Fermi surface in band 2 and band 3 are given by a number of authors (Alekseevskiĭ and Gaĭdukov 1961; Khaĭkin and Mina 1962; Mina and Khaĭkin 1963; Rayne 1963; Schirber 1963; Anderson and Gold 1965). Anderson and Gold performed an O.P.W. band structure calculation for Pb and obtained values for the parameters by fitting the results of the band structure calculation to certain of the areas of cross section of the Fermi surface determined from their de Haas–van Alphen results. This calculation further supported the view that there is no Fermi surface in band 4 in Pb. In this calculation the effects of spin–orbit coupling were found to be quite large at certain points in the Brillouin zone; for instance, one of the levels at W which is twofold degenerate in the absence of spin–orbit coupling becomes split by about 0·124 Ry when spin–orbit coupling is included. While this band structure calculation by Anderson and Gold (1965) was clearly semi-empirical it gave a set of energy bands in quite good agreement with those obtained by Loucks (1965*b*) in an *ab initio* relativistic band structure calculation for Pb. However, some Schubnikow–de Haas measurements by Tobin, Sellmyer, and Averbach (1969) suggested that there might, after all, be some pockets of Fermi surface in band 4 in Pb.

4.6. Group VB

As, Sb, Bi

[N	Nitrogen	2.(2, 3)]	—
P	Phosphorus	2.8.(2, 3)]	—
As	Arsenic	2.8.18.(2, 3)	trigonal, see below
Sb	Antimony	2.8.18.18.(2, 3)	
Bi	Bismuth	2.8.18.32.18.(2, 3)	

The crystal structure of each of the semi-metals As, Sb, and Bi belongs to the trigonal space group $R\bar{3}m$ (D^5_{3d}) with two atoms per unit cell at (x, x, x) and at $(-x, -x, -x)$. This structure can be obtained from a simple cubic structure by

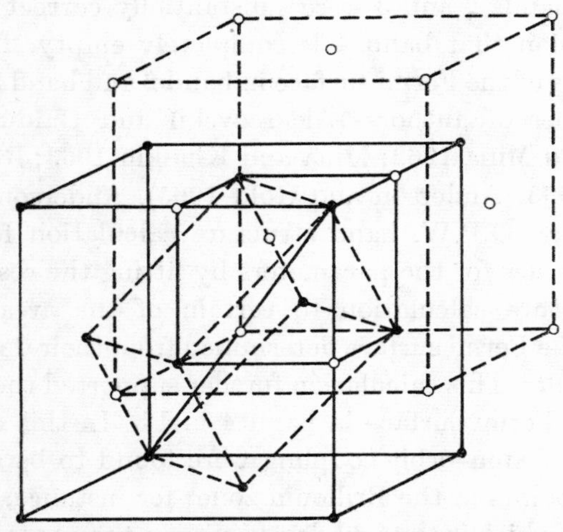

FIG. 4.26. The relationship of the structures of As, Sb, and Bi to the simple cubic lattice (Abrikosov and Fal'kovskiĭ 1962).

making slight displacements of the atoms. Suppose that we separate a simple cubic lattice into two face-centred sublattices where the two sublattices are related by a translation along the body-diagonal of the original simple cube, see Fig. 4.26. The structure of As, Sb, or Bi can then be obtained by performing a slight trigonal distortion; the extent of this distortion can be seen to be quite small from the values of the angle α in Table 4.5 which are quite close to the value of 60° for the undistorted

TABLE 4.5

Lattice parameters for As, Sb, and Bi

Coordinates: rhombohedral 2 As ($3m$ (C_{3v})): $\pm(x, x, x)$

	a (Å)	α
As	4·12	54° 10′
Sb	4·50	57° 6′
Bi	4·74	57° 14′
simple cubic	—	60°

simple cubic lattice. Alternative descriptions of these structures are possible using larger unit cells (see, for example, Smithells 1967).

Since the Bravais lattice of the crystal structure of As, Sb, or Bi is very closely related to either of the face-centred cubic sublattices shown in Fig. 4.26, the Brillouin zone of each of these structures can be obtained by a slight distortion of the face-centred cubic Brillouin zone which is shown, for example, in Fig. 1.8. The principal axis of this distorted face-centred cubic Brillouin zone is the threefold axis normal to one of the pairs of large hexagonal faces, see Fig. 4.27. Because of the differences in the lattice constants and in the values of α for the three crystal structures of As, Sb, and Bi, the dimensions and shapes of the Brillouin zones will also be slightly different for the three different elements.

Historically Bi was one of the first materials for which the free-electron energy bands and the free-electron Fermi surface

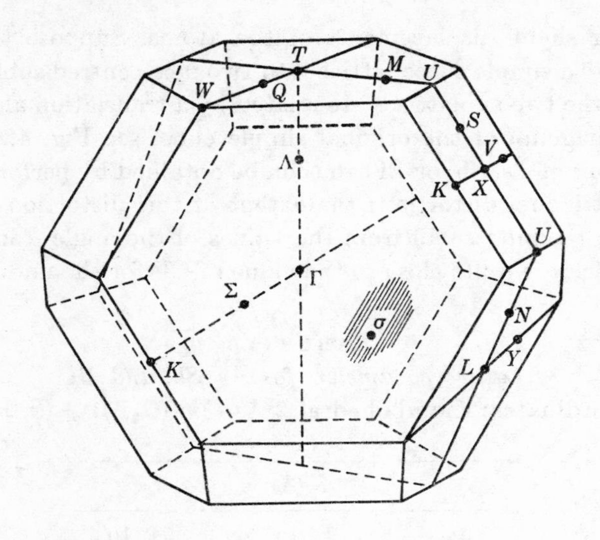

FIG. 4.27. The Brillouin zone for As, Sb, and Bi (Cohen 1961).

were investigated theoretically (Jones 1934; Mott and Jones 1936; see also Jones 1960). As well as being studied theoretically in the very early days of the investigations of the electronic properties of metals, Bi was also one of the first materials to be studied experimentally by some of the methods which have now become commonplace in Fermiology. As early as 1930 oscillations were observed in the magnetoresistance of Bi by Schubnikow and de Haas and in the diamagnetic susceptibility by de Haas and van Alphen (see Section 3.8). Moreover, no other material seems to have been studied more extensively since the discovery of these oscillatory effects. There are several reviews that contain extensive discussions of the electronic structure of Bi (Kahn and Frederikse 1959; Lax and Mavroides 1960; Boyle and Smith 1963; Shoenberg 1969*a*). Jones (1934) used a zone in extended **k**-space that corresponded to five electrons per atom; the relationship between this large zone and the conventional Brillouin zone is explained in Section 42 of the book by Jones (1960). By studying the electrical resistivity as a function of concentration of Sn impurities, Jones concluded that the number of electrons which spilled out of this large zone must be less than about 0·0013 per atom. There will be an equal number of holes in the large zone. Mott and Jones (1936) estimated the number of carriers of each sign to be about 10^{-4} per atom and more recent estimates are of the order of 10^{-5} per atom (Abelès and Meiboom 1956; Jain and Koenig 1962; Zitter 1962; Williams 1965; Bhargava 1967; Van Goor 1968). In the free-electron model these pockets of electrons would occur at a point in the large zone that corresponds to the points L or X in the conventional Brillouin zone. These pockets of electrons will be repeated by the threefold rotation axis of symmetry of the crystal. The pockets of holes are at a point in the large zone which can be shown to correspond to Γ or T in the conventional Brillouin zone (Jones 1960). It is the existence of only these relatively small pockets of electrons and holes in otherwise completely empty or full bands which leads to the low electrical conductivity of Bi and its description as a semi-metal.

This model of the Fermi surface of Bi was found to be consistent with low-field galvanomagnetic measurements (Abelès and Meiboom 1956). Shoenberg (1939, 1952*a*) concluded from the de Haas–van Alphen measurements that the three ellipsoidal pockets of electrons have one principal axis parallel to a twofold axis of symmetry but are tilted away from the trigonal axis through a small angle (see also Shoenberg and Uddin 1936; Blackman 1938; Dhillon and Shoenberg 1955). It is customary to represent each of these tilted ellipsoidal pieces of Fermi surface enclosing pockets of electrons by

$$E(\mathbf{k}) = \frac{\hbar^2}{2m}(\alpha_{11}k_x^2+\alpha_{22}k_y^2+\alpha_{33}k_z^2+2\alpha_{23}k_yk_z), \tag{4.6.1}$$

where m is the mass of a free electron and the set of orthogonal axes $Oxyz$ are oriented with the axis Oz along the threefold axis and the axis Ox along one of the twofold axes. The components of the effective mass tensor can be written in terms of the parameters α_{ij}

$$\left.\begin{aligned}\frac{m_{11}}{m} &= \frac{1}{\alpha_{11}}\\ \frac{m_{22}}{m} &= \frac{\alpha_{33}}{\alpha_{22}\alpha_{33}-\alpha_{23}^2}\\ \frac{m_{33}}{m} &= \frac{\alpha_{22}}{\alpha_{22}\alpha_{33}-\alpha_{23}^2}\\ \frac{m_{23}}{m} &= \frac{-\alpha_{23}}{\alpha_{22}\alpha_{33}-\alpha_{23}^2}\end{aligned}\right\}. \tag{4.6.2}$$

Most workers are agreed that the Fermi surface enclosing the pocket of holes consists of an ellipsoid of revolution where the axis of revolution is parallel to the trigonal axis (Pippard and Chambers 1952; Heine 1956; Aubrey and Chambers 1957). This ellipsoid of revolution is commonly represented by

$$E(\mathbf{k}) = \frac{\hbar^2}{2m}(\beta_1(k_x^2+k_y^2)+\beta_3 k_z^2) \tag{4.6.3}$$

where $\beta_1 = 1/m_{11}$ and $\beta_3 = 1/m_{33}$. Various experimental values for the effective masses of both the electrons and the holes in Bi are collected in Table 20 of Cracknell (1969*c*).

Quantum oscillations in a number of properties other than the diamagnetic susceptibility (de Haas–van Alphen effect) have been observed in Bi by many workers; these include the Hall coefficient, the electrical resistivity, the thermal capacity, the thermoelectric power, the thermal conductivity, the ultrasonic velocity, the ultrasonic attenuation, and the infra-red transmission coefficient. The results of various experiments on quantum oscillations are collected in Table 21 of Cracknell (1969*c*). Although, as a result of these and other experiments, it had been known for a long time that the pockets of electrons in Bi consist of a set of equivalent ellipsoids tilted away from the trigonal axis, the actual number (that is, three or six) of these ellipsoids and their location in the Brillouin zone was much more difficult to ascertain. Similarly, although the pockets of holes were known to be ellipsoids of revolution about the trigonal axis, there was again doubt about the number of these ellipsoids (that is, one or two) and about their location in the Brillouin zone.

The construction of the free-electron Fermi surface for Bi was studied by Harrison (1960*a*). The spherical free-electron Fermi surface in extended **k** space contains 10 electrons per unit cell of the crystal (that is, 5 electrons per atom) and this gives rise to pieces of Fermi surface in bands 2 to 8. When the actual crystal potential is considered, several of these pieces of Fermi surface disappear and on the basis of the extensive experimental evidence available it was suggested that for Bi the only pieces of Fermi surface to survive are regions of holes in band 5 and regions of electrons in band 6. The pockets of electrons that were observed experimentally can be expected to arise from the electrons of region X or L in the free-electron model. Experimentally it is known that the electron Fermi surface is elongated and is tipped slightly out of the plane perpendicular to the c-axis. It can be shown that the addition of a non-zero potential and spin–orbit coupling could not give rise to such segments at X while they might easily appear at L. The other electron Fermi surfaces are obtained by the use of the threefold rotation axis of symmetry, see Fig. 4.28. The pocket of holes in band 5 which has been observed experimentally presumably arises from the complex hole surface which is centred at Γ in band 5 in the free-electron model. However, according to the band structure calculations of Mase (1958, 1959*a*) and Cohen, Falicov, and Golin (1964), the Fermi surface of holes in band 5 would be at T rather than Γ. Harrison's arguments about the potential suggested that this piece of Fermi surface was centred somewhere between T and Γ.

Several authors have suggested the existence of a third set of carriers (heavy holes) beyond the two sets described so far on the Jones–Shoenberg picture (see Heine 1956; Brandt and Venttsel' 1958; Kalinkina and Strelkov 1958; Brandt, Dubrovskaya, and Kytin 1959; Brandt 1960; Smith 1961; Lerner 1962, 1963; Sybert, Grenier, and Reynolds 1962). However, this idea has now largely been discredited by showing that the volume

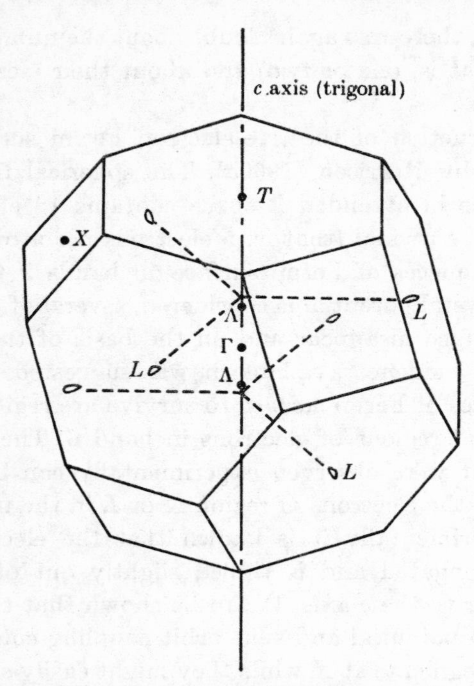

FIG. 4.28. The disposition of the electron Fermi surface pockets in Bi (Harrison 1960*a*).

of the pocket of holes in band 5 is equal to the total volume of the pockets of electrons in band 6 in the Jones–Shoenberg model, within the accuracy of the available measurements of the dimensions of these pieces of the Fermi surface (Jain and Koenig 1962; Zitter 1962). The results of an *ab initio* relativistic A.P.W. band structure calculation for Bi provided fairly clear evidence that there are pockets of holes at T and nowhere else and pockets of electrons at L and nowhere else (Ferreira 1967, 1968).

Although many experimental results showed that the tilt angle of the ellipsoidal pockets of electrons in Bi is about 6° (see, for example Edel'man and Khaĭkin 1965; Fal'kovskiĭ and Razina 1965) the sign of the tilt angle remained undetermined for some time. That is, it was not clearly established whether the vertical cross section should be represented by the ellipse indicated with the solid lines or with the broken lines in Fig. 4.29. It is convenient to establish a sign convention to describe the tilted ellipsoids and the usual convention, described by Brown, Hartman, and Koenig (1968), is such that the solid ellipsoid in Fig. 4.29 has its major axis tilted by a small positive angle from the bisectrix direction. Brown, Hartman, and

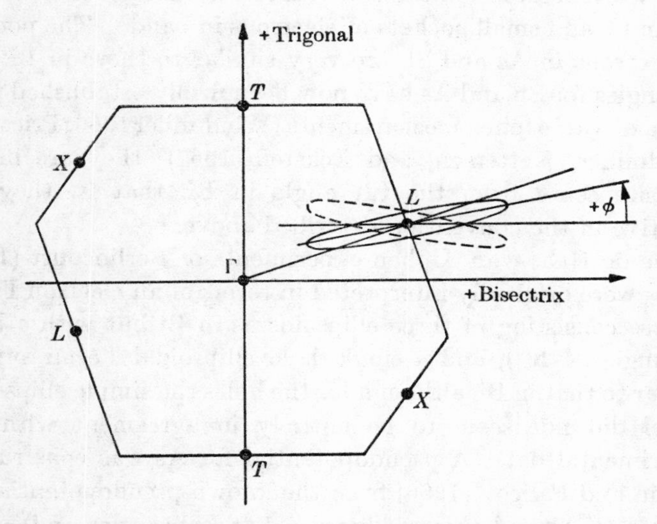

FIG. 4.29. Cross section of the Fermi surface of Bi with the cross section drawn in continuous lines for a positive tilt angle and in broken lines for a negative tilt angle (Brown, Hartman, and Koenig 1968).

Koenig (1968) studied the various pieces of experimental evidence available and also measured the de Haas–van Alphen effect for Bi and Sb specimens which were oriented by X-rays and therefore known to be in the same orientation. They found that the sign of the tilt angle in Bi is positive in the convention just described. They also found that the signs of the tilts of the electron Fermi surfaces in Bi and Sb are opposite to each other, independent of course of the actual sign convention used. Some of the ambiguities which existed previously about the sign of the tilt angle were ascribed to difficulties associated with the use of etch pits instead of X-rays by some workers when orienting their samples (see, for instance, Plate I of Boyle and Smith 1963).

Neither As nor Sb has been the object of such intensive experimental and theoretical investigation as Bi. Nevertheless, enough work has been done to establish that the Fermi surfaces of these two elements are similar to the Fermi surface of Bi; each semi-metal is compensated and has small pockets of holes in band 5 and small pockets of electrons in band 6. The pockets of electrons in As and Sb are very similar to those in Bi. The tilt angles for Sb and As have now been firmly established from de Haas–van Alphen measurements (Windmiller 1966; Priestley, Windmiller, Ketterson, and Eckstein 1967); they are in the opposite sense from the tilt angle in Bi, that is, they are negative in the convention described above.

The de Haas–van Alphen experiments of Berlincourt (1955) on As were originally interpreted in terms of an electron Fermi surface consisting of three ellipsoids as in Bi but with a large tilt angle ($-36°$), and a single hole ellipsoidal Fermi surface similar to that in Bi, although for the holes the simple ellipsoidal model did not seem to be entirely in agreement with the experimental data. A pseudopotential for As was constructed by Lin and Falicov (1966) from the known pseudopotentials of Ge and GaAs and this produced a band structure and Fermi surface for As which was in good agreement with the more recent experimental results on the de Haas–van Alphen effect, galvanomagnetic properties, cyclotron resonance, magneto-acoustic attenuation, and magnetothermal and Schubnikow–de Haas oscillations. The electron Fermi surface was still found to consist of three ellipsoids, but with a tilt angle of rather less than 10°; the parameters of this Fermi surface are given in Table 22 of Cracknell (1969*c*). The Fermi surface of holes obtained by Lin and Falicov (1966) was rather different from that of Bi in that it consisted of six pockets of holes connected together to form a 'crown' centred at T. This crown is illustrated in Fig. 4.30 and its existence has now been established by a number of experimental methods (Miziumski and Lawson 1969).

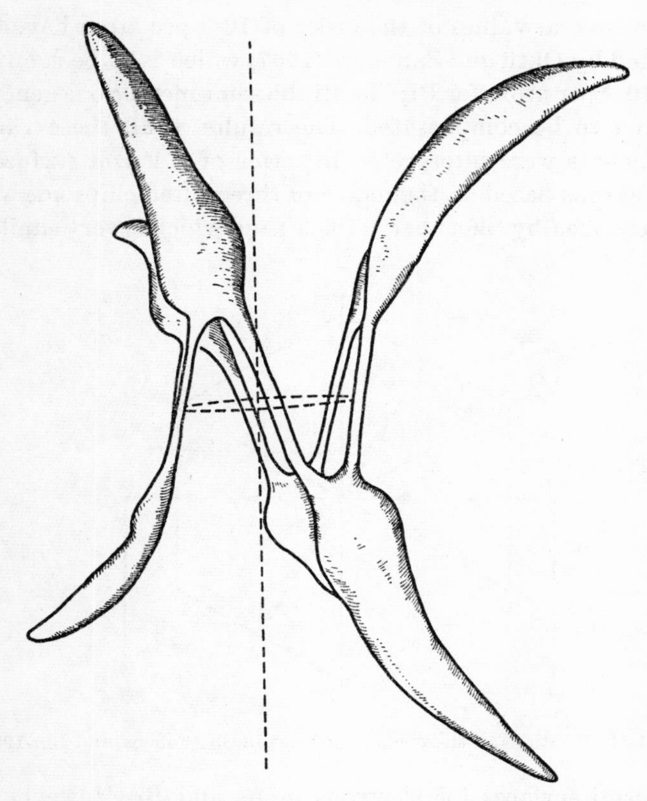

FIG. 4.30. The hole Fermi surface of As (Lin and Falicov 1966).

The amount of experimental work which has been performed on Sb is intermediate between the very large amount which has been done on Bi and the very small amount which has been done on As. Prior to the band structure calculation of Falicov and Lin (1966) there already existed a substantial number of experimental results on the de Haas–van Alphen effect, the Schubnikow–de Haas effect, cyclotron resonance, magneto-acoustic effects, anomalous skin effect, electron spin resonance, galvanomagnetic effects, and quantum oscillations in the optical reflection coefficient and in the infra-red absorption coefficient. The carrier density in Sb is slightly higher than in

Bi and several values of the order of 10^{-3} per atom have been collected by Öktü and Saunders (1967) which is to be compared with 10^{-5} per atom for Bi; like Bi the semi-metal Sb is generally assumed to be compensated. The results of all these various experiments were interpreted in terms of a Fermi surface for the electrons based on the model of three tilted ellipsoids which was described by Shoenberg (1952*a*) and which is very similar to

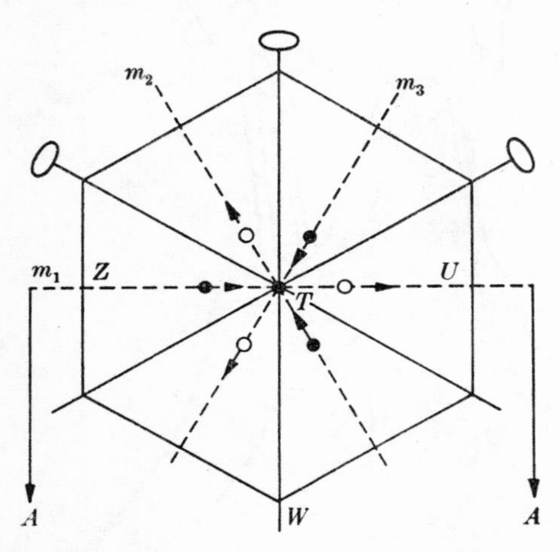

FIG. 4.31. Positions of the pockets of holes in Sb (Falicov and Lin 1966).

the Fermi surfaces for electrons in As and Bi. However, the interpretation of these results for the Fermi surface of holes was less straightforward. Some results were interpreted in terms of a Fermi surface for holes consisting of a set of three distorted ellipsoids with only a slight tilt, while other results were interpreted in terms of one ellipsoid of light holes and one ellipsoid of heavy holes.

A pseudopotential band structure calculation for Sb has been performed similar to that of Lin and Falicov (1966) for As and based on the use of the known pseudopotentials of InSb and grey Sn (Falicov and Lin 1966; Lin and Phillips 1966). On the basis of this calculation the pockets of electrons in band 6 were found to be similar to those in Bi and As, consisting of distorted ellipsoids at L with a tilt angle of $-7°$. The Fermi surface of holes in Sb obtained from the calculation of Falicov and Lin (1966) is different from that of As and that of Bi. It consists of six pockets of holes arranged around the point T which are obviously related to the six pockets of holes in As, but in Sb they are not connected together as they were in As. The positions of these pockets are illustrated in Fig. 4.31. Detailed comparisons between the dimensions of the calculated Fermi surface and various experimental results are tabulated by Falicov and Lin (1966). A compilation of other experimental values for the tilt angles for Sb is given by Öktü and Saunders (1967). The direct identification of the types of carriers as electrons or holes appropriately was performed by doping pure Sb with varying amounts of Te (donor) or Sn (acceptor) and observing the changes in the sizes of the pieces of Fermi surface by means of the de Haas–van Alphen effect (Ishizawa and Tanuma 1965; Ishizawa 1968*a*).

5

THE FERMI SURFACES OF THE METALLIC ELEMENTS. II. NON-SIMPLE METALS.

5.1. Introduction

IN CHAPTER 4 we described the shapes of the Fermi surfaces of the simple metals. In this connection we defined the term 'simple metal' as meaning a metal belonging to the s-block or p-block of the periodic table. We noticed that for many of these metals the free-electron model gives a good first approximation to the shape of the Fermi surface. The considerable success which has attended the use of pseudopotential methods in connection with the simple metals is a consequence of the relevance of the nearly-free-electron model.

In this chapter we consider the electronic band structures and the Fermi surfaces of the d-block and f-block metals, which include the transition metals and the rare-earth metals. Although for the metals of groups IB (Cu, Ag, Au) and IIB (Zn, Cd, Hg) it is possible to see some quite close relationship to the free-electron model, the Fermi surfaces of the remaining metallic elements, which are either transition metals or rare-earth metals, show very substantial departures from the predictions of the free-electron model, as we shall see in later sections. There are one or two other general points that arise in connection with the d-block and f-block metals and which did not occur in connection with the simple metals. These include the problem of distinguishing between the conduction electrons and the ion-core electrons in a given metallic element, and also the complications caused by the existence of ferromagnetic and antiferromagnetic ordering in a number of transition and rare-earth metals.

It has been a commonly accepted view for a very long time (Mott 1935; Mott and Jones 1936; Hume-Rothery and Coles

1954) that in the metals of the first transition series (Sc to Ni or Cu) the atomic 3d-electrons are in a very narrow band, and can therefore be regarded as fairly well localized, while the 4s-electrons are in a very broad band and are, therefore, not very strongly localized; this is often illustrated by the schematic density of states in Fig. 5.1. Estimates of the numbers of s- and d-electrons in the conduction band of a transition metal can be obtained from measurements of X-ray and neutron scattering

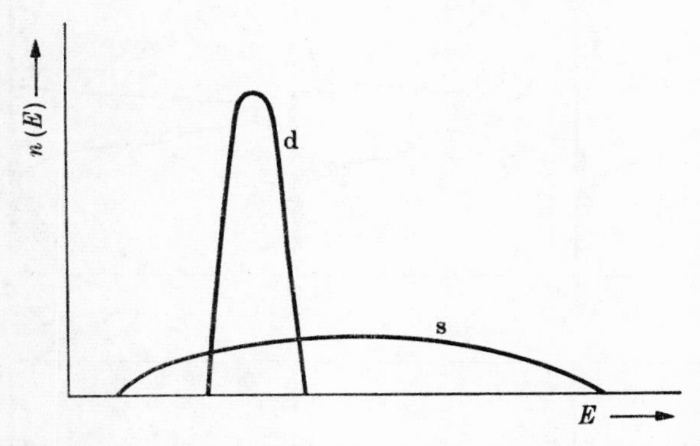

FIG. 5.1. Schematic density of states for d- and s-bands in a transition metal.

factors (see p. 143). However, the idea that an electron in a metal can be meaningfully described as an s-electron or a d-electron is an approximation which should be treated with considerable caution. Nevertheless, in so far as it is meaningful to describe the outer electrons of an atom in a metal as d- or s-electrons, this feature of narrow d-like bands and broad s-like bands is now well established and can be seen in some of the recent and quite reliable band structure calculations on these metals, see Fig. 5.2. It might then be argued that for any of these transition metals, because the d band is narrow and the d-electrons are therefore quite highly localized, it is only the s-electrons that should be regarded as the conduction electrons. However, the relative widths of the bands are much less important than the fact that the d-band and the s-band overlap; this is closely related to the chemical feature that the transition

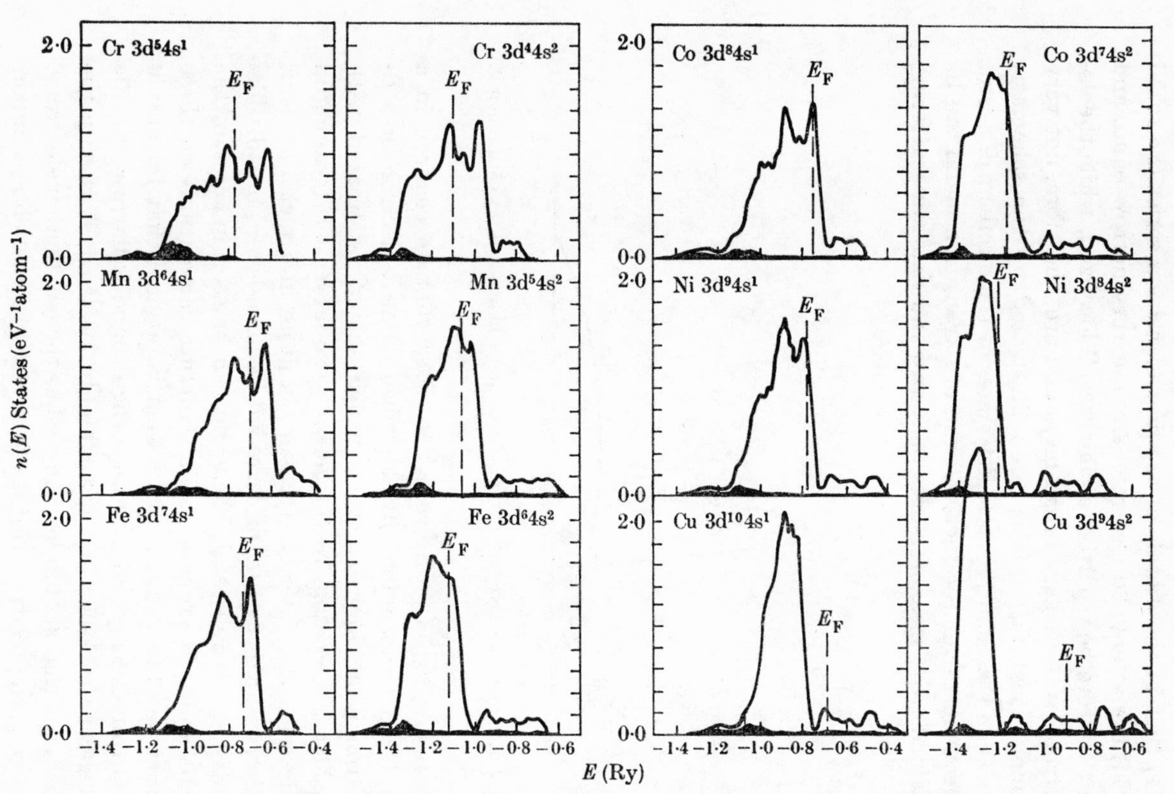

FIG. 5.2. Recent calculated density of states curves for the f.c.c. phases of the metals Cr to Cu in the two configurations s^1d^{n+1} and s^2d^n. The small darkened area represents the s-like portion of the $n(E)$ curves. (Snow and Waber 1969).

elements are notorious for exhibiting variable valence. The number of conduction electrons per atom, which is sometimes called the electron/atom ratio, e/a, in a metal of the first transition series is, therefore, given by the total number of 3d- and 4s-electrons in a free atom of that element. Similarly, in the second or third transition series the number of conduction electrons per atom is given by the total number of atomic 4d- and 5s-electrons or 5d- and 6s-electrons respectively. In the early days of work on the transition metals, band structure calculations were sometimes performed for the d-electrons alone taking no account of the hybridization with s-orbitals or other atomic orbitals at all; predictions of the shape of the Fermi surface of a transition metal based on the results of such calculations should be viewed with considerable suspicion. In the rare-earth (or lanthanide) metals, on the other hand, the situation is rather different. Variable valence is a much less prominent feature of the chemistry of the rare-earth elements than of the transition elements. The electronic structures of the atoms of the elements in the rare-earth series take the general form...$4f^n5s^25p^65d^16s^2$ or...$4f^{n+1}5s^25p^66s^2$ and the 4f-electrons are sufficiently deep that with only one or two exceptions the valence of these elements is fairly well fixed as three (Sidgwick 1950). Whether or not this means that the 4f-electrons can also be ignored in the metallic state, so that one need only consider the bands arising from the 5d- and 6s-electrons, is still an open question and we shall consider this in more detail in Section 5.10. However, what is more definitely established, and is relatively insensitive to the details of $V(\mathbf{r})$, is the fact that the 4f-bands are very narrow. For example, in their relativistic A.P.W. calculations on Yb, Johansen and Mackintosh (1970) obtained two extremely narrow sets of 4f-bands; one set consisted of four twofold degenerate bands corresponding to $j = 7/2$ and with a width of about $2{\cdot}5 \times 10^{-3}$ Ry, while the other set consisted of three twofold degenerate bands corresponding to $j = 5/2$ and with a width of about $1{\cdot}2 \times 10^{-4}$ Ry. Slightly larger widths were calculated for the 4f-bands in Ce by Mukhopadhyay and Majumdar (1969). The separation between the

$j = 5/2$ and $j = 7/2$ sets of bands was, however, quite large, namely $\sim$0·1 Ry. Even if it is assumed that the 4f-bands will not contribute appreciably to the electronic properties of the metal, nevertheless when they are partially filled they will produce a large magnetic moment. For the actinide series of metals it is the 5f-, 6d- and 7s-shells which are important (instead of the 4f-, 5d-, and 6s-shells respectively) and the 6d- and 7s-electrons which can be regarded as conduction electrons.

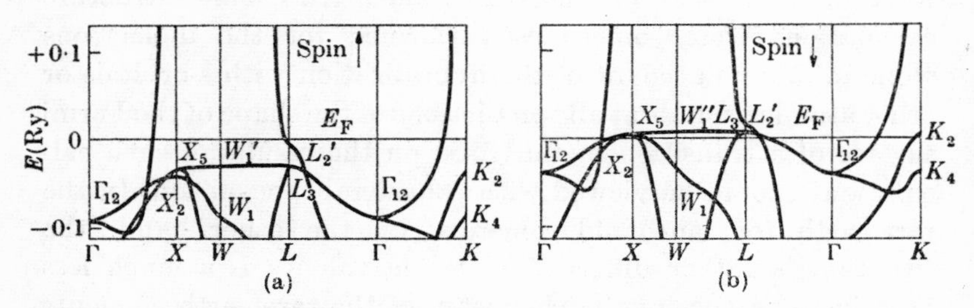

FIG. 5.3. Schematic band structure for ferromagnetic Ni. The spin-up (↑) bands are assumed to have lower energy (Tsui and Stark 1966).

In Chapter 4 we usually ignored the difference between the bands for spin-up and spin-down electrons, although for a few of the metals we did discuss the importance of spin–orbit coupling. However, if a metal undergoes a transition from a paramagnetic phase to a spontaneously ordered ferromagnetic or antiferromagnetic phase, the transition will be accompanied by a drastic change in the Fermi surface of that metal. To a first approximation the introduction of ferromagnetic ordering does not alter the shape of either the spin-up or spin-down bands. However, because of the magnetic ordering there will be an intense spontaneous magnetic field, **B**, within a specimen of the metal and this will cause a separation of $g_s\mu_B B$ between the spin-up and spin-down bands, as seen in Fig. 5.3. There will no longer be equal numbers of occupied spin-up and spin-down states and for both sets of bands the position of the Fermi level will be moved relative to the Fermi level of the paramagnetic metal. For the complicated forms of antiferromagnetic ordering that occur in Cr and in many of the rare-earth metals, the effect of the magnetic ordering on the band structure and Fermi surface is more complicated and will be discussed later in Sections 5.7 and 5.10. The origins of the magnetic phenomena are slightly different for the transition metals and the rare-earth metals. For the transition metals the magnetic moments are associated with the conduction electrons themselves, whereas for the rare-earth metals the magnetic moments arise principally from the incomplete shell of 4f-electrons which we do not regard as conduction electrons. In the transition metals it is the exchange interactions among the conduction electrons which are responsible for the magnetic ordering, whereas in the rare-earth metals it is primarily the exchange interactions between the 4f-electrons in the various ion cores, via the conduction electrons, which are responsible for the magnetic ordering (Zener 1951*a*,*b*,*c*, 1952; Ruderman and Kittel 1954; Kasuya 1956; Yosida 1957).

The appearance of a spontaneous magnetic field **B**, or, for a ferromagnetic metal, of a spontaneous magnetic moment $\mathscr{M}$, when a metal becomes magnetically ordered causes a considerable reduction in the symmetry of the metal and consequently many degeneracies in the band structure of the non-magnetic metal become lifted. The presence of a net magnetic moment, $\mathscr{M}$, in a ferromagnetic metal destroys the operation of time-reversal, θ, as a symmetry operation of the crystal. However, although the operation of time-reversal is not present on its own, θ may still occur in combination with some point-group or space-group operation as a symmetry operation of each domain of a magnetic metal. That is, one needs to consider the magnetic space group of the crystal, which is, of course, a subgroup of the grey space group of the paramagnetic metal. In recent years a considerable amount of work has been done on the theory of magnetic groups and their co-representations (for various reviews see, for example, Birss 1964; Opechowski and Guccione 1965; Cracknell 1967*a*, 1969*b*; Bradley and Davies 1968). For a ferromagnetic f.c.c. or b.c.c. metal magnetized parallel to [001], [111], or [110], it can be shown that the energy bands are non-degenerate (except for possible accidental degeneracies) all over the appropriate Brillouin zone. For ferromagnetic h.c.p. metals magnetized parallel to [0001] the sticking together of the bands all over the top hexagonal face *AHL* of the Brillouin zone in the paramagnetic metal is completely removed except at the points *A* and *L* and along the line *R*. The energy bands along the line *R*, which are fourfold degenerate in the paramagnetic state including time-reversal degeneracy, become split into two twofold degenerate bands. For ferromagnetic h.c.p. metals magnetized parallel to $[10\bar{1}0]$ or $[11\bar{2}0]$ the twofold degeneracy, due to time-reversal symmetry, all over *AHL* is lifted in general, but still survives at the points *A* and *L* and along the lines *R*, *S*, and *S′* (Falicov and Ruvalds 1968; Cracknell 1969*a*, 1970).

In addition to all the complications mentioned already there have been considerable technical problems associated with the determination of the Fermi surfaces of the transition, rare-earth, and actinide metals. On the experimental side it has, in the past, been difficult to obtain pure single-crystal specimens of many of these metals, while on the theoretical side the lack of a clear-cut division between conduction electrons and core electrons makes it very difficult to construct realistic crystal potentials $V(\mathbf{r})$ *ab initio* in these metals. Because of the uncertainties in $V(\mathbf{r})$ it is therefore quite difficult to calculate reliable band structures for these metals. Consequently, for the transition metals the use of the *rigid-band model*, or as it has sometimes been called the '*common-band model*' (Heine and Weaire 1970), has been quite popular in the past. This model is based on the assumption that, for metals with a common crystal structure, as one moves through any one of the series of transition metals the shapes of the electronic energy bands are not altered, apart from scaling due to differences in the lattice constants, and all that happens is that the number of conduction electrons per atom is changed and the position of the Fermi level changes accordingly. The rigid-band model can also be extended to describe dilute alloys by allowing departures from the integer values for the electron/atom ratio (see Section 7.2). We shall see in this chapter as we progress through the periodic table that, because of the close similarities among the various transition metals, the rigid-band model has been used quite successfully to predict the shapes of the Fermi surfaces of a number of transition metals (see Section 5.7 in particular). Consequently those transition metals which, for historical reasons only, have been studied most extensively have tended to be used as a basis for constructing a generalized transition metal band structure for all those transition metals with the same crystal structure. For any transition metal for which the shape of the Fermi surface had not been studied experimentally it was then possible, by using the generalized transition metal band structure for that crystal structure together with the known or assumed number of conduction electrons per atom, to deduce the Fermi energy and thence the shape of the Fermi surface.

5.2. Group IB

Cu, Ag, Au

Cu	Copper	2.8.(8.10).1	f.c.c.
Ag	Silver	2.8.18.(8,10).1	f.c.c.
Au	Gold	2.8.18.32.(8,10).1	f.c.c.

The electronic structure of metallic Cu is characterized by a full d-band with one electron per atom in the conduction band. It is to be expected, by comparison with the electronic structure of a free Cu atom and the fact of the variable valence of Cu, that the d-band is not far below the conduction band. Similar arguments apply to Ag and Au. The free-electron Fermi surface for a metal with the f.c.c. structure and one conduction electron per atom is shown in Fig. 1.16. This Fermi surface consists of a sphere with volume equal to half the volume of the Brillouin zone; it is completely contained within the Brillouin zone and does not touch the surface of the Brillouin zone. However, from the relationship between the electrical conductivity and the thermal conductivity of Cu, Klemens (1954) argued that the Fermi surface of Cu must be sufficiently distorted from sphericity so as to touch the Brillouin zone boundary; the most likely boundary point is L, at the centre of the hexagonal face of the Brillouin zone, since it is nearest to Γ. The sign of the thermoelectric power (Jones 1955), the magnitude of the Hall effect and the non-saturation of the magnetoresistance (Chambers 1956*a*) also indicated that the Fermi surface of Cu is multiply-connected. The shape of the Fermi surface of Cu was established quite early in the history of

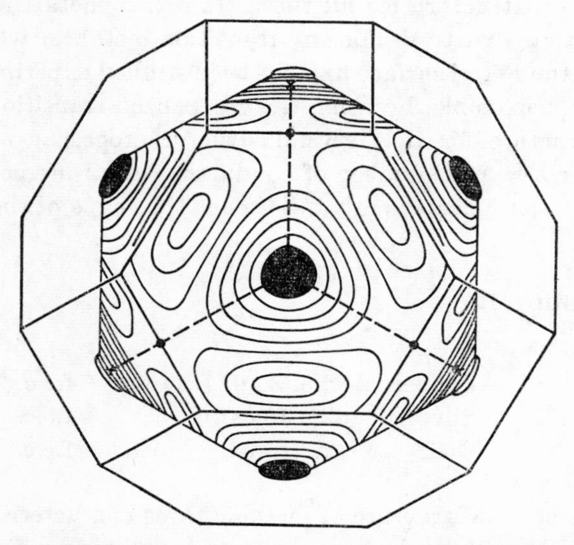

FIG. 5.4. The Fermi surface of Cu (Pippard 1957*b*).

Fermiology by Pippard (1957*b*) using the anomalous skin effect. This was before any of the other effects which give direct measurements of Fermi surface geometry, such as the de Haas–van Alphen effect, had been observed in Cu. Previous work on the electronic structure of Cu was quite sparse; it included anomalous skin effect work on polycrystalline Cu, measurements of the electronic contribution to the specific heat, and one or two band structure calculations. The Fermi surface of Cu deduced by Pippard is illustrated in Fig. 5.4. The de Haas–van Alphen effect was subsequently observed in Cu by Shoenberg (1959, 1960*a*,*b*, 1962) and these measurements confirmed the general features of the Fermi surface determined by Pippard, as also did the magnetoresistance experiments of Alekseevskiĭ and Gaĭdukov (1959*a*, *b*). Parametrization schemes to represent the Fermi surface of Cu analytically have already been discussed in Section 2.9.

Band structure calculations for Cu were performed by Burdick (1961, 1963) and Mattheiss (1964*b*) using the A.P.W. method and by Segall (1961*b*, 1962) using the Green's function method. The calculations performed by Burdick and by Segall produced band structures for Cu that were in close agreement with each other (~0·01 Ry usually), partly through having used the same potential. The dimensions of their calculated Fermi surfaces were in good semiquantitative agreement (~10 per cent) with the experimental results already mentioned for the anomalous skin effect and de Haas–van Alphen effect, as well as with magnetoacoustic geometric resonance measurements of some of the caliper dimensions (Morse 1960; Morse, Myers, and Walker 1961). Subsequently, both more refined experiments and more accurate calculations for Cu have been performed and the agreement between experiment and calculations has become very close indeed. The shape of the Fermi surface of Cu can now be regarded as known with sufficient accuracy to enable it to be used in testing theories of various electronic properties of metals, or for testing different methods for constructing potentials for use in band structure calculations. The more recent sets of de Haas–van Alphen data on Cu have been used to re-calculate the coefficients in Roaf's formula (eqn (2.9.5)) and radius vectors of the Fermi surface of Cu (to an accuracy substantially better than 1 per cent) at 5° intervals of the polar coordinates θ and ϕ are tabulated by Halse (1969); another parametrization scheme has also been applied to Cu by Zornberg and Mueller (1966).

In the free-electron model, the Fermi surfaces of Ag and Au, which are monovalent f.c.c. metals like Cu, would be spheres which are completely contained within the Brillouin zone. However, on account of the general similarity between Cu, Ag, and Au we might expect the Fermi surfaces of Ag and Au to be sufficiently distorted so as to touch the hexagonal face of the Brillouin zone as happens in Cu. The existence of open orbits in both Ag and Au was demonstrated by work on the magnetoresistance (Alekseevskiĭ and Gaĭdukov 1958, 1959*b*, 1962; Gaĭdukov 1959) so that it appeared that the Fermi surface proposed by Pippard for Cu also applies to Ag and Au as well (Priestley 1960). This idea was further supported for both Ag and Au by preliminary de Haas–van Alphen results (Shoenberg 1960*a*,*b*, 1962) and some measurements of caliper dimensions using magnetoacoustic geometric oscillations (Morse 1960; Morse, Myers, and Walker 1960, 1961; Bohm and Easterling 1962; Easterling and Bohm 1962). The determination by Roaf (1962) of the coefficients in an analytical representation of the Fermi surface for Cu described in Section 2.9 was also performed at the same time on Ag and Au. Numerous further de Haas–van Alphen measurements have been made on Ag and Au, and the angular variations of the important periods have been determined with an accuracy of about 0·1 per cent and with an estimated error of about 0·5 per cent in the absolute values of the periods. These measurements have been used in a re-determination of the coefficients in Roaf's formula for the Fermi surfaces of Ag and Au; Halse (1969) gives radius vectors of the Fermi surfaces of Ag and Au at 5° intervals in θ and ϕ.

The Fermi surfaces of Ag and Au had been determined quite accurately before many band structure calculations on these metals had been performed so that, as for Cu, these band structure calculations tend to be used to give guidance about methods for constructing suitable *ab initio* potentials or suitable pseudopotentials rather than for providing any new information about the Fermi surface topology. Although the general shapes of the Fermi surfaces of Cu, Ag, and Au are similar, it is interesting to see how the magnitude of the distortion of the spherical free-electron Fermi surface varies among these metals. The Fermi surface of Ag is least distorted from the spherical free-electron Fermi surface; consequently one would expect the neck radius to be smaller in Ag than in Cu or Au, as is indeed observed, see Fig. 5.5.

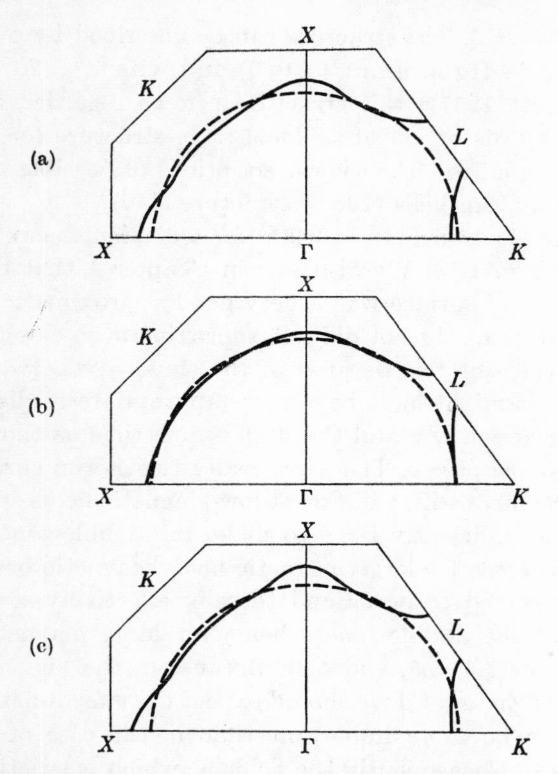

FIG. 5.5. Illustration of the relative distortions of the Fermi surfaces of (a) Cu, (b) Ag, and (c) Au (Roaf 1962).

5.3. Group IIB

Zn, Cd, Hg

Zn	Zinc	2.8.(8,10).2	h.c.p.
Cd	Cadmium	2.8.18.(8,10).2	h.c.p.
Hg	Mercury	2.8.18.32.(8,10).2	see below

The values of the c/a ratios of Zn and Cd depart considerably from the ideal value of $\sqrt{(\frac{8}{3})}$ which means that the structures are considerably more distorted than that of Mg for which the value of c/a is very close to the ideal value. Hg, which is of course a liquid at room temperature, crystallises at about $-39°C$ to form a trigonal crystal that belongs to the space group $R\bar{3}m(D_{3d}^5)$. The structure can be described by a unit cell with a single Hg atom at (0, 0, 0) and with $\alpha = 70° 44{\cdot}6'$ at 5 K (Barrett 1957); this structure can be regarded as being obtained by distortion of an ideal f.c.c. structure for which α would be equal to 60°. Other descriptions of possible unit cells are given by Smithells (1967) (structure $A10$).

The free-electron Fermi surface for a divalent h.c.p. metal is shown in Fig. 1.18. We also saw in Chapter 4 that this free-electron Fermi surface was a very good approximation to the Fermi surface of Mg but did not approximate so closely to the Fermi surface of Be. Because of the similarity between their structures it might have been more appropriate to discuss the Fermi surfaces of Zn and Cd at the same time as those of Be and Mg in Chapter 4. There are really two Fermi surfaces for each of the metals Zn and Cd; at low magnetic fields spin–orbit coupling is sufficiently large to make the double-zone scheme invalid, whereas for large magnetic fields magnetic breakdown can be expected to occur and thereby effectively restore the validity of the double-zone scheme for large magnetic fields (see Sections 1.6, 3.5, and 4.3). Because of the higher atomic numbers of Zn and Cd we should expect the effect of spin–orbit coupling to be more important than in the case of Mg (see Section 4.3). Consequently the fields at which magnetic breakdown will occur in Zn and Cd will be larger than they are in Mg. We mentioned in Chapter 1, see p. 69, the importance of the axial ratio c/a in determining whether or not the needle-shaped pockets of electrons in bands 3 and 4 are present in the free-electron model for a divalent h.c.p. metal (Harrison 1960*b*). If c/a exceeds the critical value of 1·8607 the needles will not be present. For Zn, c/a is about 1·8246 at the low temperatures usually used in Fermi surface studies, whereas for Cd c/a is about 1·886. Consequently we should expect the needles to be present in the free-electron Fermi surface for Zn but not for Cd; this feature of the Fermi surfaces of Zn and Cd which was predicted on the basis of the free-electron model has now been well established experimentally. In addition to affecting the needles, the departure of c/a from the ideal value of $\sqrt{(\frac{8}{3})}$ may also affect the connectivity of the monster in bands 1 and 2. For the very large value of c/a that is possessed by Cd, the connectivity of the arms of the monster becomes altered—see Fig. 5.6. This feature of the Fermi surface of Cd which was predicted on the basis of the free-electron model has also been established experimentally.

The early experimental work on the electronic structures of Zn and Cd has been reviewed by Harrison (1960*b*): this included

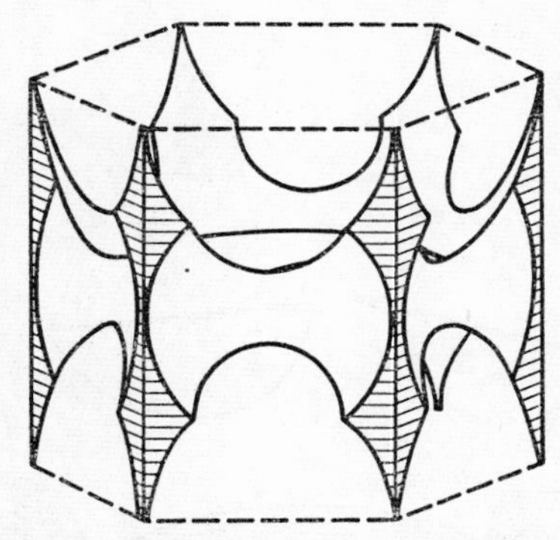

FIG. 5.6. The free-electron Fermi surface of Cd in bands 1 and 2 (Harrison 1960*b*).

Schubnikow–de Haas, cyclotron resonance, and galvanomagnetic measurements, as well as a number of de Haas–van Alphen measurements which indicated that the needles in band 3 exist for Zn but not for Cd, and in which for both metals several other oscillations were observed. An O.P.W. calculation of the band structure and Fermi surface of Zn was performed by Harrison (1962, 1963*a*) and the calculated Fermi surface was adjusted to fit the considerable number of measured de Haas–van Alphen periods of Joseph, Gordon, Reitz, and Eck (1961) and Joseph and Gordon (1962). These experimental results included measurements of periods on the needles in band 3, and the horizontal arms and diagonal arms of the monster in band 2. The volume of the needles in Zn is very small and corresponds to about 5×10^{-6} electron per atom. These experimental results also included two periods that could only be

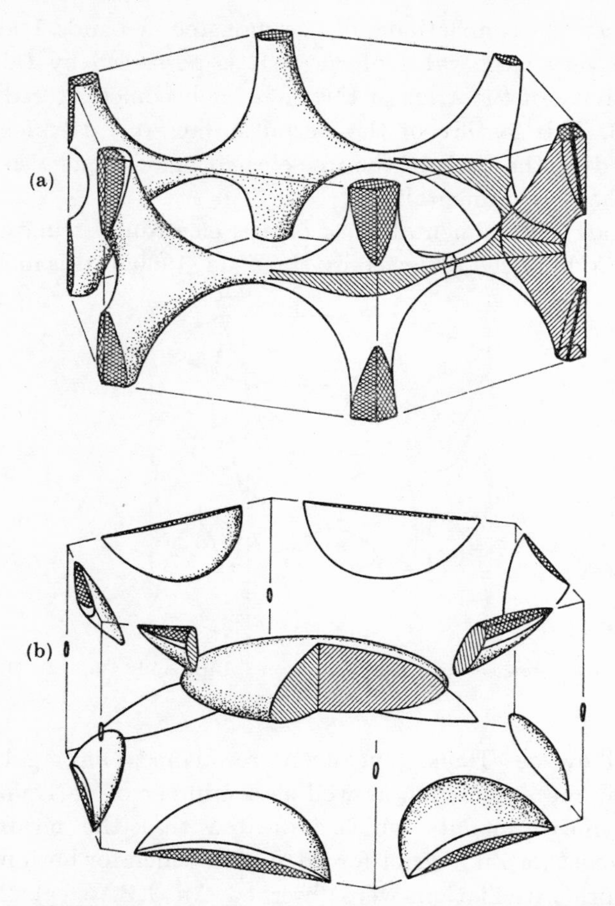

FIG. 5.7. The Fermi surface of Zn: (a) holes in bands 1 and 2 (the caps in band 1 are shown cross-hatched), (b) electrons in bands 3 and 4 (the cigars in band 4 are shown cross-hatched) (Gibbons and Falicov 1963).

explained satisfactorily by assuming that the spin–orbit coupling in Zn is sufficiently large to render the double-zone scheme invalid, at least in relatively low magnetic fields, see Fig. 5.7. A substantial amount of further experimental work on both Zn and Cd was performed before any further complete band structure calculations became available. This included work on magnetoacoustic geometric oscillations, quantum oscillations in the ultrasonic attenuation, the de Haas–van Alphen effect, galvanomagnetic properties, cyclotron resonance, and the radio-frequency size effect. A large number of linear dimensions of the Fermi surfaces of Zn and Cd are tabulated by Gibbons and Falicov (1963). From many of these experiments

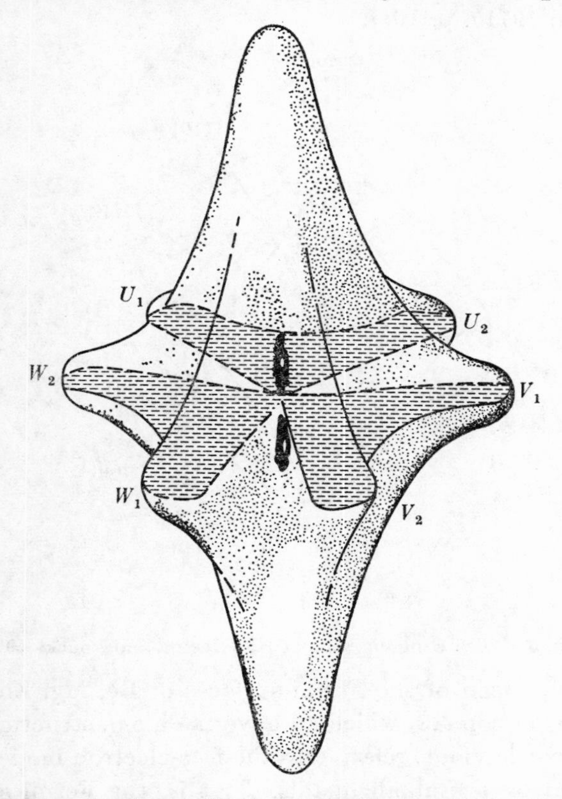

FIG. 5.8. The Fermi surface of Cd; the region of holes in band 1 formed by the dismembering of the free-electron 'monster' (Galt, Merritt, and Klauder 1965).

it became clear that the needle-shaped pockets of electrons that appear at K in band 3 for Zn do not exist in Cd and that the monster in band 2 is disconnected by the 'pinching off' of the $\langle 11\bar{2}0\rangle$ waists, see Fig. 5.8; we have already noted that both these features were predicted from the free-electron model as a consequence of the very large c/a ratio in Cd. There has been some considerable disputation as to whether or not the 'butterflies' (or 'stars') and the horizontal 'cigars' at the edges of the Brillouin zone in bands 3 and 4 which are shown in the free-electron Fermi surface for a divalent h.c.p. metal in Fig. 1.18 are present in the Fermi surfaces of Zn and Cd (for details see Cracknell 1971*b*, p. 161).

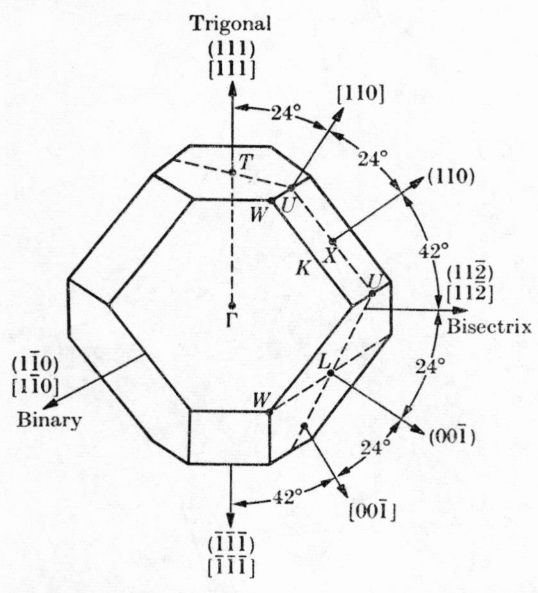

FIG. 5.9. The Brillouin zone of Hg (Keeton and Loucks 1966*b*).

A comparison of the Fermi surfaces of Be, Mg, Zn, and Cd (see also Section 4.3) which all have the h.c.p. structure reveals a similar behaviour, relative to the free-electron model, to that exhibited by the alkali metals. That is, the Fermi surface of the metal in the second row of the periodic table, namely Mg, is closest to the h.c.p. free-electron Fermi surface; with increasing atomic number beyond Mg the distortion from the free-electron Fermi surface increases. As with the alkali metals, the Fermi surface of the lightest metal, in this case Be, is also substantially distorted from the free-electron model.

Since the structure of Hg can be regarded as a distorted f.c.c. crystal, the Brillouin zone, Fig. 5.9, can be regarded as a distorted f.c.c. Brillouin zone. The labels used for the special points of symmetry are an adaptation of the labels used in the f.c.c. Brillouin zone. In the free-electron model the spherical Fermi surface intersects the six large hexagonal faces of the Brillouin zone but does not reach the remaining faces, as in Fig. 5.10. Consequently the free-electron Fermi surface consists

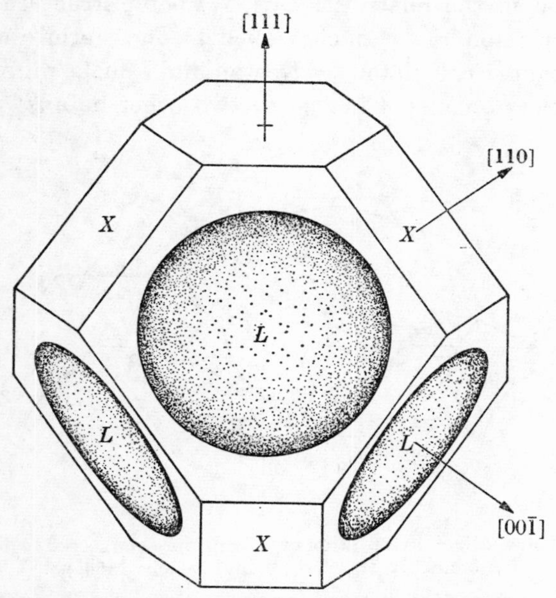

FIG. 5.10. The Fermi surface of Hg in the free-electron model: holes in band 1 (Brandt and Rayne 1965).

of a set of lens-shaped pockets of electrons in band 2 at the large hexagonal faces, together with a complicated multiply-connected Fermi surface of holes in band 1. This can be recognized as being obtained by a trigonal distortion of the free-electron Fermi surface for a divalent f.c.c. metal.

The Fermi surface of Hg has been determined only recently and its general features are similar to those of the free-electron model. The de Haas–van Alphen measurements of Brandt and Rayne (1965) were able to be interpreted in terms of orbits on the multiply-connected Fermi surface of holes in band 1, provided the spherical free-electron Fermi surface was sufficiently distorted so as to touch the rectangular faces of the Brillouin zone at the points X; confirmation of the existence of open orbits was provided by several magnetoresistance and other measurements on Hg. The contact of the Fermi surface in band 1 with the rectangular faces of the Brillouin zone was reproduced in the relativistic A.P.W. band structure calculations of Keeton and Loucks (1966*b*). The Fermi surface of holes in band 1 calculated by Keeton and Loucks differed from the free-electron model in one or two other details; occupied

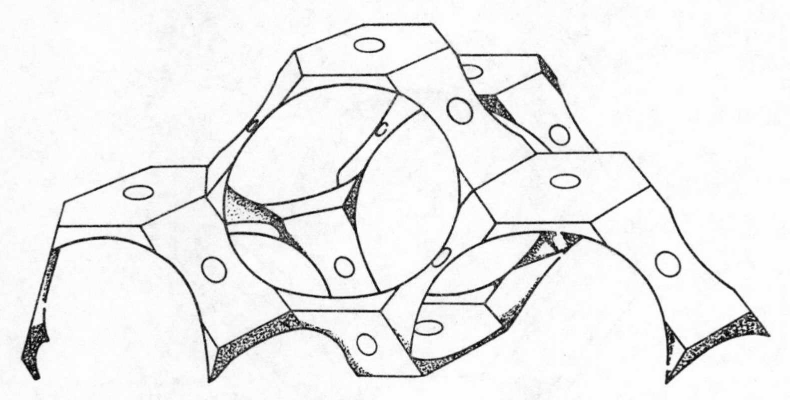

FIG. 5.11. Schematic representation of the multiply-connected region of holes in band 1 in Hg (Keeton and Loucks 1966*b*).

regions were found around the points T and K as well as around X, see Fig. 5.11. In addition to the two sets of de Haas–van Alphen periods associated with the multiply-connected Fermi surface of holes in band 1, Brandt and Rayne (1966) also observed a period associated with the lens-shaped pocket of electrons in band 2. Brandt and Rayne used their experimental results to determine the parameters in a pseudopotential band structure calculation; the resulting calculated cross-sectional areas were tabulated and compared with the experimental results (Brandt and Rayne 1966). Caliper dimensions of the Fermi surface of Hg in both band 1 and band 2 were determined from magnetoacoustic geometric resonances and the results for the principal dimensions were compared with the results of previous work (Bellessa, Reich, and Mercouroff 1969; Bogle, Coon, and Grenier 1969).

5.4. Group IIIA

Sc, Y

Sc	Scandium	2.8.(8,1).2	h.c.p.
Y	Yttrium	2.8.18.(8,1).2	h.c.p.

The lanthanides (rare-earth metals) and actinides are often classified with Sc and Y in group IIIA of the periodic table. However, because many of the rare-earth metals possess complicated magnetically ordered structures it seemed better to leave them until after the simpler ferromagnetic metals of group VIII have been discussed, rather than to discuss them with the elements Sc and Y of group IIIA which might have been more logical.

With the metals of group IIIA we enter the transition series of metals, that is, the metals which have a full 4s-, 5s-, or 6s-shell and now begin to fill up the next lower d-shell, that is, the 3d-, 4d-, or 5d-shell. In contrast with nearly all the metals that we have considered so far, both in Chapter 4 and in the previous sections of this chapter, there has been only a relatively small amount of experimental and theoretical work performed on the Fermi surfaces of the transition metals. This is because of the difficulties that are often involved in obtaining sufficiently pure samples. An extensive review of the early work on electrons in transition metals was given by Mott (1964), at which stage hardly any direct measurements of the dimensions of transition-metal Fermi surfaces had been made.

The axial ratio c/a is 1·59 for Sc and 1·58 for Y; these values depart quite substantially from the ideal h.c.p. value of $\sqrt{(\frac{8}{3})}$. Sc also exists in an f.c.c. phase at high temperatures. The free-electron Fermi surface for an h.c.p. metal of valence 3 is shown in Fig. 1.18. While the departure of c/a from the ideal value of $\sqrt{(\frac{8}{3})}$ alters the dimensions of the free-electron Fermi surface slightly, it does not lead to any drastic changes such as the complete disappearance of some parts of the Fermi surface as happens in Cd and nearly happens in Zn. Since c/a is smaller than the ideal value, the horizontal cross sections of the pockets of electrons in bands 5 and 6 can be expected to increase slightly.

The band structure, Fermi surface, and density of states of Sc have been calculated by Altmann and Bradley (1967) using

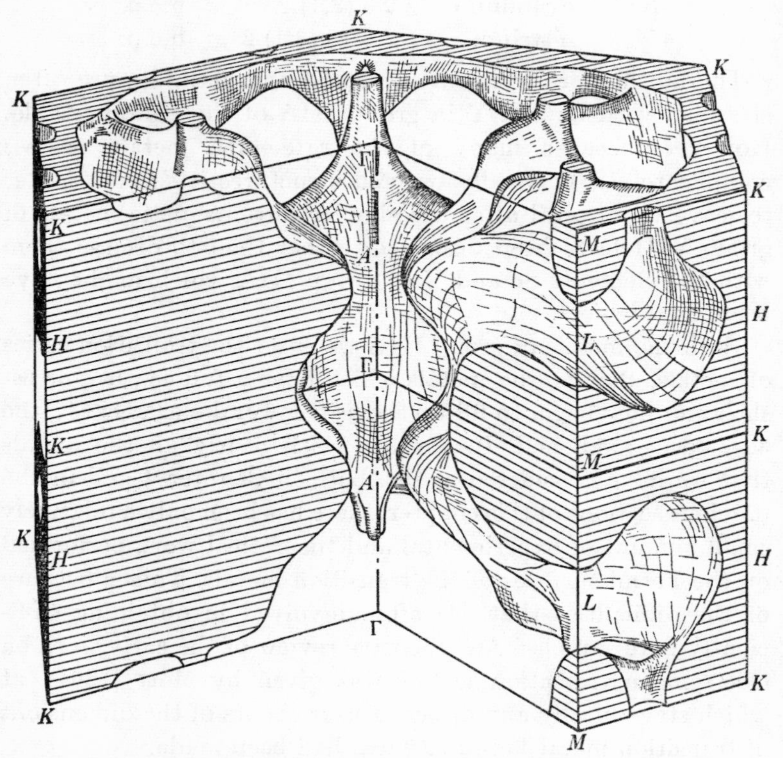

FIG. 5.12. The Fermi surface of Sc in band 3 and band 4 calculated by Altmann and Bradley (1967) (electrons shaded).

the cellular method and by Fleming and Loucks (1968) using the A.P.W. method. With the band structure and Fermi energy calculated by Altmann and Bradley, the Fermi surface is entirely in bands 3 and 4 and has the very complicated multiply-connected structure shown in Fig. 5.12. Although this complicated Fermi surface of electrons in bands 3 and 4 bears some relationship to the free-electron Fermi surface, it is very substantially distorted from the free-electron Fermi surface. As a result of this distortion holes appear at Γ, while narrow filaments parallel to k_z arise as a result of extra contacts between the top surface of the Brillouin zone and the region occupied by electrons at a point along ΓK and the region of holes at a point along MK. However, these contacts involve bands that are very close to the Fermi level and one or other of these connectivities could easily be destroyed by quite a small change in the position of the Fermi level. The Fermi surface of Sc calculated by Fleming and Loucks (1968) also consists of a multiply-connected Fermi surface in bands 3 and 4; however, many of the detailed connectivities and dimensions of the Fermi surface for Sc calculated by Fleming and Loucks were different from that obtained by Altmann and Bradley. The principal topological differences were that the narrow filaments of electrons that cross ΓK, and of holes that cross MK, vanish; these differences are most clearly seen from the section shown in Fig. 5.13. The differences between the results of these cellular and A.P.W. calculations is thought to be due to differences between the crystal potentials $V(\mathbf{r})$ used in the two calculations.

The most direct experimental evidence available so far about the shape of the Fermi surface of Sc comes from the magnetoresistance measurements of Bogod and Eremenko (1965); the B^2 law was obeyed up to the highest magnetic fields used (96 kG), from which it was concluded that the Fermi surface of Sc is multiply-connected in some complicated fashion. However, these measurements cannot be used to distinguish between the two calculated Fermi surfaces that we have described. Detailed experimental investigations of the Fermi surface of Sc are still awaited to determine the dimensions of the Fermi surface and to establish the connectivities.

For Y, similar cellular and A.P.W. band structure calculations have been performed to those that we have just described for Sc. The general features of the calculated Fermi surfaces for Y were found to be very similar to those calculated for Sc, that is,

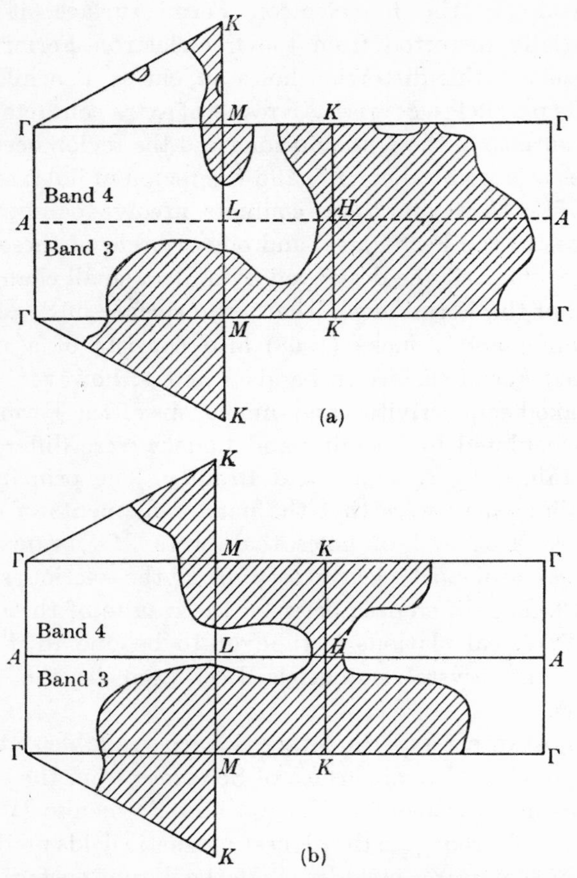

Fig. 5.13. Cross sections of the Fermi surface of Sc (with electrons shaded) calculated by (a) Altmann and Bradley (1967) and (b) Fleming and Loucks (1968).

the Fermi surface is entirely contained in bands 3 and 4 and separates a multiply-connected region of holes enclosing Γ, *A*, and *L*, and a multiply-connected region of electrons enclosing *K*, *H*, and *M*. The Fermi surface of Y calculated by Loucks (1966*b*) was almost identical with that obtained by Fleming and Loucks (1968) for Sc. The details of the dimensions and connectivities of the Fermi surface calculated by Altmann and

Bradley (1967) for Y using the cellular method were rather different from those for the Fermi surface calculated by Loucks; in particular band 4 was found to be empty at M, see Fig. 5.14.

Once again, as with Sc, there have been no direct experimental measurements of the dimensions of the Fermi surface of Y, except for some slight evidence of Kohn anomalies

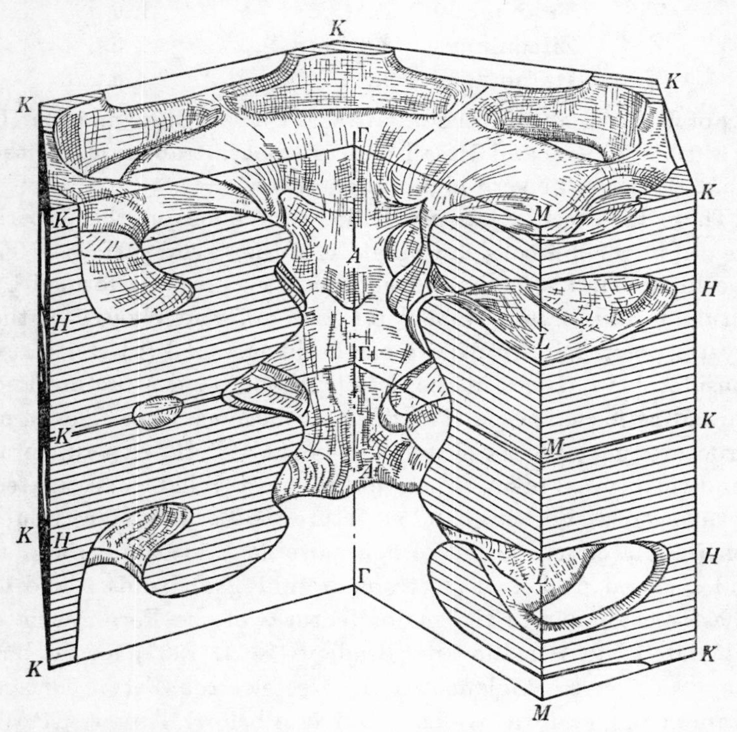

FIG. 5.14. The Fermi surface of Y in bands 3 and 4 calculated by Altmann and Bradley (1967) (electrons shaded).

(Sinha, Brun, Muhlestein, and Sakurai 1970). It is perhaps a little surprising that, although the pockets of electrons in bands 5 and 6 in the free-electron model for a trivalent h.c.p. metal are not particularly small, in both the cellular and A.P.W. calculations for each of the metals Sc and Y, bands 5 and 6 were found to be empty all over the Brillouin zone. If this is a genuine feature of the Fermi surfaces of Sc and Y, it is one indication that for transition metals the free-electron model is likely to give only a very poor first approximation to the shape of the Fermi surface.

5.5. Group IVA

Ti, Zr, Hf

Ti	Titanium	2.8.(8,2).2	h.c.p.
Zr	Zirconium	2.8.18.(8,2).2	h.c.p.
Hf	Hafnium	2.8.18.32.(8,2).2	h.c.p.

For both Ti and Zr the axial ratio c/a is 1·59, while for Hf it is 1·58. Ti, Zr, and Hf all undergo phase transitions to the b.c.c. structure at very high temperatures.

There has been relatively little work done on the experimental or theoretical determination of the Fermi surface of Ti. A cellular band structure calculation for Ti was performed by Altmann and Bradley (1967) for four different choices of the crystal potential $V(\mathbf{r})$; quite small changes in $V(\mathbf{r})$ can cause considerable changes in the details of the band structure and quite drastic changes in the calculated Fermi surface. The band structure that gave the closest agreement with the experimental bandwidth was taken to give the most reliable calculated Fermi surface. For this band structure the Fermi surface would consist of a closed pocket of holes around L in bands 3 and 4, and a closed pocket of electrons around Γ in bands 5 and 6. These are similar to the major features of the Fermi surface calculated by Altmann and Bradley (1964, 1967) for Zr but they bear no resemblance to the free-electron Fermi surface of an h.c.p. metal with valence four (see below). A new A.P.W. calculation for Ti performed by Hygh and Welch (1970) produced a Fermi surface which is illustrated in Fig. 5.15 and which is substantially different both from the results of Altmann and Bradley (1964, 1967) and from what might have been expected by extrapolation from the A.P.W. calculations for Zr performed by Loucks (1967*b*) (see below). There is still considerable doubt attached to these theoretical predictions of the Fermi surface of Ti as well as an almost complete lack of direct experimental evidence on this metal.

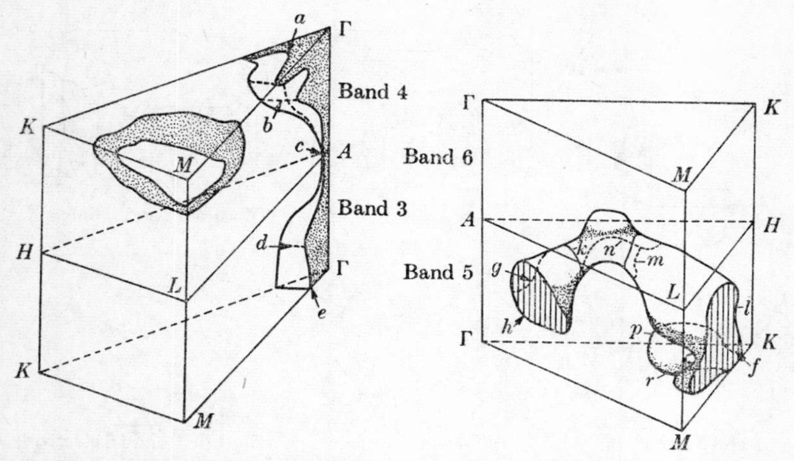

FIG. 5.15. The Fermi surface of Ti calculated by Hygh and Welch (1970), (a) holes in bands 3 and 4, and (b) electrons in bands 5 and 6.

Slightly more work has been done on the determination of the Fermi surface of Zr than has been done for Ti. The de Haas-van Alphen effect was observed in Zr by Thorsen and Joseph (1963) and five periods were observed; of these five periods, one could be followed over all angles and was therefore assumed to be associated with a closed piece of the Fermi surface. The remaining periods were present for magnetic fields in the [0001] direction and for limited inclinations to [0001] but eventually disappeared at large inclinations. The free-electron Fermi surface for an h.c.p. metal with valence four is not given with the others in Fig. 1.18; however, its general features can be determined by extrapolation from the case of a metal of valence 3 in Fig. 1.18. Bands 1 and 2 are full all over the Brillouin zone and the regions occupied by electrons in bands 3, 4, 5, and 6 can be expected to be somewhat expanded. This free-electron Fermi surface is shown in the single-zone scheme in Fig. 5.16. Thorsen and Joseph attempted to explain their de Haas–van Alphen periods in terms of a distorted version of the free-electron Fermi surface in Fig. 5.16.

A number of band structure calculations have been made for Zr using the cellular method (Altmann 1958*b*; Altmann

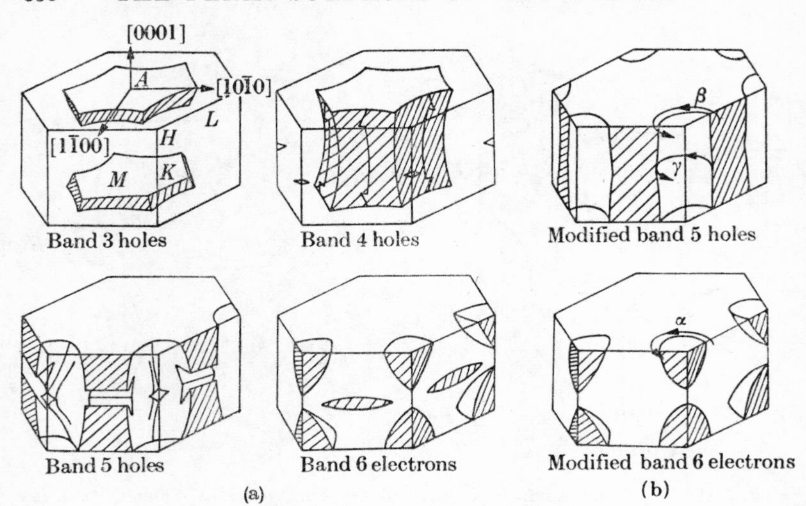

FIG. 5.16. (a) Sketches of the free-electron Fermi surface for an h.c.p. metal of valence 4, and (b) modifications of the Fermi surface in bands 5 and 6 proposed by Thorsen and Joseph (1963) to account for the observed de Haas–van Alphen results in Zr.

and Bradley 1962, 1964, 1967) and the A.P.W. method (Loucks 1967*b*). According to the cellular calculation of Altmann and Bradley (1964), bands 1 and 2 are full all over the Brillouin zone, in bands 3 and 4 in the double-zone scheme there are a number of closed regions of holes (see Fig. 5.17) and in bands 5 and 6 in the double-zone scheme there is a region occupied by electrons which is almost isolated except for very small necks across the vertical faces of the Brillouin zone (see Fig. 5.18). A small change in the Fermi surface in bands 5 and 6 was found in later calculations (Altmann and Bradley 1967) in which band 5 no longer comes below the Fermi energy anywhere along ΓM so that the narrow waist of this pocket of electrons along ΓM vanishes. This Fermi surface bears very little resemblance to the free-electron Fermi surface shown in Fig. 5.16 and was considerably more successful than the free-electron model in explaining the de Haas–van Alphen data of Thorsen and Joseph.

The A.P.W. band structure calculations performed by Loucks (1967*b*) produced a Fermi surface that was substantially

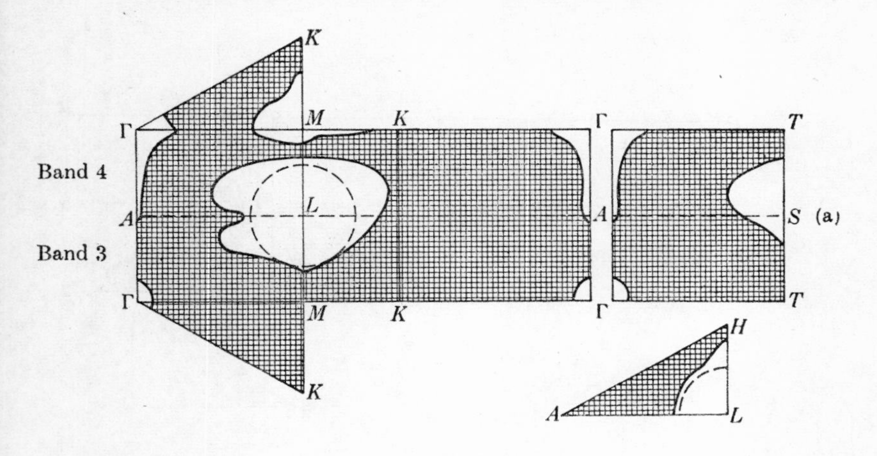

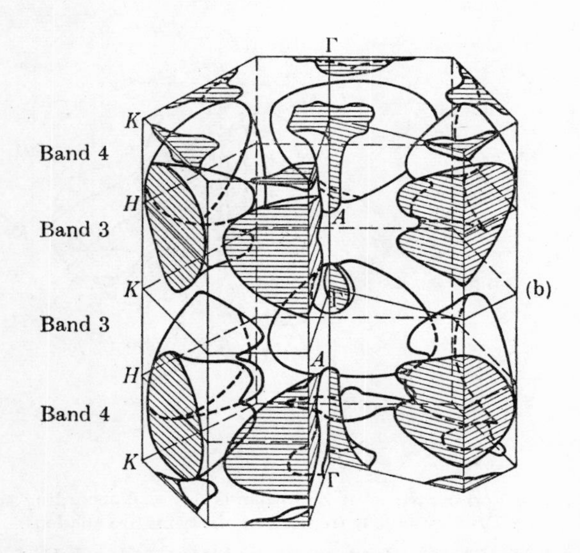

FIG. 5.17. The Fermi surface of Zr in bands 3 and 4 according to Altmann and Bradley (1964) (in (a) regions of electrons are shaded and in (b) regions of holes are shaded).

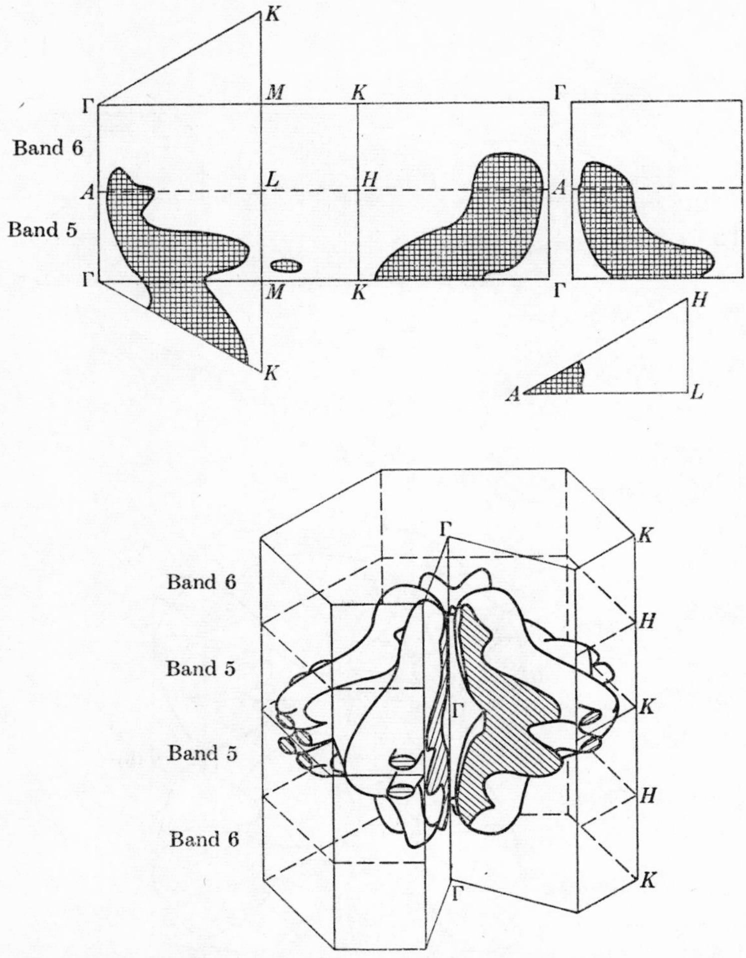

FIG. 5.18. The Fermi surface of Zr in bands 5 and 6 according to Altmann and Bradley (1964) (regions of electrons are shaded).

different from that calculated by Altmann and Bradley. The Fermi surface calculated by Loucks consisted, in the double-zone scheme, of a region of holes with axial rotational symmetry located along the [0001] axis in bands 3 and 4 (see Fig. 5.19(a)), an isolated region occupied by electrons around H in band 5, and small ellipsoidal pockets of electrons, also at H, in band 6 (see Fig. 5.19(b)). The band 6 pockets of electrons are contained

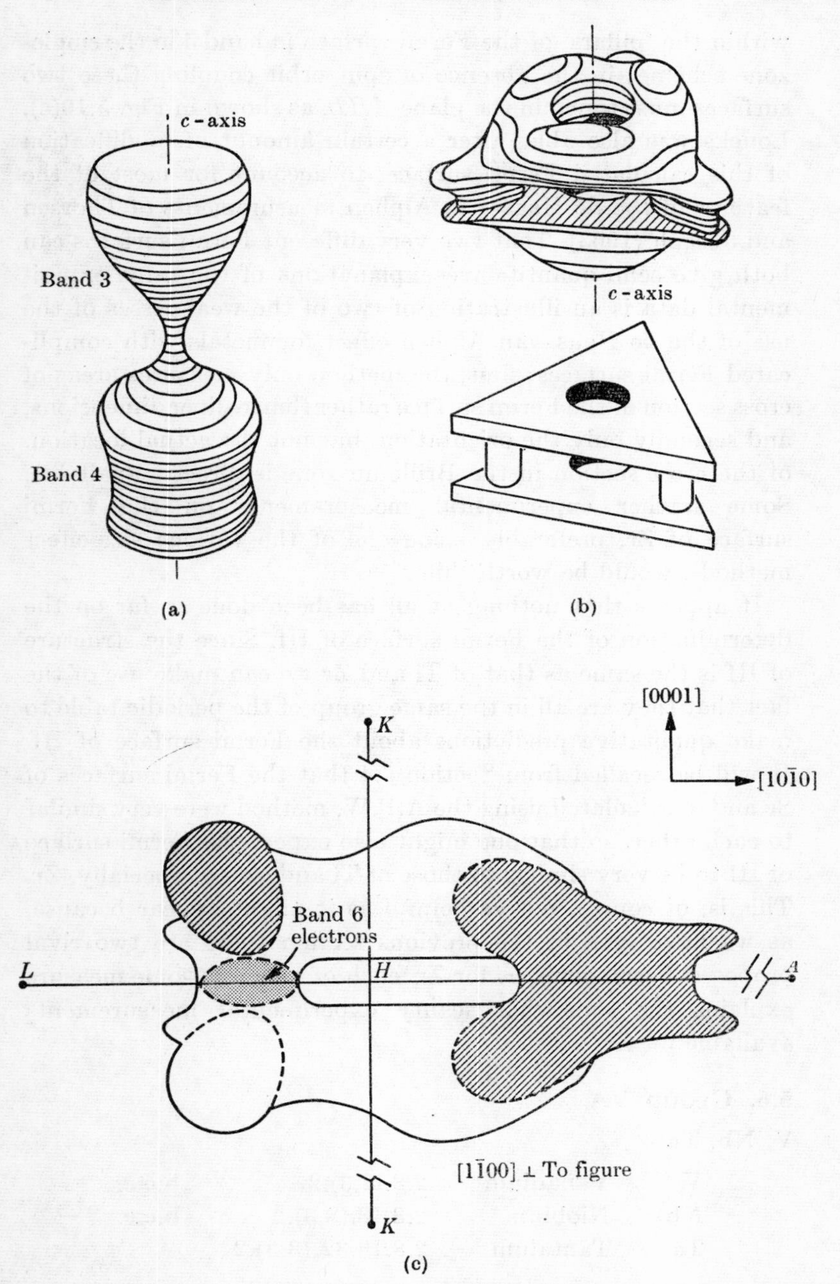

FIG. 5.19. (a) The regions of holes in bands 3 and 4 of Zr and (b) and (c) the regions of electrons in band 5 of Zr according to Loucks (1967*b*).

within the 'pillars' of the Fermi surface in band 5 in the single-zone scheme; in the absence of spin–orbit coupling these two surfaces must touch in the plane AHL, as shown in Fig. 5.19(c). Loucks was also able, after a certain amount of modification of this calculated Fermi surface, to account for most of the features of the de Haas–van Alphen measurements of Thorsen and Joseph (1963). That two very different Fermi surfaces can both give semi-quantitative explanations of the same experimental data is an illustration of two of the weaknesses of the use of the de Haas–van Alphen effect for metals with complicated Fermi surfaces; first, the method only measures areas of cross section of the Fermi surface rather than caliper dimensions, and secondly only the orientation, but not the actual location, of the cross section in the Brillouin zone is directly available. Some further experimental measurements on the Fermi surface of Zr, preferably using one of the various size-effect methods, would be worthwhile.

It appears that nothing at all has been done so far on the determination of the Fermi surface of Hf. Since the structure of Hf is the same as that of Ti and Zr we can make use of the fact that they are all in the same group of the periodic table to make qualitative predictions about the Fermi surface of Hf. It will be recalled from Section 5.4 that the Fermi surfaces of Sc and Y calculated using the A.P.W. method were very similar to each other, so that one might also expect the Fermi surface of Hf to be very similar to those of Ti and, more especially, Zr. This is, of course, not as helpful as it might appear because, as we have seen in the previous section there are two rival proposed Fermi surfaces for Zr, each of which in some measure explains the somewhat scanty experimental measurements available for Zr.

5.6. Group VA

V, Nb, Ta

V	Vanadium	2.8.(8,3).2	b.c.c.
Nb	Niobium	2.8.18.(8,3).2	b.c.c.
Ta	Tantalum	2.8.18.32.(8,3).2	

The results of preliminary A.P.W. band structure calculations for V and several other b.c.c. transition metals along the line $\Gamma\Delta H$ in the Brillouin zone performed by Mattheiss (1964*b*) suggested that, for transition metals of the same structure, a rigid-band or common-band model gave a good approximation to the calculated band structures of these metals. As we shall see in Section 5.7 the idea of a rigid-band model for transition metals had previously been used by Lomer (1962*a*, 1964*a*) in connection with predicting the shapes of the Fermi surfaces of the group VIA metals Cr, Mo, and W using the results of the A.P.W. band structure calculations performed by Wood (1962) for Fe. The general features of these proposed Fermi surfaces were supported by the results of a direct A.P.W. calculation for W performed by Mattheiss (1965). Rigid-band-model arguments were then applied by Mattheiss to the calculated band structure and Fermi surface for W to predict the general features of the Fermi surfaces of the metals V, Nb, and Ta in group VA. Three slightly different Fermi surfaces for group VA metals were obtained by Mattheiss using Wood's results for Fe and two sets of bands obtained from A.P.W. calculations using two slightly different potentials in the band structure calculations for W. These predictions are only semi-quantitative and are not able to distinguish between the three metals, V, Nb, and Ta. This proposed Fermi surface is illustrated in Fig. 5.20 and consists of

(i) an inner region of holes at Γ in band 2 (Fig. 5.20(a)),
(ii) a multiply-connected set of hole tubes in band 3 along the $\langle 100 \rangle$ (or $\Gamma\Delta H$) directions, which has been described as a 'jungle gym' (Mattheiss 1970) (Fig. 5.20(b)), and
(iii) ellipsoidal pockets of holes in band 3 at N (Fig. 5.20(b)).

The predictions based on the band structure of W suggest that the ellipsoids at N are isolated whereas those based on the band structure of Fe suggest that these ellipsoids at N are joined to the multiply-connected tubular region by necks along ΓN. Further work was needed to distinguish between these two possibilities. In the $\langle 100 \rangle$ and $\langle 110 \rangle$ planes the inner hole surface

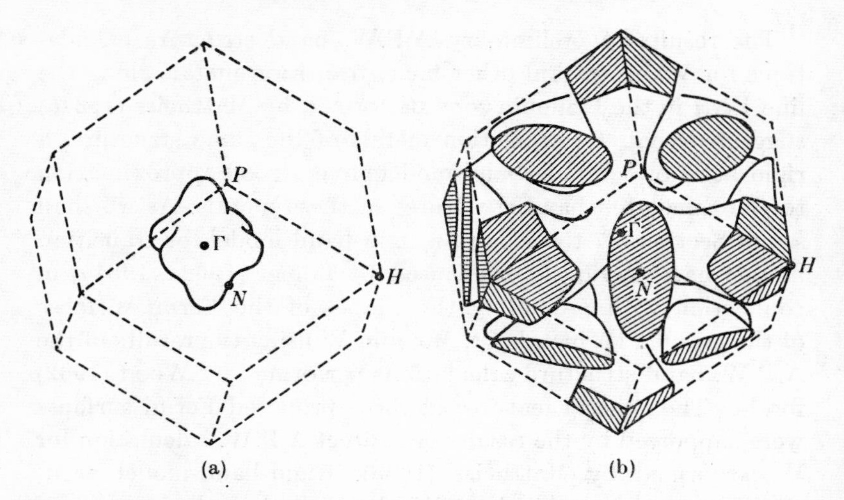

FIG. 5.20. The Fermi surface proposed by Mattheiss (1965) for group VA metals (V, Nb, and Ta), (a) closed region of holes around Γ, and (b) the 'jungle-gym' hole tubes and the ellipsoidal pockets of holes at N.

touches the multiply-connected tubular surface; the degeneracies that give rise to these contacts will be removed by the inclusion of spin–orbit coupling, although they may be restored, in V and Nb at least, by magnetic breakdown.

There is still in the Fermi surface proposed in Fig. 5.20 for group VA metals a faint resemblance to the free-electron model for a b.c.c. metal with five conduction electrons per atom. By extrapolation from the case of a b.c.c. metal of valence four shown in Fig. 1.17 we should expect the free-electron Fermi surface for a b.c.c. metal of valence five to have a region of holes in band 2 centred at Γ and with roughly the shape of an octahedron, quite fat arms of electrons in band 3 pointing from P towards Γ and with the pockets of electrons in band 4 somewhat enlarged. The region of holes in band 2 is not unlike the shape shown in Fig. 5.20(a) while in Fig. 5.20(b) the electrons are concentrated around the points P. Since band 2 is rather more full than in the free-electron model it is not too surprising that the pockets of electrons in band 4 in the free-electron model have vanished.

An A.P.W. band structure calculation for V was performed by Mattheiss (1964*b*) along only one line in the Brillouin zone. Preliminary A.P.W. calculations by Anderson, McCaffrey and Papaconstantopoulos (1969) have been reported using two different expressions for the exchange contribution to the potential; tentative comparisons were made with preliminary de Haas–van Alphen results which had been ascribed to ellipsoidal pockets of holes at N. Little experimental evidence on V is available at present but it seems likely from the magnetoresistance data of Nelson, Stanford, and Schmidt (1968) that the Fermi surface described above and illustrated in Fig. 5.20 is fairly close to the actual Fermi surface of V. Several workers have reported the existence of anomalies in the electrical resistivity and the magnetic susceptibility that have been ascribed to the onset of magnetic ordering in V; however, no direct evidence of the existence of magnetic ordering in V has been obtained (for references see Cracknell 1971*b*, p. 187).

The results of the band structure calculation of Deegan and Twose (1967) on Nb, using the O.P.W. method, as well as the work of Sharp (1969*a*, *b*) on the observation of Kohn anomalies in Nb seemed to indicate that in Nb the pockets of holes at N are closed and are not connected together by necks along ΓN. The Fermi surfaces of Nb and Ta have been investigated experimentally principally by using the de Haas–van Alphen effect and also by some galvanomagnetic measurements. Open orbits were detected in Nb parallel to the four fold axes as well as to the three fold and two fold axes. Thorsen and Berlincourt (1961*b*) observed two de Haas–van Alphen periods in each of the metals Nb and Ta. The two periods observed in Nb were both later ascribed to the ellipsoidal pockets of holes at N in band 3 by Scott, Springford. and Stockton (1968*a*, *b*) who also observed other periods which were ascribed to the multiply-connected tubular region of holes in band 3. The anisotropy of the galvanomagnetic properties of Nb was found to be very similar to that of Ta, and for each of them the anisotropy of the magnetoresistance could be satisfactorily interpreted in terms of open orbits associated with the multiply-connected Fermi surface in band 3 proposed by Mattheiss. New A.P.W. band structure calculations for Nb and Ta, including relativistic effects for Ta, have been performed by Mattheiss (1970). For Nb no significant differences from the Fermi surface illustrated in Fig. 5.20 were obtained and, so long as spin–orbit coupling is neglected, the sheet of Fermi surface around Γ in band 2 and the multiply-connected jungle gym Fermi surface in band 3 are degenerate in the $\langle 220 \rangle$ and $\langle 100 \rangle$ symmetry planes. In the case of Ta spin–orbit coupling removes these degeneracies but the gaps which are introduced are quite small and can easily escape observation as a result of magnetic breakdown. Otherwise the Fermi surface of Ta calculated by Mattheiss (1970) was also very similar to that shown in Fig. 5.20. Extensive new measurements of cross sections of the ellipsoids at N and the jungle gym for both Nb and Ta were made by Halloran, Condon, Graebner, Kunzler, and Hsu (1970) using both the de Haas–van Alphen effect and magnetothermal oscillations. A detailed comparison between the calculated and measured extremal areas of cross section for both Nb and Ta is given by Mattheiss (1970) and in most cases the agreement between theory and experiment is to within 5 per cent and neither in Nb nor in Ta were oscillations observed experimentally that could be ascribed to the predicted region of holes around Γ in band 2. However, in further de Haas–van Alphen measurements on Nb by Scott and Springford (1970) some oscillations were observed which were ascribed to orbits involving the region of holes around Γ in band 2; it is now fairly certain that this piece of the Fermi surface does actually exist although Scott and Springford proposed a slight distortion of the octahedron from the shape given by Mattheiss. We have now seen that the details of the Fermi surface of Nb are fairly well established and that it seems that the Fermi surface of Ta resembles that of Nb quite closely.

5.7. Group VIA

Cr, Mo, W

Cr	Chromium	2.8.(8,5).1	b.c.c., see below
Mo	Molybdenum	2.8.18.(8,5).1	b.c.c.
W	Tungsten	2.8.18.32.(8,4).2	b.c.c.

At very high temperatures (above 1840°C) Cr exhibits the f.c.c. structure; this is called the β-phase of Cr while the low-temperature b.c.c. structure is called the α-phase. Other electrolytic or unstable forms of Cr and W exist but they do not concern us here. Cr undergoes a transition to an anti-ferromagnetically ordered state at a Néel temperature, T_N, of 311·5 K. Below T_N the structure is not that of a simple two-sublattice antiferromagnet but one in which the spin polarization $\mathbf{P}(\mathbf{r})$ of each sublattice varies sinusoidally with a wave-vector $\mathbf{Q}$, that is, there is a spin-density wave (S.D.W.) (Overhauser 1962) in which

$$\mathbf{P}(\mathbf{r}) = \mathbf{P}(\mathbf{Q}) \exp(i\mathbf{Q}.\mathbf{r}). \tag{5.7.1}$$

References to the experimental evidence for this magnetic structure are cited by Koehler, Moon, Trego, and Mackintosh (1966). One of the sublattices consists of the Cr atoms at the corners of the unit cells and the other consists of the Cr atoms at the centres of these unit cells. The amplitude of the magnetic moment, or spin polarization $\mathbf{P}(\mathbf{Q})$, on one sublattice is exactly opposite to that on the other sublattice. In general the orientation of $\mathbf{P}(\mathbf{Q})$, that is, the polarization of the spin-density wave, may take any value relative to $\mathbf{Q}$, and $\mathbf{Q}$ in turn can take any arbitrary value, so that the wavelength of the spin-density wave need not be commensurate with the crystal lattice of the metal. For Cr, if the temperature is lower than T_N but above a second transition temperature, T_S, of 115 K, $\mathbf{P}(\mathbf{Q})$ is perpendicular to $\mathbf{Q}$, that is, the spin-density wave has transverse polarization; below T_S the polarization becomes longitudinal, that is, $\mathbf{P}(\mathbf{Q})$ is parallel to $\mathbf{Q}$. T_S is called the 'spin flip' temperature (see, for example, Matsumoto, Sambongi, and Mitsui (1969), and Meaden, Rao, Loo, and Sze (1969)). In Cr the wave vector $\mathbf{Q}$ of the spin-density wave orients itself along the $\langle 100 \rangle$ type directions and its magnitude is given by

$$\mathbf{Q} = \tfrac{1}{2}\mathbf{G}_0(1-\delta) \tag{5.7.2}$$

where $\mathbf{G}_0$ is the basic reciprocal lattice vector in the [100] direction and δ varies with temperature from about 0·035 near T_N to about 0·05 at very low temperatures.

The first successful suggestion for the Fermi surfaces of Cr, Mo, and W was based on the use of the rigid-band model in conjunction with the results of the A.P.W. band structure calculation performed for Fe by Wood (1962). Lomer (1962*a*) observed that in previous band structure calculations for a number of f.c.c. d-block metals (Cu, Ag, Fe, and Ni) the general appearance of the various band structures was quite similar although the actual widths of the bands might differ considerably. It therefore seemed that in the almost complete absence, at that time, of direct band structure calculations for Cr, Mo, and W, a good approximation to the band structures and Fermi surfaces of these metals could be obtained by using the rigid-band model. For the b.c.c. structure which is possessed by Cr, Mo, and W there were very few band structure calculations available at that time for any d-block metals nearby in the periodic table and Lomer based his arguments almost exclusively on the band structure calculation by Wood (1962) for Fe. An interpolation scheme was used to determine the value of the Fermi energy corresponding to six electrons per atom and the Fermi surface shown in Fig. 5.21 was obtained. This Fermi surface consists of

(i) in band 3, a large closed region of holes around H,
(ii) in band 3, a smaller closed region of holes around N,
(iii) in band 4, a large closed region of electrons around Γ, which resembles an octahedron with knobs on its corners and which is sometimes described as an electron 'jack', and
(iv) in band 5, a set of small electron pockets, or 'lenses', located at a position on ΓH so as to be within the necks that connect the knobs to the body of the electron jack.

While this Fermi surface was constructed for a metal of valence 6 and could therefore be expected to apply to all three metals Cr, Mo, and W, it must be realized that for Cr it can only be expected to apply to the high-temperature non-magnetic phase and that for W it may be severely modified by spin–orbit

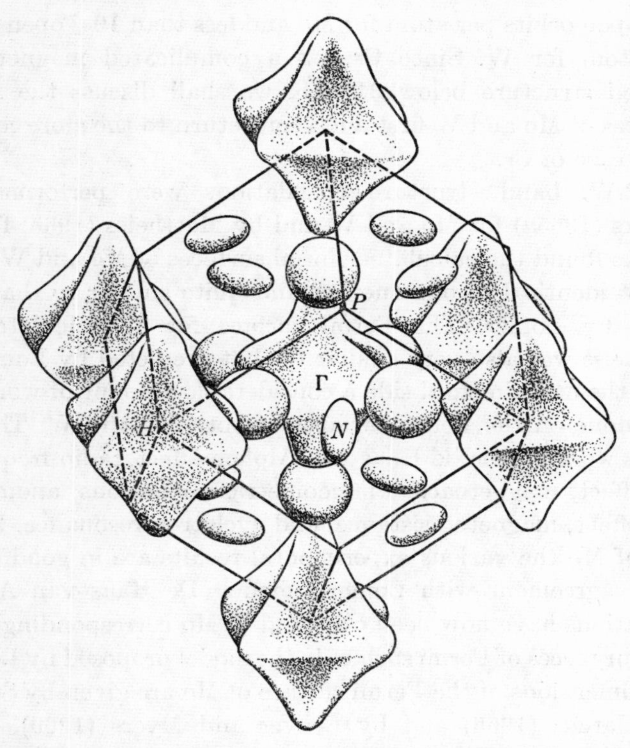

FIG. 5.21. The Fermi surface of Mo, W, and paramagnetic Cr proposed by Lomer (Mattheiss 1965).

coupling (Mattheiss and Watson 1964, 1965). The prediction of Fermi surfaces for group VIA metals consisting only of closed surfaces is consistent with the results of the magnetoresistance work on Mo and W (Fawcett 1961*a*, 1962; Alekseevskiĭ, Egorov, Karstens, and Kazak 1962). If the two sheets of Fermi surface enclosing the electron jack and the region of holes at *H* in Lomer's model touch at a point along the line Γ*H*, this might be expected to give rise to an infinitesimally small number of open orbits; this number could become finite and produce an appreciable effect on the high-field magnetoresistance if magnetic breakdown occurs. Fawcett and Reed (1964*a*) showed from their magnetoresistance measurements on Mo and W in fields up to 83 kG that the Fermi surfaces of these metals support less than

10^{-4} open orbits per atom for Mo and less than 10^{-7} open orbits per atom for W. Since Cr has a complicated magnetically ordered structure below 311·5 K we shall discuss the Fermi surfaces of Mo and W first and then return to the more complicated case of Cr.

A.P.W. band structure calculations were performed by Loucks (1965*a*) for Mo and W and by Mattheiss (1965) for W. Loucks found the calculated Fermi surfaces of Mo and W to be almost identical to one another and quite similar to that proposed by Lomer. The Fermi surface for W calculated by Mattheiss was also very similar to that predicted by Lomer.

On the experimental side a considerable amount of work has been done on the Fermi surfaces of both Mo and W. This includes work on the de Haas-van Alphen effect, radio-frequency size effect, magnetoacoustic geometric oscillations, anomalous skin effect, magnetoresistance, and cyclotron resonance. In the case of Mo the various experimental results are in good qualitative agreement with Lomer's model. De Haas–van Alphen oscillations have now been observed in Mo corresponding to all the four pieces of Fermi surface in the model proposed by Lomer. The dimensions of the Fermi surface of Mo are given by Sparlin and Marcus (1966) and by Leaver and Myers (1969). From radio-frequency size effect measurements the separation between the electron jack and each of the hole octahedra along the $\langle 100 \rangle$ directions was estimated to be $(7{\cdot}5 \pm 1)$ per cent of ΓH (Walsh and Grimes 1964, Cleveland and Stanford 1970). This quite large separation was taken to imply that the importance of spin–orbit coupling in Mo metal is considerably greater than had previously been thought to be the case.

Because of the importance of spin–orbit coupling in W, an *ab initio* relativistic A.P.W. calculation for W was performed by Loucks (1965*b*, 1966*a*). Loucks found that for W both the electron lenses along ΓH in band 5 and the holes at N in band 3 disappeared, and the size of the electron jack had been reduced. Later de Haas–van Alphen work revealed no trace of the electron lens pockets in band 5 in W. There is experimental evidence that although the pockets of holes at N in W do not actually disappear completely as predicted by the relativistic A.P.W. calculation, they are, nevertheless, very much reduced from their sizes in Lomer's model and in the results of the non-relativistic A.P.W. band structure calculations. For instance the de Haas–van Alphen measurements of Sparlin and Marcus (1966) showed that the linear dimensions of the pocket of holes at N in band 3 in W are about half the linear dimensions of the corresponding pocket in Mo. The reduction in size of the electron jack and of the hole octahedron at H is also supported by several other sets of de Haas–van Alphen results and by the anomalous skin effect results of Fawcett and Griffiths (1962). The parameters of the Fermi surface of W are given by Sparlin and Marcus (1966) and by Girvan, Gold, and Phillips (1968). Thus we see that the inclusion of relativistic effects generally decreases the sizes of all the pieces of the calculated Fermi surface of W and leads to the complete disappearance of the small electron lens in band 5 along ΓH.

By analogy with the elements in the previous groups in the periodic table we should expect the Fermi surfaces of the elements in the fourth and fifth periods, namely Cr and Mo, to be very similar to one another. Thus the Fermi surface for group VIA metals which was originally proposed by Lomer and supported by the non-relativistic A.P.W. calculations of Loucks (1965*a*), and which was found to give a good description of the Fermi surface of Mo, can be expected also to apply to non-magnetic Cr, that is, to Cr at temperatures above T_N. Slight differences from Lomer's model for the Fermi surface of non-magnetic Cr were obtained in the non-relativistic A.P.W. calculations of Loucks (1965*a*): in Cr the pockets of holes at N in band 3 disappeared and the sizes of the knobs on the electron jack were also reduced. However, direct experimental measurements of Fermi surface features at such high temperatures are also impossible to obtain.

Experimental work related to the determination of the Fermi surface topology of Cr has, of necessity, been carried out almost exclusively below T_N and the results show that there is considerable departure from the Lomer model, presumably as a consequence of the magnetic ordering. It is therefore necessary to consider what effect the presence of an oriented spin-density wave will have on the Fermi surface. One would expect to find some magnetostrictive distortion along the direction of **Q**; however, when a single crystal of Cr is cooled below T_N it has several different choices of $\langle 100 \rangle$ type directions so that antiferromagnetic domains will be formed in the crystal, where in different domains the magnetic ordering may occur with different directions, such as [100] and [010], for **Q**. Because of the domains any magnetostrictive distortions will tend to cancel out and any departure from cubic symmetry will be difficult to detect. However, if a single crystal of Cr is cooled below T_N in the presence of a magnetic field parallel to the [100] direction then it may be possible to produce a single-domain single-crystal specimen of Cr with **Q** parallel to [100], in which case the magnetostrictive distortion in the direction of **Q** will reduce the symmetry of the crystal from cubic to tetragonal. Evidence of tetragonal symmetry in samples of field-cooled antiferromagnetic Cr has been found in several de Haas–van Alphen and magnetoresistance experiments (Arko and Marcus 1964; Montalvo and Marcus 1964; Watts 1964*a*,*c*; Graebner and Marcus 1966, 1968); the magnitude of this distortion of the crystal from cubic symmetry is too small to be detected directly in X-ray diffraction measurements (Werner, Arrott, and Kendrick 1966; Combley 1968) but the existence of the distortion was detected directly by Lee and Asgar (1969) using electrical resistance strain gauges. Since the presence of an oriented axial vector **P**(**r**), the spin polarization, reduces the symmetry of the Cr crystal, it will therefore reduce quite substantially the number of degeneracies in the electronic band structure (Falicov and Ruvalds 1968; Cracknell 1969*a*, 1970). Any magnetostrictive distortions will generally be found to be compatible with this very much reduced symmetry of the crystal in its magnetically ordered phase.

It was noticed by Lomer (1962*a*) that the magnitude of **Q** in eqn (5.7.2) was just of the size that corresponds to the separation between opposite surfaces of the jack at Γ and the region of holes around H in the Fermi surface of the non-magnetic form of Cr. Similar relationships between the wave vector $\mathbf{Q}$ of a periodic magnetic structure and the 'nesting' of portions of the Fermi surface occur in a number of the rare-earth metals (see Section 5.10). The theoretical reasons for this are discussed by Herring (1966) and Roth, Zeiger, and Kaplan

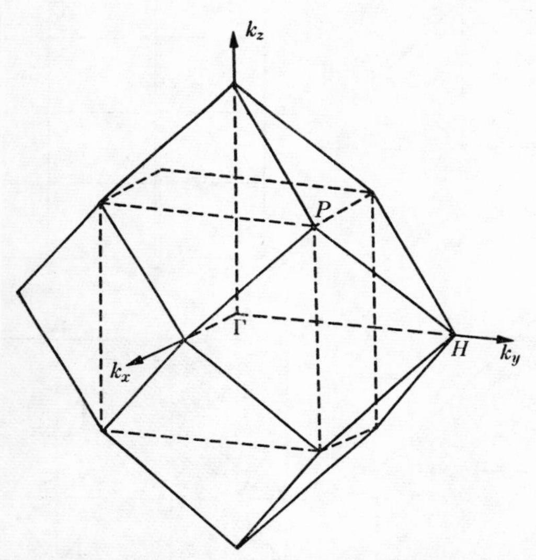

FIG. 5.22. The Brillouin zones of non-magnetic Cr (continuous lines) and of antiferromagnetic Cr (broken lines) (Cracknell 1971*b*).

(1966). The term 'nesting' is employed to describe the situation in which two approximately parallel sheets of the Fermi surface can be made to coincide by displacing either of them through an appropriate distance in $\mathbf{k}$-space (Andersen and Loucks 1968). When Cr is cooled below T_N and becomes antiferromagnetically ordered, the size of the unit cell will be doubled because the atoms at the centres of the conventional b.c.c. unit cells are no longer equivalent to the atoms at the corners of the conventional unit cells. The volume of the Brillouin zone is therefore halved, see Fig. 5.22; in particular, the points that were H points on the corners of the Brillouin zone now become Γ points

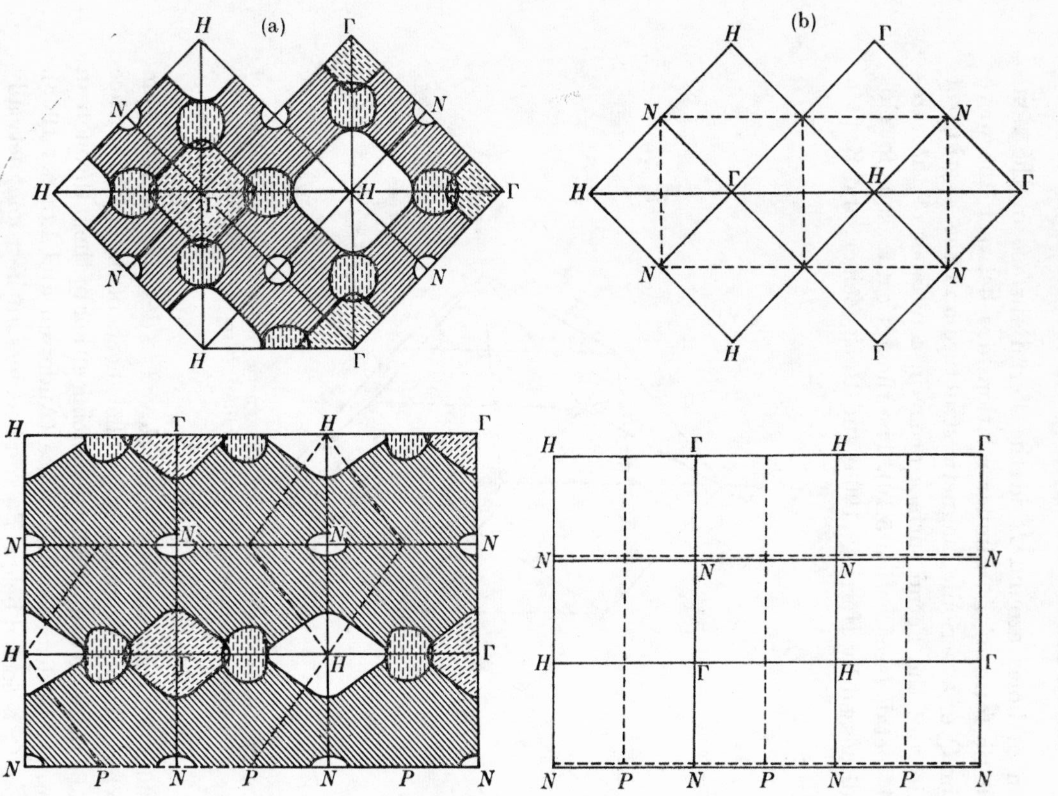

FIG. 5.23. (a) ⟨100⟩ and ⟨110⟩ cross sections of the Fermi surface of group VIA metals proposed by Lomer (1964*b*); the contact between the electron and hole surfaces along ΓH can be expected to be removed by spin–orbit coupling in W at least. (b) Brillouin zone boundaries for the antiferromagnetic phase of Cr indicated (by broken lines) on the cross sections shown in (a).

at the centre of the new Brillouin zone for the antiferromagnetic phase.

The relationship of the new Brillouin zone boundaries to some of the principal cross sections of the Fermi surface of non-magnetic Cr is shown in Fig. 5.23. Because the points H and Γ now coincide, the hole octahedron at H becomes transferred to Γ, and becomes nearly filled up using the electrons from the octahedral body of the jack and the electron lens in band 5,

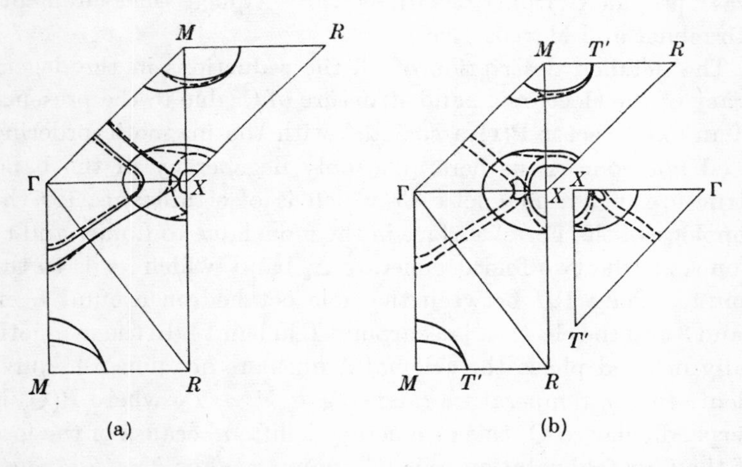

FIG. 5.24. Fermi surface cross sections for Cr according to Asano and Yamashita (1967) for two slightly different choices of E_F (solid lines are for the antiferromagnetic phase and broken lines for the paramagnetic phase).

leaving only the approximately spherical pockets of electrons that were the knobs on the corners of the jack. This leads to an unoccupied region corresponding to a thin shell of which the inner surface is that of the body of the original electron jack and with its outer surface formed by the surface of the original octahedron. Indeed one such surface was found in the band structure calculations for antiferromagnetic Cr performed by Asano and Yamashita (1967) using the Green's function method, see Fig. 5.24(a). Quite a small change in the choice of the Fermi energy could cause most of this thin shell to vanish except for the small regions near X as shown by Fig. 5.24(b). It can therefore be seen qualitatively how, of the electron jack at Γ, the hole octahedron at H, and the electron lens in band 5 along ΓH in the paramagnetic phase, all that is likely to remain in the magnetic phase is two pockets of electrons around X. The holes at N simply become folded back to give two pockets of holes in bands 5 and 6 at the same point, which happens, however, to be labelled M in the magnetic Brillouin zone. Evidence for the disappearance of the electron jack and the hole octahedron was provided from de Haas–van Alphen measurements (Graebner and Marcus 1968).

The detailed description of all the reductions in the degeneracy of the electronic band structure of Cr due to the presence of an axial vector $\mathbf{P}(\mathbf{r})$ associated with the magnetic ordering need not concern us here. The only degeneracy in the band structure of non-magnetic Cr which is of significance for the topology of the Fermi surface in the model due to Lomer and to Loucks is the two fold degenerate Δ_5 band, which leads to the contact along ΓH between the hole octahedron around H in band 3 and the electron jack around Γ in band 4. In the magnetically ordered phase the original Δ lines are not now all equivalent. In the temperature range $T_S \leqslant T \leqslant T_N$ where $\mathbf{P}(\mathbf{Q})$ is perpendicular to $\mathbf{Q}$, this degeneracy is lifted because of the loss of the four fold rotation axis of symmetry. For $T \leqslant T_S$, when $\mathbf{P}(\mathbf{Q})$ is parallel to $\mathbf{Q}$, the four fold rotation axis of symmetry returns for the Δ lines parallel to $\mathbf{Q}$ but is not present for the other Δ lines normal to $\mathbf{Q}$. However, unless the splitting of the Δ_5 band due to the magnetic ordering is quite large, one can expect that in practice magnetic breakdown will restore the original connectivity.

Although in principle a considerable number of degeneracies in the band structure of paramagnetic Cr can be expected to be lifted as a result of the reduction in the symmetry due to magnetic ordering, nevertheless in practice the magnetostrictive distortions are so small that the changes in the potential $V(\mathbf{r})$ can be expected to be small. The changes in the energies $E(\mathbf{k})$, as a result of changes in $V(\mathbf{r})$ induced by the magnetostrictive distortion of the crystal, can therefore also be expected to be quite small. However, for a few wave vectors **k** in the Brillouin zone that have special relationships to the wave vector **Q** of the spin-density wave, the effect of the presence of the spin-density wave becomes important. The presence of the spin-density wave means that in the crystal there is now a second periodic potential, due to the spin-density wave, superposed on the original periodic potential of the crystal lattice; these two periodic potentials are generally incommensurate.

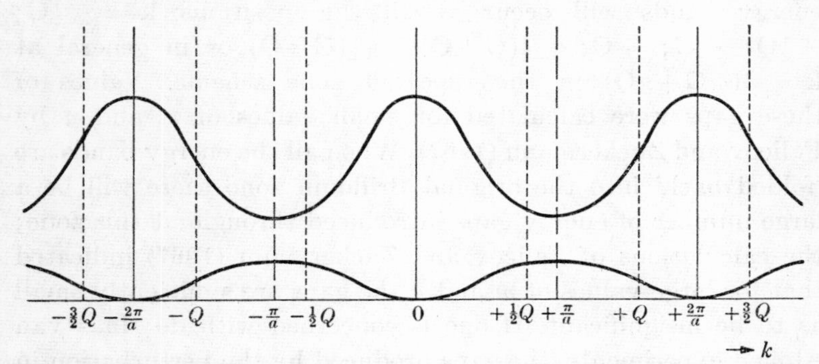

FIG. 5.25. The band structure of a hypothetical one-dimensional metal with magnetic ordering vector **Q** (Cracknell 1971*b*).

This can be illustrated by considering a hypothetical one-dimensional crystal, see Fig. 5.25. An electron is free to 'jump' by adding integer multiples of the reciprocal lattice vector **G** to its wave vector since

$$E(\mathbf{k}) = E(\mathbf{k}+m\mathbf{G}) \tag{5.7.3}$$

for all **k**, when m is an integer. The reciprocal lattice vector associated with the potential due to the spin-density wave is **Q** so that a second set of repeated zones with width $|\mathbf{Q}|$ appear, as shown in Fig. 5.25. The potential due to the spin-density wave is not strong enough to impose on all the wave vectors **k** its own periodicity condition.

$$E(\mathbf{k}) = E(\mathbf{k}+n\mathbf{Q}), \tag{5.7.4}$$

where n is an integer. However, for those particular wave vectors for which this condition is accidentally satisfied by the original energy bands, the interaction of the spin-density wave with the translational motion of the electron is sufficiently strong to cause the electron to jump by some integral multiple of $\mathbf{Q}$ since no appreciable energy change is involved. A particularly simple case is of course when $\mathbf{k} = -\frac{1}{2}\mathbf{Q}$ when the condition $E(\mathbf{k}) = E(\mathbf{k}+\mathbf{Q})$ will be satisfied because $E(\mathbf{k}) = E(-\mathbf{k})$. If one evaluates the perturbation in the energy bands due to the spin-density wave, then gaps in the energy bands will occur at all the positions $\mathbf{k} = \pm\frac{1}{2}\mathbf{G}$; $\pm\frac{1}{2}\mathbf{Q}$; $\pm\mathbf{G}$; $\pm\mathbf{Q}$; $\pm\frac{1}{2}(\mathbf{G}+\mathbf{Q})$; $\pm\frac{1}{2}(\mathbf{G}-\mathbf{Q})$ or in general at $\mathbf{k} = \frac{1}{2}(m\mathbf{G}+n\mathbf{Q})$ in the repeated zone scheme. Values of these gaps were calculated for small values of m and n by Falicov and Zuckermann (1967). When all the energy bands are 'folded back' into the original Brillouin zone there will be a large number of energy gaps introduced throughout this zone; the calculations of Falicov and Zuckermann (1967) indicated that for large values of m and n the gaps are sufficiently small as to be insignificant. If one is concerned with de Haas–van Alphen experiments, the gaps produced by the perturbation in the potential caused by the spin-density wave are liable to be small enough for magnetic breakdown to occur across them. The correct pictorial description of the electron orbits in antiferromagnetic Cr in a magnetic field can therefore be obtained by considering the Fermi surface cross section of the non-magnetic metal, and allowing electrons to 'jump' by wave vectors $\mathbf{Q}$ as well as by the usual $\mathbf{G}$ (Arko, Marcus, and Reed 1966, 1968, 1969; Graebner and Marcus 1968; Wallace and Bohm 1968).

5.8. Group VIIA

Mn, Tc, Re

Mn	Manganese	2.8.(8,5).2	see below
Tc	Technetium	2.8.18.(8,6).1	h.c.p.
Re	Rhenium	2.8.18.32.(8,5).2	h.c.p.

Mn exists in several forms most of which have quite complicated structures. α-Mn, which is stable at temperatures up to

742°C, has a complicated cubic structure which belongs to the space group $I\bar{4}3m(T_d^3)$, with lattice constant $a = 8{\cdot}90$ Å (Smithells 1967), and has 58 atoms in the conventional unit cell or 29 atoms in the fundamental unit cell. At low temperatures α-Mn becomes antiferromagnetic with a Néel temperature of 100 K (Shull and Wilkinson 1953; Kasper and Roberts 1956; Arrott and Coles 1961). The other forms of Mn, β-Mn, γ-Mn, and δ-Mn, are stable in various temperature ranges between 742°C and the melting point (1244°C), although one or more of these forms may be metastable below 742°C. We shall not discuss these high-temperature forms of Mn any further, except to note that Fletcher (1969) has calculated the band structure of γ-Mn.

Since the structure of Mn is so complicated there are obvious difficulties in the way of discussing the band structure and Fermi surface of this metal. The number of atoms per unit cell is an order of magnitude larger than we have encountered in any other metal so far; even the fundamental unit cell for Ga, which was considered in Section 4.4, only contains four atoms. The volume of the unit cell in Mn is 352·5 Å^3, whereas the volume of the fundamental unit cell of Cr, which is next to Mn in the periodic table and has the ordinary b.c.c. structure, is only 12·1 Å^3. Correspondingly the linear dimensions of the Brillouin zone of Mn are only 32·5 per cent of the corresponding linear dimensions of the Brillouin zone of Cr, and the volume is only 3·4 per cent of the volume of the Brillouin zone of Cr. This means that any piece of the Fermi surface of Mn, unless it is multiply-connected, must be very small indeed and may, therefore, prove difficult to measure experimentally, particularly if one uses techniques such as the de Haas–van Alphen effect which involve cross-sectional areas rather than linear dimensions. From the point of view of band structure calculations for Mn, if one assumes, as we have done in connection with the neighbouring metals V and Cr in the periodic table, that the 4s-electrons and all the 3d-electrons are in the conduction band, then with 29 atoms in the fundamental unit cell this means that there must be the equivalent of $29 \times \frac{7}{2} = 101\frac{1}{2}$ full bands.

To obtain a meaningful Fermi surface by making a realistic band structure calculation involving so many bands is therefore a difficult task, while the existence of the antiferromagnetic ordering below 100 K is a further complication. One might try to simplify the problem by ascribing some or all of the five 3d-electrons in Mn to the ion cores and thereby reduce the number of conduction electrons to be considered; however, experience with neighbouring metals in the periodic table suggests that this is not likely to be very realistic. In view of the above difficulties which hamper both experimental and theoretical determinations of the Fermi surface of Mn it is not very surprising that little has been attempted so far.

Since Tc has no stable isotopes it is not surprising that its Fermi surface has not been studied experimentally. There is no obvious reason why band structure calculations for Tc should not be performed but, again, nothing appears to have been done so far in this connection. Since the structure of Tc is the same as that of Re there is every reason to suppose that the Fermi surface of Tc is very similar to that of Re, which has been determined quite accurately, except that one would expect spin–orbit coupling effects in Tc to be less marked than in Re.

Unlike the cases of the other elements of group VIIA—Mn and Tc—the Fermi surface of Re has been studied quite extensively. A relativistic A.P.W. band structure calculation for Re was performed by Mattheiss (1966) which produced a Fermi surface that consisted of five sheets. Closed regions occupied by holes were found in bands 5, 6, and 7, and a closed region occupied by electrons in band 9. In band 8 Mattheiss obtained a Fermi surface enclosing a region occupied by electrons and corresponding roughly to a cylinder with its axis directed parallel to [0001]. The closed hole surfaces obtained by Mattheiss in bands 5 and 6 are illustrated in Fig. 5.26(a) and (b) and are quite similar to the two surfaces which had previously been deduced from the de Haas–van Alphen data by Joseph and Thorsen (1964) and supported by magnetoacoustic results (Jones and Rayne 1965*a*,*b*; Testardi and Soden 1967) shown in

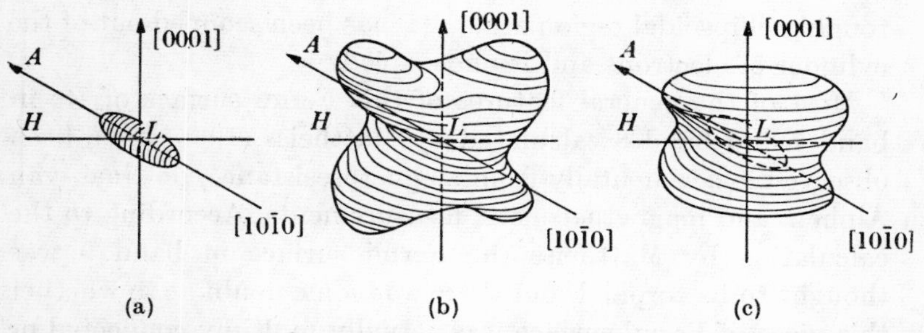

FIG. 5.26. The Fermi surface of Re: (a) holes in band 5, (b) holes in band 6 and (c) the ellipsoid and dumb-bell model of Joseph and Thorsen (Mattheiss 1966).

Fig. 5.26(c). A comparison between the calculated and measured extremal areas of cross section was given by Mattheiss (1966); the agreement between the calculated and measured dimensions was particularly good in the case of the ellipsoid in band 5. The larger pieces of Fermi surface are illustrated in Fig. 5.27. The closed pockets of holes in band 7 are located at L and the multiply-connected vertical cylinder of electrons passes through Γ and A; these two regions touch along the line AL. A small

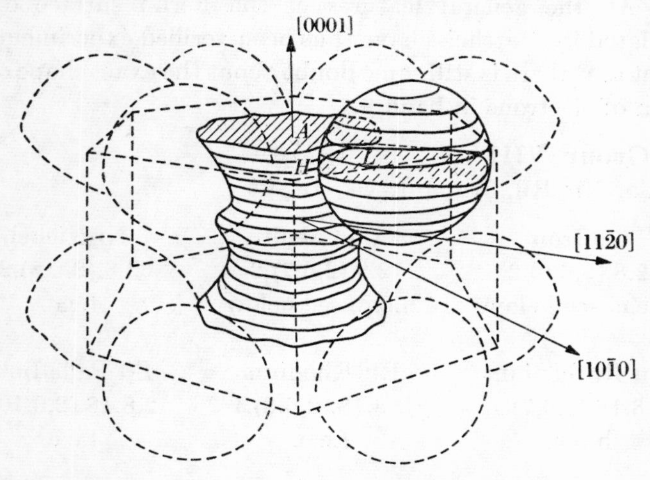

FIG. 5.27. The Fermi surface of Re, holes in band 7 and electrons in band 8 (Mattheiss 1966).

roughly ellipsoidal region around Γ has been scooped out of the cylinder of electrons and is unoccupied.

Most of the general features of this Fermi surface of Re in band 7 and band 8 calculated by Mattheiss (1966) have been observed experimentally from magnetoresistance, de Haas–van Alphen, and magnetoacoustic measurements. According to the calculation by Mattheiss the Fermi surface in band 9 was thought to be toroidal, but there was some doubt as to whether this piece of Fermi surface was actually multiply-connected or not since its shape is very sensitive to small changes in the Fermi energy. Thorsen, Joseph, and Valby (1966) observed two de Haas–van Alphen periods which they assigned to this piece of Fermi surface in band 9, but they were unable to detect several other periods that were predicted by Mattheiss from the Fermi surface in band 9. They suggested some slight modifications to the shape of this torus to explain their results; they proposed that the torus has the approximate shape of six spherical balls connected together to form a ring and that these balls are deformed so as to be concave towards the inside of the ring. All the general features of the Fermi surface of Re calculated by Mattheiss have thus been verified experimentally, except that there is still some doubt about the exact shape of the region of electrons in band 9.

5.9. Group VIII

Fe, Co, Ni; Ru, Rh, Pd; Os, Ir, Pt

Fe Iron	Co Cobalt	Ni Nickel
2.8.(2,6,6).2	2.8.(2,6,7).2	2.8.(2,6,8).2
b.c.c. see below	h.c.p. see below	f.c.c.
Ru Ruthenium	Rh Rhodium	Pd Palladium
2.8.18.(2,6,7).1	2.8.18.(2,6,8).1	2.8.18.(2,6,10).
h.c.p.	f.c.c.	f.c.c.
Os Osmium	Ir Iridium	Pt Platinum
2.8.18.32.(2,6,6).2	2.8.18.32.(2,6,9).	2.8.18.32.(2,6,9).1
h.c.p.	f.c.c.	f.c.c.

It is convenient to divide this collection of elements into smaller sets and both straightforward horizontal and vertical division have been used quite commonly. A particularly attractive scheme is to consider together the three elements in the first row, which are all ferromagnetic, and then to divide the remaining six elements vertically into three sets of two metals each (Sidgwick 1950); we shall follow this scheme.

The b.c.c. structure of Fe (α-Fe) is only stable below 900°C and above 1400°C; between these two temperatures the f.c.c. form (γ-Fe) is stable. The h.c.p. structure of Co (α-Co) is only stable below about 390°C and above this temperature this metal exhibits the f.c.c. structure (β-Co). Some of the other metals also exhibit different structures under unusual conditions.

5.9.1. Fe, Co, Ni

The electronic properties of Fe, Co, and Ni have been studied very extensively over the years. Fe was one of the metals for which band structure calculations were performed in the very early days (Greene and Manning 1943; Manning 1943) and a large number of calculations have been performed on these three metals since then. These metals are of particular interest because they exhibit ferromagnetism. However, many of the early band structure calculations for these metals completely ignored the fact of the ferromagnetic ordering; that is, the Hamiltonian, $\mathscr{H}$, used in calculating the band structure contained no terms attributable to the large internal magnetic field which must exist inside each domain of a sample of the metal below its Curie temperature. It is then assumed that the only effect of the internal field **B** is to cause a uniform separation of $g_s\mu_B B$ to appear between the spin-up and spin-down bands at each wave vector **k**. This gives a good first approximation to the band structure of one of these metals. However, even if the band structure has been calculated sufficiently accurately it is also necessary to know the value of the energy difference, $g_s\mu_B B$, between the spin-up and spin-down electrons with the same **k** and in the same band before one can determine the

Fermi energy and hence the shape of the Fermi surface. Because $g_s\mu_B B$ is quite large, the topological features of the Fermi surface for spin-up electrons will be completely different from those of the Fermi surface for spin-down electrons. Although the actual magnitudes of any changes in the band structure due to the inclusion of magnetic terms in the Hamiltonian, $\mathscr{H}$, may be quite small, the inclusion of these extra terms may cause quite drastic reductions in the essential degeneracies of a ferromagnetic metal, as we have already mentioned in Section 5.1. These reductions in the degeneracies of the band structure may have important repercussions on the possible connectivities of the Fermi surface.

Several measurements of the X-ray and neutron scattering factors for the ferromagnetic transition metals have been made (Weiss and De Marco 1958; Nathans, Shull, Shirane, and Andresen 1959; Weiss and Freeman 1959; Batterman, Chipman, and De Marco 1961; Sirota and Olekhnovich 1961; Shull 1963) with the object of investigating the spatial distribution of the charge density and spin density of the conduction electrons (see also Section 2.12). For example Weiss and De Marco (1958) deduced that Ni possesses $(9{\cdot}7 \pm 0{\cdot}3)$ d-electrons per atom, while Weiss and Freeman (1959) deduced that there were 5·0 3d-electrons with spin up and 4·4 3d-electrons with spin down. Such deductions have normally been made on the basis of atomic wave functions, although the use of atomic wave functions in calculating scattering factors is unrealistic because, as is well known, no conduction electron energy band in a metal is ever pure s-like or pure d-like but always consists of some admixture of these and other atomic states. One or two calculations of scattering factors have now also been made using more realistic wave functions for the electrons in a metal (Hodges, Lang, Ehrenreich, and Freeman 1966). Scattering factor measurements, of course, do not give direct information about the shape of the Fermi surface of a metal. Other methods have also been used for investigating the polarization of the electrons in a ferromagnetic metal. For example, if there is an appreciable net polarization of the electrons with the Fermi energy there should be a change in the Fermi energy with applied magnetic field; this appears as a change in the contact potential and can be measured relative to a non-magnetic metal (Walmsley 1962, Belson 1966).

Doubts used to be expressed as to whether the methods commonly used in measuring the shapes of Fermi surfaces could be used for ferromagnetic metals (Pippard 1960*c*, Shoenberg 1960*b*), however, with the magnetoresistance measurements of Fawcett and Reed (1962) on Ni it became apparent that direct information about the shape of the Fermi surface of a ferromagnetic metal could be determined in the usual way. Shortly after this Anderson and Gold (1963) unequivocably observed conventional de Haas–van Alphen oscillations in Fe although, as might be expected, the oscillations were found to be periodic in $|\mathbf{B}|^{-1} = |\mu_0(\mathbf{H}+\mathbf{M}_s)|^{-1}$ rather than in just $|\mathbf{H}|^{-1}$, where $\mathbf{M}_s$ is the saturation magnetization (see also Kittel 1963*a*). Since then a number of other direct measurements related to the Fermi surfaces of Fe, Co, and Ni have been made using the standard methods. There are substantial difficulties in the way of preparing suitable single-domain single-crystal specimens of Fe and Co because of the phase transitions which occur between ordinary temperatures and the melting points, and the specimens are commonly in the form of whiskers.

Band structure calculations on Fe have been performed by a number of workers using a variety of methods; of these we have already mentioned in Section 5.7 the A.P.W. calculation by Wood (1962) that was used by Lomer in connection with the group VIA metals Cr, Mo, and W using the rigid-band model. In the Green's function calculation by Wakoh and Yamashita (1966) proper account was taken of the different crystal potentials $V(\mathbf{r})$ that apply to spin-up and spin-down electrons; separate band structures and densities of states were then calculated for the spin-up and spin-down electrons. As a result of their calculations Wakoh and Yamashita obtained 5·1 electrons per atom in the spin-up bands and 2·9 electrons per atom in the spin-down bands: the corresponding calculated

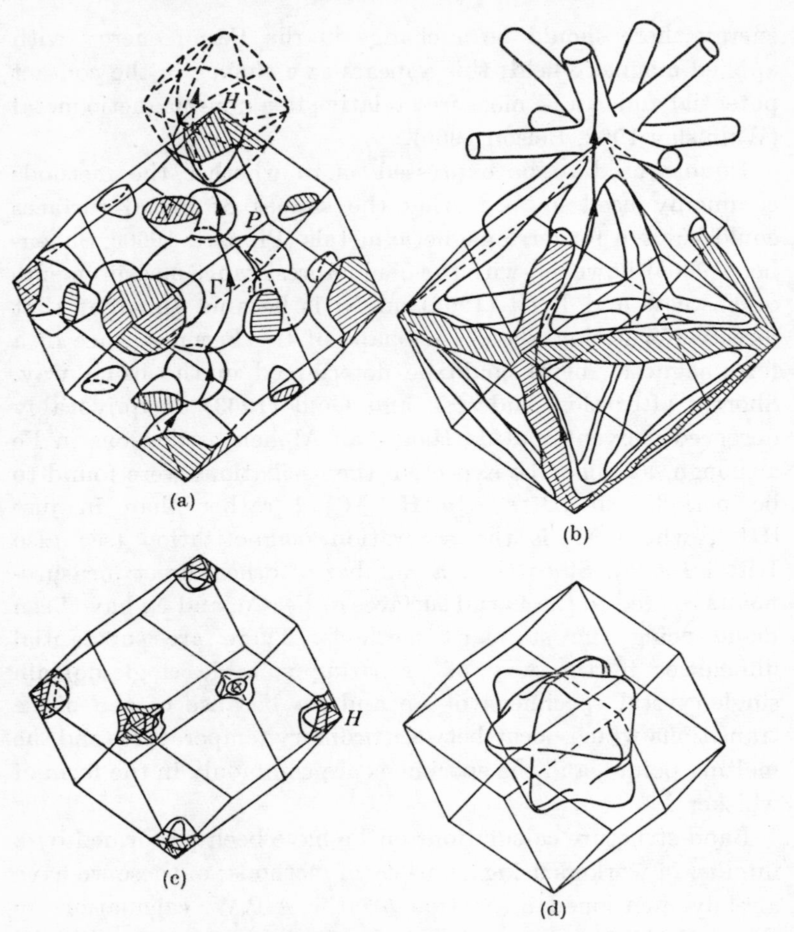

Fig. 5.28. The Fermi surface of Fe calculated by Wakoh and Yamashita (1966), (a) spin-down (electrons), (b) and (c) spin-up (holes), and (d) spin-up (electrons).

Fermi surface is shown in Fig. 5.28. The Fermi surface for the spin-down bands shown in Fig. 5.28(a) is almost identical to the Fermi surfaces of the group VIA metals; this is not entirely surprising because in the group VIA metals there are three electrons per atom in the spin-up or spin-down bands, and this is very close to the number of electrons per atom (2·9) in the spin-down bands of Fe obtained by Wakoh and Yamashita. The

Fermi surface for the spin-up bands shown in Fig. 5.28(b), (c), and (d) is quite different from this and consists of two pockets of holes at H (Fig. 5.28(c)), a multiply-connected tubular region of holes (Fig. 5.28(b)), and an isolated region of electrons at Γ (Fig. 5.28(d)). The Fermi surface shown in Fig. 5.28 is capable of supporting open orbits in certain directions and these have been detected experimentally in various magnetoresistance measurements; examples of these open orbits are illustrated, respectively, in Figs. 5.28(a) and 5.28(b). Because of the experimental difficulties associated with the preparation of good samples of pure single-crystal Fe, relatively few de Haas–van Alphen measurements exist for Fe. Anderson and Gold (1963) observed two de Haas–van Alphen periods that were assigned to the lens along ΓH of the spin-up Fermi surface.

Since Co has an atomic structure of ...$3d^7 4s^2$, and there are two atoms per unit cell in the h.c.p. form which is stable at ordinary temperatures, there will be the equivalent of 18 bands to be considered, and for a band structure calculation this represents no mean undertaking; however, band structure calculations have been performed by Hodges and Ehrenreich (1968) and by Wakoh and Yamashita (1970). On the experimental side there is the difficulty mentioned before of preparing pure single-domain single-crystal specimens of Co because of the phase transition at about 390°C. No measurements of properties that give direct information about the shape of the Fermi surface of Co appear to have been made so far. The band structure calculation by Wakoh and Yamashita (1970) for paramagnetic Co was based on the Green's function method and the Fermi surface was predicted by assigning 10·56 and 7·44 electrons per unit cell to the majority and minority spin bands, respectively. This Fermi surface, which is illustrated in Figs. 5.29 and 5.30, consists of:

(i) for the spin-up electrons a large sphere in the double-zone scheme, centred at Γ and elongated in the k_z direction. In the single-zone scheme this reduces to a smaller closed and roughly spherical Fermi surface around Γ, together with a cylinder in the k_z direction; and

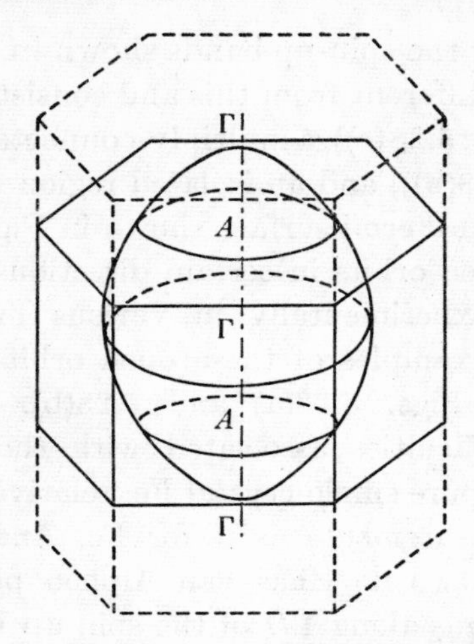

FIG. 5.29. The Fermi surface for spin-up electrons in Co in the double-zone scheme (Wakoh and Yamashita 1970).

(ii) for the spin-down electrons there are several pieces of Fermi surface including cylinders of electrons directed parallel to the k_z direction, with their axes along ML lines and connected to one another at K points, one set of pockets centred at L, another set of smaller pockets near to L, and three closed pockets around Γ. The three concentric surfaces at Γ separate, respectively, bands 9 and 10, bands 10 and 9, and bands 9 and 8 as one moves out from Γ towards the surface of the Brillouin zone.

Apart from a few de Haas–van Alphen frequencies (Batallan and Rosenman 1970) there is only indirect experimental evidence available related to the Fermi surface of Co, and this involves the value of γ and photoemission measurements of the density of states.

It is easier to prepare pure single-crystal specimens of Ni than it is for Fe and Co because of the absence of inconvenient phase transitions in Ni; consequently more experimental work has been done in connection with the Fermi surface of Ni than has been done in Fe and Co. Direct information concerning the shape of the Fermi surface in ferromagnetic Ni has come

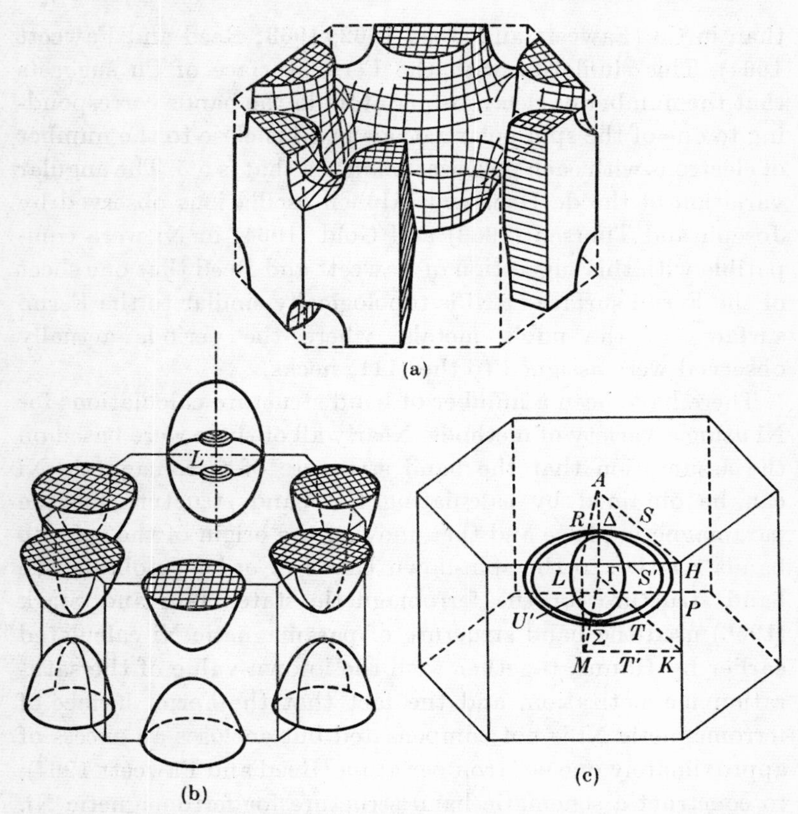

FIG. 5.30. The Fermi surface for spin-down electrons in Co calculated by Wakoh and Yamashita (1970) (a) the multiply-connected region of electrons, (b) closed pockets at L (the larger surfaces enclose holes and the smaller surfaces may be hole or electron surfaces according to the potential), (c) closed pockets at Γ (the smallest surface encloses holes and the other two are electron-like surfaces).

primarily from measurements of the magnetoresistance and of the de Haas–van Alphen effect. The existence of open orbits in Ni in a number of orientations was demonstrated by magnetoresistance measurements; the results could be explained by assuming that one sheet of the Fermi surface of ferromagnetic Ni is similar to the Fermi surface of Cu and consists of a sphere pulled out to touch the ⟨111⟩ faces of the Brillouin zone, although in Ni the contact areas are considerably smaller than in Cu (Fawcett and Reed 1962, 1963; Reed and Fawcett 1964). This similarity with the Fermi surface of Cu suggests that the number of electrons per atom in the bands corresponding to one of the spin polarizations in Ni is close to the number of electrons with each polarization in Cu, that is 5·5. The angular variation of the de Haas–van Alphen oscillations observed by Joseph and Thorsen (1963) and Gold (1964) in Ni were compatible with this suggestion of Fawcett and Reed that one sheet of the Fermi surface of Ni is topologically similar to the Fermi surfaces of the noble metals, where the periods actually observed were assigned to the $\langle 111 \rangle$ necks.

There have been a number of band structure calculations for Ni using a variety of methods. Nearly all of them were based on the assumption that the band structure of ferromagnetic Ni can be obtained by calculating the band structure for the paramagnetic phase and then moving the origin of the spin-up bands relative to the spin-down bands in order to obtain the band structure for the ferromagnetic state. Tsui and Stark (1966) used the band structure of paramagnetic Ni calculated earlier by Hanus, together with the known value of the saturation magnetization, and the fact that the Fermi surface of ferromagnetic Ni is not compensated but encloses an excess of approximately one electron per atom (Reed and Fawcett 1964), to construct a schematic band structure for ferromagnetic Ni. In this schematic band structure the multiply-connected sheet of Fermi surface which we have already described is assumed to be for the spin-up electrons and this is the only sheet of Fermi surface for the spin-up bands. There are then several sheets of Fermi surface for the spin-down bands, namely two large closed electron pockets centred at Γ in bands 5 and 6, one small hole pocket at L in band 4, and two small hole pockets at X in bands 3 and 4 (Tsui 1967). All these pieces of Fermi surface, except the small pocket of holes at L, were also obtained in the A.P.W. band structure calculation of Connolly (1967). Cross sections of all these pieces of Fermi surface are shown in Fig. 5.31; this is quite similar to the Fermi surfaces deduced earlier by Ehrenreich, Philipp, and Olechna (1963) (based on the same

A.P.W. results of Hanus), by Phillips (1964), and by Wakoh and Yamashita (1964). The Fermi surface calculated by Wakoh and Yamashita is illustrated in Fig. 5.32. Further de Haas–van Alphen periods were also observed, and were assigned to an ellipsoidal pocket of holes which was assumed to be one of the two pockets of holes at X (Tsui and Stark 1966;

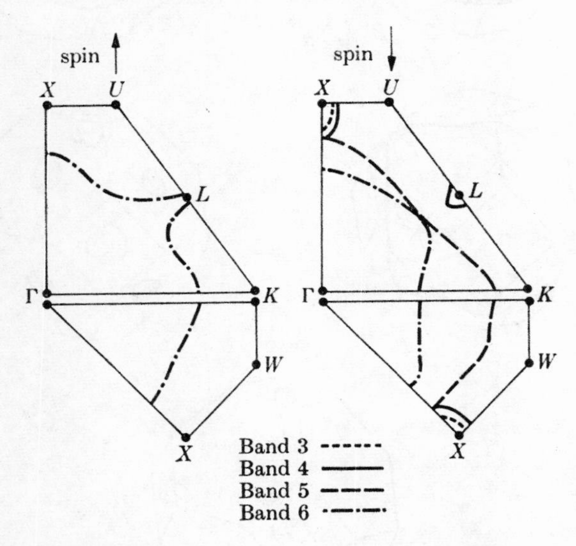

FIG. 5.31. Cross sections of the Fermi surface of Ni (Tsui 1967).

Hodges, Stone, and Gold 1967; Stark and Tsui 1968). This surface was parametrized by the expression given in eqn (2.9.6). When the band structure calculated using the *combined interpolation scheme* (see Section 2.7) was compared with the shape of the roughly ellipsoidal Fermi surface at the point X determined experimentally, it was concluded by Hodges, Stone, and Gold that the importance of spin–orbit coupling in Ni is quite considerable; this is a little surprising in view of the relatively low atomic number of Ni. The angular dependence of the amplitude of the de Haas–van Alphen oscillations associated with this pocket of holes at X as observed by Tsui (1967) was also rather strange, see Fig. 5.33. However, rather

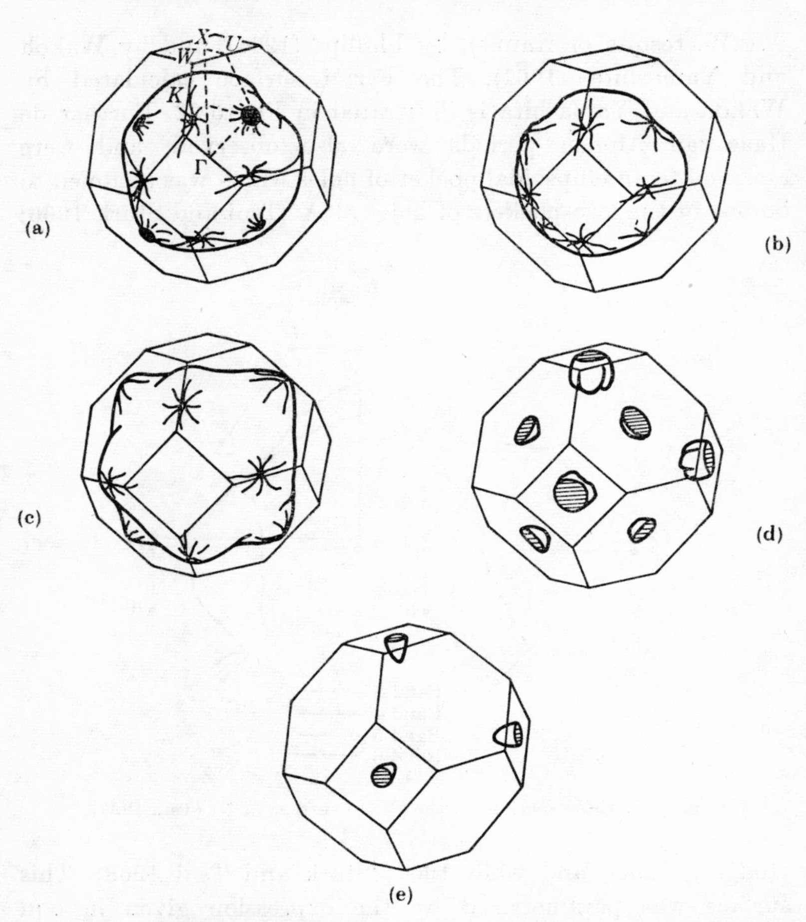

FIG. 5.32. The Fermi surface of ferromagnetic Ni calculated by Wakoh and Yamashita (1964), (a) spin-up (electrons), (b) and (c) spin-down (electrons at Γ), (d) and (e) spin-down (holes at L and X).

than assuming a large spin–orbit interaction in a metal of such low atomic number, it is possible to show that both the shape of this pocket of holes at X, as well as the curious angular dependence of the amplitude of the de Haas–van Alphen oscillations, can be explained in terms of the lifting of many of the band structure degeneracies by the internal magnetic field, together with the existence of magnetic breakdown (for details

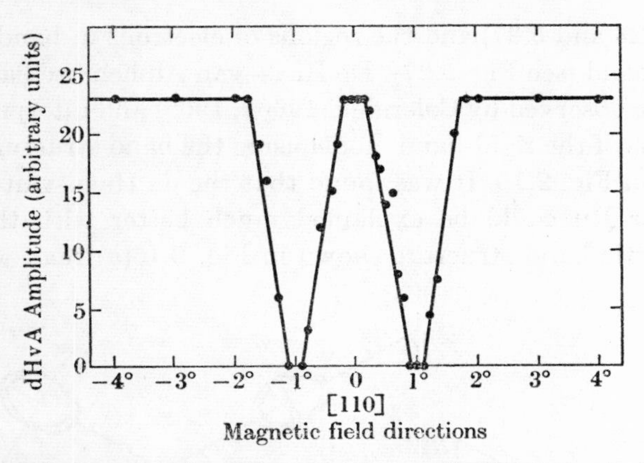

FIG. 5.33. The angular dependence of the amplitude of the de Haas–van Alphen oscillations associated with the pocket of holes at X in Ni (Tsui 1967).

of the explanation see Falicov and Ruvalds 1968; Ruvalds and Falicov 1968; Cracknell 1969*a*, 1970). No de Haas–van Alphen oscillations corresponding to a second pocket of holes at X, or to a pocket of holes at L were observed experimentally (Tsui and Stark 1966; Hodges, Stone, and Gold 1967; Tsui 1967).

5.9.2. Ru, Os

Ru and Os have the same structure (h.c.p.) as Re and do not possess any spontaneous magnetization. In the absence of any band structure calculations for Ru or Os it is tempting to use the rigid-band model and the relativistic A.P.W. band structure calculated by Mattheiss (1966) for Re. This can be expected to give a good description of the Fermi surface of Os, which differs from Re by only one in its atomic number, so that spin–orbit coupling effects will be very similar in the two metals. This approach will be slightly less appropriate to Ru because of the difference between the importance of spin–orbit coupling in Ru and Re. From the band structure of Re shown in Fig. 2.15 it can be seen that it is only at Γ, L, and H that there are any bands just above the Fermi energy for Re. If one electron per atom is added to the band structure shown in Fig. 2.15, we can expect the regions of holes in bands 5, 6, and 7 to contract (see

Figs. 5.26 and 5.27) and the regions of electrons in bands 8 and 9 to expand (see Fig. 5.27). De Haas–van Alphen oscillations in Ru were observed by Coleridge (1966*a*, 1969) and interpreted on the basis of the rigid-band model using the band structure of Re shown in Fig. 2.15. It was found that the de Haas–van Alphen data for Ru could be explained much better with the non-relativistic band structure shown in Fig. 2.15(a) than with the

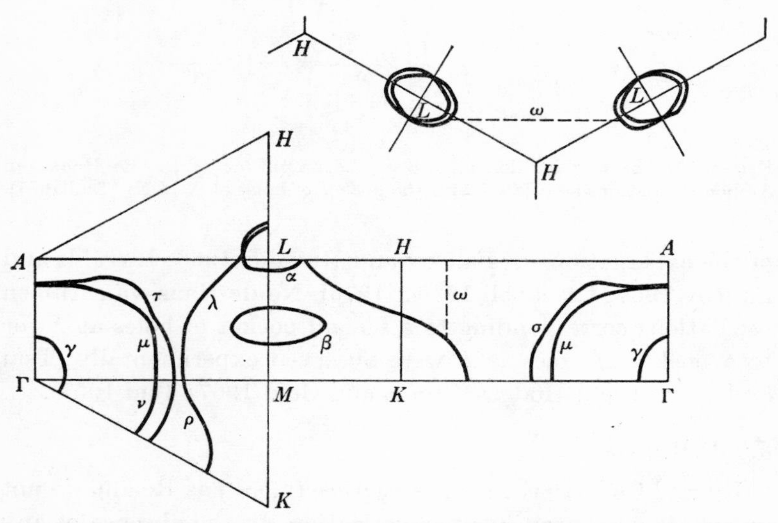

Fig. 5.34. Cross sections of the Fermi surface proposed for Ru by Coleridge (1969).

relativistic band structure shown in Fig. 2.15(b); this is presumably because the atomic number of Ru is lower than that of Re. On the basis of this non-relativistic band structure the Fermi surface proposed by Coleridge for Ru consisted of:

(i) a small region of holes (γ) around Γ and another small region of holes (β) on the line LM, and

(ii) two large closed electron surfaces (μ, ν; ρ, σ, τ) centred at Γ, and one large multiply-connected electron surface (α, λ, ω) that touches the small pockets of holes around L.

The principal cross sections of all these sheets of the proposed Fermi surface of Ru are illustrated in Fig. 5.34.

5.9.3. Rh, Ir

Extensive de Haas–van Alphen measurements were performed on Rh by Coleridge (1965, 1966*b*) and detailed measurements of the dimensions of a number of sheets of the Fermi surface were obtained. Since there were, at that time, no band structure calculations available for Rh, Coleridge employed the rigid-band model and used the band structure of Ni, which was the nearest available f.c.c. metal in the periodic table for which the band structure had been calculated (see above). On the basis of this band structure, but with one less electron per atom, and with one or two quite small *ad hoc* modifications, Coleridge deduced that there were in all five closed sheets of Fermi surface in Rh; detailed dimensions of four of these five sheets agreed rather nicely with the experimental results. The Fermi surface of Rh proposed by Coleridge (1966*b*) is illustrated in Fig. 5.35.

Ab initio relativistic A.P.W. band structure calculations for the metals Rh and Ir, which both have the f.c.c. structure,were performed by Andersen and Mackintosh (1968). As one would expect, relativistic effects were found to be more important in the heavier metal, Ir. For Rh the calculated Fermi surface possessed the same general features that had been suggested previously by Coleridge, namely two large closed electron surfaces at Γ, two closed hole surfaces at X and one small closed hole surface at L. For Ir the calculated Fermi surface is very similar to that of Rh, except that the small pocket of holes at L has vanished. Extremal areas of cross section of these calculated Fermi surfaces for Rh and Ir are tabulated by Andersen and Mackintosh (1968) and for Rh the agreement between experiment and theory is quite close. For Ir the experimental cross-sections of the two pockets of holes at X have been measured using the de Haas–van Alphen effect by Grodski and Dixon (1969) and Hörnfeldt (1970).

5.9.4. Pd, Pt

Magnetoresistance measurements revealed the existence of open orbits in $\langle 100 \rangle$ directions in Pt and similar results were also obtained later for Pd. The sign of the Hall coefficient indicates

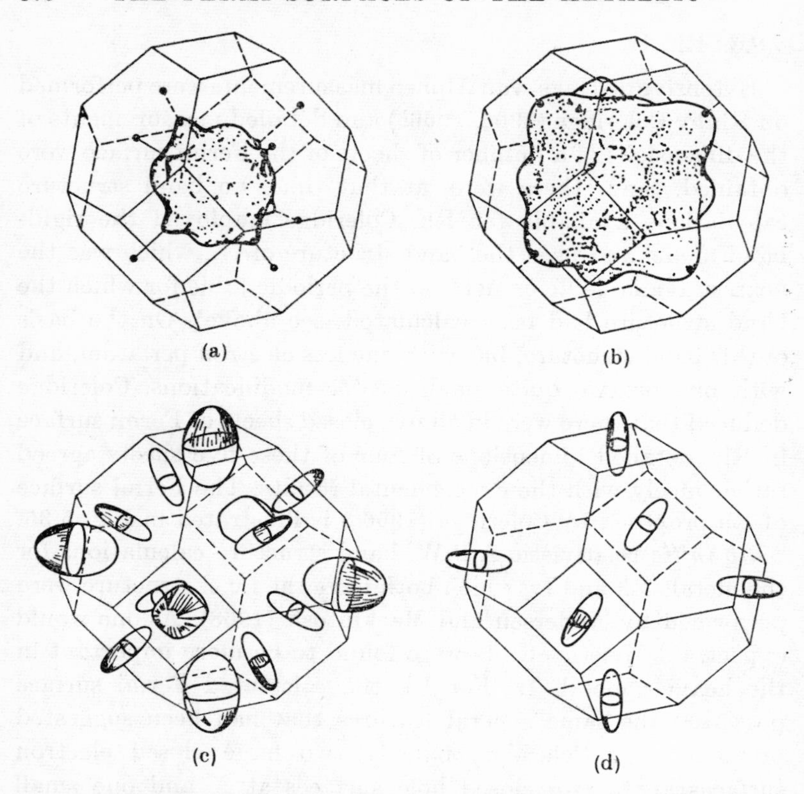

FIG. 5.35. The Fermi surface of Rh proposed by Coleridge (1966b), showing (a) and (b) regions of electrons and (c) and (d) pockets of holes.

that the multiply-connected Fermi surface encloses a region of holes (Alekseevskiĭ, Karstens, and Mozhaev 1964). De Haas–van Alphen measurements of the Fermi surface of Pd (Vuillemin and Priestley 1965, Vuillemin 1966) were interpreted on the basis of the rigid-band model using the band structures calculated by Segall (1962) for Cu and by Mattheiss (1964b) for non-magnetic Ni; this approach suggested that the Fermi surface of Pd consists of

(i) a large closed region of electrons centred at Γ,

(ii) small closed ellipsoidal regions of holes at X, and

(iii) a multiply-connected region of holes passing through the points X and W. (See Fig. 5.36.)

The general features of this proposed Fermi surface were supported by the results of the A.P.W. calculations of Mueller, Freeman, Dimmock, and Furdyna (1970). Of these three surfaces, the dimensions of the region of electrons at Γ and the small pockets of holes at X were measured in considerable detail

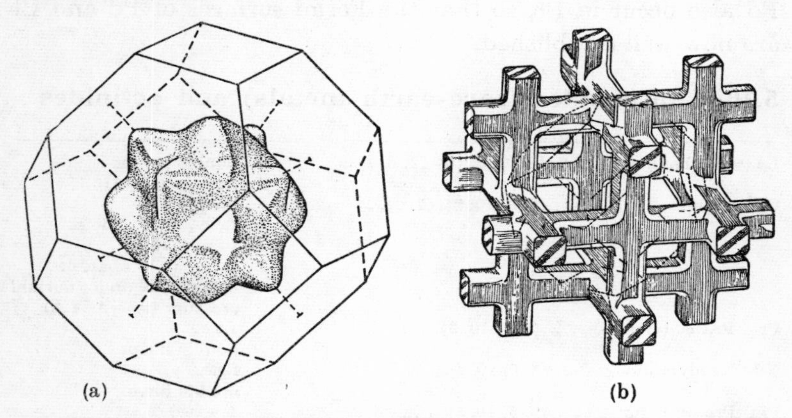

FIG. 5.36. The Fermi surface of Pd (a) electrons at Γ (b) multiply-connected tubes of holes (Mueller, Freeman, Dimmock, and Furdyna 1970).

by Vuillemin and Priestley. The region of electrons at Γ is roughly spherical in shape with bumps along [111] and [100]. Although no de Haas–van Alphen periods directly attributable to the multiply-connected hole region were observed, the general topological features of this surface are consistent with the results of the magnetoresistance measurements mentioned above.

The Fermi surface of Pt has been found to be very similar to that of Pd. In measurements of the de Haas–van Alphen effect in Pt, Stafleu and de Vroomen (1965) demonstrated the existence of small ellipsoidal pockets which were interpreted as being regions of holes at X because of their close similarity to those previously found in Pd. Ketterson, Priestley, and Vuillemin (1966) found two further sets of de Haas–van Alphen periods in Pt. One of these sets of periods could be associated with a Fermi surface that was very similar to the region of electrons around Γ in Pd which was described above. The other set of oscillations was tentatively assigned to a multiply-connected region of holes similar to that just described for Pd and the existence of which in Pt as well was indicated by galvano-magnetic measurements. More recent de Haas–van Alphen and magnetoacoustic measurements and band structure calculations showed that all the three sheets of Fermi surface that occur in Pd also occur in Pt, so that the Fermi surfaces of Pd and Pt are now well established.

5.10. Lanthanides (rare-earth metals) and actinides

La	Lanthanum	2.8.18.(2, 6, 10).(2, 6, 1).2	α	double h.c.p.
			β	f.c.c., above 613 K
Ce	Cerium	......(2, 6, 10, 2).(2, 6).2	β	h.c.p.
			γ	f.c.c., above 263 K ($a = 5{\cdot}16$ Å)
			α	f.c.c., at 15 kbar, or below 95 K under normal pressure ($a = 4{\cdot}84$ Å)
Pr	Praseodymium	......(2, 6, 10, 3)......		f.c.c.
				double h.c.p.
Nd	Neodymium	......(2, 6, 10, 4)......		f.c.c.
				double h.c.p.
Pm	Promethium	......(2, 6, 10, 5)......		?
Sm	Samarium	......(2, 6, 10, 6)......	α	$R\bar{3}m$ (D^5_{3d}), below 1190 K
			β	b.c.c. 1190 K to 1345 K
Eu	Europium	......(2, 6, 10, 7)......		b.c.c.
Gd	Gadolinium	......(2, 6, 10, 7).(2, 6, 1).2		h.c.p.
Tb	Terbium	......(2, 6, 10, 9).(2, 6).2		h.c.p.
Dy	Dysprosium	......(2, 6, 10, 10)......		h.c.p.
Ho	Holmium	......(2, 6, 10, 11)......		h.c.p.
Er	Erbium	......(2, 6, 10, 12)......		h.c.p.
Tm	Thulium	......(2, 6, 10, 13)......		h.c.p.
Yb	Ytterbium	......(2, 6, 10, 14)......		h.c.p.
Lu	Lutetium	......(2, 6, 10, 14).(2, 6, 1).2		h.c.p.
Ac	Actinium	2.8.18.32.(2, 6, 10).(2, 6, 1).2		f.c.c.
Th	Thorium	2.8.18.32.(2, 6, 10).(2, 6, 2).2		f.c.c.
			ξ	orthorhombic
Pa	Protactinium	2.8.18.32.(2, 6, 10, 2).(2, 6, 1).2		$I4/mmm$ (D^{17}_{4h})
U	Uranium	2.8.18.32.(2, 6, 10, 3).(2, 6, 1).2	α	$Cmcm$ (D^{17}_{2h})
etc.		2.8.18.32.(2, 6, 10, n).(2, 6, 1).2		

where n = (atomic number −89).

In the double hexagonal close-packed structure, which is possessed by La, Pr, and Nd, there are four atoms in the unit cell located at (0,0,0), ($\frac{1}{3}$, $\frac{2}{3}$, $\frac{1}{4}$), (0, 0, $\frac{1}{2}$), and ($\frac{2}{3}$, $\frac{1}{3}$, $\frac{3}{4}$) where the coordinates are referred to $\mathbf{t}_1 = a\mathbf{i}$, $\mathbf{t}_2 = \frac{1}{2}a\mathbf{i}+\frac{1}{2}(\sqrt{3})a\mathbf{j}$,

$\mathbf{t}_3 = c\mathbf{k}$. Fleming, Liu, and Loucks (1968) related the fact of the existence of this double h.c.p. structure to one of the features of their calculated Fermi surfaces of these metals (see below). The transition between the two phases γ- and α- of Ce, which both have the same structure (f.c.c.) but different values of the lattice constant (or between the h.c.p. β-phase of Ce and the α-phase), is thought to correspond to the emptying of the 4f-shell by the raising of its 4f electrons into the conduction band (Itskevich 1962). The positions of the four U atoms in the unit cell of α-U are $[000; \frac{1}{2}\frac{1}{2}0]+0y\frac{1}{4}$ and $[000; \frac{1}{2}\frac{1}{2}0]+0\bar{y}\frac{3}{4}$, where $y = 0{\cdot}105$. There are several other U structures.

Many of the rare-earth metals between La and Lu exhibit various magnetically ordered structures, some of which are ferromagnetic and some of which are antiferromagnetic. The existence of these magnetic phases was first suspected from the existence of anomalies in various macroscopic properties such as the specific heat, thermal expansion, magnetic susceptibility, electrical resistivity, and magnetoresistance, but has now been established from neutron diffraction measurements. The magnetic structures of many of these metals are summarized in Table 5.1. Some of the structures are simple ferromagnetic structures with all the individual magnetic moments parallel to the c-axis (Gd) or in the plane normal to the c-axis (Tb and Dy). Actually the magnetization in Gd is only parallel to the c-axis between 294 K and 232 K; below 232 K Gd remains ferromagnetic but the direction of the magnetization moves away from the c-axis to a maximum deviation of about 65° near 180 K and then moves back to within about 32° of the c-axis at low temperatures (Cable and Wollan 1968). More complicated conical ferromagnetic structures are found in Ho and Er in which the magnetic moment of each individual ion has the same c-component, while the components normal to the c-axis are arranged helically, see Fig. 5.37(a). In each plane normal to the c-axis the magnetic moments are all parallel, that is, each individual sheet is ferromagnetic. If the components of the spins in Fig. 5.37(a) parallel to the c-axis vanish, the structure will have no net magnetic moment and it is described as a

TABLE 5.1

Magnetic structures of rare-earth metals

Element	*Atomic number*	*Structure*
Ce	58	antiferromagnetic $T_N \fallingdotseq 4{\cdot}2$ K
Pr	59	*double h.c.p.:* sinusoidal antiferromagnetic $T_N \fallingdotseq 24$ K, sub-lattice magnetization normal to *c*-axis *f.c.c.:* ferromagnetic, $T_C \fallingdotseq 8{\cdot}7$ K
Nd	60	as Pr but for *double h.c.p.* with $T_N \fallingdotseq 19$ K and for *f.c.c.* with $T_C \fallingdotseq 29$ K
Eu	63	helical antiferromagnetic $T_N \fallingdotseq 91$ K, axis parallel to cube edge
Gd	64	ferromagnetic $T_C = 294$ K (see text)
Tb	65	ferromagnetic in basal plane $T_C = 218$ K helical antiferromagnetic $T_N = 230$ K
Dy	66	ferromagnetic in basal plane $T_C = 86$ K helical antiferromagnetic $T_N = 179$ K
Ho	67	conical ferromagnetic with moment along *c*-axis $T_C = 19$ K helical antiferromagnetic $T_N = 133$ K
Er	68	conical ferromagnetic with moment along *c*-axis $T_C = 20$ K sinusoidally modulated antiferromagnetic $T_N = 80$ K
Tm	69	4, 3, 4, 3 ferrimagnetic at 4 K changing to sinusoidal antiferromagnetic with $T_N = 56$ K

'helical antiferromagnetic' structure (Eu, Tb, Dy, and Ho). At 4 K Tm has a 4, 3, 4, 3 ferrimagnetic structure, see Fig. 5.37(b), which changes progressively, as the temperature is raised, into a sinusoidal antiferromagnet with no resultant magnetic moment, see Fig. 5.37(c). Sinusoidal antiferromagnetism is also exhibited by Pr and Nd but for these two metals the direction of the spins is normal, instead of parallel, to the *c*-axis. If an external magnetic field is applied to an antiferromagnetic metal there may be a critical field above which the ordering becomes ferromagnetic; this has been observed, for example, in Dy, Ho, and Er (Behrendt, Legvold, and Spedding 1958; Flippen 1964). We shall see below that there is quite an

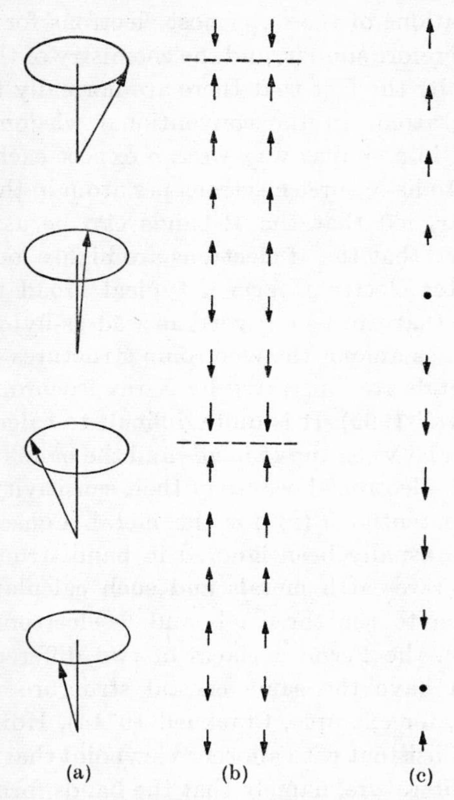

FIG. 5.37. Some types of magnetic ordering (a) conical ferromagnetism, (b) 4, 3, 4, 3 ferrimagnetism, and (c) sinusoidal antiferromagnetism (Cracknell 1971*b*).

intimate relationship between the type of magnetic ordering that occurs in a rare-earth metal and the Fermi surface of that metal.

Relatively little work has so far been performed on direct measurements of the dimensions of the Fermi surfaces of the rare-earth metals, but quite a large number of indirect measurements have been made. As we have already mentioned at the beginning of this chapter, the electronic structure of a rare-earth atom, outside a Xe core, is $4f^n5s^25p^65d^16s^2$ or $4f^{n+1}5s^25p^66s$, where $n =$ (atomic number $-$ 57) and goes from 0 to 14.

The configurations of the outermost electrons for each of these metals are therefore similar, and the chemistry of these elements is dominated by the fact that there are normally three valence electrons per atom. In the conventional wisdom it is often assumed that in a similar way we can expect each of the rare-earth metals to have three electrons per atom in the conduction band. It is argued that the 4f-bands can be assumed to be very narrow so that the 4f-electrons are highly localized, while the three outer electrons form a typical broad metallic conduction band that can be regarded as a 5d-6s-hybrid band; the close similarities among the electronic structures of the heavy rare-earth metals are illustrated by X-ray isochromat measurements (Bergwall 1965). It is quite difficult to calculate directly the separation between the 4f-bands and the bands derived from the 5d- and 6s-electrons because of their sensitivity to errors in the crystal potential $V(\mathbf{r})$ for the metal. Consequently, the 4f-levels have usually been ignored in band structure calculations for the rare-earth metals and such calculations confine their attention to the three 5d- and 6s-electrons. To a first approximation the Fermi surfaces of two different rare-earth metals which have the same crystal structure will then be identical (see, for example, Cracknell 1971*a*). However, this is not entirely consistent with another viewpoint that is commonly found in the literature, namely that the bands formed from the atomic 4f-levels in a solid rare-earth metal are assumed to be widely separated from the 5d- and 6s-bands, except perhaps in Ce (Gupta and Loucks 1969). There is an inconsistency between these two approaches for the following reason. Suppose that we consider free atoms of different rare-earth elements. In so far as it is meaningful to represent the ground states of these atoms in terms of individual electronic states, the results of atomic structure calculations show that the pattern of atomic energy levels must be similar to that shown in Fig. 5.38. The ten 5d-levels must be assumed split in such a manner that one level is below the 4f-levels and the remaining 9 are above the 4f-levels. As we proceed from La with no 4f-electrons, to Lu with 14 4f-electrons, so the position of the highest occupied level

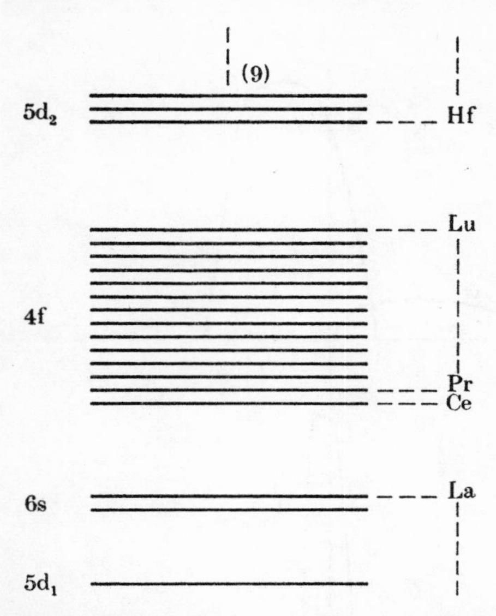

FIG. 5.38. Schematic representation of the individual-particle electron energy levels of a rare-earth atom. The positions of the highest occupied levels in various atoms are indicated.

moves up steadily through the set of 4f-levels. This individual-electron picture must be treated with caution because it is the 5d- and 6s-electrons and not the 4f-electrons that are responsible for the chemical properties of the rare-earth elements. In order to account for the chemical properties of the rare-earth metals we need to postulate that once n 4f-levels are occupied these n-levels move down so as to be below the $5d_1$- and 6s-levels, whereas the remaining $(14-n)$ 4f-levels stay above the $5d_1$- and 6s-levels. In the metallic state each of the levels shown in Fig. 5.38 is spread out into a band and it is convenient to replace Fig. 5.38 by densities of states for the bands derived from the 4f-, 5d-, and 6s-atomic states, see Fig. 5.39. One cannot glibly say, as it is sometimes suggested, that with the exception of Ce the 4f-band is narrow and widely separated from the Fermi energy and therefore the 4f-band can be ignored. If this were the case and the 4f-band were a long way below the Fermi energy there would have to be $(n+3)$ (or 14, whichever is the smaller) occupied 4f-states per atom for all the

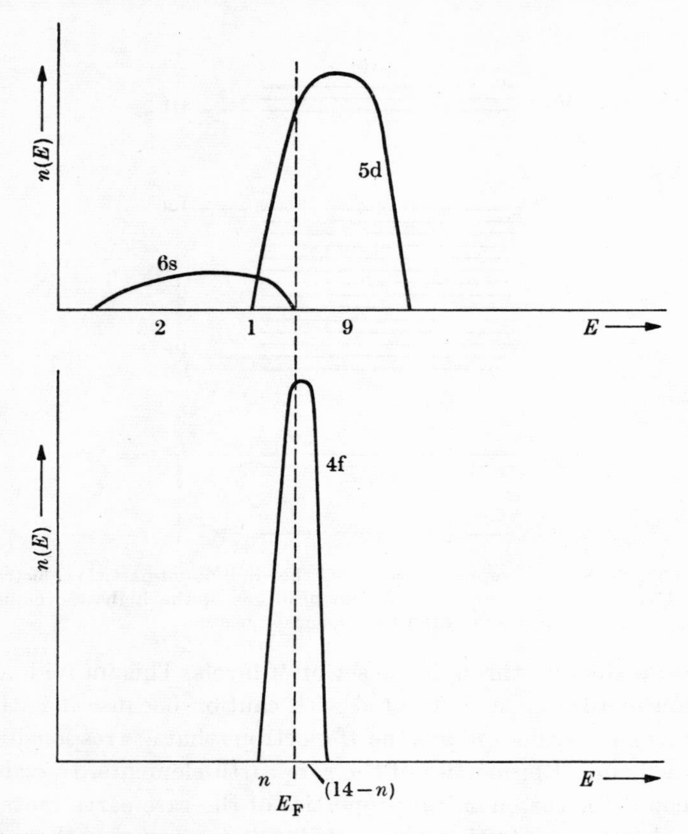

FIG. 5.39. Very schematic densities of states for 4f-, 5d-, and 6s-bands in rare-earth metals, indicating the number of electrons on each side of the Fermi energy.

rare-earth metals, instead of n. On the other hand, if the 4f-band were a long way above the Fermi energy the number of occupied 4f-states per atom for any given rare-earth metal would be fewer than n. To preserve the conventional approach of considering only the 5d- and 6s-electrons it would be necessary to postulate the following process as n increases from 0 to 14 as we pass from one rare-earth metal to the next: the shapes and relative positions of the 5d- and 6s-densities of states remain fixed relative to each other and to the Fermi energy, while the 4f-band moves down relative to the 5d- and 6s-bands, so that the area of the 4f-density of states that comes below the Fermi energy is always exactly equal to n, as in Fig. 5.39. To say the least, this seems a little unlikely. It is probably fair to draw an analogy between the present-day treatment of the 4f-levels in the rare-earth metals and the treatment of the 3d- or 4d-levels in transition metals that was common about a decade earlier. In the past, attempts were often made to calculate band structures for the transition metals ignoring the possibility of hybridization between the 3d- and 4s- (or 4d- and 5s-) levels but, as we have seen earlier, this approach had to be abandoned. The account of the Fermi surfaces of the rare-earth metals given in this section is based on the assumption that the hybridization between the $4f^n$-electrons and the $5d^1 6s^2$-electrons can be neglected; future investigations may render this account obsolete. The argument given in this paragraph and illustrated by Figs. 5.38 and 5.39 is based on the...$4f^n$...$5d^1$...configuration; its details, but not the general principles, would have to be modified for the other, and more common,...$4f^{n+1}$...configuration.

For the free atoms of the rare-earth elements there are one or two exceptions to the electronic structure with valence 3 which we have described previously; Ce, Pr, Tb, and probably Nd can be tetravalent while Sm, Eu, Yb, and perhaps Tm can be divalent. The reasons for the departures from valence 3 in these particular elements are discussed in numerous chemistry books (see, for example, Sidgwick 1950). The α-phase of metallic Ce is thought to be associated with the increase of the number of conduction electrons per atom from three to four at the expense of the 4f-shell which thereby becomes empty (Itskevich 1962; Likhter and Venttsel' 1962). Yb becomes divalent by using one of its outermost electrons to complete the 4f-shell, while Eu becomes divalent by using one of its outermost electrons to form an exactly full 4f-shell which is also very stable. Yb is therefore rather different from most of the other rare-earth metals (Lounasmaa 1963). Yb is soft, its density (7·0 gm cm^{-3}) is lower than the densities of the neighbouring elements Tm (9·3 gm cm^{-3}) and Lu (9·8 gm cm^{-3}) and its crystal structure is f.c.c. rather than h.c.p. That Yb has fewer conduction electrons per atom is indicated by the very low value of its electronic specific heat; γ for Yb is only $(2{\cdot}92 \pm 0{\cdot}01)$ mJ mol^{-1} K^{-2} (Lounasmaa 1963) compared with values of about 10 mJ mol^{-1} K^{-2} for most other rare-earth metals. The positions of the 4f-levels in metallic Yb, which on the above argument can all be expected to be full, have been investigated by X-ray photoemission measurements (Hagström, Hedén, and Löfgren 1970); two sets of 4f-levels, separated as a result of spin–orbit coupling, were found at $(1{\cdot}4 \pm 0{\cdot}4)$ eV and $(2{\cdot}7 \pm 0{\cdot}4)$ eV below the Fermi energy.

If we are prepared to make the assumption which is commonly made that, with one or two exceptions, the rare-earth metals can be regarded as being composed of trivalent positive ions in a sea of three conduction electrons per atom, we would expect the Fermi surfaces of the quite large number of rare-earth metals which have the h.c.p. structure to be very similar to the Fermi surfaces of Sc and Y which were described in Section 5.4, while the Fermi surface of Eu, which is divalent and b.c.c., should be very similar to that of Ba, and the Fermi surface of Yb, which is divalent and h.c.p., should be very similar to those of Zn and Cd. Recent calculations, mostly using the A.P.W. method, do indeed show that the Fermi surfaces of the heavy rare-earth metals and of Y are very much alike (Loucks 1966*b*). Non-relativistic A.P.W. band structure calculations have been performed for Ce, Gd, and Tm, while relativistic A.P.W. band structure calculations have been performed for a large number of the rare-earth metals. As one might expect by analogy with the transition metals, the energy bands for the 5d- and 6s-electrons—which are the conduction electrons—depart very considerably from the free-electron energy bands. With three conduction electrons per atom and two atoms per unit cell in the h.c.p. structure the equivalent of three bands must be full. The Fermi surfaces in bands 3 and 4 in the double-zone scheme for Gd and Tm, calculated non-relativistically, were almost identical and are shown in Fig. 5.40. Although there are some differences from the Fermi surface of Y described in Section 5.4, the general features are quite similar to those of the Fermi surface calculated for Y; that is, the Fermi surface is entirely in bands 3 and 4 and, in the double-zone scheme, is multiply-connected in a rather

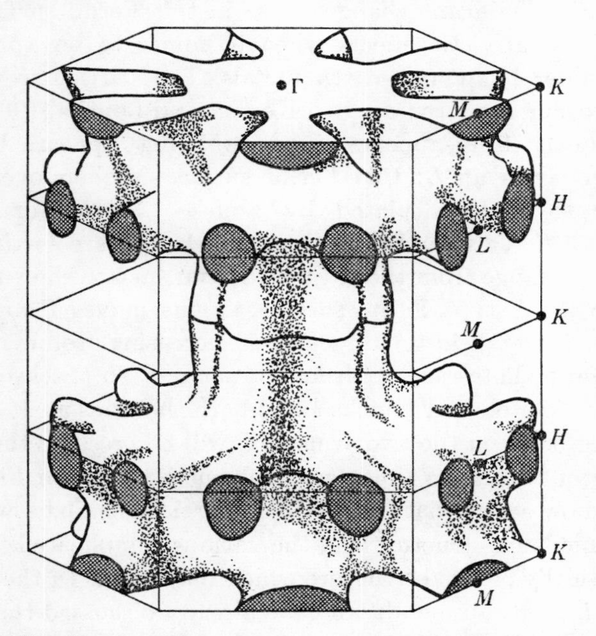

Fig. 5.40. The Fermi surface of Tm, in the double-zone scheme, calculated by Freeman, Dimmock, and Watson (1966) (holes shaded).

complicated fashion. Since the atomic numbers of the rare-earth metals are quite large it is natural to expect that spin–orbit coupling effects will be quite significant. In particular, many degeneracies in the electronic band structure at the special points of symmetry in the Brillouin zone will be lifted (see, for example, Keeton and Loucks 1966*a* on Lu); the lifting of these degeneracies means that the double-zone scheme will no longer be appropriate for the description of the Fermi surfaces of those metals. In their relativistic A.P.W. calculations Keeton and Loucks (1968) found that the inclusion of relativistic effects does not alter the calculated Fermi surfaces very significantly from the results of non-relativistic calculations, whereas for the heavier rare-earth metals the inclusion of relativistic effects does produce significant effects. Therefore, while Fig. 5.40 probably describes fairly accurately the Fermi surface of any of the lighter rare-earth metals that happen to have the h.c.p. structure, significant changes arise in this Fermi surface as a result of relativistic effects when it comes to be applied to the heavier h.c.p. rare-earth metals. The particular changes which occur are, according to Keeton and Loucks (1968), that the arm at M disappears and a 'webbing' appears between the two arms at L; this Fermi surface is then even more similar to that calculated by Loucks (1966*b*) for Y, see Section 5.4. We should thus expect for h.c.p. rare-earth metals a steady change from the Gd-type Fermi surface shown in Fig. 5.40 to the Y-type Fermi surface, as one moves through the rare-earth metals in the direction of increasing atomic number. For intermediate rare-earth metals such as Tb (Jackson 1969) and Dy (Keeton and Loucks 1968) both the arms at M and the webbing between the two arms at L will be present; the Fermi surface of Tb in the single-zone scheme is shown in Fig. 5.41.

It is now appropriate to consider the relationship between the electronic band structure and the various complicated forms of magnetically ordered structure which may exist. In the section on the iron-group metals we have already discussed the effects of ferromagnetic ordering on the band structure and Fermi surface of a metal. There is no change in the size of the Brillouin zone, but many of the essential degeneracies that existed in the paramagnetic phase will be removed by the introduction of ferromagnetic ordering. However, if relativistic effects were included in the band structure calculations for the paramagnetic phase there would be very few essential degeneracies left even before the magnetic ordering is introduced (see, for example, Keeton and Loucks 1966*a*). The effect of ferromagnetic ordering on the shapes of the bands can be expected to be quite small. However, as we saw in connection with the iron-group metals, the introduction of ferromagnetic ordering gives rise to a large

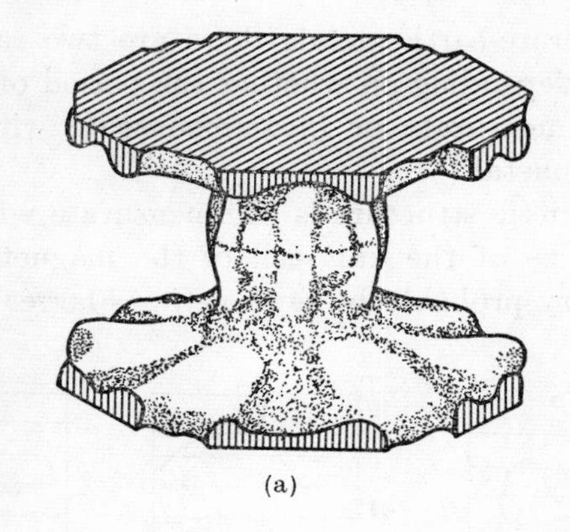

(a)

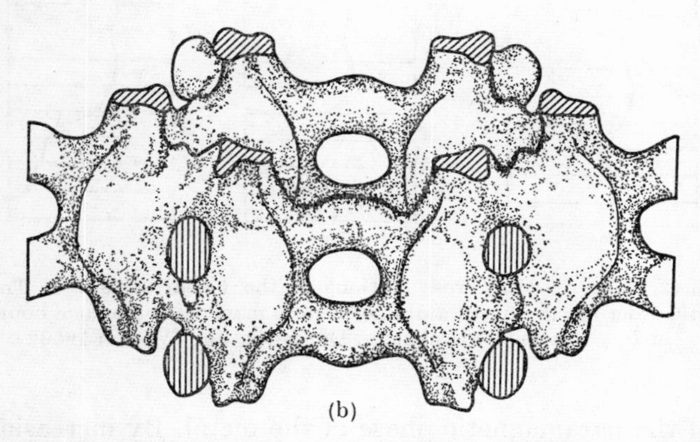

(b)

FIG. 5.41. The Fermi surface of Tb shown in the single-zone scheme, (a) holes in band 3, and (b) electrons in band 4 (Jackson 1969).

magnetic induction which causes a large separation of $g_s\mu_B B$ between the spin-up and spin-down bands and this in turn causes drastic changes in the shape of the Fermi surface. Completely different Fermi surfaces will then apply to the spin-up and spin-down bands. Little work appears to have been done for the rare-earth metals on the estimation of the magnitude of this separation and the study of the consequent shapes of the Fermi surfaces for the spin-up and spin-down bands. For the other more complicated forms of magnetic ordering which can occur in the rare-earth metals there are two rather different possibilities, depending on whether the period of the magnetic order is commensurate or incommensurate with the crystal lattice of the metal.

If the magnetic structure is commensurate with the crystal lattice, the size of the unit cell of the magnetically ordered metal will very probably be several times larger than the unit

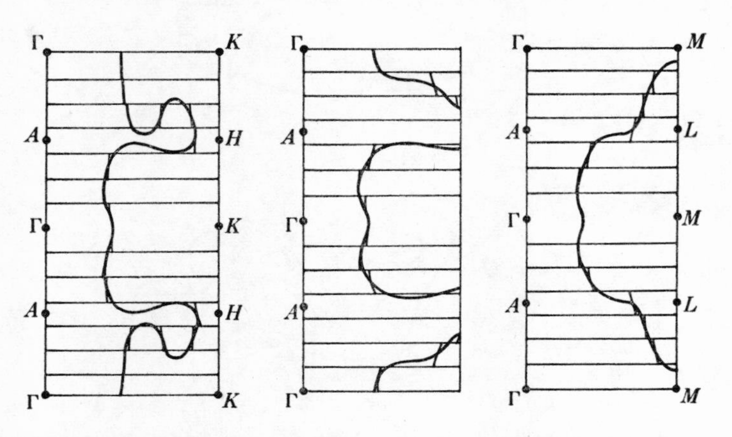

FIG. 5.42. Some vertical cross sections of the Fermi surface of Tm with magnetic ordering; the horizontal lines denote magnetic superzone boundaries at $k_z = \pm n(2\pi/7c)$ (Freeman, Dimmock, and Watson 1966).

cell of the paramagnetic phase of the metal. By increasing the size of the unit cell of the metal the size of the Brillouin zone will be correspondingly reduced. For example, between 40 K and 50 K, Tm has a sinusoidal antiferromagnetic structure in which the magnitudes of the magnetic moments, which in each horizontal layer are ferromagnetically aligned in the z direction, vary sinusoidally with a period of $7c$. The size of the unit cell is therefore increased by a factor of seven and new *superzone boundaries* appear in the reciprocal lattice at $k_z = \pm n(2\pi/7c)$. These new superzone boundaries are shown in Fig. 5.42; this figure now represents an extended zone scheme for the antiferromagnetic structure while the true Brillouin zone only extends from $k_z = -(2\pi/7c)$ to $k_z = +(2\pi/7c)$ and the entire band structure and Fermi surface can be folded back into this region. A particularly suitable metal for studying the effect of magnetic superzone boundaries is Dy because between 86 K and 178 K it exhibits a helical structure, with no magnetic moment parallel to the c-axis, and if a sufficiently large magnetic field is applied normal to c this will effect a change to simple ferromagnetic ordering so that the superzone boundaries will disappear; it is therefore possible to study directly the effects of the controlled appearance and disappearance of the superzone boundaries (Wilding and Lee 1965).

If the period of the magnetic ordering is incommensurate with the crystal lattice of the metal the situation is very similar to that described in Section 5.7 in connection with antiferromagnetic Cr, and we may expect to find some relationship between the wave vector $\mathbf{Q}$ associated with the magnetic ordering and the separation between some important points on the Fermi surface; in the case of Cr the wave vector $\mathbf{Q}$ corresponds to the separation along the $\langle 100 \rangle$ axes between the octahedron of holes around H and the body of the electron jack around Γ. There is a qualitative difference between the cases of Cr and of the rare-earth metals in that the magnetic moments in Cr are associated with the conduction electrons themselves, whereas the magnetic moments in the rare-earth metals are generally thought not to be associated with the conduction electrons, but with the incomplete shell of 4f-electrons which are conventionally not regarded as conduction electrons. The ideas of the relationship between the spin-density waves in antiferromagnetic Cr and the Fermi surface of that metal were extended to the case of the magnetic rare-earth metals by Mackintosh (1962) and Miwa (1963) and discussed in considerable detail by Elliott and Wedgwood (1963, 1964) and Herring (1966). Much of the initial interest was concerned with attempting to explain the temperature dependence of the electrical resistivity of rare-earth metals and, in particular, the anomalies which occur at the transition temperatures. The discussion has been extended to the case of the electrical resistivity, thermal conductivity, and Seebeck effect in Tm using the calculated Fermi surface shown in Fig. 5.40 (Freeman, Dimmock, and Watson 1966; Edwards and Legvold 1968). The shapes of the Fermi surfaces of the rare-earth metals are very closely linked with the origins of the magnetic ordering which may occur. Keeton and Loucks (1968) noticed that there appears to be a close correlation between the wave vector **Q** associated with the magnetic ordering, and the thickness in the *c*-axis direction of the webbing which appears between the arms near L in the hole Fermi surface of h.c.p. rare-earth metals heavier than Gd. The values of **Q** determined experimentally from neutron diffraction work are compared in Table 5.2 with the thickness of the webbing calculated by Keeton and

TABLE 5.2

Comparison of the magnetic wave vector (in units of π/c) as determined from experiment and from minimum separation of the webbing

	Y	Lu	Er	Dy	Gd
Experiment	0·54	0·53	0·57	0·49	0
Theory	0·49	0·45	0·54	0·46	0

(After Keeton and Loucks 1968).

Loucks (1968) and illustrated in Fig. 5.43. The webbing is absent from the Fermi surface of Gd and, in practice, Gd is observed not to have an antiferromagnetic phase. Similar work on Tb (Harris 1969, Jackson 1969, Jackson and Doniach 1969) led to calculated values of **Q** that were in semiquantitative agreement with the experimentally determined value of 0·224 π/c (Yosida and Watabe 1962). Assuming exchange interactions of the Ruderman–Kittel–Kasuya–Yosida (R.K.K.Y.) type between the electrons in the 4f-shells of different ion cores via the conduction electrons (Ruderman and Kittel 1954; Kasuya 1956; Yosida 1957), it is possible to calculate the total energy of the system. The value of the total energy will be dependent on the value of **Q** so that the actual value of **Q** can be found by minimization. This was done by Elliott and Wedgwood (1964) for Dy, Ho, Er, and Tm using the free-electron Fermi surface, by Watson, Freeman, and Dimmock (1968) for a number of the heavy rare-earth metals, and by Evenson and Liu (1968, 1969) for Gd, Dy, Er, and Lu using the Fermi surfaces calculated by Keeton and Loucks (1968). Evenson and Liu obtained values of **Q** for Dy, Er, and Lu that were in

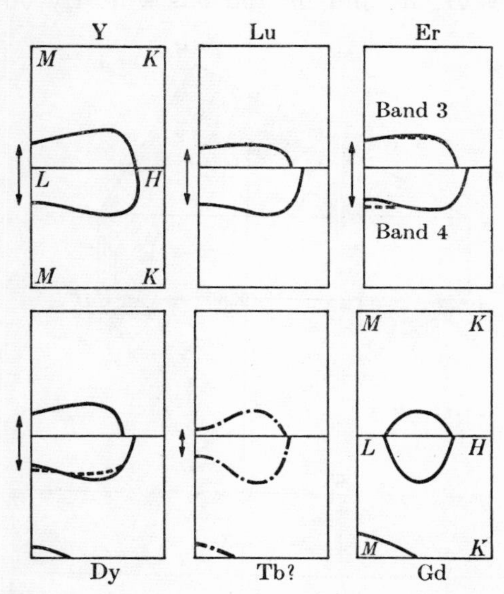

FIG. 5.43. Cross sections of various Fermi surfaces showing the webbing feature (Keeton and Loucks 1968).

quite good agreement with the observed **Q** values for the antiferromagnetic ordering in these metals.

Having discussed in considerable detail the Fermi surfaces of those rare-earth metals that possess the h.c.p. structure we now turn to the remaining rare-earth metals which possess other structures. Fleming, Liu, and Loucks (1968) used the relativistic A.P.W. method to calculate the band structures and Fermi surfaces of the metals La, Pr, and Nd which have the double h.c.p. structure. Since there are four atoms per unit cell there will have to be the equivalent of six full bands. The calculated Fermi surface for La consists of

(i) a nearly circular cylinder of holes centred along ΓA in band 5, and another cylinder of holes centred along ΓA in band 6 which tapers from a nearly hexagonal cross section in the ΓKM plane to a circular cross section in the AHL plane;

(ii) a multiply-connected region of electrons in band 7 which encloses H, K, and M and has a nearly circular cross

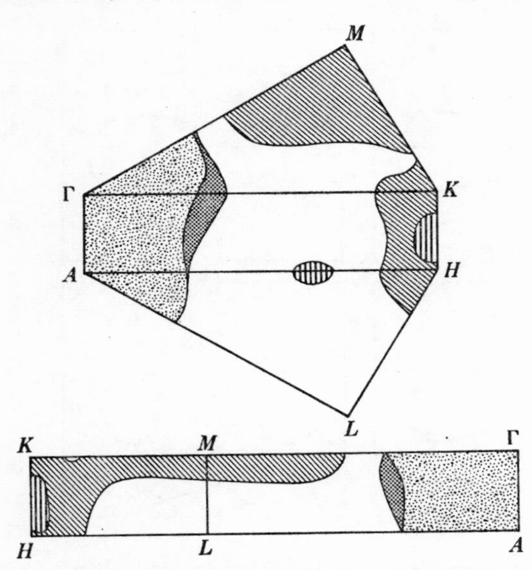

FIG. 5.44. Cross sections of the Fermi surface of La (Fleming, Liu, and Loucks 1968).

section near H but spreads out to produce a shallow shelf centred at M in the ΓKM plane; and

(iii) small ellipsoidal pockets of electrons along AH in band 7 and along HK in band 8.

Cross sections of this calculated Fermi surface for La are illustrated in Fig. 5.44.

The Fermi surfaces of Pr and Nd can be expected to be similar to that of La. La, of course, is non-magnetic but Pr and Nd do exhibit magnetic ordering. For Pr and Nd the wave vector **Q** associated with the magnetic ordering is normal to the c-axis

(Cable, Moon, Koehler, and Wollan 1964; Moon, Cable, and Koehler 1964) and was assumed by Fleming, Liu, and Loucks (1968) to be associated with the shelf-like part of the multiply-connected region of electrons in band 7. We have already seen (Figs. 5.40 and 5.41) that in the heavy rare-earth metals (beyond Gd) which have the ordinary h.c.p. structure, there are very flat pieces of Fermi surface normal to the c-axis and approximately half-way between the central plane and the top surface of the Brillouin zone. These flat pieces can be eliminated, thereby reducing the total energy, by the introduction of a periodic potential with period $2c$; this causes new superzone boundaries to appear at $\pm\frac{1}{2}\pi/c$. Fleming, Liu, and Loucks (1968) suggested that the reason for the existence of the double h.c.p. structure was because it produces just such a potential with a periodicity of $2c$ in the z-direction and thereby lowers the energy of the system. Relativistic band structure calculations for the high-temperature f.c.c. phases of La and Pr have been performed by Myron and Liu (1970). Their predicted Fermi surface was entirely contained in band 2 and was multiply-connected in a rather complicated way. It bore no obvious resemblance to the Fermi surface of Al, a simple metal with the same structure and valence. The only remaining rare-earth metals not covered by the above discussion of the h.c.p. and double h.c.p. structures are Pm, Sm, Eu, Yb, and the f.c.c. phases of La, Ce, and Pr. Nothing is known about the Fermi surfaces of Pm, Sm, and the f.c.c. phases of La, Ce, and Pr.

We have already mentioned that Yb commonly has a valence of two rather than three because it prefers to form a complete 4f-shell. Yb metal is then an h.c.p. metal with two electrons per atom and one can therefore expect that its band structure and Fermi surface will resemble quite closely the band structures and Fermi surfaces of Zn and Cd (Tanuma, Datars, Doi, and Dunsworth 1970). The de Haas–van Alphen frequencies obtained by Tanuma, Datars, Doi, and Dunsworth indicated the existence of hyperboloidal regions on the Fermi surface of Yb. Earlier de Haas–van Alphen and magnetoresistance measurements had been interpreted on the assumption, which now seems to be incorrect, that it is the f.c.c. phase of Yb that is stable.

Eu is similar to Yb in having only two conduction electrons per atom rather than three; in this case it is because one extra electron is used to make the 4f-shell exactly half full. Since Eu has the b.c.c. structure it is not unreasonable to suppose that the band structure and Fermi surface of Eu should be very similar to that of Ba which has been discussed in Section 4.3. Experimental evidence for the similarity between the band structures of Ba and Eu was found in the very close similarities between the optical reflectance and transmission spectra of these

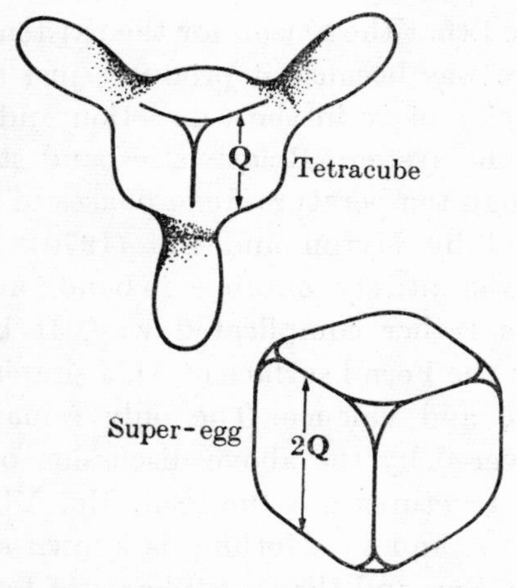

FIG. 5.45. The Fermi surface of Eu (Andersen and Loucks 1968).

two metals (Müller 1965, 1966; Schüler 1965) and a comparison of the fusion curves and compressibility data also suggests this relationship (Jayaraman 1964). This similarity between the Fermi surfaces of Eu and Ba was confirmed by the relativistic A.P.W. calculation of Andersen and Loucks (1968) on Eu in which the Fermi surface of Eu was found to be almost identical to that of Ba; that is, it consists of a pocket of holes at P, described as a 'tetracube', and a pocket of electrons at H, described as a 'super-egg'. These two pieces of Fermi surface are illustrated in Fig. 5.45, and Andersen and Loucks

(1968) found that with $n = \frac{13}{4}$, eqn (4.3.1) described quite closely the shape of their calculated region of electrons at H. They associated the wave vector $\mathbf{Q}$ of the helical magnetic ordering with the nesting of the tetracube with itself, see Fig. 5.45. This gives a value of $\mathbf{Q}$ quite close to the experimental value. It is an interesting coincidence that the opposite faces of the super-egg are separated by almost exactly $2\mathbf{Q}$.

Since the existence of a partially filled 4f-shell in most of the rare-earth metals results in the existence of a ferromagnetic or antiferromagnetic structure for most of these metals at low temperatures, one would expect to be able to detect the existence of spin waves in these metals at low temperatures. Spin waves in magnetically ordered rare-earth metals have indeed been detected both directly by inelastic neutron scattering and also indirectly from their contributions to such properties as the specific heat, thermal conductivity, and electrical resistivity. In the very simple theories of spin waves the spin-wave contribution to the specific heat is proportional to $T^{\frac{3}{2}}$ for a ferromagnet and to T^3 for a simple antiferromagnet (Van Kranendonk and Van Vleck 1958). The calculated and measured values of γ suggest that the phonon enhancement of γ in the rare-earth metals is of comparable magnitude to that in the transition metals. Since many of the rare-earth metals are magnetically ordered there is also the possibility of the enhancement of γ by electron–magnon interactions in these metals (Nakajima 1967).

It is not very surprising that relatively little work has been done so far on the Fermi surfaces of the actinides because most of them do not occur naturally and nearly all of them are radioactive. The electronic structures of the free atoms of the actinides are very similar to those of the lanthanides except that it is now the 5f-, 6d-, and 7s-shells rather than the 4f-, 5d-, and 6s-shells which are of importance. By analogy with the lanthanides we may suppose that the conduction bands of the actinides are derived from the 6d- and 7s-electrons and that, in general, the partially filled 5f-levels can be neglected. Although this is the approach that is actually used, the same caution that we advised in connection with the neglect of the 4f-levels in the lanthanides (see pp. 382–5) should also be exercised in connection with the 5f-levels in the actinides. If the 5f-levels are neglected, the Fermi surface of any of the actinides can be expected to be similar to that of the appropriate iso-electronic transition metal with the same crystal structure (if that exists).

The group-theoretical analysis of the Brillouin zone and a study of the free-electron energy bands in α-U was performed by Jones (1960) and Suffczyński (1960); however, from our experience with the lanthanides and with the transition metals, together with the fact that the atomic number of U is so high, it seems unlikely that the free-electron model will give even a first approximation to the shape of the Fermi surface of U. Relativistic A.P.W. band structure calculations for Ac and Th, which have the f.c.c. structure, were performed by Keeton and Loucks (1966*a*). According to Keeton and Loucks the 5f-shell is empty in both Ac, which is analogous to La, and in Th, which is similar to Ce; there are therefore three conduction electrons per atom in Ac ($\ldots 6d^1 7s^2$) and four conduction electrons per atom in Th ($\ldots 6d^2 7s^2$). The first direct experimental work on the Fermi surface of Th was performed by Thorsen, Joseph, and Valby (1967) using the de Haas–van Alphen effect. Their results were interpreted as indicating the existence of a nearly spherical piece of Fermi surface situated at the centre of the Brillouin zone, and a set of six closed segments located along the ⟨100⟩ axes at X points. These results were not in agreement with the band structure calculations mentioned above. The reason for this is explained by Gupta and Loucks and is connected with the position of the 5f-levels. The difficulty was overcome in a new relativistic A.P.W. band structure calculation for Th performed by Gupta and Loucks (1969) and, apart from the increased importance of relativistic effects, the calculated band structure showed a considerable resemblance to the band structures of the f.c.c. transition metals. The shape of the calculated Fermi surface is illustrated in Fig. 5.46. This consists of

(i) a hole surface around Γ shaped like a rounded cube (a 'super-egg'),

(ii) a dumb-bell shaped hole surface with triangular ends centred at L and directed along the line ΓL, and

(iii) regions of electrons on the line ΓK shaped like a pair of lungs or a woman's breasts.

The existence of the regions of holes around Γ is consistent with the de Haas–van Alphen results of Thorsen, Joseph, and Valby (1967). The de Haas–van Alphen periods that were assigned by

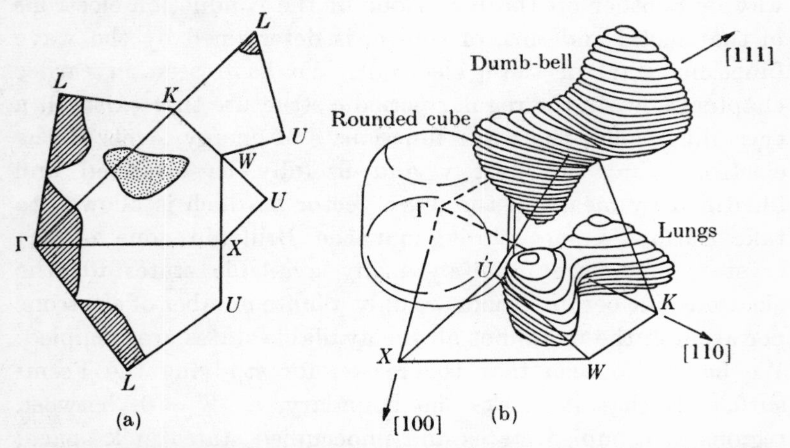

FIG. 5.46. The Fermi surface of Th calculated by Gupta and Loucks (1969), (a) cross sections (holes shaded, electrons dotted), and (b) sketch.

Thorsen, Joseph, and Valby to an ellipsoidal pocket at X can readily be reassigned to the dumb-bell surface or the lung surface shown in Fig. 5.46. An extensive set of further de Haas–van Alphen measurements on Th by Boyle and Gold (1969) lend strong support to the essential correctness of the shape of the calculated Fermi surface for Th shown in Fig. 5.46. The experimental orbits on the lung surfaces were in fairly good agreement with those calculated from the Fermi surface shown in Fig. 5.46, but the experimental orbits assigned to the other two surfaces suggested that the calculated dumb-bell is slightly too small while the calculated super-egg at Γ is slightly too large.

6
ELECTRON INTERACTIONS

6.1. Introduction

NEARLY all the physical properties of a metal depend in some way or another on the behaviour of the conduction electrons in that metal and this, of course, is determined by the wave functions $\psi_{\mathbf{k}}(\mathbf{r})$ of these electrons. We have seen in earlier chapters how, in the regular periodic structure that exists in a crystalline solid, the wave functions and energy levels of the electrons can conveniently and usefully be classified and identified by means of the wave vector $\mathbf{k}$ which is allowed to take various values throughout the Brillouin zone of the crystal. There are infinitely many available states for the electrons but because there are only a finite number of electrons per atom in the metal not all the available states are occupied. We have also seen that the reason for studying the Fermi surface is that it marks the boundary, at $T = 0$, between regions of occupied states and unoccupied states in $\mathbf{k}$ space; since $kT \ll E_{\mathrm{F}}$ for all reasonable temperatures that are normally encountered in the laboratory, relatively few electrons will stray across this boundary.

There are two important assumptions that we have employed in all the previous chapters of this book. The first assumption concerns the *independent-particle model*, namely we assume that a meaningful description of the behaviour of the system of the conduction electrons in a metal can be obtained in terms of the behaviour of the individual electrons, which can be studied by means of the one-electron Schrödinger equation. The interactions between the electrons are either neglected completely or, at best, are included in the crystal potential in some rather unsatisfactory *ad hoc* manner. The second assumption is that the positive ions in the metal are rigidly fixed at certain sites in each unit cell of the crystal. That is, the thermal vibrations of the ions in the metal have been neglected. These motions of the ions will cause changes in the potential so that $V(\mathbf{r})$ now becomes a function of time, t. The lattice vibrations can be resolved into normal modes; the quanta of vibrational energy in these normal modes are quasiparticles which are called *phonons* and the most convenient way to consider the effect of the motions of the ions on the behaviour of the conduction electrons is to think in terms of interactions between the electrons and the phonons. The fact that the theoretical methods described in Chapter 2 and the experimental methods described in Chapter 3 have been so successful in connection with the study of the electronic structures of the metals described in Chapters 4 and 5 indicates that these two assumptions give a very good first approximation to the truth. However, having seen that the Fermi surface is a meaningful concept and that its shape has been determined for most of the metallic elements, it is now appropriate that we should re-examine both these two assumptions in a little more detail.

Several interesting questions are excluded by the treatment that we have given so far in this book based on the assumptions mentioned at the beginning of the previous paragraph. For example, we have excluded any consideration of the factors that determine which is the stable structure exhibited by any given metal. That is, we have not considered the details of the cohesive forces in a metal. Closely related to the question of the cohesive forces in a metal is the problem of obtaining quantitative explanations of the elastic properties of a metal. Other interesting omissions are several other features of the properties of a metal that are connected with the interactions of the conduction electrons, either with other conduction electrons or with the ion cores.

It is relatively easy to give a qualitative description of the forces that are responsible for holding a metal together, but it is notoriously difficult to obtain satisfactory quantitative numerical results for these forces and for the cohesive energy of a metal. A metal is considered to be an array of positively charged ions of large mass, together with a sea of electrons which are negatively charged and highly mobile. The total energy of the metal will involve contributions from the repulsive forces among the positively charged ions, the repulsive forces among the conduction electrons, and the attractive forces between the ion cores and the conduction electrons. In principle, by assuming that the ion cores in a metal are arranged in some fixed regular array compatible with one of the crystallographic space groups, it would be possible to calculate the wave functions of all the conduction electrons in the metal for that given crystal structure. Quantitative expressions for all the various contributions to the total energy could then be evaluated. By repeating this procedure for various different assumed crystal structures for a given metal it would be possible to determine which of these structures possessed the lowest total energy, and therefore to predict the structure that this metal would be expected to exhibit under ordinary circumstances. Attempts to perform such calculations were made in the very early days of band structure calculations (Wigner and Seitz 1933, 1934) and have been made from time to time since then. Reviews of the early work have been given by Wigner and Seitz (1955), Löwdin (1956), and Mott (1962).

It is only very recently that band structure calculations have become sufficiently accurate to enable anything approaching quantitative success to be achieved in calculating cohesive energies of metals. Such calculations are often now based on the use of pseudopotential methods. We shall mention just one or two examples. Schneider and Stoll (1967*b*) calculated the total energy at 0 K for each of the five alkali metals and for each of the three important common possible structures, b.c.c., f.c.c., and h.c.p., using a model pseudo-potential determined from the measured phonon dispersion curves for Na and K and from the elastic constants for Li, Rb, and Cs. The results are shown in Table 6.1. These calculations are in agreement with the experimental results (Li and Na h.c.p. at very low temperatures, others b.c.c.) with one exception. Schneider and Stoll found that for potassium at 0 K the h.c.p. structure has the lowest

TABLE 6.1

Relative energies of b.c.c., f.c.c., and h.c.p. structures of alkali metals at 0 K.

	Li	Na	K	Rb	Cs
b.c.c.	20·47	13·88	0·76	0	0
f.c.c.	1·84	1·14	1·60	1·76	5·03
h.c.p.	0	0	0	0·12	3·37

Energies are in units of 10^{-16} erg ion^{-1} (Schneider and Stoll 1967*b*).

energy; similar results for potassium were also obtained by Pick (1967) and Shaw (1969). This means either that the calculations are inaccurate, possibly because of the fact that the d-levels in potassium were ignored, or else that potassium is expected to undergo a transition to the h.c.p. structure at very low temperatures, below those used in previous structure determinations, i.e. below 1·2 K. Similar calculations by Heine (1968) for the unusual orthorhombic structure exhibited by Ga have already been mentioned in Section 4.4, while considerable success has also been achieved for the noble metals Cu, Ag, and Au (Gubanov and Nikulin 1966; Nikulin 1966; Nikulin and Trzhaskovskaya 1968) and in explaining the general features of the structural trends among the transition metals (Brewer 1968; Deegan 1968; Pettifor 1970). An extensive account of recent work on the calculation of the cohesive energies of metals has been given by Heine and Weaire (1970).

The quantitative explanation of the elastic properties of metals in terms of their known band structures is at a more rudimentary stage than the explanation of the crystallographic structures of the metals in terms of the cohesive energies. The early stages of the work on elasticity were described by Jones (1949); since then there have been several isolated calculations for various groups of metals.

In the rest of this chapter we shall devote some space to the discussion of electron–electron interactions and of electron–phonon interactions and to the consequences of these interactions. We have already encountered one or two direct consequences of electron–electron interactions and electron–phonon interactions in Chapter 3, namely in connection with the electronic specific heat (see Section 3.2) and the Kohn effect (see Section 3.3.2). There are several other properties that depend on these interactions.

We have seen in Chapter 3 that there are certain experiments that can be performed in which it is possible to isolate the contribution from a certain small group of electrons on the Fermi surface of the metal and so to make a measurement that gives direct information about the shape of the Fermi surface, either in the form of a caliper radius or an extremal area of cross section. We discussed the physics of such phenomena in some detail. We also saw in Chapter 3 that there are a number of other phenomena in which the magnitude of some measurable quantity arises as a result of contributions from the electrons all over the Fermi surface, and we discussed the using of such properties as, for example, the surface impedance of a metal in the anomalous skin effect regime, for the purpose of obtaining information about the shape of the Fermi surface of the metal. These methods have the disadvantage that it is necessary to postulate a shape for the Fermi surface of the metal, calculate the expected magnitude of the quantity that is being measured, and then vary the shape of the postulated Fermi surface until the closest possible agreement between calculations and experiment is obtained. In the present chapter our purpose is slightly different. We now assume that in our study of a given metal the shape of the Fermi surface of that metal has been determined, probably as a result of the efforts of several different workers using a variety of the methods described in Chapters 2 and 3. We can, therefore, turn our attention to the problem of seeing how it is possible to use the known Fermi surface of a given metal to help in explaining the behaviour of the physical properties of that metal. There is inevitably some arbitrariness in the division of the discussion of the indirect properties of a metal between Chapter 3 and the present chapter. This division is merely a matter of convenience. In Chapter 3 we included those indirect properties that have, at least at some time in the past, been used to determine the shape of the Fermi surface of some metal. In the present chapter we shall be concerned with properties that depend indirectly on the shape of the Fermi surface but which have hardly ever been used in practice in attempts to obtain detailed geometrical information about the Fermi surface of any metal.

At its lowest level we can say that if the detailed shape of the Fermi surface of a particular metal is known, it is possible to predict the behaviour of all the various properties that we considered in Chapter 3. This might not be a particularly profitable pastime because several of these properties would probably have been used to determine the shape of the Fermi surface of this metal in the first place. However, there are many properties of a metal, including one or two of those discussed in Chapter 3, which while they would make very blunt and insensitive instruments for measuring the fine details of the shape of the Fermi surface are, nevertheless, sufficiently sensitive to the details of the band structure and Fermi surface of the metal as to make it worthwhile for us to discuss them further in this chapter.

6.2. Electron–electron interactions and plasma oscillations

We have mentioned the scattering of conduction electrons by collision with other conduction electrons in one or two connections already. For example, we deliberately excluded them from our discussions of band structure calculations in Chapter 2. In Section 3.2 we noted that electron–electron interactions lead to an extra contribution to γT, the electronic specific heat of a metal (see Table 3.3). In this section and the next we shall discuss the subject of the interactions between the electrons in a metal, concentrating in this section on the effects of these interactions on the band structure while retaining, as far as possible, the independent-particle approximation, and concentrating in the next section on a very simple phenomenological approach to many-body effects associated with electron–electron interactions.

Because the conduction electrons in a metal are Fermions the Pauli exclusion principle requires that no two electrons shall occupy the same eigenstate of the system simultaneously. We assume that this requirement has already been satisfied in any calculation of the band structure, density of states, or Fermi surface of a metal. However, there are further restrictions on the motions of the conduction electrons in a metal. In the methods described in Chapter 2 for the calculation of the band structure of a metal we made use of the independent-particle approximation in which it is assumed that the motion of a given electron does not depend on the motions of all the other electrons. In the one-electron Schrödinger equation (1.2.3) the electron was assumed to move in a potential $V(\mathbf{r})$ which was constructed from a distribution of positive ion cores and some 'smear' of electric charge associated with the conduction electrons. In this independent-particle approximation the probability densities associated with two electrons can be calculated completely independently from the expressions $|\psi_{\mathbf{k}_1}(\mathbf{r}_1)|^2$ and $|\psi_{\mathbf{k}_2}(\mathbf{r}_2)|^2$ respectively, and there is a finite probability that the two electrons will both be in the same small volume element $d\mathbf{r}$ at $\mathbf{r}$. However, when two electrons approach each other very closely the Coulomb electrostatic force between them, which is proportional to $1/|\mathbf{r}_1-\mathbf{r}_2|^2$, becomes very large indeed and will tend to keep them apart. There is, consequently, a correlation between the positions of the electrons; if one electron is at $\mathbf{r}_1$ it will tend to keep other electrons away from $\mathbf{r}_1$ to a much greater extent than would be suggested by the 'smear' of charge density associated with the first electron. This gives rise to an extra contribution to the total energy of the system of the conduction electrons in a metal which is called the *correlation energy*. Its effect on the band structure of the metal can be accounted for, approximately at least, by the addition of an extra contribution to the potential $V(\mathbf{r})$ used in the one-electron Schrödinger equation. The treatment of the correlation energy owes much of its development to the work of Bohm and Pines, and has been described in detail in a series of papers by these authors (Bohm and Pines 1951, 1953; Pines and Bohm 1952; Pines 1953, 1955) although there was some significant earlier work on the subject of the correlations between electrons by Wigner (1934, 1938). In the work of Bohm and Pines the correlated motion of the electrons is considered right from the start. The correlated motion corresponds to collective oscillations of the whole system of the conduction electrons and is very much akin to the plasma oscillations which are familiar in discharges in gases. It was found that for distances which are large compared with the average separation between the electrons, the system behaves collectively and is most suitably described by a set of harmonic oscillators which represent the plasma oscillations. By analogy with phonons and magnons, the quanta of energy of these plasma oscillations are called *plasmons*.

Although a gaseous plasma and a metal are superficially very different, there are several similarities. In a plasma the electron density is of the order of 10^{18} m^{-3} while in a metal it is of the order of 10^{29} m^{-3}. In a metal the positive ions are arranged at specified sites in a crystallographic lattice and undergo only very small vibrations about their equilibrium positions. In a plasma, although the positive ions are free to move, their masses are so very much greater than the masses of the electrons that, in any collective oscillation, the displacements of the positive ions will be correspondingly very much smaller than the displacements of the electrons in order to conserve momentum. Both the plasma and the metal therefore consist of a set of positive ions which are almost stationary and an assembly of highly mobile itinerant electrons. Each system is, of course, electrically neutral overall. Moreover, the fact that the electrons are highly mobile means that if an excess of positive or negative charge appears in some locality in one of these systems, the electrons will very quickly re-arrange themselves so as to maintain electrical neutrality locally as well. Any Coulomb force between electric charges will therefore be screened by these rapid movements of the conduction electrons. This tendency of the electrons to re-arrange themselves to restore electrical neutrality is of course opposed by the random thermal motions of the electrons. The magnitude of the electrostatic force arising from a local excess charge, $-e$, will be proportional to $1/r^2$, so that for large distances the effect of $-e$ is dominated by the thermal fluctuations, whereas for small distances, the presence of $-e$ dominates over the thermal fluctuations. It is therefore convenient to introduce a characteristic distance, called the *Debye length*, which will indicate the distance beyond which the effect of a local excess charge $-e$ will no longer be felt. The actual value of the Debye length λ_D will depend on the temperature and on n_0, the number of electrons per unit volume in the system. It is possible to obtain an expression for λ_D by using Poisson's equation as follows (Debye and Hückel 1923). If a point charge $-e$ is introduced into the plasma at $\mathbf{r} = 0$ it will give rise to a change $\delta\rho$ in the number of electrons per unit volume and to an effective potential ϕ where

$$\nabla^2\phi = \frac{e\,\delta(\mathbf{r})}{\varepsilon_0} + \frac{e\,\delta\rho}{\varepsilon_0}. \tag{6.2.1}$$

If we assume that $\delta\rho$ is given by using the Boltzmann distribution then for small ϕ we have

$$\begin{aligned}\delta\rho &= n_0\exp(e\phi/kT) - n_0\\ &\doteqdot n_0\frac{e\phi}{kT}.\end{aligned} \tag{6.2.2}$$

Therefore substituting into equation (6.2.1) gives

$$\left[\nabla^2 - \frac{n_0 e^2}{\varepsilon_0 kT}\right]\phi = \frac{e\,\delta(\mathbf{r})}{\varepsilon_0}, \tag{6.2.3}$$

so that

$$\phi = \left(\frac{-e}{4\pi\varepsilon_0 r}\right)\exp(-r/\lambda_\mathrm{D}), \tag{6.2.4}$$

where

$$\lambda_\mathrm{D} = \left(\frac{\varepsilon_0 kT}{n_0 e^2}\right)^{\frac{1}{2}}. \tag{6.2.5}$$

The expression for ϕ shows that if $r \ll \lambda_\mathrm{D}$ the potential is almost the same as that due to a point charge q in free space, i.e. the charge is hardly screened at all, but that if $r \gg \lambda_\mathrm{D}$ the charge $-e$ is very effectively screened. From the actual expression for λ_D in eqn (6.2.5) we see that if T is increased λ_D will increase and the effectiveness of the screening is reduced, while if n_0 is increased λ_D will decrease and the effectiveness of the screening will be increased. For a typical plasma with $n_0 \sim 10^{18}\ m^{-3}$ and $kT \sim 3$ eV corresponding to quite a high temperature, λ_D will be about 10^{-5} m. It is not immediately obvious that this classical argument should be at all relevant to a metal which ought to be described by quantum mechanics. However, if one substitutes into eqn (6.2.5) values that are of the correct order of magnitude for a metal: $n_0 \sim 10^{29}\ m^{-3}$ and $kT \sim 1/40$ eV for a metal at room temperature, λ_D would come to about 3×10^{-13} m. Since this is substantially smaller than the average separation between the conduction electrons in a metal it can be taken to indicate that for most of the time a conduction electron is unaware of the motions of the other electrons and therefore the independent-particle model is a good first approximation to the behaviour of the conduction electrons in a metal.

By analyzing the density fluctuations in an electron gas it is possible to show that the electrons in a plasma are capable of both collective and independent-particle behaviour (for details see Pines and Bohm 1952, Pines 1955). For phenomena that involve distances which are longer than the Debye length, the electrons behave collectively and their motions are best described by a set of simple harmonic oscillations which are the plasma oscillations. For phenomena that involve distances which are shorter than the Debye length, the motions of the electrons are best described as those of a system of individual particles moving nearly independently and interacting rather weakly by means of a screened Coulomb potential. These conclusions were originally based on classical arguments but it was found that similar conclusions could be obtained from quantum-mechanical arguments, although it is necessary to obtain an expression for the quantum-mechanical analogue of the Debye length. When this quantity was determined it was found to be somewhat shorter than the average separation of the electrons in the metal. We do not propose to go into the details of this quantum-mechanical treatment here; the interested reader should refer to the accounts given in the various papers by Bohm and Pines. Having performed this analysis it is possible, in principle at least, to calculate the contributions to various measurable physical properties of a metal as a result of the electron correlation arising from the Coulomb electrostatic interactions. Calculations have been made of the effect which the correlation between the motions of the electrons in the alkali metals will have on the cohesive energy, the band width, the density of states, the electronic specific heat, the electrical conductivity, the thermoelectric power, the Knight shift, and the magnetic susceptibility of these metals (for details see the review by Pines 1955). Although these calculations give an indication of the size of the effects of electron–electron interactions on these various properties, one should not attach too much significance to the actual numbers quoted because of various approximations made in the numerical calculations, and because the theory of Bohm and Pines only includes one of the two contributions to electron–electron interactions, namely the Coulomb forces of repulsion, and does not include the forces of attraction between the electrons arising from the exchange of virtual phonons.

We have mentioned the possibility of the existence of plasma oscillations involving the conduction electrons in a metal. It is instructive to consider the frequency of these plasma oscillations. In the specimen of plasma with the rather idealized shape of a rectangular box illustrated in Fig. 6.1, the electrons are all assumed to be displaced a distance ξ relative to the fixed positive ions, so that charge densities of $\pm en_0\xi$ per unit area of cross section appear at each end of the piece of plasma. The electric field $\mathbf{E}$ within the plasma can be determined by the use of Gauss' theorem: $\epsilon_0\epsilon_L E = en_0\xi$, where ϵ_L is the dielectric constant of the lattice of positive ions. The electric field gives rise to a restoring force on each electron so that, from Newton's second law of motion, the equation of motion of the electrons in the plasma is

$$m\frac{\mathrm{d}^2\xi}{\mathrm{d}t^2} = -\frac{e^2n_0\xi}{\varepsilon_0\varepsilon_L}, \tag{6.2.6}$$

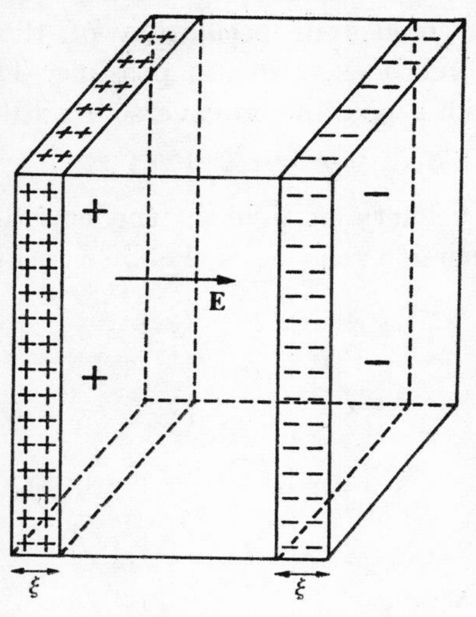

FIG. 6.1.

where we neglect the possibility of collisions between the electrons. The angular frequency, ω_p, of the plasma oscillations is therefore given by

$$\omega_p = \left(\frac{n_0 e^2}{\epsilon_0 \epsilon_L m}\right)^{\frac{1}{2}}. \tag{6.2.7}$$

Substitution of a typical value of n_0 for a gaseous plasma into eqn (6.2.7) leads to a value of ω_p of the order of 10^{12} rad s^{-1}. Although one cannot expect to obtain an accurate expression for the plasma frequency in a metal from the classical argument outlined above, nevertheless eqn (6.2.7) does give an order-of-magnitude estimate of the frequency of plasma oscillations in a metal. Substituting a typical value of n_0 for a metal gives a value of ω_p of the order of 10^{16} rad s^{-1}. The energy $\hbar\omega_p$ of a plasmon in a metal is therefore of the order of 6·6 eV which is larger than the Fermi energy of a metal (typically $\sim$5 eV); consequently very few plasmons are spontaneously excited under ordinary circumstances in a metal.

In the mode of oscillation which we have just described all the electrons in the plasma are moving in phase. That is, we have a $\mathbf{k} = 0$ mode of oscillation of the system. However, other wave-like modes of oscillation of the plasma, with non-zero wave vector **k**, are also possible. The angular frequency ω of such a possible transverse excitation is given by

$$\omega^2 = c^2k^2 + \omega_p^2, \tag{6.2.8}$$

where c is the velocity of light in the crystal lattice of the metal. This dispersion relation is sketched in Fig. 6.2. For any

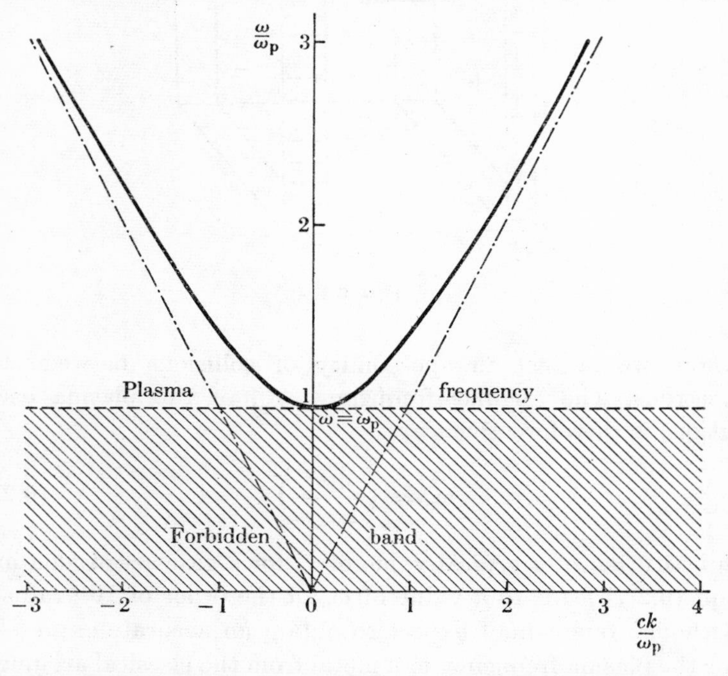

FIG. 6.2. Sketch of the dispersion relation for transverse waves in a single-component plasma in the absence of a magnetic field (Baynham and Boardman 1971).

value of ω below ω_p the value of k given by eqn (6.2.8) is imaginary so that we have the familiar situation that it is not possible to transmit electromagnetic waves through a metal when $\omega < \omega_p$. Since we have seen that for a metal ω_p is likely to be of the order of 10^{16} rad s^{-1}—which is almost the frequency of very soft X-rays—a metal will be opaque to all electromagnetic radiation except X-rays and γ-rays. It is this non-penetration of a metal at radio-frequencies except within the skin depth δ, which we have noted in connection with cyclotron resonance in Chapter 3. However, it was pointed out by Konstantinov and Perel' (1960) and Aigrain (1960) that if a magnetic field is applied to a solid metal or semiconductor, it becomes possible for electromagnetic waves of low frequency to propagate within the solid along the direction of the applied magnetic field. It would not be appropriate to become involved in a lengthy discussion of the theory behind this phenomenon and we shall simply note the results. In spite of the importance of quantum mechanics in the theory of electrons in metals, much of the theory behind magnetoplasma waves can actually be presented in classical terms; it was first developed in connection with gaseous plasmas and only much later applied to solid-state plasmas (for further details see Baynham and Boardman 1971; Kaner and Skobov 1971). There are two kinds of waves to be considered, namely *helicons* and *Alfvén waves.*

Suppose that we consider a plasma with only one kind of carrier present; these carriers may be either electrons or holes. Then it is possible to show that in addition to the usual high-frequency excitations that are possible in the absence of the external magnetic field there is another set of low-frequency excitations corresponding to circularly polarized waves propagating along the direction of the applied magnetic field. These waves are called *helicons* and they were first detected experimentally in a metal by Bowers, Legendy, and Rose (1961). Their importance from the point of view of Fermi surface measurements has already been mentioned in Chapter 3 in connection with Doppler-shifted cyclotron resonance (see p. 208). It is intuitively reasonable to expect that these waves should be circularly polarized because the applied magnetic field will cause the carriers to execute cyclotron motion. The frequency of the helicon mode with wave vector **k** for a one-component plasma in the absence of collisions is given by solving the equation

$$\frac{c^2k^2}{\omega^2} = \epsilon_{\mathrm{L}} - \frac{\omega_p^2}{\omega(\omega_c + \omega)}, \tag{6.2.9}$$

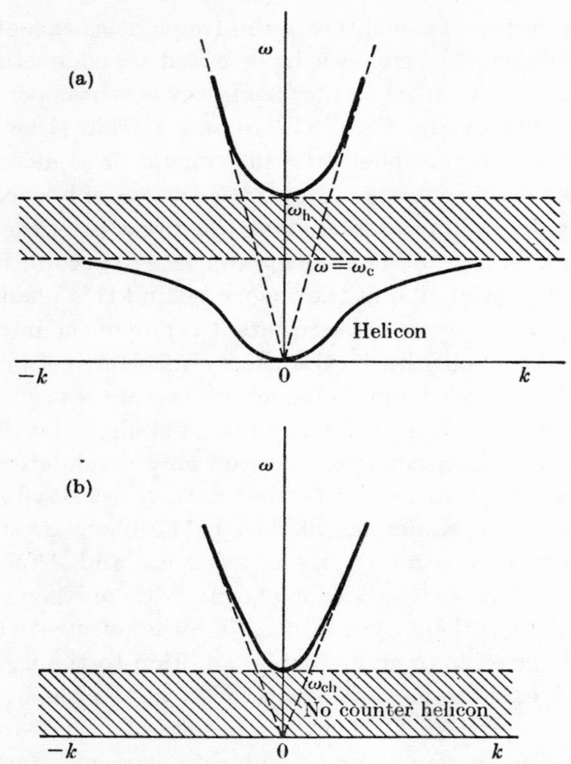

Fig. 6.3. Sketch of the dispersion relations for transverse waves propagating along an external field $\mathbf{B}_0$ in a collisionless one-component plasma for two possible senses of circular polarization (a) with propagating helicon mode, and (b) with non-propagating counter-helicon mode.

$$\omega_h = \tfrac{1}{2}\omega_c\{(1+4\omega_p^2/\omega_c^2)^{\frac{1}{2}}+1\} \quad \text{and} \quad \omega_{ch} = \tfrac{1}{2}\omega_c\{(1+4\omega_p^2/\omega_c^2)^{\frac{1}{2}}-1\}.$$

(Baynham and Boardman 1971).

and the resulting dispersion relation is sketched in Fig. 6.3(a). The sense of the circular polarization is not arbitrary but is determined by the direction of **B** and the sign of the carriers. For the opposite sense of circular polarization there will be no low-frequency propagating mode (see Fig. 6.3(b)) and the dispersion relation is obtained by solving the equation

$$\frac{c^2k^2}{\omega^2} = \epsilon_L + \frac{\omega_p^2}{\omega(\omega_c-\omega)}\,. \tag{6.2.10}$$

The non-propagating wave is sometimes called the *counter-helicon* mode. Although for the counter-helicon mode in a collisionless plasma k is purely imaginary and we have an evanescent wave, when collisions are included k becomes complex and we have a heavily-damped, fast, propagating wave. The helicon mode has been known in atmospheric physics—in which context it is called the *whistler*—for a long time (Barkhausen 1919), but its significance in the physics of metals has only been appreciated more recently.

In the discussion of helicons so far we have assumed that we are concerned with a solid-state plasma containing only one kind of carriers: either electrons or holes. However, in many metals the carriers will include both electrons and holes. In the absence of any external magnetic field, eqn (6.2.8) can be generalized to cover the situation in which more than one species of carrier is present, namely

$$\omega^2 = c^2k^2 + \sum_{\alpha} \omega_{p\alpha}^2, \tag{6.2.11}$$

where $\omega_{p\alpha}$ is the plasma frequency for the species of carriers labelled by α. Therefore, as in the single-carrier case, it will not be possible to have propagating electromagnetic waves in the plasma except at very high frequencies, that is, unless $\omega^2 > \sum_{\alpha} \omega_{p\alpha}^2$. However, if an external magnetic field is applied to a solid-state plasma that contains both electrons and holes then helicon-mode propagation is possible at low frequencies just as it is in the single-carrier case. As in the single-carrier situation there will be one circularly polarized excitation that is a propagating wave, while the opposite circular polarization will correspond to an evanescent wave. The sense of the circular polarisation of the propagating mode will depend on whether it is the electrons or the holes that are the majority carriers. It is in the special case when the metal is compensated, that is, when there are equal numbers of electrons and holes present, that the *Alfvén wave* replaces the helicon mode as the form of electromagnetic wave that can propagate in the two-component solid-state plasma in the metal. The Alfvén wave mode differs from the helicon mode in that, at least at low frequencies and for propagation parallel to **B**, it is a plane polarized wave rather than a circularly polarized wave. Qualitatively this is not too surprising because if we imagine ourselves passing from the situation of excess electrons to the situation of excess holes, the propagating mode will change from one sense of circular polarization to the opposite sense of circular polarization. Alfvén waves also differ from helicons in that in the low frequency region ($\omega \ll \omega_p$) the dispersion relation is linear rather than parabolic, that is we have $\omega \propto k$ rather than $\omega \propto k^2$.

6.3. Landau's theory of a Fermi liquid

We have, so far, discussed electron–electron interactions primarily in connection with the effects which they produce on the potential $V(\mathbf{r})$ in a metal and therefore on the band structure of the metal. By changing the band structure and the density of states, and consequently the shape of the Fermi surface, the inclusion of electron–electron interactions may make significant changes in the theoretical estimates of any of the properties of a metal which have been described in Chapter 3 of this book. For example, the contributions of electron–electron interactions to the electronic specific heats of a number of simple metals are listed in Table 3.3. In the Bohm–Pines treatment which was mentioned in the previous section, it was shown that there are two extreme possibilities: completely independent motions of the electrons and completely correlated motion of the electrons in a plasma oscillation in which all the electrons oscillate together in simple harmonic motion with a single common frequency. The real situation is assumed, not unreasonably, to lie somewhere between these two extremes. Having established that the correlation between the motions of the electrons leads to the screening of any charge q inserted into the metal, it is then a common procedure to assume that correlation effects can be adequately accounted for by making appropriate modifications to the potential $V(\mathbf{r})$ to account for this screening of the charges of the ions in the metal. This adjusted $V(\mathbf{r})$ is then used in a band structure calculation giving energy eigenvalues $E(\mathbf{k})$ in which it is said that correlation effects have been included. This is not the whole story. In regarding the interactions between the conduction electrons as simply contributing an additional 'correlation' term to $V(\mathbf{r})$ and then calculating $E(\mathbf{k})$ in the usual way, the validity of the independent-particle approximation has not really been seriously questioned because all the methods of band structure calculation described in Chapter 2 are based on solving the one-electron Schrödinger equation (1.2.3) by some special technique or another. However, there is no *a priori* reason for assuming the validity or usefulness of the independent-particle approximation and, since we are now prepared to consider the fact that the conduction electrons in a metal interact with each other, we are really concerned with a complicated many-body system. During the last decade or so very substantial progress has been made in the formal study of many-body systems of this type, without assuming at the outset that each wave function of the complete system can be separated into a number of independent wave functions corresponding to single-particle wave functions (see, for example, Nozières 1964; March, Young, and Sampanthar 1967; Ziman 1969*b*; Fetter and Walecka 1971).

Although the formal many-body treatments are very worthy and certainly provide a more rigorous and satisfying quantum-mechanical description than the independent-particle approximation which we have used so far, it is rather difficult to apply the results in practice except for one or two very simple metals like Na and K (Rice 1968). It is therefore perhaps not inappropriate to consider the theory developed by Landau (1956, 1957, 1958) for treating a collection of interacting Fermions which, although it is frankly phenomenological in its approach, does show a remarkable insight into the physics of the many-body system of the conduction electrons in a metal. At this stage we make no particular assumptions about the physical origins of the interactions between the electrons, but simply note that the principal force between two electrons is the electrostatic Coulomb repulsion between the electrons which will be balanced, at least partially, by the attractive force due to the exchange of virtual phonons between the electrons (see Section 6.5). This second contribution was not included at all in the Bohm–Pines treatment mentioned in Section 6.2. If one neglects the interactions between the conduction electrons in a metal this corresponds to regarding the electrons as a gas of free, or nearly free, electrons. If we slowly 'turn on' the interactions between the electrons, their behaviour becomes less and less similar to that of a gas and more like that of a liquid. Hence the term *Fermi liquid* is often applied to an assembly of Fermions in which the interactions between the particles are not ignored. The theory of a Fermi liquid which is due to Landau was originally developed phenomenologically without too much regard for the microscopic structure of the system. The theory was later shown to be compatible with the more formal quantum-mechanical treatment of this many-body system by perturbation theory, so that in addition to being useful Landau's theory is also 'respectable' (see, for example, Chapter 6 of Nozières 1964).

Landau's Fermi liquid theory was originally applied to liquid ^{3}He rather than to the conduction electrons in a metal. One of the great successes of the theory was in dealing with the various collective excitations—particularly zero sound and first sound—in liquid ^{3}He. There are some differences worth noting between the case of liquid ^{3}He and the conduction electrons in a metal. First of all, liquid ^{3}He really is a liquid, while it is rather more artificial to regard the conduction electrons in a metal as a liquid. Whereas the atoms of ^{3}He are electrically neutral, the conduction electrons in a metal are negatively charged; this will involve differences of detail in the theory. In liquid ^{3}He the Fermi surface is spherical in the absence of interactions between the atoms; switching on the interactions will therefore not alter the shape of the Fermi surface and it will remain spherical. The application of Fermi liquid theory to the conduction electrons in a metal is not so simple. Even in the absence of interactions between the electrons the Fermi surface of a metal is anisotropic, except in the free-electron approximation, so that the switching on of the interactions can be expected to distort the Fermi surface.

The theory of a Fermi liquid developed by Landau involves the introduction of the idea of a quasi-particle in a system of many Fermions, and in the particular application with which we are concerned these Fermions are electrons. So far in this book we have used the individual-particle approximation in describing the conduction electrons in a crystalline metal. In this approximation the solutions of the one-particle Schrödinger equation are characterized by a wave vector $\mathbf{k}$ and an energy $E(\mathbf{k})$. These are the eigenstates of the individual particles. To specify an eigenstate of the whole system it is enough to specify by means of a distribution function, $n(\mathbf{k})$, those states which are occupied and those states which are unoccupied. As long as the electrons are assumed to be non-interacting the total energy of the system will just be given by

$$E = \sum_{\mathbf{k}} E(\mathbf{k}) n(\mathbf{k}), \tag{6.3.1}$$

and if a change, $\delta n(\mathbf{k})$, is made in the distribution function the corresponding change, δE, in the energy of the whole system is given by

$$\delta E = \sum_{\mathbf{k}} E(\mathbf{k}) \delta n(\mathbf{k}). \tag{6.3.2}$$

At absolute zero the distribution function is the step function

$$n_0(\mathbf{k}) = \begin{cases} 1 & E(\mathbf{k}) < E_{\mathrm{F}} \\ 0 & E(\mathbf{k}) > E_{\mathrm{F}}. \end{cases} \tag{6.3.3}$$

Now we suppose that the interactions between the electrons are introduced progressively in an adiabatic manner (or 'slowly turned on') and we proceed from the case of a 'gas' to the case of a 'liquid'. It is then assumed that as the magnitudes of the interactions are increased so the eigenstates of the original non-interacting situation evolve continuously into the eigenstates of the real system. That this assumption is not too unreasonable may be justified by remembering the steady development of eigenstates that occurs in ordinary time-dependent perturbation theory; nevertheless it remains an assumption that, strictly, requires more formal justification. The entities which are associated with these eigenstates are described as *quasi-particles*. These quasi-particles are equal in number to N, the number of electrons in the system, and they also still obey Fermi–Dirac statistics. One can envisage a quasi-particle in the following way. Suppose we start with a system of non-interacting electrons in the ground state characterized by the distribution $n_0(\mathbf{k})$. If we introduce an extra electron with wave vector $\mathbf{k}$ and then introduce the electron–electron interactions, we shall obtain an eigenstate, with wave vector $\mathbf{k}$, of the system with electron–electron interactions included. The quasi-particle with wave vector $\mathbf{k}$ can therefore be regarded as an individual electron that is 'clothed' by its interactions with all the other electrons. As described by Landau, the quasi-particle can be considered as a particle in a self-consistent field of surrounding particles. It would be dangerous to assume that all the properties of a particle in such surroundings are the same as those of the bare particle, and describing it as a quasi-particle helps to emphasise this fact. Of course, in the limit of extremely weak interactions the behaviour of these quasi-particles will become very similar to that of the ordinary or 'bare' electrons. In seeking to understand the quasi-particle as a collective excitation of the whole system it may be helpful to draw an analogy with the lattice vibrations of a crystalline metallic or non-metallic solid, remembering of course that an analogy should never be pursued too far. One can, if one so wishes, describe the vibrations of the atoms in a crystalline solid in terms of the displacements of all the various atoms but, because of the strong coupling that exists between one atom and its neighbours, such a description would be very difficult to handle. It is much easier not to consider the individual displacements of the various atoms in the solid but to discuss the vibrations of the crystal in terms of the normal modes of vibration of the crystal. These normal modes behave as *non-interacting* harmonic oscillators which are very much easier to handle (see Section

6.4). In a similar way we can expect that when we study the behaviour of a system of interacting conduction electrons in a metal it will be easier and more profitable to discuss the situation not in terms of the motions of the individual electrons but in terms of the motions of some other quasi-particles or collective excitations of the system.

One very important feature of these quasi-particles is that for a quasi-particle with some given value of $\mathbf{k}$ there is no unique value for the energy $E(\mathbf{k})$ of the quasi-particle because this will depend on which other quasi-particles are actually present. If the quasi-particle with wave vector $\mathbf{k}$ is the only quasi-particle present in the system, then its energy will have some value $E_0(\mathbf{k})$ which could, in principle at least, be determined. However, if other quasi-particles with wave vectors $\mathbf{k}'$ are also present, the energy of the quasi-particle with wave vector $\mathbf{k}$ will be altered as a result of the interactions with the other quasi-particles. The energy $E(\mathbf{k})$ of the quasi-particle can then be written as

$$E(\mathbf{k}) = E_0(\mathbf{k})+\sum_{\mathbf{k}'} f(\mathbf{k}, \mathbf{k}')\delta n(\mathbf{k}'), \tag{6.3.4}$$

where the *interaction function* (or *correlation function*) $f(\mathbf{k}, \mathbf{k}')$ represents the interaction energy between two quasi-particles with wave vectors $\mathbf{k}$ and $\mathbf{k}'$. The second term in eqn (6.3.4) therefore represents the total interaction energy between the quasi-particle of wave vector $\mathbf{k}$ and all the other quasi-particles that are actually present as indicated by $\delta n(\mathbf{k}')$. Since the quasi-particles obey Fermi–Dirac statistics the distribution function $\delta n(\mathbf{k}')$ must itself involve the energies $E(\mathbf{k}')$ of the quasi-particles with wave vector $\mathbf{k}'$ via the Fermi–Dirac distribution function $1/[\{\exp(E(\mathbf{k}')-E_{\mathrm{F}})/kT\}+1]$. Since the $E(\mathbf{k}')$ in turn depend on which quasi-particles are actually present, the determination of the energy $E(\mathbf{k})$ of the quasi-particle $\mathbf{k}$ via eqn (6.3.4) is a complicated problem in self-consistency. The fact that the interaction function $f(\mathbf{k}, \mathbf{k}')$ is unknown for most metals is a further complication. The equation that replaces eqn (6.3.2) to give the energy relative to the ground state of the whole system of quasi-particles is

$$\delta E = \sum_{\mathbf{k}} E_0(\mathbf{k})\delta n(\mathbf{k}) + \tfrac{1}{2}\sum_{\mathbf{k}}\sum_{\mathbf{k}'} f(\mathbf{k}, \mathbf{k}')\delta n(\mathbf{k})\delta n(\mathbf{k}'), \quad (6.3.5)$$

where the first term represents the sum, over all the quasi-particles present, of the energy that each of the quasi-particles would have if it were present on its own, and the second term represents the interaction energies between the quasi-particles. The factor of $\frac{1}{2}$ in the second term is to prevent the interaction between the pair of quasi-particles $\mathbf{k}$ and $\mathbf{k}'$ from being counted twice. Equation (6.3.5) can be regarded as the sum $\sum_{\mathbf{k}} E(\mathbf{k})\delta n(\mathbf{k})$ over the energies $E(\mathbf{k})$ given by eqn (6.3.4) for those quasi-particles that are actually present, provided the sum over $\mathbf{k}$ is restricted in such a way that the interaction $f(\mathbf{k}, \mathbf{k}')$ is not counted twice in the implied double summation over $\mathbf{k}$ and $\mathbf{k}'$.

So far a quasi-particle has been identified by its wave vector $\mathbf{k}$ and energy $E(\mathbf{k})$. The dynamics of the quasi-particle will be characterized by the expression for the velocity

$$\mathbf{v}_{\mathbf{k}\alpha} = \frac{1}{\hbar}\frac{\partial E(\mathbf{k})}{\partial k_\alpha}, \quad (6.3.6)$$

where the suffix α indicates the three components of $\mathbf{v}_{\mathbf{k}}$, and $E(\mathbf{k})$ is the quasi-particle energy given by eqn (6.3.4). Equation (6.3.6) is a simple extension of eqn (1.3.17). We have not mentioned the spins of the electrons so far in this section. Since we use the wave vector $\mathbf{k}$ and the spin $\boldsymbol{\sigma}$ of an electron to classify the eigenstates in the independent-particle approximation, we can also use $\mathbf{k}$ and $\boldsymbol{\sigma}$ to classify the quasi-particle states. So long as there is no magnetic field present $E(\mathbf{k})$ will not depend on $\boldsymbol{\sigma}$, so we can take spin into account by simply using the pair $(\mathbf{k}, \boldsymbol{\sigma})$ in place of $\mathbf{k}$ in the discussion given so far in this section.

Now, since we have specified the momentum $\hbar\mathbf{k}$, the energy $E(\mathbf{k})$, and the velocity $\mathbf{v}_{\mathbf{k}}$ of each quasi-particle, and since the quasi-particles are known to obey Fermi–Dirac statistics, it is possible to derive expressions for many of the measurable macroscopic properties of a metal. By comparison of these expressions with the expressions for the same quantities obtained in the independent-particle approximation (see Chapter 3) it is possible to study the effect of electron–electron interactions. However, when the theory was introduced by Landau (1956, 1957, 1958) the detailed form of the interaction function $f(\mathbf{k}, \boldsymbol{\sigma}; \mathbf{k}', \boldsymbol{\sigma}')$ was unknown, so that one had to try to use what little experimental information was available for the purpose of determining some of its features. Recently a start has been made on using standard many-body techniques to determine the interaction function *ab initio* (see Rice 1968). It is useful to expand the interaction function $f(\mathbf{k}, \boldsymbol{\sigma}; \mathbf{k}', \boldsymbol{\sigma}')$ in terms of some convenient set of basis functions and the expansion that is conventionally used is in terms of Legendre polynomials in the angle $\theta_{\mathbf{kk}'}$ between the wave vectors $\mathbf{k}$ and $\mathbf{k}'$. In the absence of an external magnetic field and in the absence of spin–orbit coupling, $f(\mathbf{k}, \boldsymbol{\sigma}; \mathbf{k}', \boldsymbol{\sigma}')$ can be separated into singlet and triplet contributions so that

$$f(\mathbf{k}, \boldsymbol{\sigma}; \mathbf{k}', \boldsymbol{\sigma}') = \sum_{l=0}^{\infty} (f_l^s + f_l^a \boldsymbol{\sigma} \cdot \boldsymbol{\sigma}') P_l(\cos \theta_{\mathbf{kk}'}) \tag{6.3.7}$$

and it is conventional to introduce the coefficients

$$A_l = \frac{m^* k_{\mathrm{F}}}{\pi^2(2l+1)} f_l^s \tag{6.3.8}$$

and

$$B_l = \frac{m^* k_{\mathrm{F}}}{\pi^2(2l+1)} f_l^a \tag{6.3.9}$$

as the unknown parameters in the interaction function $f(\mathbf{k}, \boldsymbol{\sigma}; \mathbf{k}', \boldsymbol{\sigma}')$. For the conduction electrons in a given metal the Fermi-liquid coefficients A_l and B_l can either be regarded as adjustable parameters that have to be determined experimentally, or else attempts can be made to calculate them using many-body techniques, the details of which are beyond the scope of this book. Some calculated values of these coefficients for Na are given in Table 6.2 where both the Coulomb electrostatic repulsion and the attraction by the exchange of virtual phonons are included; the various different theoretical estimates, I – V, of the coefficients are based on various different

TABLE 6.2

The experimental values and five different sets of theoretical values for the Landau parameters in Na

	Experiment	Theory (I)	(II)	(III)	(IV)	(V)
A_0		−0·62	−0·64	−0·66	−0·45	−0·17
A_1		+0·12	+0·11	+0·10	+0·04	+0·03
A_2	−0·05±0·01	−0·03	−0·04	−0·03	−0·01	+0·006
A_3	0·0±0·005	+0·004	+0·005			
B_0	−0·18±0·03	−0·14	−0·17	−0·22	−0·17	−0·17
B_1	+0·05±0·04	+0·01	−0·005	−0·02	−0·02	+0·03
B_2	0·0±0·05	−0·01	−0·02	0·00	+0·01	+0·006
B_3		+0·000	+0·001			
m^*/m	1·24±0·02	1·26	1·21	1·15	1·19	1·17

(Rice 1968).

combinations of treatments of these two contributions to the electron–electron interactions. In principle, measurements of any physical property to which there is a contribution arising from electron–electron interactions should yield some information about the coefficients A_l and B_l in the expansion of the interaction function $f(\mathbf{k}, \boldsymbol{\sigma}; \mathbf{k}', \boldsymbol{\sigma}')$; however, in some cases, such as the electronic specific heat (see Section 3.2) or the electrical resistivity (see Section 6.6) the electron–phonon interactions are much more important than the electron–electron interactions. The observation of high-frequency plasma waves and of spin-wave-like excitations can be used in the determination of the experimental values of the coefficients A_l and B_l (Walsh and Platzman 1965; Platzman and Walsh 1967; Platzman and Wolff 1967; Schultz and Dunifer 1967; Dunifer, Schultz, and Schmidt 1968; Platzman, Walsh, and Foo 1968).

6.4. Phonon dispersion relations

The student's first encounter with any attempt at a quantitative discussion of the vibrations of the atoms in a solid is usually in connection with the specific heat. As the temperature is raised so the amplitude, and hence the energy, of the vibrations in the solid increases and the specific heat is defined as the derivative, with respect to temperature, of the internal energy of the solid. At high temperatures the molar specific heat of a solid is approximately constant and takes the value of $3R$, where R is the gas constant; this is the well-known law of *Dulong and Petit*. However, at low temperatures departures from Dulong and Petit's law occur, see Fig. 6.4. The first more

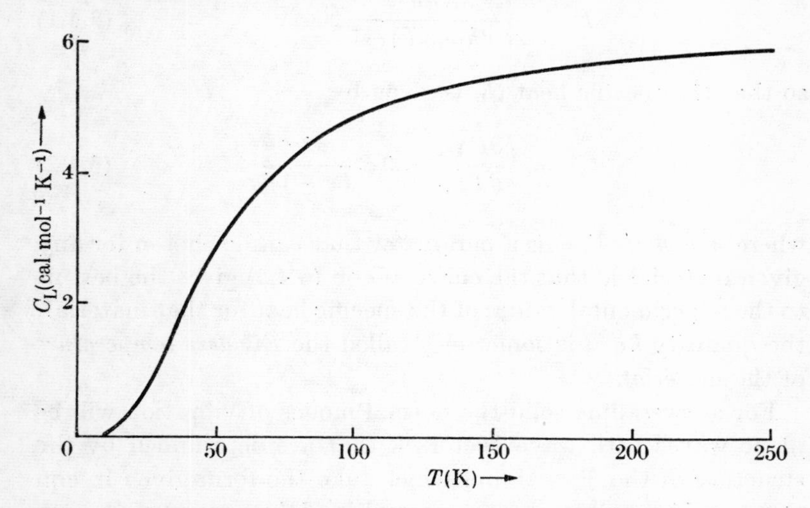

FIG. 6.4. The specific heat of Ag illustrating the departure from the law of Dulong and Petit at low temperatures.

or less successful attempt to explain the form of the curve of the specific heat as a function of temperature was due to Einstein (1907). Since each of the N atoms in a solid has three vibrational degrees of freedom, there are $3N$ vibrational degrees of freedom for the whole specimen. These vibrations can be resolved into $3N$ normal modes so that it is possible to consider the system as a collection of $3N$ harmonic oscillators, each normal mode corresponding to one of these oscillators. In the absence, at that time, of any experimental or theoretical indication as to the frequencies of all these $3N$ normal modes Einstein made the simplest possible assumption, namely that the frequencies of all the oscillators take the same value $\omega/2\pi$. The quantum of energy which is associated with each of these oscillators, and which is called a *phonon*, is then equal to $\hbar\omega$, so that the number of phonons that one expects to be present for each oscillator at a temperature T is given by the Bose-Einstein distribution function $1/\{\exp(\hbar\omega/kT)-1\}$. Since the normal modes are non-interacting, the total energy E is given by

$$E = \frac{3N\hbar\omega}{\exp(\hbar\omega/kT)-1}, \tag{6.4.1}$$

so that the specific heat C_V is given by

$$C_V = \left(\frac{\partial E}{\partial T}\right)_V = 3R\,\frac{x^2e^x}{(e^x-1)^2} \tag{6.4.2}$$

where $x = \hbar\omega/kT$. ω is a parameter that can be chosen for any given material so that the curve of eqn (6.4.2) gives the best fit to the experimental values of the specific heat for that material; the quantity $\hbar\omega/k$ is sometimes called the *Einstein temperature* of the material.

For a crystalline solid the normal modes of vibration will be plane waves with wave vectors **k** that are determined by the structure of the crystal and which take the form given in eqn (1.2.12). These allowed wave vectors **k** of the normal modes are exactly the same as the allowed wave vectors that we have been using previously in connection with electrons. That is, although so far we have always used Bloch's theorem in connection with the eigenfunctions of the Hamiltonian that describes the behaviour of the conduction electrons in a metal, the theorem also applies to a Hamiltonian that describes the lattice vibrations of a crystalline solid or, indeed, to a Hamiltonian that describes any quasi-particle excitation in a crystalline solid. Therefore the wave vector **k** as described in Sections 1.2 and 1.3 is a useful quantity for labelling the normal modes of vibration of a crystalline solid. Some authors use the symbol **q** for the wave vector of a phonon and use the symbol **k** only for the wave vector of an electron. In spite of the reasonably good agreement of eqn (6.4.2) with experimental results the assumption made by Einstein that the $3N$ normal modes of a specimen of a solid all have the same frequency is clearly unrealistic. The next improvement in the understanding of the vibrations of a solid is due to Debye (1912). Instead of assuming that all the normal modes have the same frequency, Debye regarded the normal modes as standing waves in a homogeneous elastic medium, the behaviour of which was completely specified by the moduli of elasticity of the crystal as measured macroscopically; the fact that a solid is not, on an atomic scale, a continuous medium was ignored. This means that the velocity v of each wave of a given polarization is determined by the elastic moduli of the crystal and, in particular, is independent of the wavelength, so that $v = \omega/|\mathbf{k}|$ or

$$\omega = v\,|\mathbf{k}|, \tag{6.4.3}$$

and the normal modes are regarded as stationary sound waves in the medium. After the Einstein theory in which ω was assumed to be a constant, this is the next simplest possible dispersion relation for the phonons in the lattice, namely a straight line dependence on $|\mathbf{k}|$. The other important approximation that is made in Debye's theory is in the method used to count the normal modes. Instead of using eqn (1.2.12) as the definition of $\mathbf{k}$, based on the microscopic structure of the crystal, $\mathbf{k}$ was also defined macroscopically. If the specimen is in the form of a cube and we apply Born–von Kármán periodic boundary conditions to it, the allowed values of k_x, k_y, and k_z, the three components of the wave vector $\mathbf{k}$ must each be an integer multiple of $2\pi/L$. In the vector space of $\mathbf{k}$ the allowed wave vectors $\mathbf{k}$ terminate on the points of a simple cubic lattice with lattice spacing equal to $2\pi/L$, so that the volume associated with each allowed wave vector $\mathbf{k}$ is $(2\pi/L)^3$. Since there are three possible polarizations for each wave vector $\mathbf{k}$, it was assumed that the $3N$ normal modes of the crystal can be assigned to the N wave vectors contained within a sphere of radius k_D where

$$\tfrac{4}{3}\pi k_\mathrm{D}^3 = N\left(\frac{2\pi}{L}\right)^3 \tag{6.4.4}$$

so that

$$k_\mathrm{D} = \frac{(6N\pi^2)^{\frac{1}{3}}}{L}. \tag{6.4.5}$$

It is this replacing of the proper Brillouin zone by a sphere of radius k_D that constitutes the second major approximation in the Debye treatment.

For a long time little was known about the forces that exist between the atoms in a solid, and consequently the Debye treatment has been used very extensively. It was only with the introduction of inelastic neutron scattering techniques that detailed studies of the lattice vibration spectra of materials were made on a large scale. It then became important to consider the theory of the lattice vibrations of a crystalline solid from a microscopic point of view. The motion of the nuclei in a crystal will affect the motion of the electrons and *vice versa*. To handle simultaneously the motions of both the nuclei and the electrons would be an extremely difficult problem. However, it is possible to separate the treatment of the nuclear vibrations from the behaviour of the electrons. The electrons are very much lighter than the nuclei and therefore they respond very quickly to any changes in the distribution of charge due to the motions of the nuclei. This leads to the *adiabatic approximation*, or the *Born–Oppenheimer approximation*, in which one assumes that at any time the electrons move as if the nuclei were frozen in their instantaneous positions. The nuclear vibrations can then be separated from the motions of the electrons, and the phonons can be treated as reasonably independent entities. We have found it convenient to separate the electrons in a metal into core electrons and conduction electrons. The core electrons, being strongly bound to the nucleus, will move with the nucleus as it vibrates so that we can regard each complete positive ion as vibrating, rather than just each nucleus. There are several difficulties in the way of calculating the lattice vibration frequencies of a crystal even in Newtonian mechanics, without introducing quantum mechanics at all. The restoring forces that act on a displaced atom in a crystal are not known exactly. Even if these forces were known exactly there would still be the problem of solving $3rN$ coupled equations of motion, where N is the number of unit cells in the crystal and r is the number of atoms per unit cell. These equations are of course simultaneous second-order differential equations. The decoupling of this set of equations is achieved by replacing the vibrations of the individual nuclei by the normal modes of vibration of the crystal.

The situation in which the restoring forces experienced by a nucleus or ion in a solid involve only expressions that are linear in the displacements of the atoms is called the *harmonic approximation*; this approximation leads to a system of non-interacting phonons. The one-dimensional problem of the vibrations of a linear chain of identical masses separated by identical springs obeying Hooke's law is considered in numerous textbooks. The study of the normal modes of vibration of a monatomic crystal in the harmonic approximation consists of the adaptation of this problem to three dimensions (Born and Huang 1954; Cochran 1965; Cochran and Cowley 1967; Maradudin and Vosko 1968). Suppose that $u_\alpha(l\kappa)$ denotes the α Cartesian component of the displacement of the atom labelled by κ in the unit cell labelled by l. The potential energy, in the harmonic approximation, can be written as

$$\Phi = \Phi_0+\tfrac{1}{2}\sum_{\alpha\beta}\sum_{ll'}\sum_{\kappa\kappa'}\Phi_{\alpha\beta}(l\kappa;l'\kappa')u_\alpha(l\kappa)u_\beta(l'\kappa') \qquad (6.4.6)$$

where $\Phi_{\alpha\beta}(l\kappa;l'\kappa') = \partial^2\Phi/(\partial x_\alpha(l\kappa)\partial x_\beta(l'\kappa'))$ evaluated at the equilibrium positions. The $\Phi_{\alpha\beta}(l\kappa;l'\kappa')$ are then the force constants between the atoms specified by l, κ and l', κ'. κ (or κ') runs from 1 to r, the number of atoms in the fundamental unit cell of the crystal, and l (or l') runs from 1 to N, the number of unit cells in the crystal. The $x_\alpha(l\kappa)$ are the Cartesian components of the vector $\mathbf{x}(l\kappa)$ which specifies the position of the atom specified by l and κ; $\mathbf{r}(l\kappa)$, the vector which specifies the equilibrium position of this atom, can be written as

$$\mathbf{r}(l\kappa) = \mathbf{r}(l)+\mathbf{r}(\kappa), \qquad (6.4.7)$$

where $\mathbf{r}(l)$ specifies the origin of the unit cell labelled by l. The notation that is used in connection with eqn (6.4.6) is somewhat cumbersome but we use it because it is the conventional notation used in work on lattice dynamics. It is often convenient to introduce the *dynamical matrix* $D_{\alpha\beta}(l\kappa;l'\kappa')$ which is related to $\Phi_{\alpha\beta}(l\kappa;\, l'\kappa')$ by

$$D_{\alpha\beta}(l\kappa;\, l'\kappa') = \frac{1}{\sqrt{(M_\kappa M_{\kappa'})}}\,\Phi_{\alpha\beta}(l\kappa;\, l'\kappa') \qquad (6.4.8)$$

where M_κ and $M_{\kappa'}$ are the masses of the atoms κ and κ' respectively. The force in the α direction on the l, κ atom is given by

$$-\{\partial\Phi/\partial u_\alpha(l\kappa)\} = -\sum_{l'\kappa'\beta} \Phi_{\alpha\beta}(l\kappa;\, l'\kappa')u_\beta(l'\kappa'),$$

so that the corresponding equation of motion is

$$M_\kappa \ddot{u}_\alpha(l\kappa) = -\sum_{l'\kappa'\beta} \Phi_{\alpha\beta}(l\kappa;\, l'\kappa')u_\beta(l'\kappa'). \qquad (6.4.9)$$

The mass M_κ can be removed from these equations by using the reduced displacements $w_\alpha(l\kappa)$ given by

$$w_\alpha(l\kappa) = (\sqrt{M_\kappa})u_\alpha(l\kappa) \qquad (6.4.10)$$

when eqn (6.4.9) becomes

$$\ddot{w}_\alpha(l\kappa) = -\sum_{l'\kappa'\beta} D_{\alpha\beta}(l\kappa;\, l'\kappa')w_\beta(l'\kappa'). \qquad (6.4.11)$$

We therefore have in either eqn (6.4.9) or eqn (6.4.11) $3rN$ coupled equations of motion. The standard technique which can be used in solving these equations involves making a transformation to a set of normal coordinates. As in the one-dimensional case the displacements of the atoms in one of these normal modes corresponds to wave-like displacements of the atoms so that if the system is vibrating in one of its normal modes we can write

$$w_\alpha(l\kappa) = \varepsilon_\alpha(\kappa \mid \mathbf{k}j)\exp[\mathrm{i}\{\mathbf{k}\,.\,\mathbf{r}(l\kappa) - \omega_j(\mathbf{k})t\}] \qquad (6.4.12)$$

where j runs from 1 to $3r$ and is used to distinguish between the $3r$ normal modes for each $\mathbf{k}$, and $\omega_j(\mathbf{k})$ is the frequency of the normal mode. For a vibration in the normal mode with frequency $\omega_j(\mathbf{k})$ we can substitute from eqn (6.4.12) into eqn (6.4.11), when we obtain

$$\omega_j(\mathbf{k})^2\epsilon_\alpha(\kappa \mid \mathbf{k}j) = \sum_{\beta\kappa'} D_{\alpha\beta}(\kappa\kappa' \mid \mathbf{k})\epsilon_\beta(\kappa' \mid \mathbf{k}j) \qquad (6.4.13)$$

where

$$D_{\alpha\beta}(\kappa\kappa' \mid \mathbf{k}) = \sum_{l'} D_{\alpha\beta}(l\kappa;\, l'\kappa')\exp[\mathrm{i}\mathbf{k}\,.\,\{\mathbf{r}(l'\kappa') - \mathbf{r}(l\kappa)\}] \qquad (6.4.14)$$

and $D_{\alpha\beta}(\kappa\kappa' \mid \mathbf{k})$ is called the *Fourier-transformed dynamical matrix*. For each value of $\mathbf{k}$ the set of $3r$ simultaneous eqns (6.4.14) only have non-trivial solutions for certain frequencies $\omega_j(\mathbf{k})$ for which

$$\det |\omega_j(\mathbf{k})^2 \delta_{\kappa\kappa'} \delta_{\alpha\beta} - D_{\alpha\beta}(\kappa\kappa' \mid \mathbf{k})| = 0. \quad (6.4.15)$$

There will then be $3r$ values of $\omega_j(\mathbf{k})$ which satisfy eqn (6.4.15) and once the force constants $\Phi_{\alpha\beta}(l\kappa; l'\kappa')$ are given, the determination of the $3r$ frequencies $\omega_j(\mathbf{k})$ for each value of $\mathbf{k}$ is simply a mechanical process.

The dispersion relations of $\omega_j(\mathbf{k})$ as a function of $\mathbf{k}$, for the normal modes of the lattice vibrations, which are commonly called the *phonon dispersion relations*, are obtained by the solution of eqn (6.4.15). For any real crystal there is always a set of three, possibly degenerate, branches of the phonon dispersion relations for which $\omega_j(\mathbf{k})$ takes the form similar to eqn (6.4.3)

$$\omega_j(\mathbf{k}) = v_j \, |\mathbf{k}| \quad (6.4.16)$$

where v_j is some constant. That is, for these three branches the lattice vibrations are just the same as stationary sound waves in the crystal; consequently these three branches of the dispersion relations are referred to as the *acoustic modes* of vibration of the crystal. The remaining $(3r-3)$ branches are referred to as the optic modes. Eqn (6.4.16) is of the same form as eqn (6.4.3) which was used in the Debye theory. Although eqn (6.4.3) describes only some of the branches of the phonon dispersion relations, and even these branches only near $\mathbf{k} = 0$, nevertheless the Debye theory of the specific heat of a solid met with considerable success because it is the acoustic modes near $\mathbf{k} = 0$ which make the main contributions to the specific heat of a solid.

The quantities $\epsilon_\alpha(\kappa \mid \mathbf{k}j)$ which are the eigenvectors of the Fourier-transformed dynamical matrix specify the relative amplitudes of the displacements of the various atoms in the unit cell of the crystal for the normal mode specified by j. The normal modes of a crystal are often labelled group-theoretically in a manner similar to that outlined for electronic energy bands in Section 1.4. It is possible to identify the space-group representation to which a given normal mode belongs by studying the transformation of the displacements under the symmetry operations of the space group of the crystal (see Maradudin and Vosko 1968; Warren 1968).

The problem of determining the frequencies of the $3r$ normal modes for each wave vector $\mathbf{k}$ in the harmonic approximation has been reduced formally to the determination of the eigenvalues of the Fourier-transformed dynamical matrix $D_{\alpha\beta}(\kappa\kappa' \mid \mathbf{k})$. However, the determination *ab initio* of the force constants $\Phi_{\alpha\beta}(l\kappa; l'\kappa')$ for any real crystal is a far from trivial undertaking, so that the aim of calculating the phonon dispersion relations of a real crystal from first principles is difficult to realize. In practice most of our knowledge of the detailed form of the frequencies of the normal modes of vibration of a crystal has come from direct experimental measurements of the dispersion relations, $\omega_j(\mathbf{k})$ versus $\mathbf{k}$, from the inelastic scattering of either X-rays or neutrons. It is then possible to regard the force constants as parameters that can be adjusted until the best fit to the observed dispersion relations is obtained.

We now give a more formal approach to the problem of determining the eigenfunctions and eigenvalues of the Hamiltonian that describes the lattice vibrations of a crystal. Suppose that $p_\alpha(l\kappa)$ denotes the α Cartesian component of the momentum of the atom, or nucleus, with displacement $u_\alpha(l\kappa)$ from its equilibrium position. Then using the potential in eqn (6.4.6) and neglecting the constant term Φ_0 we can write the Hamiltonian that describes the vibrations of the crystal in the harmonic approximation, neglecting electronic excitations, as

$$\mathscr{H} = \sum_{\alpha,l,\kappa} \frac{p_\alpha(l\kappa)^2}{2M_\kappa} + \\ + \tfrac{1}{2} \sum_{\alpha\beta} \sum_{ll'} \sum_{\kappa\kappa'} \Phi_{\alpha\beta}(l\kappa; l'\kappa') u_\alpha(l\kappa) u_\beta(l'\kappa'). \qquad (6.4.17)$$

In the quantum mechanical treatment of a single one-dimensional harmonic oscillator with operators x and p for the position and momentum coordinates, it is often convenient to work in terms of the boson creation and annihilation operators a^+ and a defined by

$$\begin{aligned} a^+ &= (2m\hbar\omega)^{-\frac{1}{2}}(p+im\omega x) \\ a &= (2m\hbar\omega)^{-\frac{1}{2}}(p-im\omega x), \end{aligned} \tag{6.4.18}$$

where m is the mass of the particle and ω is the frequency of the oscillations. The Hamiltonian in eqn (6.4.17) describes a system of coupled harmonic oscillators and we have already seen that by a transformation to the normal coordinates of the system the vibrations of a crystal can be considered in terms of a set of non-interacting, or de-coupled, harmonic oscillators with frequencies $\omega_j(\mathbf{k})$. It is therefore possible to introduce a pair of operators corresponding to a^+ and a for each of the normal modes of vibration of the crystal. To identify these operators we proceed as follows. The displacement $u_\alpha(l\kappa)$ for a general vibration of the system can be expanded in terms of the displacements of the $l\kappa$ atom in the normal modes so that, by analogy with eqn (6.4.12) and ignoring the time dependence, we may write

$$u_\alpha(l\kappa) = (NM_\kappa)^{-\frac{1}{2}} \sum_j \sum_{\mathbf{k}} \epsilon_\alpha(\kappa \mid \mathbf{k}j) Q_j(\mathbf{k}) \exp\{-i\mathbf{k} \,.\, \mathbf{r}(l\kappa)\} \tag{6.4.19}$$

where $Q_j(\mathbf{k})$ is a coefficient that expresses the extent to which the mode j, $\mathbf{k}$ participates in the vibration. If this expression is substituted into the potential energy term in the Hamiltonian in eqn (6.4.17) we obtain

$$\begin{aligned} &\frac{1}{2N} \sum_{\alpha\beta} \sum_{ll'} \sum_{\kappa\kappa'} D_{\alpha\beta}(l\kappa;\, l'\kappa') \times \\ &\quad \times \left[\sum_j \sum_{\mathbf{k}} \epsilon_\alpha(\kappa \mid \mathbf{k}j) Q_j(\mathbf{k}) \exp\{-i\mathbf{k} \,.\mathbf{r}(l\kappa)\} \right] \times \\ &\quad \times \left[\sum_{j'} \sum_{\mathbf{k}'} \epsilon_\beta(\kappa \mid \mathbf{k}'j') Q_{j'}(\mathbf{k}') \exp\{-i\mathbf{k}' \,.\, \mathbf{r}(l'\kappa')\} \right] \\ &\qquad = \tfrac{1}{2} \sum_j \sum_{\mathbf{k}} \omega_j(\mathbf{k})^2 Q_j(\mathbf{k}) Q_j(-\mathbf{k}). \end{aligned} \tag{6.4.20}$$

where we have used eqns (6.4.13) and (6.4.14) and the orthogonality of the normal modes. The form of this expression suggests that we should also expand the momenta $p_\alpha(l\kappa)$ with an expression similar to that given in eqn (6.4.19). We therefore write

$$p_\alpha(l\kappa) = (M_\kappa/N)^{\frac{1}{2}}\sum_j\sum_{\mathbf{k}} \pi_\alpha(\kappa \mid \mathbf{k}j)P_j(\mathbf{k})\exp\{-i\mathbf{k}\,.\,\mathbf{r}(l\kappa)\}. \tag{6.4.21}$$

The Hamiltonian in eqn (6.4.17) therefore becomes

$$\mathscr{H} = \tfrac{1}{2}\sum_j\sum_{\mathbf{k}}\{P_j(\mathbf{k})P_j(-\mathbf{k})+\omega_j(\mathbf{k})^2Q_j(\mathbf{k})Q_j(-\mathbf{k})\}. \tag{6.4.22}$$

This equation gives the Hamiltonian $\mathscr{H}$ in terms of the normal coordinates of the system and it can be regarded as the Hamiltonian of a system of non-interacting, or de-coupled, harmonic oscillators, with each oscillator characterized by a particular set of values of j and $\mathbf{k}$. We can define a pair of creation and annihilation operators $a^+_{j\mathbf{k}}$ and $a_{j\mathbf{k}}$ for each normal mode by

$$\left.\begin{aligned} a^+_{j\mathbf{k}} &= \{2\hbar\omega_j(\mathbf{k})\}^{-\frac{1}{2}}\{\omega_j(\mathbf{k})Q_j(-\mathbf{k})-iP_j(\mathbf{k})\} \\ a_{j\mathbf{k}} &= \{2\hbar\omega_j(\mathbf{k})\}^{-\frac{1}{2}}\{\omega_j(\mathbf{k})Q_j(\mathbf{k})+iP_j(-\mathbf{k})\}\end{aligned}\right\} \tag{6.4.23}$$

which are similar to the operators a^+ and a defined for a single harmonic oscillator in eqn (6.4.18). Equation (6.4.22) can be re-written in terms of the operators $a^+_{j\mathbf{k}}$ and $a_{j\mathbf{k}}$:

$$\mathscr{H} = \sum_j\sum_{\mathbf{k}}(a^+_{j\mathbf{k}}a_{j\mathbf{k}}+\tfrac{1}{2})\hbar\omega_j(\mathbf{k}). \tag{6.4.24}$$

The commutation rules for the operators $a^+_{j\mathbf{k}}$ and $a_{j\mathbf{k}}$ can be determined by using eqns (6.4.19), (6.4.21), and (6.4.23) and the known commutation rules for the operators $p_\alpha(l\kappa)$ and $u_\alpha(l\kappa)$.

Whether one prefers to think in terms of a classical equation of motion such as that given in eqn (6.4.9), or in terms of the rather more sophisticated quantum-mechanical formulation of the problem in terms of the operators $a^+_{j\mathbf{k}}$ and $a_{j\mathbf{k}}$, the problem of the calculation of the phonon frequencies $\omega_j(\mathbf{k})$ for a given crystal reduces to solving the same determinantal or secular equation (6.4.15). The more sophisticated quantum-mechanical argument that we have just outlined takes us no farther than we were before in relation to the really difficult part of the calculation of $\omega_j(\mathbf{k})$, namely the determination of the force constants $\Phi_{\alpha\beta}(l\kappa; l'\kappa')$. It is not our intention to enter into any lengthy description of the possible methods that can be used in the determination of these force constants and we shall just indicate the general features of what is involved (for further details see, for example, Cochran 1965; Cochran and Cowley 1967; Donovan and Angress 1971). The earliest attempts to determine *ab initio* values of the force constants were made by trying to relate them to the observed values of the macroscopic elastic constants of the crystal in question. It is then necessary to restrict the range of the interactions that are included because there is only a (fairly small) finite number of independent elastic constants of a crystal. However, it is a fairly general feature that to obtain a realistic set of phonon dispersion relations for a crystal it is necessary to use the values of many more of the force constants $\Phi_{\alpha\beta}(l\kappa; l'\kappa')$ than it would be possible to determine from the values of the macroscopic elastic constants of the crystal. One could regard all these force constants as adjustable parameters that have to be determined by fitting the calculated dispersion relations to the results of inelastic neutron scattering measurements. At the opposite extreme, if the wave functions of all the atoms in a solid were known it would be possible to calculate the force constants $\Phi_{\alpha\beta}(l\kappa; l'\kappa')$ *ab initio* and, indeed, it would even be possible to include terms that go beyond the harmonic approximation—that is to include *anharmonic interactions*. However, as we mentioned in Section 6.1, at present there is insufficient information available about the wave functions of atoms in solids to make this a practical approach.

It is possible to avoid both of these extremes by making some use of the knowledge of the physical nature of the forces that hold a given crystal together. The conventional treatment of these forces varies according to the crystal in question: whether it be an inert-gas solid, an ionic solid, a covalent solid, a molecular crystal or a metal. For an inert-gas solid it is common to assume that the force between two atoms in the solid is given by the usual semi-empirical expression for van der Waals forces involving powers of the distance between the two atoms. For an ionic crystal the principal contributions to the force constants can be separated into (i) the electrostatic interactions between the charged ions in the crystal which can be regarded as moving point charges, and (ii) repulsive forces that arise due to the distortions of the charge clouds of the outer electrons on each ion as the ions move relative to each other. The electrostatic forces are quite long-range forces but the repulsive forces arise principally between nearest-neighbour ions in the crystal; the repulsive force between two ions at $\mathbf{x}(l\kappa)$ and $\mathbf{x}(l'\kappa')$ is often represented by a term proportional to $\exp(-|\mathbf{x}(l\kappa)-\mathbf{x}(l'\kappa')|/\rho)$ in the potential. When the Coulomb forces are regarded as being between point charges we have the *rigid-ion model* which has been used in calculations of the phonon dispersion relations of a number of alkali halides (Kellermann 1940; Iona 1941; Karo 1959, 1960). However, the rigid-ion model is not entirely satisfactory and better calculated phonon dispersion relations for an ionic crystal can be obtained by taking into account the distortions of the ions when evaluating the Coulomb contributions to the force constants; this is done, for example, in the *shell model* (Dick and Overhauser 1958; Hanlon and Lawson 1959; Woods, Cochran, and Brockhouse 1960). Like an ionic solid, a metal too contains a collection of positive and negative charges. However, as we have noted in Section 6.2, the masses of the electrons are very small compared with the masses of the positive ions. The electrons can therefore move very much faster and so they very quickly readjust themselves to any changes in the positions of the positive ions. It might be a reasonable first approximation to consider the repulsive electrostatic force between a pair of positive ions as a force between a pair of point charges, as in the rigid-ion model for ionic crystals. However, to determine the electrostatic force either between a positive ion and a conduction electron or between two positive ions screened by conduction electrons, it is necessary to know the positions of the conduction electrons and it is not adequate to regard the conduction electrons as a uniform distribution of negative charge. Moreover, to determine the positions of the conduction electrons it is necessary to know the wave functions of the conduction electrons which it is, of course, part of the object of band structure calculations to determine. Accurate wave functions for the conduction electrons in a metal are notoriously difficult to determine from a band structure calculation because they are more sensitive than the energies $E_j(\mathbf{k})$ to the choice of the crystal potential in the metal. However, we have seen in earlier chapters that, at least for many simple metals, the use of a suitable pseudopotential gives a very good approximation to the band structure of a metal. For many metals the parameters in a model pseudopotential have been determined by fitting the calculated band structures and Fermi surfaces to the Fermi surfaces that have been determined experimentally for those metals. Since it is possible to express in terms of the pseudopotential those contributions to the force constants involving the conduction electrons, it is possible to use such an empirically determined model pseudopotential to calculate the phonon frequencies $\omega_j(\mathbf{k})$ for any given $\mathbf{k}$ and hence to determine the complete phonon dispersion relations. Further details of the principles involved in such calculations are given, for example, by Harrison (1966, 1970). Pseudopotential calculations of the phonon dispersion relations have now been performed for a number of the simple metals (Vosko, Taylor, and Keech 1965; Animalu, Bonsignori, and Bortolani 1966; Schneider and Stoll 1966*a*,*b*, 1967*a*; Ho and Ruoff 1967).

In view of the difficulties that arise when one tries to calculate the force constants and hence the phonon frequencies it is not surprising that a considerable amount of effort has been expended in connection with the experimental determination of the phonon dispersion relations in solids. The energies of the phonons that can be excited in a crystal can be determined experimentally by studying the inelastic interactions of the crystal with electromagnetic radiation or with beams of particles. By far the most successful experimental technique, which can in principle be applied all over the Brillouin zone and which has been used for a large number of crystals, is by the inelastic scattering of neutrons. However, there are other methods which should not be overlooked: infra-red absorption, Raman scattering, and X-ray scattering. In all these methods one makes use of the laws of conservation of energy and of momentum where the energy of a phonon is $\hbar\omega_j(\mathbf{k})$ and the momentum of the phonon is $\hbar\mathbf{k}$. Long wavelength (that is, small $\mathbf{k}$) acoustic phonon modes can also be studied by means of ultrasonic techniques; a review of such measurements has been given by Huntington (1958). An extensive discussion of the importance of the microscopic description of lattice vibrations in obtaining a detailed understanding of many of the thermal properties of a solid is given by Wallace (1972).

A great deal of information on phonon dispersion curves has been obtained from neutron and X-ray scattering experiments (for example Bacon 1962; Donovan and Angress 1971). Slow thermal neutrons are particularly useful since they have wavelengths of the order of 1 Å, which is comparable to the interatomic spacing. Hence large momentum transfers can easily be observed. Moreover, the kinetic energy of the thermal neutrons is about the same as the phonon energies (0·01–0·1 eV) so that any interactions between the neutrons and the phonons can easily be detected. The energy of an X-ray, however, is of the order of 10 keV so that the energy transferred to the phonons would be quite a negligible fraction of the X-ray photon energy.

In a solid a slow neutron can be scattered elastically, during which process no energy is transferred to the nuclei. Such scattering is therefore not of interest to us in connection with the determination of the phonon dispersion relations of a solid, although information about the structure of a crystal can be determined from elastic neutron scattering measurements. In the inelastic scattering of a neutron the nuclei receive energy from the neutron and transfer it to the crystal vibrations (creation of a phonon) or transfer energy from the crystal vibrations to the neutron (annihilation of a phonon). The total scattering cross section σ can be expressed as a sum of two terms

$$\sigma = \sigma_{\text{coh}} + \sigma_{\text{incoh}}. \tag{6.4.25}$$

σ_{coh} is the cross section for scattered neutrons which interfere with each other coherently and σ_{incoh} is the cross section for incoherent scattering. Non-zero nuclear spins and the presence of different isotopes of an element contribute to σ_{incoh}. It is the coherent (and inelastic) scattering which is of interest in the present connection.

In an inelastic collision the laws of the conservation of energy and of momentum must be satisfied, where of course it must be remembered that the momentum of a phonon is indeterminate up to the addition of a term $\hbar\mathbf{G}_n$ where $\mathbf{G}_n$ is any reciprocal lattice vector of the crystal. If $\hbar\mathbf{k}_1$ and $\hbar\mathbf{k}_2$ are the momenta of a neutron before and after inelastic scattering by a crystal then the conservation of momentum requires that

$$\hbar\mathbf{k}_1 - \hbar\mathbf{k}_2 = \sum_i \hbar\mathbf{k}_i, \tag{6.4.26}$$

where the $\mathbf{k}_i$ are the wave vectors of the phonons created in the crystal by the scattering event. If E_1 and E_2 are the energies of the incident and scattered neutron then the conservation of energy requires that

$$E_1 - E_2 = \sum_i \hbar\omega_j(\mathbf{k}_i), \tag{6.4.27}$$

where $\omega_j(\mathbf{k}_i)$ are the frequencies of the phonons created in the crystal. Fortunately scattering processes involving one phonon are very much more probable than those involving more than one so that eqns (6.4.26) and (6.4.27) simplify to

$$\hbar\mathbf{k}_1 - \hbar\mathbf{k}_2 = \hbar\mathbf{k}_i \tag{6.4.28}$$

and

$$E_1 - E_2 = \hbar\omega_j(\mathbf{k}_i). \tag{6.4.29}$$

Phonon spectra can therefore be measured by performing neutron scattering experiments and studying the one-phonon coherent cross section (for further details of the theory see, for example, Marshall and Lovesey 1971). In these experiments a beam of monochromatic neutrons of known wavelength and energy is allowed to fall on a crystal at a defined angle. The wavelength and energy of the neutrons after they have been scattered through a known angle are then measured. From these measurements and the conservation eqns (6.4.28) and (6.4.29), points on the dispersion curve $\omega_j(\mathbf{k})$ can be determined. The wave vector $\mathbf{k}_i$ determined from eqn (6.4.28)

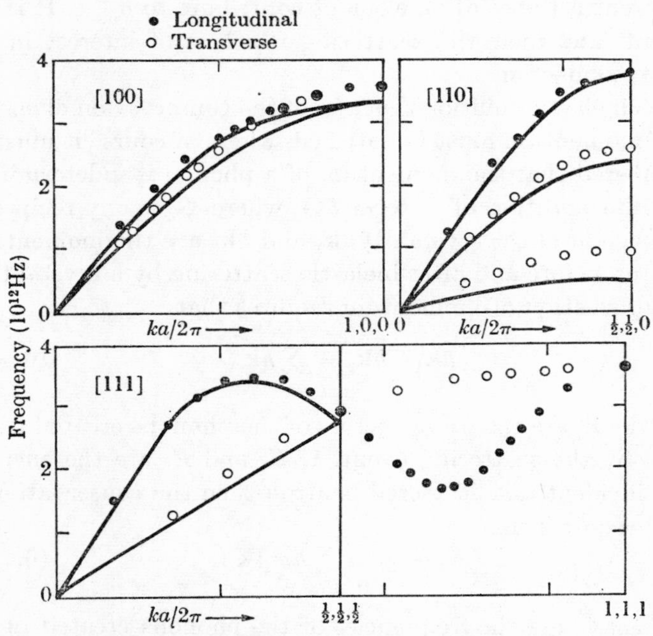

FIG. 6.5. Phonon dispersion relations for Na, showing experimental values of Woods, Brockhouse, March, and Bowers (1962) (dots and open circles) and the results of the calculations of Toya (1958) (continuous curves)

may lie outside the first Brillouin zone but it can always be brought within the first Brillouin zone by the addition or subtraction of some reciprocal lattice vector $\mathbf{G}_n$. The phonon dispersion relations have now been determined for many metals using inelastic neutron scattering and we give just two examples: the phonon dispersion curves for Na and K are shown in Figs. 6.5 and 6.6.

Although inelastic neutron scattering is a very versatile

technique for the study of phonon spectra, it has certain disadvantages. In the first place, the equipment needed to perform the experiments is very bulky and very expensive and must of necessity be situated at a nuclear reactor. Large specimens (about 50 cc) have to be used, because of the weakness of the interaction between the thermal neutrons and the

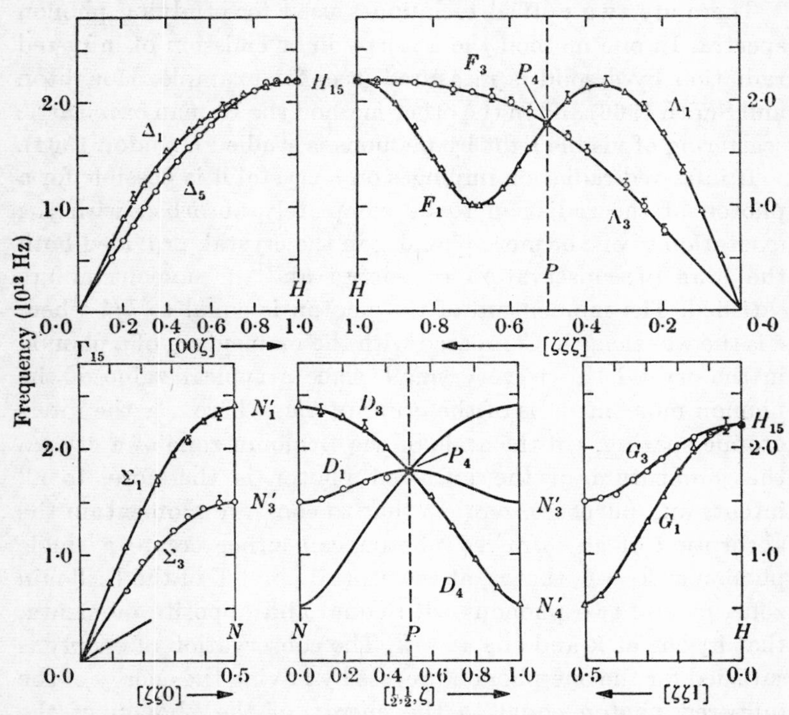

FIG. 6.6. The phonon dispersion relations for K as determined with the aid of inelastic neutron scattering (Cowley, Woods, and Dolling 1966).

solid. It therefore takes a long time to perform the experiments and the equipment must be automated and very reliable. Since the object of the experiment is to determine the coherent scattering cross-section, the neutron absorption cross-sections and the incoherent scattering cross-sections must be comparatively small. However, inelastic neutron scattering experiments do yield $\omega_j(\mathbf{k})$ throughout the Brillouin zone whereas some of the other methods yield less complete results. Since X-ray photon energies are much greater than the phonon energies, these cannot be measured directly. The phonon spectra are obtained indirectly through intensity measurements (Cochran 1965). These experiments are very difficult, although good results have been obtained by Walker (1956) for Al.

There are two optical techniques used for studying phonon spectra. In one method the absorption or emission of infra-red radiation by a solid is measured (see, for example, Houghton and Smith 1966) and in the other method the Raman or inelastic scattering of visible light by phonons is studied (Loudon 1964).

If infra-red radiation impinges on a crystal it is possible for a photon of the radiation to be completely absorbed with the excitation of one or more phonons in the crystal, provided both the laws of conservation of energy and of momentum are satisfied. The momentum of the photon is equal to h/λ where λ is the wavelength. Compared with the momentum of a phonon in the crystal this is very small, since a typical value of the phonon momentum is of the order of h/a, where a is the inter-atomic spacing. On the scale of the Brillouin zone of a crystal the momentum of the infra-red photon is therefore, to all intents and purposes, zero, so that to conserve momentum the absorption of an infra-red photon can either create a single phonon at $\mathbf{k} = 0$, that is, at the central point Γ of the Brillouin zone, or else two phonons with equal and opposite momenta, that is, one at $\mathbf{k}$ and one at $-\mathbf{k}$. The conservation of energy is satisfied for the one-phonon process by having the energy of the infra-red photon equal to the energy of the phonon of the lattice vibration that is excited at Γ, that is,

$$\hbar\omega_0 = \hbar\omega_1(0). \tag{6.4.30}$$

For the two-phonon process the energy of the infra-red photon is equal to the total energy of the two phonons produced, so that

$$\hbar\omega_0 = \hbar\omega_1(\mathbf{k})+\hbar\omega_2(-\mathbf{k}). \tag{6.4.31}$$

For many crystals the phonon spectrum at $-\mathbf{k}$ is the same as that at **k**, so that equation (6.4.31) simplifies to

$$\hbar\omega_0 = \hbar\omega_1(\mathbf{k}) + \hbar\omega_2(\mathbf{k}). \tag{6.4.32}$$

If one used a source of infra-red radiation containing a continuous spread of frequencies one might therefore expect to find heavy absorption of the infra-red radiation for the whole range of phonon frequencies in the crystal; however, there are very strong selection rules which mean that the excitation of very many of the phonons that are allowed by the conditions on the conservation of energy and momentum is actually forbidden by the selection rules. For example, it is possible to show that the two-phonon process can *only* occur for wave vectors **k** that correspond to special points of symmetry in the Brillouin zone; for all other wave vectors in the Brillouin zone the two-phonon process is forbidden. Even at the special points of symmetry there may be further selection rules. In principle, it would also be possible for a single photon to be completely absorbed with the creation of more than two phonons, although the probability of such a process actually occurring would be so small that the phenomenon would be unobservable in practice. It is possible to remove the restriction to $\mathbf{k} = 0$ phonons in one-phonon absorption by the introduction of defects or impurities. Although this does not enable the complete dispersion relations to be determined, because values of **k** are not measured, it does give a rather direct method for studying the density of states. Thus, to a certain extent, infra-red absorption and neutron scattering are complementary techniques for the study of lattice vibrations in a crystal.

The second method which is similar to the infra-red absorption and which also only gives information about the phonon energies at the special points of symmetry in the Brillouin zone is by Raman scattering. The difference between the Raman effect and the infra-red absorption that we have just described is that in the Raman effect only part of the energy of the incident photon is absorbed and used to create one or more phonons in the crystal; the remaining energy then re-emerges as a photon of lower energy, that is of longer wavelength. Another possibility is that the incident photon may acquire extra energy as a result of the destruction of one or more phonons, so that the photon that emerges has more energy, that is shorter wavelength, than the incident photon. With the invention of the laser there has been a considerable re-awakening of experimental interest in Raman spectroscopy. As in the case of infra-red absorption the conditions of conservation of energy and of momentum must be satisfied. If ω_1 and ω_2 are the frequencies of the incident and emerging photons we have for the conservation of energy condition

$$\hbar\omega_1 = \hbar\omega_2 \pm \sum_i \hbar\omega_j(\mathbf{k}_i), \tag{6.4.33}$$

where $\omega_j(\mathbf{k}_i)$ and $\mathbf{k}_i$ are the frequencies and wave vectors of the phonons created or annihilated in the process. Since the momentum of each of the photons is negligible compared with the momentum of a typical phonon in the crystal, the conservation of momentum condition is

$$\sum_i \hbar\mathbf{k}_i = 0. \tag{6.4.34}$$

If only one phonon is created or annihilated eqn (6.4.34) will only be satisfied at $\mathbf{k} = 0$, that is at the centre point, Γ, of the Brillouin zone, so that eqn (6.4.33) leads to

$$\hbar\omega_1 = \hbar\omega_2 \pm \hbar\omega_1(0), \tag{6.4.35}$$

which is analogous to eqn (6.4.30) where the + sign corresponds to the creation of a phonon and the − sign to the annihilation of a phonon. If two phonons are produced eqn (6.4.34) requires that if one phonon is at $\mathbf{k}$ the other is at $-\mathbf{k}$ so that

$$\hbar\omega_1 = \hbar\omega_2 + \hbar\omega_1(\mathbf{k}) + \hbar\omega_2(-\mathbf{k}). \tag{6.4.36}$$

There are in fact some selection rules which show that this two-phonon Raman scattering will only occur at wave vectors $\mathbf{k}$ which correspond to special points of symmetry in the Brillouin zone. If a phonon is annihilated rather than created the sign attached to the appropriate phonon frequency in eqn

(6.4.36) will be reversed. The phonon spectrum at $-\mathbf{k}$ will be the same as the phonon spectrum at $+\mathbf{k}$, unless the crystal is of very low symmetry, so that eqn (6.4.36) simplifies to

$$\hbar\omega_1 = \hbar\omega_2 + \hbar\omega_1(\mathbf{k}) + \hbar\omega_2(\mathbf{k}). \tag{6.4.37}$$

Again in principle it should be possible for Raman scattering involving three, or more, phonons to occur if one considers only the requirements of the conservation of energy and the conservation of momentum. However, in practice, the probability that such a process involving more than two phonons will occur is so small as to make the process unobservable experimentally. For the process of Raman scattering involving phonons at the various points of symmetry in the Brillouin zone there will be a set of selection rules that are similar, but not identical, to the selection rules for straightforward infra-red absorption or emission.

6.5. Electron-phonon interactions

The proper quantum-mechanical treatment of the physics of a pure crystalline specimen of a metal would be a very complicated many-body problem involving all the nuclei and all the electrons in the metal. For each atom of atomic number Z in the metal it will be necessary to specify $(Z+1)$ pairs of position and momentum coordinates, one pair for the nucleus and one pair for each of the electrons. We can therefore formally write the Hamiltonian for the system of nuclei and electrons constituting a metal as

$$\mathscr{H} = \sum_{\alpha}\sum_{l,\kappa}\frac{p_\alpha(l\kappa)^2}{2M_\kappa} + \frac{1}{2}\sum_{l,\kappa}\sum_{\substack{l',\kappa' \\ (l,\kappa \neq l',\kappa')}}\frac{(Ze)^2}{|\mathbf{x}(l\kappa)-\mathbf{x}(l'\kappa')|} +$$

$$+\sum_{i}\frac{\mathbf{p}_i^2}{2m} + \frac{1}{2}\sum_{i}\sum_{\substack{j \\ i\neq j}}\frac{e^2}{|\mathbf{r}_i-\mathbf{r}_j|} - \sum_{l,\kappa}\sum_{i}\frac{Ze^2}{|\mathbf{x}(l\kappa)-\mathbf{r}_i|} \tag{6.5.1}$$

where $\mathbf{p}(l\kappa)$ and $\mathbf{x}(l\kappa)$ denote the momentum and position coordinates of the nuclei, $\mathbf{p}_i$ and $\mathbf{r}_i$ represent the momentum and position coordinates of the ith electron, and the summations over i and j are over all the electrons in the crystal. Although any given nucleus is moving it is not free to move throughout the entire crystal but is confined to a region in the vicinity of some fixed point in the crystal. We have found it convenient to separate the electrons in a metal into two categories, the localized ion-core electrons and the highly mobile conduction electrons. Although for certain metals there may be some disputation as to exactly which electrons are conduction electrons and which electrons are ion-core electrons, this division is generally very useful and we shall retain it. As in the previous section we can therefore regard the core electrons as moving together with the appropriate nuclei, so that instead of having to specify $(Z+1)$ sets of position and momentum coordinates for each atom in the metal we now only have to specify $(s+1)$ sets of coordinates, one set for the ion core and one set for each of the s conduction electrons contributed to the metal by this atom. Thus for metallic La, with $Z = 57$ and $s = 3$, we have reduced the number of sets of coordinates that need to be specified from $58N$ to $4N$, where N is the number of atoms in the sample under consideration. The symbol Z in eqn (6.5.1) can therefore be replaced by s. Although we have reduced the number of coordinates in the Hamiltonian in eqn (6.5.1) this does not really achieve any significant simplification of our many-body problem because it is still not feasible to determine the eigenvalues and eigenfunctions of this Hamiltonian directly.

So far in this book we have always made some kind of simplifying assumptions before attempting to find the eigenvalues and eigenfunctions of the Hamiltonian given in eqn (6.5.1). It is possible to re-arrange eqn (6.5.1) so as to give a little more insight into the approximations that we have used previously and into the interactions between the lattice vibrations and the translational motions of the electrons. We can use the fact that the displacements of the ions are small and expand the last term on the right-hand side of eqn (6.5.1) as a Taylor series about the equilibrium configuration of the ions.

Rewriting this term as $\sum_{i}\sum_{l,\kappa} V(\mathbf{r}_i-\mathbf{x}(l\kappa))$ and expanding about $\mathbf{r}(l\kappa)$ we obtain

$$-\sum_{l,\kappa}\sum_{i}\frac{se^2}{|\mathbf{x}(l\kappa)-\mathbf{r}_i|}=\sum_{l,\kappa}\sum_{i}V(\mathbf{r}_i-\mathbf{x}(l\kappa))$$

$$=\sum_{l,\kappa}\sum_{i}V(\mathbf{r}_i-\mathbf{r}(l\kappa))+\sum_{l,\kappa}\sum_{i}\mathbf{u}(l\kappa)\,.\,\nabla V(\mathbf{r}_i-\mathbf{r}(l\kappa)) \quad (6.5.2)$$

where $\mathbf{u}(l\kappa)$ is the displacement of the l, κ ion in the notation of the previous section. Equation (6.5.1) can therefore be rewritten as

$$\mathscr{H}=\sum_{\alpha}\sum_{l,\kappa}\frac{p_\alpha(l\kappa)^2}{2M_\kappa}+\tfrac{1}{2}\sum_{\substack{l,\kappa\\(l,\kappa\neq l',\kappa')}}\sum_{l',\kappa'}\frac{(se)^2}{|\mathbf{x}(l\kappa)-\mathbf{x}(l'\kappa')|}+$$

$$+\sum_{i}\frac{\mathbf{p}_i^2}{2m}+\tfrac{1}{2}\sum_{\substack{i\\(i\neq j)}}\sum_{j}\frac{e^2}{|\mathbf{r}_i-\mathbf{r}_j|}+\sum_{l,\kappa}\sum_{i}V(\mathbf{r}_i-\mathbf{r}(l\kappa))+$$

$$+\sum_{l,\kappa}\sum_{i}\mathbf{u}(l\kappa)\,.\,\nabla V(\mathbf{r}_i-\mathbf{r}(l\kappa)). \quad (6.5.3)$$

We now collect the terms on the right-hand side of eqn (6.5.3) into three sets

$$\mathscr{H}_{\mathrm{e}}=\sum_{i}\frac{\mathbf{p}_i^2}{2m}+\tfrac{1}{2}\sum_{\substack{i\\i\neq j}}\sum_{j}\frac{e^2}{|\mathbf{r}_i-\mathbf{r}_j|}+\sum_{l,\kappa}\sum_{i}V(\mathbf{r}_i-\mathbf{r}(l\kappa)) \quad (6.5.4)$$

$$\mathscr{H}_{\mathrm{n}}=\sum_{\alpha}\sum_{l,}\frac{p_\alpha(l\kappa)^2}{2M_\kappa}+\tfrac{1}{2}\sum_{\substack{l,\kappa\\(l,\kappa\neq l',\kappa')}}\sum_{l',\kappa'}\frac{(se)^2}{|\mathbf{x}(l\kappa)-\mathbf{x}(l'\kappa')|} \quad (6.5.5)$$

$$\mathscr{H}_{\mathrm{I}}=\sum_{l,\kappa}\sum_{i}\mathbf{u}(l\kappa)\,.\,\nabla V(\mathbf{r}_i-\mathbf{r}(l\kappa)) \quad (6.5.6)$$

so that

$$\mathscr{H}=\mathscr{H}_{\mathrm{e}}+\mathscr{H}_{\mathrm{n}}+\mathscr{H}_{\mathrm{I}}. \quad (6.5.7)$$

The expression $\mathscr{H}_{\mathrm{e}}$ involves only the position and momentum coordinates of the conduction electrons and the equilibrium positions, $\mathbf{r}(l\kappa)$, of the ion cores. $\mathscr{H}_{\mathrm{e}}$ is just a sum over i of one-electron operators and is the Hamiltonian which was used as the basis of the various methods described in Chapter 2 for the calculation of non-relativistic band structures. The expression $\mathscr{H}_{\mathrm{n}}$ involves only the position and momentum coordinates of the ion cores and does not involve the conduction-electron coordinates at all; it includes terms involving the kinetic energy of the vibrations of the ion cores and the potential energy associated with the electrostatic forces between all possible pairs of ion cores. The expression $\mathscr{H}_{\mathrm{I}}$ involves the displacements of the ion cores from their equilibrium positions as well as the coordinates of the conduction electrons. $\mathscr{H}_{\mathrm{I}}$ therefore describes the interactions between the conduction electrons and the ion cores. Since we have found it convenient to study the vibrations of the ion cores in terms of phonons, which we can regard as quasi-particles, the expression $\mathscr{H}_{\mathrm{I}}$ can be regarded as describing the interaction between the electrons and the phonons. If it were not for the presence of $\mathbf{r}_i$, the position coordinates of the conduction electrons in $\mathscr{H}_{\mathrm{I}}$, it would be possible to achieve a separation of variables in the eigenvalue equation based on

$$\mathscr{H} = (\mathscr{H}_{\mathrm{e}})+(\mathscr{H}_{\mathrm{n}}+\mathscr{H}_{\mathrm{I}}) \tag{6.5.8}$$

into one equation involving only the coordinates of the electrons and one equation involving only the coordinates of the ion cores. The fact that such a separation is not possible shows that the procedures adopted in Chapter 2 and in Section 6.4 are only approximations. If we make the assumption that the ion cores are permanently frozen in their equilibrium positions then $\mathbf{p}(l\kappa)$ and $\mathbf{u}(l\kappa)$ are zero so that $\mathscr{H}_{\mathrm{n}}$ reduces to a constant:

$$\tfrac{1}{2}\sum_{l,\kappa}\sum_{\substack{l',\kappa' \\ (l,\kappa \neq l',\kappa')}} (se)^2/|\mathbf{r}(l\kappa)-\mathbf{r}(l'\kappa')|,$$

and $\mathscr{H}_{\mathrm{I}}$ vanishes so that, apart from a constant term the total Hamiltonian $\mathscr{H}$ of the metal reduces to $\mathscr{H}_{\mathrm{e}}$, which is the Hamiltonian used in the non-relativistic band structure calculations described in Chapter 2. Once the band structure has been calculated the wave functions of the conduction electrons can be used to determine the probability distribution of the $\mathbf{r}_i$ to substitute into $\mathscr{H}_{\mathrm{I}}$ so that $(\mathscr{H}_{\mathrm{n}}+\mathscr{H}_{\mathrm{I}})$ becomes a function only of the coordinates of the ion cores. It is then possible to solve the problem of determining the eigenstates of $(\mathscr{H}_{\text{n}}+\mathscr{H}_{\text{I}})$; this is the approach adopted in Section 6.4 in connection with calculating the lattice vibration frequencies of a metal.

We have chosen to study the vibrations of the lattice of a metal in terms of eigenstates of the Hamiltonian in eqn (6.4.17) which we have seen can be classified by wave vectors $\mathbf{k}$ and which can be occupied by the quasi-particles which are known as phonons. It is therefore instructive to consider the electron-ion interaction Hamiltonian $\mathscr{H}_{\text{I}}$ to see how it is related to these eigenstates and to the eigenstates of the Hamiltonian, $\mathscr{H}_{\text{e}}$, for the conduction electrons. $\mathscr{H}_{\text{I}}$ has been written as

$$\mathscr{H}_{\text{I}} = \sum_{l,\kappa} \sum_{i} \mathbf{u}(l\kappa) \,.\, \nabla V(\mathbf{r}_i - \mathbf{r}(l\kappa)). \tag{6.5.6}$$

We can make use of eqns (6.4.19) and (6.4.23) to express the components of $\mathbf{u}(l\kappa)$ in terms of the phonon creation and annihilation operators $a_{j\mathbf{k}}^{+}$ and $a_{j\mathbf{k}}$

$$u_{\alpha}(l\kappa) = \sum_{j}\sum_{\mathbf{k}} \left\{\frac{2\hbar}{NM_{\kappa}\omega_j(\mathbf{k})}\right\}^{\frac{1}{2}} (a_{j\mathbf{k}}+a_{j-\mathbf{k}}^{+}) \times$$
$$\times\, \epsilon_{\alpha}(\kappa \mid \mathbf{k}j) \exp\{-i\mathbf{k} \,.\, \mathbf{r}(l\kappa)\}. \tag{6.5.9}$$

For brevity we shall write

$$q_j(\mathbf{k}) = \left\{\frac{2\hbar}{NM_{\kappa}\omega_j(\mathbf{k})}\right\}^{\frac{1}{2}} (a_{j\mathbf{k}}+a_{j-\mathbf{k}}^{+}). \tag{6.5.10}$$

We can see from the form of the $\mathbf{u}(l\kappa)$ factor in the operator $\mathscr{H}_{\text{I}}$ that the effect of $\mathscr{H}_{\text{I}}$ on a given eigenstate of the lattice-vibration problem will be either to create or to destroy a phonon in that state. The other factor, $\nabla V(\mathbf{r}_i - \mathbf{r}(l\kappa))$, can be regarded as acting on the electronic states. After some manipulation, with which we do not wish to obscure the present discussion (for details see Schrieffer 1964; Schultz 1964) it is possible to express $\mathscr{H}_{\text{I}}$ as

$$\mathscr{H}_{\text{I}} = \tfrac{1}{2}N \sum_{j}\sum_{\mathbf{k}}\sum_{\mathbf{k}_1}\sum_{\mathbf{k}_2} q_{j\mathbf{k}} \mathscr{I}_{j\mathbf{k}}(\mathbf{k}_2, \mathbf{k}_1) c_{\mathbf{k}_2}^{+} c_{\mathbf{k}_1} \tag{6.5.11}$$

where $c^{+}_{\mathbf{k}_2}$ creates an electron in the state with wave function $\psi_{\mathbf{k}_2}(\mathbf{r})$, and $c_{\mathbf{k}_1}$ destroys an electron in the state with wave function $\psi_{\mathbf{k}_1}(\mathbf{r})$, and $\mathscr{I}_{j\mathbf{k}}(\mathbf{k}_2, \mathbf{k}_1)$ is a coupling constant of the form given by

$$\mathscr{I}_{j\mathbf{k}}(\mathbf{k}_2, \mathbf{k}_1) = \epsilon_\alpha(\kappa \mid \mathbf{k}j)\int \psi^*_{\mathbf{k}_2}(\mathbf{r})\nabla V(\mathbf{r})\psi_{\mathbf{k}_1}(\mathbf{r})\, d\mathbf{r}. \quad (6.5.12)$$

The coupling constant $\mathscr{I}_{j\mathbf{k}}(\mathbf{k}_2, \mathbf{k}_1)$ indicates the strength of the scattering process involving the scattering of an electron from state $\psi_{\mathbf{k}_1}(\mathbf{r})$ to state $\psi_{\mathbf{k}_2}(\mathbf{r})$ with the creation or annihilation of a phonon of frequency $\omega_j(\mathbf{k})$. Each set of values of $\mathbf{k}$, $\mathbf{k}_1$, and $\mathbf{k}_2$ in eqn (6.5.11) must satisfy the conservation of momentum, so that

$$\hbar\mathbf{k}_1 = \hbar\mathbf{k}_2 \pm \hbar\mathbf{k} + \hbar\mathbf{G}_n \quad (6.5.13)$$

where $\mathbf{G}_n$ is a reciprocal lattice vector defined by eqn (1.2.11). If eqn (6.5.13) is satisfied by $\mathbf{G}_n = 0$ we have a normal process and if $\mathbf{G}_n \neq 0$ we have an *umklapp* process. However, there is no unique distinction between normal processes and umklapp processes as their definitions depend on the choice of the first Brillouin zone, which is not unique.

In Chapter 2 we have described various methods of calculating the electronic band structure of a metal assuming that the positions of the ion cores are known and in Section 6.4 we have described the calculation of the frequencies of the normal modes of vibration assuming that the spatial distribution of the conduction electrons is already known. In Section 6.4 we have seen that in calculating the phonon dispersion relations in a metal, realistic values of the force constants $\Phi_{\alpha\beta}(l\kappa; l'\kappa')$ between the ions or nuclei can only be determined theoretically if one has a detailed knowledge of the spatial distributions of the conduction electrons in the metal; this means that one requires to know the wave functions, or at least some pseudo-wave functions, of the conduction electrons. In its turn this means that some calculation of the band structure of the metal must have been performed in order to determine these wave functions or pseudo-wave functions. In calculating the band structure and the wave functions for the conduction electrons in a metal using any of the methods described in Chapter 2 it is necessary to assume that the positions of all the nuclei are known. Since these nuclei are vibrating, the crystal potential $V(\mathbf{r})$ varies with time and therefore the band structure and the conduction-electron wave functions will also vary with time. If one makes use of the adiabatic (or Born–Oppenheimer) approximation and assumes that the nuclei are frozen in their instantaneous positions, one can calculate the band structure and the conduction-electron wave functions at one given time. At a later time the positions of the nuclei will have changed so that one will need to calculate the band structure again. Thus we have a vicious circle or, as it is often called, a problem in self-consistency; to calculate the frequencies of the normal modes of vibration we need to know the wave functions of the conduction electrons, but these wave functions depend on the positions of the nuclei which, of course, are vibrating about their equilibrium positions in a manner which is, as yet, undetermined. The approach which we have adopted in Chapter 2 and in Section 6.4 was to calculate "time-averaged" values of the band structure assuming that the nuclei are permanently frozen in their equilibrium positions.

We have already considered one aspect of electron–phonon interactions in Section 6.4. In that section we noted that it was only possible to obtain a quantitatively accurate description of the lattice vibrations of a metal if one used force constants that were evaluated by including a realistic treatment of the spatial distributions of the conduction electrons. We now have to consider the effect of the vibrations of the ion cores on the electronic band structure and, consequently, on the Fermi surface of the metal. At least at the rather low temperatures that are used in most direct measurements of the Fermi surface of a metal, it seems to be the case that the effect of the motions of the ion cores on the band structure is rather insignificant. That this gives a good first approximation to the true description of the behaviour of the conduction electrons in a metal is justified by all the wealth of the experimental results described in Chapters 3–5 which it has been possible to interpret successfully in terms of band structures and Fermi surfaces calculated on this basis. It is perhaps a little surprising that the effect of the lattice vibrations on the energy eigenvalues $E_j(\mathbf{k})$ should be relatively insignificant, when the effect of the spatial distribution of the conduction electrons on the phonon energies $\hbar\omega_j(\mathbf{k})$ was rather important. However, it has to be remembered that from the point of view of altering the potential used in the calculation of the band structure of a metal it is not so much the magnitudes of the frequencies $\omega_j(\mathbf{k})$ or even the variations of these frequencies with $\mathbf{k}$ that are important, but rather it is the amplitudes of the vibrations of the ions that are significant. Because of the large masses of the nuclei, the amplitudes of the vibrations of the ion cores are very small compared with the distances between neighbouring ions in the metal. Moreover, the amplitudes of the vibrations also depend on the number of phonons present in each normal mode, which is determined by the Bose–Einstein distribution function. Consequently at very low temperatures the amplitudes of the vibrations of the ion cores will be particularly small. Therefore, since the conduction electrons in a metal spend most of their time at distances far removed from the nuclei where the crystal potential only varies relatively slowly with distance, the effect of the lattice vibrations on the crystal potential and therefore on the band structure is rather small.

Although the direct effect of electron–phonon interactions on the electron energy bands and on the shape of the Fermi surface is rather small there are other ways in which the electron–phonon interactions are important in connection with the behaviour of the conduction electrons in a metal. We have already had occasion to mention one example of this in Chapter 3 in connection with the electronic contribution to the specific heat of a metal. The separation of the specific heat of a non-magnetic metal in Section 3.2 into two terms

$$C = \alpha T^3 + \gamma T \qquad (3.2.6)$$

where $\gamma = \frac{1}{3}\pi^2 k^2 n(E_F)$ was based on the assumption that the total energy of a metal at temperature T could be obtained as a sum of the energies of all the occupied eigenstates of the electronic Hamiltonian $\mathscr{H}_{\mathrm{e}}$ and of the energies of all the phonons present in the metal at T. The contribution of the electron–phonon interactions, represented by $\mathscr{H}_{\mathrm{I}}$, to the total energy, and therefore also to the specific heat, was ignored. It is possible to show that the electron–phonon interactions also contribute another term to the low temperature specific heat that is also proportional to T (see Eliashberg 1962; Krebs 1963; Prange and Kadanoff 1964; Abrikosov, Gor'kov, and Dzyaloshinskiĭ 1965, pp. 188–9). If this additional term is written in the form $\lambda_S\{\frac{1}{3}\pi^2k^2n(E_{\mathrm{F}})\}T$ the whole linear term in the specific heat of the metal will be $(1+\lambda_S)\{\frac{1}{3}\pi^2k^2n(E_{\mathrm{F}})\}T$. If the linear term in the specific heat of a metal has been determined experimentally, and if $n(E_{\mathrm{F}})$ is known accurately, it is possible to determine the factor $(1+\lambda_S)$ which is sometimes referred to as the *thermal effective mass*. While λ_S may be quite small, less than 0·2 for Na, K, Rb, and Cs (see Table 3.3), it may become quite significant, say of the order of 1 or 2, for some of the heavier metals. It is also possible to calculate λ_S using the electron–phonon interaction Hamiltonian $\mathscr{H}_{\mathrm{I}}$ and the known, or assumed, wave functions or pseudo-wave functions for the conduction electrons; the values of λ_S given in Column 4 of Table 3.3 were the results of such calculations performed by Ashcroft 1965, Ashcroft and Wilkins 1965, and Ashcroft and Lawrence 1968. Table 3.3 suggests the existence of at least order-of-magnitude agreement between the calculated and measured values of λ_S, although it is difficult to make detailed comparisons because of the presence of the electron–electron contribution to the specific heat as well. Strictly speaking λ_S is not a constant but should be written as $\lambda_S(T)$, a function of temperature. The quantity $\lambda_S(T)$ has been calculated by Grimvall (1969) using the expressions given by Prange and Kadanoff (1964). The results are shown in Fig. 6.7. It can also be shown that there is a similar, but not identical, electron–phonon correction factor $(1+\lambda_c(T))$ for the cyclotron resonance

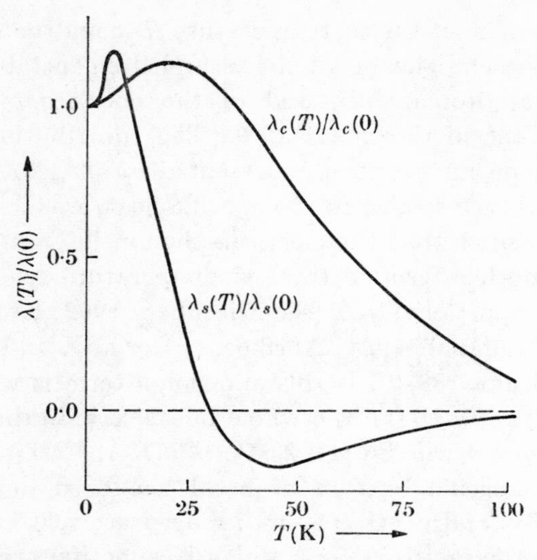

FIG. 6.7. Temperature dependence of the enhancement factors $\lambda_c(T)$ for the cyclotron mass and $\lambda_s(T)$ for the electronic specific heat in Pb (Grimvall 1969).

effective mass. Values of $\lambda_c(T)$, also calculated by Grimvall (1969) on the basis of the expressions given by Prange and Kadanoff (1964), are shown in Fig. 6.7.

Another aspect of the thermal properties of a metal that has received some attention from the theoretical point of view is the thermal expansion coefficient. The treatment of thermal expansion is naturally related to the problem of the cohesive energy and the crystal structure which we discussed in Section 6.1. In describing the calculation of the cohesive energy in Section 6.1, we assumed that the ion cores were at rest. For a given crystal structure the equilibrium values of the lattice constants could, in principle, be calculated by varying the lattice constants and determining the particular values for which the total energy is a minimum. If one relaxes the restriction that the ion cores are at rest, then it is possible to include the effect of the lattice vibrations and to calculate the equilibrium values of the lattice constants as a function of temperature and therefore to calculate the thermal expansion coefficient. However, it is very difficult to perform quantitative calculations on the basis of the above procedure; the formal theory is complicated (Leibfried and Ludwig 1961; Wallace 1970, 1972; Donovan and Angress 1971) and relatively few calculations have been attempted for real metals (Collins 1966; Fomin 1966; Gallina and Omini 1966; Wallace 1968; Srivastava and Singh 1970).

The corrections to the thermal and cyclotron effective masses which we have just mentioned are examples of many-body corrections to quantities that, in a first approximation, can be considered quite adequately by neglecting electron–phonon interactions completely. In this connection electron–phonon interactions are therefore relatively unimportant and are only of interest to the enthusiast or the perfectionist. However, there are many other properties in which it is necessary to consider electron–phonon interactions even in a first approximation to a quantitative explanation of these properties, indeed in some cases if it were not for the existence of electron–phonon interactions the property in question would not exist at all. The properties that we have in mind include all the transport properties of a metal and the phenomenon of superconductivity which occurs in many metals. In view of the importance of electron–phonon interactions in these connections it would be desirable to evaluate the coupling constants $\mathscr{I}_{j\mathbf{k}}(\mathbf{k}_2, \mathbf{k}_1)$; however, it can be seen from eqn (6.5.12) that this involves using several quantities which we have seen at various stages are notoriously difficult to determine accurately, namely the electronic wave functions, the crystal potential and the normal-mode eigenvectors. It is also difficult to imagine an experiment in which it would be possible to identify and measure the individual coupling constants $\mathscr{I}_{j\mathbf{k}}(\mathbf{k}_2, \mathbf{k}_1)$. In most experiments the measurable quantities depend on summations of these interactions over a large, and possibly infinite, number of sets of states $\mathbf{k}$, $\mathbf{k}_1$, and $\mathbf{k}_2$. All that one can then hope to do is to construct the coupling constants $\mathscr{I}_{j\mathbf{k}}(\mathbf{k}_2, \mathbf{k}_1)$, based on the use of eqn (6.5.12) and then use these values in the calculation of some macroscopic property of a metal; the answer can then be compared with the results of measurements of the property in question. The choice of assumptions that one makes about the various factors entering into the integrand on the right-hand side of eqn (6.5.12) is a matter of taste and expediency. After the comparison with experimental results these assumptions can be modified and the calculation repeated to see if better agreement can be obtained.

One of the most obvious ways of studying electron–phonon interactions in a metal experimentally is by pumping a beam of phonons, in the form of a sound wave or preferably an ultrasonic wave, into the metal and by studying the attenuation of the beam as it passes through the metal. We have, of course, had occasion in Chapter 3 to discuss the attenuation of ultrasonic waves in a metal in the presence of a magnetic field, but we are now concerned with the problem of the attenuation of ultrasonic waves in the absence of a magnetic field. The contribution of the conduction electrons to the attenuation of the ultrasonic wave can be obtained by determining the transition probability for the process illustrated in Fig. 6.8, namely the absorption of a phonon of wave vector $\mathbf{k}$ by a conduction electron in a state with wave function $\psi_{\mathbf{k}_1}(\mathbf{r})$ which is thereby scattered into a state with wave function $\psi_{\mathbf{k}_2}(\mathbf{r})$. The potential that is responsible for the scattering of the phonons is, of course, the electron–phonon interaction term, $\mathscr{H}_{\mathrm{I}}$, in the Hamiltonian of the metal. Using eqns (6.5.10) and (6.5.11) we can write $\mathscr{H}_{\mathrm{I}}$ as

$$\mathscr{H}_{\mathrm{I}} = \sum_{j}\sum_{\mathbf{k}}\sum_{\mathbf{k}_1}\sum_{\mathbf{k}_2} D_{j\mathbf{k}}(\mathbf{k}_2, \mathbf{k}_1)(a_{j\mathbf{k}}+a^{\dagger}_{j-\mathbf{k}})c^{\dagger}_{\mathbf{k}_2}c_{\mathbf{k}_1} \qquad (6.5.14)$$

where

$$D_{j\mathbf{k}}(\mathbf{k}_2, \mathbf{k}_1) = \left\{\frac{\hbar}{2NM_{\kappa}\omega_j(\mathbf{k})}\right\}^{\frac{1}{2}} \mathscr{I}_{j\mathbf{k}}(\mathbf{k}_2, \mathbf{k}_1). \qquad (6.5.15)$$

Inspection of eqn (6.5.14) reveals two things. First, the total scattering cross section for the ultrasonic wave with wave vector $\mathbf{k}$ involves a summation of contributions from the process illustrated in Fig. 6.8 for all possible occupied electronic states $\mathbf{k}_1$ and all possible empty electronic states $\mathbf{k}_2$, subject of course to the conservation of energy and the conservation of momentum. Secondly, we can see from eqn (6.5.14) that $\mathscr{H}_{\mathrm{I}}$

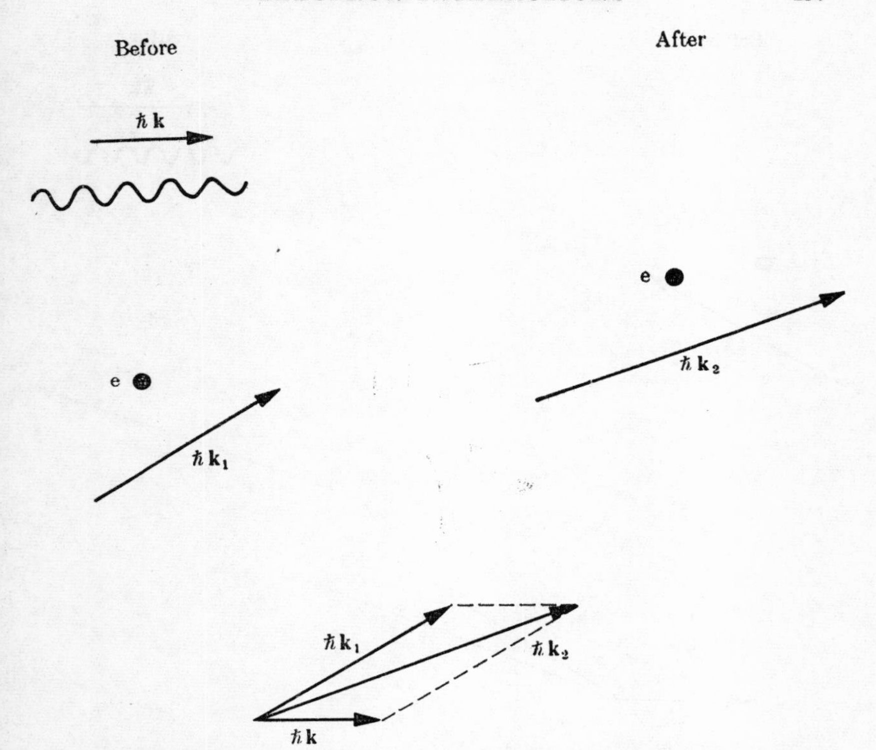

FIG. 6.8. Diagram representing the absorption of a phonon with wave vector **k** by an electron with wave vector $\mathbf{k}_1$.

includes a factor $(a_{j\mathbf{k}}+a^{+}_{j-\mathbf{k}})$. The operator $a_{j\mathbf{k}}$ involves the annihilation of a phonon of wave vector **k** as illustrated in Fig. 6.8. The presence of $a^{+}_{j-\mathbf{k}}$ indicates that in calculating the absorption coefficient we must also include scattering processes between electronic states $\mathbf{k}_1$ and $\mathbf{k}_2$ that involve the creation of a phonon as illustrated in Fig. 6.9. The fact that Fig. 6.9 creates a phonon with wave vector $-\mathbf{k}$ rather than **k** is still relevant to the absorption of the beam of phonons with wave vector **k**; having created phonons of $-\mathbf{k}$ they will now cancel with quanta of the propagating wave to produce quanta of energy of the stationary wave of wave vector **k** (or $-\mathbf{k}$), which is, of course, just one of the thermal vibrations of the solid. A quantum of energy of wave vector **k** can therefore be removed

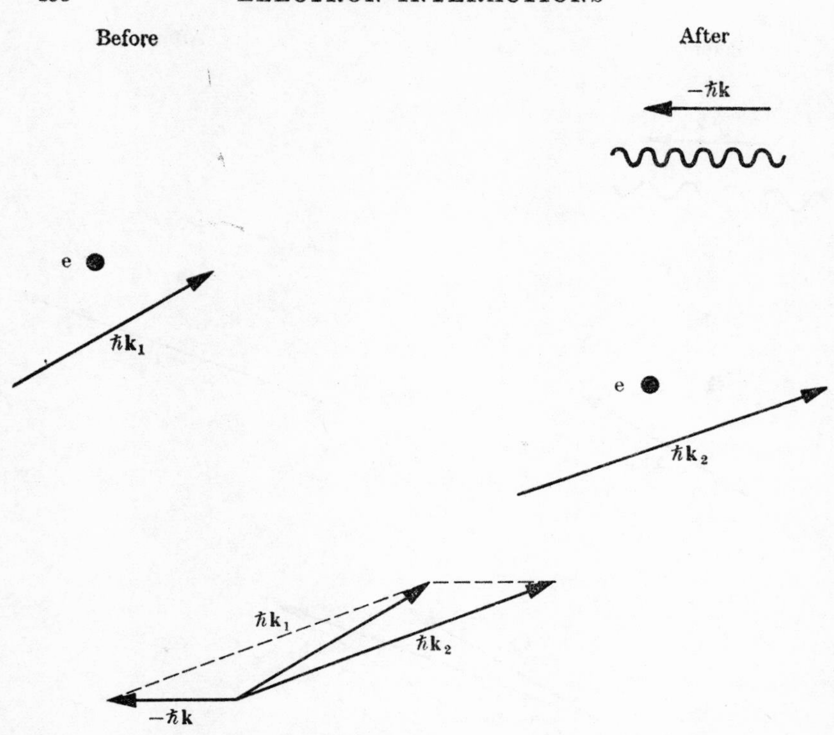

FIG. 6.9. Diagram representing the emission of a phonon with wave vector $-\mathbf{k}$ by an electron with wave vector $\mathbf{k}_1$.

from the ultrasonic beam by either of the processes illustrated in Figs. 6.8 and 6.9. If we separate $\mathscr{H}_\mathrm{I}$ into two parts, $\mathscr{H}_1$ and $\mathscr{H}_2$ involving the terms in $a_{j\mathbf{k}}$ and $a^+_{j-\mathbf{k}}$, respectively, the rate of absorption of phonons by the two processes will be given by W_1 and W_2 where

$$W_1 = \frac{2\pi}{\hbar}\,|\langle\psi_{1f}\,|\mathscr{H}_1|\,\psi_{1i}\rangle|^2 \tag{6.5.16}$$

and

$$W_2 = \frac{2\pi}{\hbar}\,|\langle\psi_{2f}\,|\mathscr{H}_2|\,\psi_{2i}\rangle|^2. \tag{6.5.17}$$

ψ_{1i} and ψ_{1f} are formal expressions for the wave functions for the initial and final states in the first process and ψ_{2i} and ψ_{2f} are similar expressions for the second process. Each wave function in these matrix elements is the total wave function of the whole system of both electrons and phonons.

The expressions for W_1 and W_2 in eqns (6.5.16) and (6.5.17) will involve summations over all possible sets of values of $\mathbf{k}$, $\mathbf{k}_1$, and $\mathbf{k}_2$. However, the summation over $\mathbf{k}$ is eliminated by using a single value of $\mathbf{k}$ in the ultrasonic wave (i.e. a single frequency) and the summations over $\mathbf{k}_1$ and $\mathbf{k}_2$ can be restricted drastically as a result of selection rules based on the conservation of energy and momentum. The values of $\mathbf{k}$ that can be achieved ultrasonically are generally quite small so that, remembering that we are dealing with acoustic phonons, the phonon energy is also quite small. Therefore, the conservation of energy will require that only electrons near the Fermi surface can participate in the scattering of the phonons. Moreover, the conservation of momentum (eqn (6.5.13)) will require that only pairs of $\mathbf{k}_1$ and $\mathbf{k}_2$ that differ by $\mathbf{k}$ will contribute to the attenuation. Therefore, by using a single value of $\mathbf{k}$ the attenuation will be due only to those electrons on a belt of the Fermi surface tangential to the direction of propagation of the ultrasonic waves. If one is prepared to make very drastic assumptions about the form of the coupling constants $\mathscr{I}_{j\mathbf{k}}(\mathbf{k}_2, \mathbf{k}_1)$ and about the crystal wave functions it is possible to evaluate analytically W_1 and W_2, and hence the attenuation coefficient $\alpha_{\mathbf{k}}$ (see, for example, Blount 1959; Morse 1959; Chambers 1960; Pippard 1960*b*; Berre and Olsen 1965; Fate 1968). However, our experience in connection with the electronic structures of real metals in Chapters 4 and 5 suggests that such analytical expressions are unlikely to be very meaningful except, perhaps, for the very simplest metals.

There are two other points that ought to be mentioned when considering ultrasonic attenuation and electron–phonon interactions. First, when an ultrasonic wave of wave vector $\mathbf{k}$ is propagating through a specimen, the number of phonons present with this particular wave vector is so large that the displacements of the lattice associated with the travelling wave of wave vector $\mathbf{k}$ are very large compared with the displacements that occur in the thermal vibrations of the crystal. Such large displacements mean that the local strain in the metal is large and this strain will cause changes in the eigenstates of the electronic Hamiltonian, $\mathscr{H}_e$, and corresponding changes in the shape of the Fermi surface. This strain is, of course, not homogeneous but varies from point to point in the crystal. The matrix elements that are involved in the evaluation of the ultrasonic attenuation coefficient therefore involve electronic states that are different from the electronic states of the metal in the absence of the ultrasonic wave (Pippard 1960*c*; Sham and Ziman 1963). Secondly, eqns (6.5.16) and (6.5.17) are based on the use of perturbation theory. This is only valid provided the energy levels of the particles or quasi-particles involved in the scattering process are well-defined, that is, they are not blurred out by the uncertainty principle. We have already mentioned this condition in Section 3.1 in connection with the electronic states. Rewriting eqn (3.1.3) in terms of the wave vector $\mathbf{k}_i$ $(i = 1, 2)$ and mean free path λ for the electrons we have

$$k_i \lambda \gg 2\pi, \tag{6.5.18}$$

which means that the mean free path of an electron must be greater than its wavelength. The phonon states must also be well-defined and one can argue (see Ziman 1960, p. 213) that therefore the condition in eqn (6.5.18) must hold for all the wave vectors involved in the scattering process, including the wave vector $\mathbf{k}$ of the phonon. That is, the wavelengths of the electrons and of the phonons must be less that λ the electronic mean free path if the perturbation treatment used to obtain eqns (6.5.16) and (6.5.17) is to be valid. Therefore, in addition to eqn (6.5.18) we have

$$k\lambda \gg 2\pi \tag{6.5.19}$$

where $\mathbf{k}$ is the phonon wave vector. Consider the orders of magnitude involved: for an ultrasonic wave of frequency 10 MHz and velocity $\sim$3,000 ms^{-1}, k will be of the order of 20,000 m^{-1} so that condition (6.5.19) will only be satisfied if the electronic mean free path is considerably in excess of

5×10^{-5} m, or 50 μm. The temperature of the metal must therefore be quite low to obtain such values of λ. If the condition $k\lambda \gg 2\pi$ is not satisfied, a more sophisticated treatment of the electron–phonon interactions will be needed.

Before passing on to a fairly general discussion of transport properties in the next section it is appropriate to devote a few words to the rather special phenomenon of superconductivity without which no discussion of electron–phonon interactions would be complete. There is now a vast literature on this subject and we shall not attempt to give any extensive discussion either of the phenomenological aspects of superconductivity or of the very considerable amount of theoretical work that has now been devoted to it. We shall simply mention one or two aspects that seem to be particularly closely related to the main theme of this book; for further details the interested reader should consult some of the numerous more specialized works on the subject (for example, Shoenberg 1952*b*; Lynton 1962; Blatt 1964; Schrieffer 1964; Rickayzen 1965; Kuper 1968; Galasiewicz 1970).

In Sections 6.2 and 6.3 we always discussed electron–electron interactions either from a phenomenological point of view or else in terms of the correlation of the motions of the electrons as a result of direct electrostatic forces of repulsion between the electrons. What we did not discuss was the possibility of indirect interactions between the electrons via the system of the ion cores. That is, an electron may collide with an ion core and lose some of its energy to the vibrations of the lattice and a second electron may subsequently also undergo a collision with one of the ion cores and gain an equal amount of energy from the lattice vibrations; this transfer of energy leads to an attractive or repulsive force between the two electrons. The situation is much more conveniently handled in terms of the normal modes of the vibrations of the ion cores, in which case we can regard the force as being associated with the emission of a phonon by the first electron and its subsequent re-absorption by the second electron, see Fig. 6.10. Two points should be noticed, first the emission of the phonon by the first electron

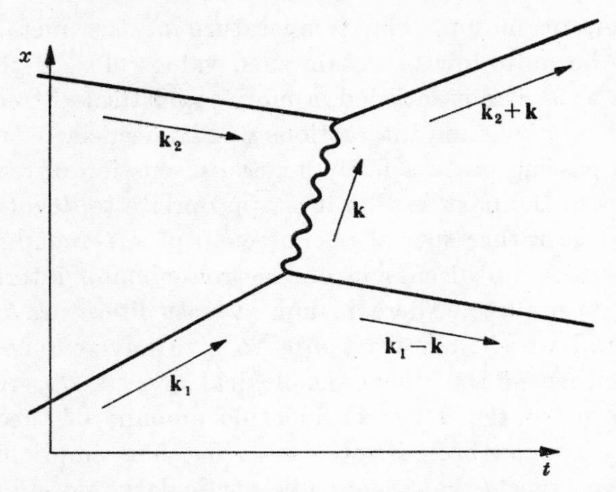

FIG. 6.10. Diagram illustrating electron–electron interaction via the exchange of a phonon in a solid; straight lines represent paths of electrons and the wavy line represents the phonon.

need not precede the re-absorption by the second electron but could occur at a later time, and secondly the phonon that is exchanged is normally a virtual phonon, that is we do not require energy to be conserved separately for the emission and re-absorption processes but only for the complete process involving both emission and re-absorption. The formulation of the interaction between electrons by the exchange of phonons was developed by Fröhlich (1950, 1952) who suggested that this interaction is responsible for the existence of superconductivity. The Hamiltonian which describes the Fröhlich interaction can be obtained as follows.

The Hamiltonian given in eqn (6.5.1) has previously been re-written as

$$\mathscr{H} = \mathscr{H}_{\mathrm{e}} + \mathscr{H}_{\mathrm{n}} + \mathscr{H}_{\mathrm{I}}, \tag{6.5.7}$$

and we now rewrite it again as

$$\mathscr{H} = \mathscr{H}_0 + \mathscr{H}_{\mathrm{I}} \tag{6.5.20}$$

where $\mathscr{H}_0 = \mathscr{H}_n + \mathscr{H}_e$. $\mathscr{H}_0$ contains all the terms that do not involve electron–phonon interactions. We have already expressed $\mathscr{H}_{\mathrm{I}}$ in terms of creation and annihilation operators for both the electrons and the phonons

$$\mathscr{H}_{\mathrm{I}} = \sum_{j}\sum_{\mathbf{k}}\sum_{\mathbf{k}_1}\sum_{\mathbf{k}_2} D_{j\mathbf{k}}(\mathbf{k}_2, \mathbf{k}_1)(a_{j\mathbf{k}} + a^{+}_{j-\mathbf{k}})c^{+}_{\mathbf{k}_2}c_{\mathbf{k}_1} \tag{6.5.14}$$

where $D_{j\mathbf{k}}(\mathbf{k}_2, \mathbf{k}_1)$ is simply related to the coupling constant $\mathscr{I}_{j\mathbf{k}}(\mathbf{k}_2, \mathbf{k}_1)$. Since the values of the coupling constants are generally unknown it was assumed by Fröhlich that they were independent of the electronic states specified by $\mathbf{k}_1$ and $\mathbf{k}_2$ and depend only on the phonon state involved. We then have

$$\mathscr{H}_{\mathrm{I}} = \sum_{j}\sum_{\mathbf{k}}\sum_{\mathbf{k}_1}\sum_{\mathbf{k}_2} D_{j\mathbf{k}}(a_{j\mathbf{k}} + a^{+}_{j-\mathbf{k}})c^{+}_{\mathbf{k}_2}c_{\mathbf{k}_1} \tag{6.5.21}$$

where the $D_{j\mathbf{k}}$ are independent of $\mathbf{k}_1$ and $\mathbf{k}_2$. We have previously interpreted $\mathscr{H}_{\mathrm{I}}$ as describing the scattering of an electron from the state specified by $\psi_{\mathbf{k}_1}(\mathbf{r})$ to the state specified by $\psi_{\mathbf{k}_2}(\mathbf{r})$ with the emission or absorption of a phonon. However, this change in the distribution of the phonons will in turn alter the distribution of the electrons. Since this interaction is of second order, it does not appear explicitly in eqn (6.5.7). Fröhlich showed that by performing a suitable canonical transformation on the Hamiltonian $\mathscr{H}$ it is possible to obtain a term of the form

$$\mathscr{H}' = \sum_{j}\sum_{\mathbf{k}}\sum_{\mathbf{k}_1}\sum_{\mathbf{k}_2} \frac{D^2_{j\mathbf{k}}}{E(\mathbf{k}_1) - E(\mathbf{k}_1 - \mathbf{k}) - \hbar\omega_j(\mathbf{k})} \times c^{+}_{\mathbf{k}_2+\mathbf{k}}c_{\mathbf{k}_2}c^{+}_{\mathbf{k}_1-\mathbf{k}}c_{\mathbf{k}_1} \tag{6.5.22}$$

(for details see Fröhlich 1952). For simplicity, since very little is usually known about the coupling constants, it is quite common to write $\mathscr{H}'$ simply as

$$\mathscr{H}' = -V\sum_{j}\sum_{\mathbf{k}}\sum_{\mathbf{k}_1}\sum_{\mathbf{k}_2} c^{+}_{\mathbf{k}_2+\mathbf{k}}c_{\mathbf{k}_2}c^{+}_{\mathbf{k}_1-\mathbf{k}}c_{\mathbf{k}_1}. \tag{6.5.23}$$

$c^{+}_{\mathbf{k}_1-\mathbf{k}}c_{\mathbf{k}_1}$ describes the annihilation of an electron in the initial state $\psi_{\mathbf{k}_1}(\mathbf{r})$, and the creation of an electron in the final state $\psi_{\mathbf{k}_1-\mathbf{k}}(\mathbf{r})$, i.e. the scattering of an electron from state $\mathbf{k}_1$ to $\mathbf{k}_1 - \mathbf{k}$ with the emission of a phonon with wave vector $\mathbf{k}$. Similarly, $c^{+}_{\mathbf{k}_2+\mathbf{k}}c_{\mathbf{k}_2}$ describes the scattering of the second electron with the absorption of a phonon. The term $\mathscr{H}'$ in eqn (6.5.23) therefore describes the process illustrated in Fig. 6.10 in which there is an attraction or repulsion between two electrons as a result of the exchange of a virtual phonon. Strictly we should also include in $\mathscr{H}'$ the complex conjugate of the operator $c^{+}_{\mathbf{k}_2+\mathbf{k}}c_{\mathbf{k}_2}c^{+}_{\mathbf{k}_1-\mathbf{k}}c_{\mathbf{k}_1}$ which would then describe the second version of the process in Fig. 6.10, namely the absorption of a phonon by the first electron followed by the subsequent re-emission of a similar phonon by the second electron.

There is an electrostatic force of repulsion between two electrons in free space. For two conduction electrons in a metal this electrostatic repulsion still survives although it will be screened to some extent by the presence of all the other charges in the metal. What we have seen in the previous paragraph is that for two itinerant electrons in a solid there is an additional interaction via the lattice vibrations. This interaction is depicted diagrammatically in Fig. 6.10 and is described by a term in the Hamiltonian given in eqn (6.5.22). This term will represent a repulsion or an attraction depending on the relative magnitudes of the quantities in the denominator. In a normal metal these electron–electron interactions, via the exchange of phonons, will contribute to the electrical resistivity of the metal but do not cause anything spectacular to occur.

Experimental evidence in favour of Fröhlich's suggestion that the electron–phonon interactions play a significant role in connection with superconductivity was provided by the *isotope effect.* If one considers different isotopes of the same metal the only significant change in the Hamiltonian $\mathscr{H}$ for the metal will be the alteration in the nuclear mass M_{κ}. This will alter $\mathscr{H}_{\mathrm{n}}$, and therefore also the phonon frequencies $\omega_j(\mathbf{k})$, but it will not have any significant effect on the other parts of the Hamiltonian $\mathscr{H}$ of the metal. Consequently, if there are any differences between the parameters (critical temperature T_c, critical magnetic field H_c) of the superconducting transition for samples of different isotopes of the same metal this can be taken as a strong indication of the importance of the lattice vibrations in connection with the phenomenon of superconductivity. According to Fröhlich's predictions the critical magnetic field for the disappearance of superconductivity should depend on the isotopic mass M_κ of the metal according to

$$H_c \propto M_\kappa^{-\frac{1}{2}}, \tag{6.5.24}$$

(for details see Fröhlich 1950) and therefore, from the empirical law of corresponding states, the superconducting transition temperature should obey

$$T_c \propto M_\kappa^{-\frac{1}{2}} \tag{6.5.25}$$

as well. This dependence of H_c and T_c on M_κ is called the *isotope effect* and experimental confirmation of eqns (6.5.24) and (6.5.25) was first provided by work on Hg (Maxwell 1950; Reynolds, Serin, Wright, and Nesbitt 1950) and Sn (Allen, Dawton, Bär, Mendelssohn, and Olsen 1950; Allen, Dawton, Lock, Pippard, and Shoenberg 1950).

The mere fact of having shown that electron–phonon interactions are important in connection with the existence of superconductivity, and having obtained the formal expression for $\mathscr{H}'$ in eqn (6.5.23), does not provide us with a complete microscopic theory of the phenomenon of superconductivity. Significant progress in this direction came with the work of Bardeen, Cooper, and Schrieffer (1957*a*, *b*), commonly referred to as the *B.C.S. theory*, only after several other attempts based on the use of the Fröhlich interaction term $\mathscr{H}'$ in perturbation theory had failed. It would not be appropriate to become involved in the details of the B.C.S. theory in this book. We simply note that an essential pre-requisite of this theory is the idea that pairs of conduction electrons in a metal can combine together to form bound states. These bound states are called *Cooper pairs* (Cooper 1956) and the force that is responsible for producing these bound states is the interaction by exchange of phonons described by eqn (6.5.23). The two electrons that are involved in forming a Cooper pair have equal and opposite wave vectors $\mathbf{k}$ and $-\mathbf{k}$. Once Cooper pairs can be formed, by the exchange of virtual phonons, they form a superconducting ground state that is separated by an *energy gap* from the excited states in which one or more pairs are broken; there is abundant experimental evidence for the existence of the superconducting energy gap (see, for example, the summary given by Kuper 1968, pp. 27–34). It is the existence of the energy gap that allows the persistence of the non-dissipative current which characterises superconductivity. Thus we have the curious fact that electron–phonon interactions—which usually lead to an important contribution to the electrical resistance of a metal—are also responsible for the existence of superconductivity when all resistance disappears. The B.C.S. theory makes use of the variational principle, with a suitable Hamiltonian and a trial wave function for the superconducting ground state, in which it is assumed that all the electrons in the metal have been formed into Cooper pairs. In principle, therefore, if one knows both the wave functions for the conduction electrons and the eigenstates of the Hamiltonian for the lattice vibrations for a given metal, it should be possible to calculate the energy of the electron–electron interactions via the exchange of phonons and then to see whether this is attractive and large enough to exceed the Coulomb repulsion; if so the metal would be expected to exhibit superconductivity and the critical temperature T_c can be determined from the B.C.S. theory. Such calculations have now been performed for a few metals (Carbotte and Dynes 1967, 1968, on Na, K, Al, and Pb; Allen, Cohen, Falicov, and Kasowski 1968, on Zn and Cd; Allen and Cohen 1969 on sixteen simple metals and the alkaline–earth metals Ca, Sr, and Ba), but the uncertainties involved are such that the success or failure of the results of these calculations tells us more about the reliability or otherwise of the band structure and phonon dispersion relations used in the calculation than about the superconducting properties of the metal in question. In practice the determination of whether or not a given material can exhibit superconductivity is a problem that still has to be solved empirically, while the B.C.S. expression for T_c,

$$T_c = \Theta\exp\{-1/n(E_{\mathrm{F}})V\} \tag{3.2.2}$$

where V is the constant in eqn (6.5.23) representing the interaction between the electrons, is more commonly used with a measured value of T_c to determine V than with a theoretical value of V to predict T_c.

A question that is of particular importance to us in this book is to consider the effect on the electronic band structure and Fermi surface of a metal when that metal becomes superconducting, and indeed even to ask whether the Fermi surface exists at all for a metal in the superconducting state. The Fermi surface is a surface in the abstract space based on the reciprocal lattice vectors $\mathbf{g}_1$, $\mathbf{g}_2$, and $\mathbf{g}_3$. The usefulness of this particular abstract space hinges on the existence of a periodic potential in the metal and the consequent relevance of Bloch's theorem and the usefulness of $\mathbf{k}$ as a good quantum number for classifying the eigenstates of particles, or quasiparticles, in a solid. The formation of Cooper pairs and the onset of superconductivity does not destroy the periodicity of the crystal structure and therefore does not destroy the usefulness of $\mathbf{k}$ as a good quantum number. The Cooper pairs will, of course, obey Bose–Einstein statistics rather than Fermi–Dirac statistics, but this does not imply that the concept of the Fermi surface is no longer relevant in the superconducting state. Cooper pairs can only be formed involving those one-electron states which are actually occupied by electrons, and the total number of conduction electrons is fixed for the superconducting state just as it is for the normal metal. The probability that a given one-electron state is occupied, and therefore available to form a Cooper pair, is consequently of some importance and it is in this connection that the concept of the Fermi surface is still valid and useful in the superconducting state, assuming that the constituent one-electron states of a Cooper pair are still identifiable after the pair has been formed (for a more complete discussion of the validity of the concept of the Fermi surface in the presence of electron interactions see Falicov and Heine 1961; Mattuck 1967, p. 175; Hedin and Lundqvist 1969). Of course the onset of superconductivity may alter the shapes of the bands of one-electron states and therefore also the shape of the Fermi surface. Complete pairing of all the electrons, as required in the B.C.S. ground state, is possible because the band structure of any metal must possess inversion symmetry, that is, $E_j(-\mathbf{k}) = E_j(\mathbf{k})$, as a result of time-reversal symmetry, even if the space inversion is not a symmetry operation of the crystal itself (see, for example, Bradley and Cracknell 1970).

We can perhaps draw an analogy with the case of a metal that exhibits a transition to a magnetically ordered state. As we have seen in Chapter 5 there was a time (about 1960) when it was fashionable in certain quarters to doubt whether the Fermi surface existed for a ferromagnetic metal such as Fe. Nevertheless in due course it was found that there was no reason why the concept of the Fermi surface should not be used for a magnetic metal, although the magnetic ordering may cause quite drastic changes in the topology of the Fermi surface. In a similar way there have been people who have said either that the Fermi surface does not exist for a superconductor or that the Fermi surface is not a useful concept in the case of a superconductor. It is true that, in general, a detailed knowledge of the shape of the Fermi surface is more difficult to obtain for the superconducting state of a metal than for the normal state and also that, once it has been obtained, such information is likely to be of less use than the corresponding information for a normal metal. Nevertheless it seems likely that as the accuracy of quantitative calculations based on the microscopic theory of superconductivity become more refined, a detailed knowledge of the Fermi surfaces of individual metals in the superconducting state will become more important.

There is also the problem of the usefulness of the Fermi surface as a concept for a superconducting metal because many of the methods described in Chapter 3 for the determination of the shape of the Fermi surface cannot be used in the case of a superconductor, while some of the properties that one often seeks to explain in terms of the Fermi surface simply do not exist for a superconductor. The existence of the Meissner effect, the expulsion of the magnetic field from a body when it becomes superconducting, means that none of those methods which make use of a magnetic field can be used to investigate the Fermi surface of a superconducting metal. This unfortunately eliminates all those methods which have been particularly successful in obtaining direct measurements of the

Fermi surfaces of normal metals. Of the few methods that are not impossible for a superconducting metal, the one which has actually been used is the study of the angular correlations of γ-rays produced by the annihilation of positrons in a metal (Briscoe, Beardsley, and Stewart 1966; Dekhtyar 1969). Although these experiments were not entirely conclusive they did seem to indicate that, apart from the formation of Cooper pairs, the momentum distribution of the individual one-electron states in a metal was not very greatly affected by the transition to the superconducting state.

In this section we have discussed electron–phonon interactions in some detail. It should also be mentioned that in a magnetically-ordered metal there will also be interactions between the translational motions of the conduction electrons and the ordered array of magnetic moments in the metal, whether these magnetic moments are associated with the conduction electrons themselves or with the core electrons. These interactions are most conveniently studied in terms of *electron–magnon interactions* where the magnons are quasi-particles which are the quanta of spin wave energy in the magnetic metal. In many ways the behaviour of the magnons is analogous to that of the phonons; for example, they give rise to an electron–magnon enhancement of the coefficient γ in the electronic specific heat that may be of the same order of magnitude as the electron–phonon enhancement of γ (Bennemann 1967; Davis and Liu 1967; Nakajima 1967).

6.6. Transport properties

One important group of properties of a solid is the transport properties, of which we have had occasion to mention one or two examples in Section 3.7. The simplest of these are the electrical conductivity and the thermal conductivity but there are also a number of more complicated transport properties, involving the transfer of electricity or of heat under various configurations of external conditions. Central to any quantitative discussion of the transport properties of a solid is Boltzmann's equation (6.6.1) (see, for example, the extensive discussion in the book by Ziman 1960). To prevent our discussion from becoming too abstract we use the particular example of the electrical conductivity to illustrate the general principles involved; the discussion could then be adapted to the consideration of other transport properties and would proceed in a similar way although some of the details would be different.

If an external electric field, $\mathbf{E}$, is applied to a specimen of a metal there will be a re-distribution of electric charge within the metal and, if the specimen is part of an electrical circuit, an electric current will flow, the electric current density, $\mathbf{j}$, being related to $\mathbf{E}$ by $\mathbf{j} = \sigma\mathbf{E}$, where σ is the conductivity of the metal. If the carriers of electric charge within the metal were completely free they would go on accelerating as long as the external field was applied, so that the electric current in a piece of wire connected to a battery would go on increasing indefinitely. This is, of course, not observed in practice because of the scattering of the carriers by the other electrons, by the thermal vibrations of the lattice, and by defects or impurities in the specimen. A steady state is rapidly reached in which the carriers acquire a mean drift velocity, $\bar{\mathbf{v}}$, the magnitude of which depends on the magnitude of the external electric field and on the cross sections for the various scattering mechanisms involved. Any calculation from first principles of a transport coefficient such as the electrical conductivity of a metal will require a proper study of the various scattering processes which affect the carriers in the metal, which are, of course, the conduction electrons.

We suppose that there exists a distribution function $f_{\mathbf{k}}(\mathbf{r}, t)$ such that, at time t, the number of carriers with wave vector $\mathbf{k}$ in the small element of volume $d\mathbf{r}$ at $\mathbf{r}$ is $f_{\mathbf{k}}(\mathbf{r}, t)\, d\mathbf{r}$. There are several ways in which the passage of time will affect the function $f_{\mathbf{k}}(\mathbf{r}, t)$. Even in the absence of the external influence, $f_{\mathbf{k}}(\mathbf{r}, t)$ will suffer changes because the carriers are in constant motion with velocities $\mathbf{v}_{\mathbf{k}}$ and because the carriers are constantly being scattered; these two changes in $f_{\mathbf{k}}(\mathbf{r}, t)$ may be called $(\partial f_{\mathbf{k}}/\partial t)_{\text{diff}}$, for the diffusion of the carriers and $(\partial f_{\mathbf{k}}/\partial t)_{\text{scatt}}$, for the scattering of the carriers. In the absence of any external field these will balance each other and there will be a dynamic equilibrium set up in which $f_{\mathbf{k}}(\mathbf{r}, t)$ takes the value $f^0_{\mathbf{k}}$ given by the Fermi–Dirac distribution function

$$1/(\exp[\{E(\mathbf{k})-E_{\mathrm{F}}\}/kT]+1).$$

When the external field is applied there is a third influence causing $f_{\mathbf{k}}(\mathbf{r}, t)$ to change and which we write as $(\partial f_{\mathbf{k}}/\partial t)_{\text{field}}$. Although there is, of course, no equilibrium when the external field is applied, there is nevertheless a steady state set up and the Boltzmann equation expresses the fact that in this steady state there is no change in the number of carriers with wave vector $\mathbf{k}$ in the volume $\mathrm{d}\mathbf{r}$. That is

$$\left(\frac{\partial f_{\mathbf{k}}}{\partial t}\right)_{\text{diff}}+\left(\frac{\partial f_{\mathbf{k}}}{\partial t}\right)_{\text{scatt}}+\left(\frac{\partial f_{\mathbf{k}}}{\partial t}\right)_{\text{field}}=0, \tag{6.6.1}$$

which is *Boltzmann's equation*. If this equation is solved for $f_{\mathbf{k}}(\mathbf{r}, t)$ it is then possible to use the solution in the expression

$$\mathbf{j}=\int e\mathbf{v}_{\mathbf{k}}f_{\mathbf{k}}(\mathbf{r}, t)\,\mathrm{d}\mathbf{k} \tag{6.6.2}$$

to calculate the current density $\mathbf{j}$ and hence the electrical conductivity σ. This is easier said than done for a number of reasons. As we mentioned in Section 3.7 it is only for cubic crystalline materials and for isotropic non-crystalline materials that σ is a scalar; otherwise σ is a tensor of rank two so that the details of the solution of Boltzmann's equation and the determination of σ will depend on the relative orientation of $\mathbf{E}$ and the crystallographic axes of the sample. Even more serious than this is the difficulty that the expressions for some of the terms in eqn (6.6.1) are rather complicated and make the equation very difficult to solve.

It is quite easy to determine the expression for $(\partial f_{\mathbf{k}}/\partial t)_{\text{diff}}$. The net flow of carriers into the volume $\mathrm{d}\mathbf{r}=\mathrm{d}x\mathrm{d}y\mathrm{d}z$ is given by considering the flow of particles across all the six faces of

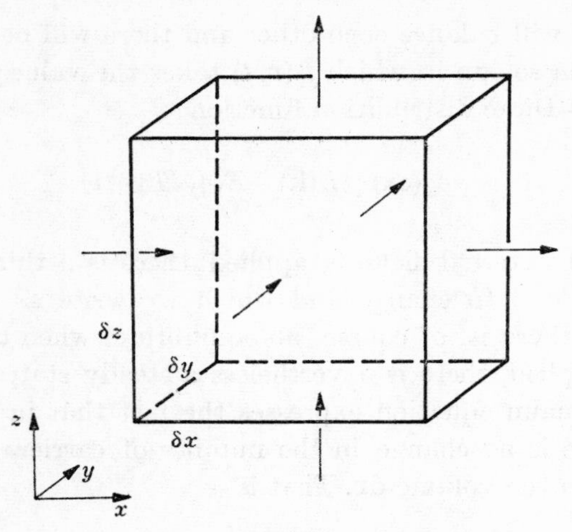

FIG. 6.11.

the box shown in Fig. 6.11 and is equal to

$$
\begin{aligned}
&\mathbf{v}_{\mathbf{k}x}\delta y\delta z\{f_{\mathbf{k}}(x,y,z,t)-f_{\mathbf{k}}(x+\delta x,y,z,t)\}+\\
&+\mathbf{v}_{\mathbf{k}y}\delta z\delta x\{f_{\mathbf{k}}(x,y,z,t)-f_{\mathbf{k}}(x,y+\delta y,z,t)\}+\\
&+\mathbf{v}_{\mathbf{k}z}\delta x\delta y\{f_{\mathbf{k}}(x,y,z,t)-f_{\mathbf{k}}(x,y,z+\delta z,t)\}\\
&\quad=-\left\{\mathbf{v}_{\mathbf{k}x}\frac{\partial f_{\mathbf{k}}}{\partial x}+\mathbf{v}_{\mathbf{k}y}\frac{\partial f_{\mathbf{k}}}{\partial y}+\mathbf{v}_{\mathbf{k}z}\frac{\partial f_{\mathbf{k}}}{\partial z}\right\}\delta x\delta y\delta z\\
&\quad=-\mathbf{v}_{\mathbf{k}}\,.\,\nabla f_{\mathbf{k}}\delta x\delta y\delta z\\
&\quad=\left(\frac{\partial f_{\mathbf{k}}}{\partial t}\right)_{\mathrm{diff}}\delta x\delta y\delta z
\end{aligned}
\tag{6.6.3}
$$

so that

$$
\left(\frac{\partial f_{\mathbf{k}}}{\partial t}\right)_{\mathrm{diff}}=-\mathbf{v}_{\mathbf{k}}\,.\,\nabla f_{\mathbf{k}}. \tag{6.6.4}
$$

The contribution $(\partial f_{\mathbf{k}}/\partial t)_{\mathrm{scatt}}$ is difficult to evaluate. We restrict ourselves to the case of elastic scattering. The probability that a carrier in state $\mathbf{k}_1$ is scattered into a state $\mathbf{k}_2$ depends on the probability that state $\mathbf{k}_1$ is occupied, the probability that state $\mathbf{k}_2$ is unoccupied, and the intrinsic transition probability $Q_{\mathbf{k}_1\mathbf{k}_2}$ which depends on the quantum-mechanical matrix element $V_{\mathbf{k}_1\mathbf{k}_2}=\int\psi^*_{\mathbf{k}_2}(\mathbf{r})V(\mathbf{r})\psi_{\mathbf{k}_1}(\mathbf{r})\,d\mathbf{r}$ where $V(\mathbf{r})$ is the operator describing the potential associated with the scattering mechanism.

The probability that a carrier is scattered from $\mathbf{k}_1$ to $\mathbf{k}_2$ in the element of volume $\mathrm{d}\mathbf{r}$ is therefore given by $\{f_{\mathbf{k}_1}(1-f_{\mathbf{k}_2})Q_{\mathbf{k}_1\mathbf{k}_2}\}\,\mathrm{d}\mathbf{r}$. We have to subtract from this the probability that a carrier is scattered back from $\mathbf{k}_2$ into $\mathbf{k}_1$. Moreover, there are many different states $\mathbf{k}_2$ into which a carrier with wave vector $\mathbf{k}_1$ may be scattered, so that the total net probability that a carrier with wave vector $\mathbf{k}_1$ within the volume element $\mathrm{d}\mathbf{r}$ will be scattered into any other state is

$$\left[\int\{f_{\mathbf{k}_1}(1-f_{\mathbf{k}_2})-f_{\mathbf{k}_2}(1-f_{\mathbf{k}_1})\}Q_{\mathbf{k}_1\mathbf{k}_2}\,\mathrm{d}\mathbf{k}_2\right]\mathrm{d}\mathbf{r} = \left(\frac{\partial f_{\mathbf{k}_1}}{\partial t}\right)_{\text{scatt}}\mathrm{d}\mathbf{r} \tag{6.6.5}$$

so that

$$\left(\frac{\partial f_{\mathbf{k}_1}}{\partial t}\right)_{\text{scatt}} = \int\{f_{\mathbf{k}_1}(1-f_{\mathbf{k}_2})-f_{\mathbf{k}_2}(1-f_{\mathbf{k}_1})\}Q_{\mathbf{k}_1\mathbf{k}_2}\,\mathrm{d}\mathbf{k}_2. \tag{6.6.6}$$

In the presence of external electric and magnetic fields the rate of change of momentum, $\mathrm{d}(\hbar\mathbf{k})/\mathrm{d}t$, of any one of the carriers, will be equal to the external force on that carrier as a result of Newton's second law of motion so that

$$\hbar\frac{\mathrm{d}\mathbf{k}}{\mathrm{d}t} = e(\mathbf{E}+\mathbf{v}_{\mathbf{k}}\wedge\mathbf{B}). \tag{6.6.7}$$

In the presence of the external fields, $f_{\mathbf{k}}(\mathbf{r},t)$ is therefore undergoing some changes as a result of the movement of the wave vectors $\mathbf{k}$ of the particles in reciprocal space with a velocity $\mathrm{d}\mathbf{k}/\mathrm{d}t$ given by eqn (6.6.7). $(\partial f_{\mathbf{k}}/\partial t)_{\text{field}}$ is therefore given by the expression (6.6.4) with $\mathbf{v}_{\mathbf{k}}$ replaced by $\mathrm{d}\mathbf{k}/\mathrm{d}t$ and ∇ replaced by

$$\nabla_{\mathbf{k}}\left(=\left(\frac{\partial}{\partial k_x},\frac{\partial}{\partial k_y},\frac{\partial}{\partial k_z}\right)\right)$$

so that

$$\left(\frac{\partial f_{\mathbf{k}}}{\partial t}\right)_{\text{field}} = -\dot{\mathbf{k}}\,.\,\nabla_{\mathbf{k}}f_{\mathbf{k}}. \tag{6.6.8}$$

The expressions for $(\partial f_{\mathbf{k}}/\partial t)_{\text{diff}}$, $(\partial f_{\mathbf{k}}/\partial t)_{\text{scatt}}$, and $(\partial f_{\mathbf{k}}/\partial t)_{\text{field}}$ given by eqns (6.6.4), (6.6.6), and (6.6.8) respectively can then be substituted into eqn (6.6.1) giving

$$-\mathbf{v}_{\mathbf{k}_1}\cdot\nabla f_{\mathbf{k}_1}+\int\{f_{\mathbf{k}_1}(1-f_{\mathbf{k}_2})-f_{\mathbf{k}_2}(1-f_{\mathbf{k}_1})\}Q_{\mathbf{k}_1\mathbf{k}_2}\,d\mathbf{k}_2 -\dot{\mathbf{k}}_1\cdot\nabla_{\mathbf{k}_1}f_{\mathbf{k}_1}=0. \quad (6.6.9)$$

There are several features which contribute to making it far from trivial to solve this equation, Boltzmann's equation. First, the term which arises from $(\partial f_{\mathbf{k}_1}/\partial t)_{\text{scatt}}$ involves the quantities $Q_{\mathbf{k}_1\mathbf{k}_2}$ which can only be evaluated if the wave functions $\psi_{\mathbf{k}_1}(\mathbf{r})$ and $\psi_{\mathbf{k}_2}(\mathbf{r})$ (all $\mathbf{k}_2$) of the carriers are known. Secondly, eqn (6.6.9) is not just a simple differential equation but an integro-differential equation. Finally, it is not unreasonable to assume that $f_{\mathbf{k}}$ does not depart too much from $f_{\mathbf{k}}^0$ its value in the absence of the external field, where,

$$f_{\mathbf{k}}^0=\frac{1}{\exp[\{E(\mathbf{k})-E_{\mathrm{F}}\}/kT]+1}. \quad (6.6.10)$$

That is, one writes

$$f_{\mathbf{k}}=f_{\mathbf{k}}^0+g_{\mathbf{k}} \quad (6.6.11)$$

and rewrites eqn (6.6.9) in terms of $g_{\mathbf{k}}$ and $f_{\mathbf{k}}^0$ (for details of the re-arrangement see Ziman 1964*a*, p. 181). However, apart from the fact that $f_{\mathbf{k}}^0$ is a function of the temperature T, one can only use the expression (6.6.10) for $f_{\mathbf{k}}^0$ if $E(\mathbf{k})$ is known, that is, if the band structure of the metal has already been determined. Thus, to calculate the electrical conductivity, or any other transport property, by a proper solution of Boltzmann's equation we find that it is necessary to know the wave functions $\psi_{\mathbf{k}}(\mathbf{r})$ and energies $E(\mathbf{k})$ for all the electronic states in the metal before Boltzmann's equation can even be set up and then, when the equation has been set up, it is difficult to solve because it is an integro-differential equation. Even if all the necessary information were available, which very often is not the case, this process would be extremely complicated and time consuming so that, in practice, Boltzmann's equation is usually solved at one of a number of possible levels of approximation depending on the extent of the information available about the carriers in the material and their interactions, on the patience and enthusiasm of the worker in question, and also on what purpose lay behind solving the problem in the first place.

So far in this section we have not devoted much attention to the nature of the physical mechanisms which contribute to the scattering of the conduction electrons in a metal and which therefore contribute to the electrical resistivity of the metal; these processes appear in Boltzmann's equation as contributions to the potential $V(\mathbf{r})$ that is used in evaluating the matrix elements $V_{\mathbf{k}_1\mathbf{k}_2}$. A conduction electron with wave vector $\mathbf{k}_1$ may be scattered by collisions with other electrons, by collisions with lattice defects, by collisions with impurity atoms if they are present in the metal, and by collisions with the ion cores; the last of these mechanisms can most conveniently be described in terms of collisions with phonons, the quanta of lattice vibrational energy of the metal. It is commonly assumed that the resistivity, ρ, of a metal can be written as

$$\rho = \rho_1 + \rho_2 + \rho_3 + \dots \tag{6.6.12}$$

where ρ_1, ρ_2, ρ_3,... are the individual contributions that would be obtained by considering each of the various scattering mechanisms to be acting alone in the metal in the absence of all the others. This assumption which is commonly called *Matthiessen's law* (Matthiessen 1862) is very widely used although there are some circumstances, which we do not propose to discuss here, in which the rule breaks down (see Ziman 1960).

Assuming that Matthiessen's law is at least qualitatively correct, it is desirable to know the relative magnitudes of the various contributions to the scattering of the carriers; for those of the scattering mechanisms which are most important it is then necessary to be able to determine the details of their contributions to the intrinsic transition probabilities $Q_{\mathbf{k}_1\mathbf{k}_2}$. Of the four scattering mechanisms mentioned it is possible to eliminate the impurity scattering by considering the ideal case of a pure metal. The contribution to the electrical resistivity due to the scattering of the conduction electrons by point defects and by dislocations is also small. Of the two remaining mechanisms the one that is dominant under most conditions in a metal is the scattering by the phonons. The contribution of electron–electron scattering to the electrical resistivity of a metal is very small and was not observed experimentally until quite recently (Yaqub and Cochran 1965; White and Tainsh 1967; Anderson, Peterson, and Robichaux 1968; Garland and Bowers 1968); this contribution to the resistivity varies as T^2. We have already discussed electron–phonon interactions and electron–electron interactions quite extensively in the earlier sections of this chapter. The dominance of the scattering by phonons means that for a proper treatment of Boltzmann's equation for electrical conduction in a metal it is necessary to have a complete knowledge not only of the band structure of the metal but also of the phonon dispersion relations. For a simple metal this is not quite as bad as it may appear at first sight because the same pseudopotential that can be used to describe the band structure of the metal (see Section 2.6) can also be used to describe the phonon dispersion relations (see Section 6.4). Since the scattering of an electron with the creation or annihilation of a phonon is not an example of elastic scattering, the expression for $(\partial f_{\mathbf{k}_1}/\partial t)_{\text{scatt.}}$ given in eqn (6.6.6) does not apply because the probability of scattering will now also depend on the number of phonons actually present with wave vector $\mathbf{k}$ and on the intrinsic transition probability that an electron which is known to be in state $\mathbf{k}_1$ will be scattered into a state $\mathbf{k}_2$, which is known to be empty, with the associated creation or annihilation of a phonon of wave vector $\mathbf{k}$. This transition probability will be determined by using a term of the form of $\mathscr{H}_{\text{I}}$ in eqn (6.5.14) to construct matrix elements similar to those in eqns (6.5.16) and (6.5.17) where the initial and final state wave functions describe the lattice vibrations as well as the motions of the conduction electrons. The requirement of the conservation of energy will mean that it is only those electrons which are close to the Fermi surface, and therefore also close to unoccupied states, which will make significant contributions to the total scattering cross section. The importance of the total area, A, of the Fermi surface can be seen from the simplified expression for the conductivity which applies to a cubic metal and which we have already mentioned in Chapter 3

$$\sigma = \frac{e^2}{12\pi^3\hbar}\,\lambda A. \tag{3.9.2}$$

The scattering of an electron by a phonon must also satisfy the conservation of momentum so that we have to remember to include both normal processes and umklapp processes in the scattering. The theoretical treatment of umklapp processes is rather more involved than that of normal processes (see, for example, Ziman 1960, Chapter 9; Rosenberg 1963, Chapter 4) while the relative importance of either process would depend on the detailed shape of the Fermi surface.

In view of the several difficulties involved it is therefore not surprising that various simplifying assumptions have been made in attempts to calculate the electrical resistivity of a metal *ab initio*. A considerable number of calculations of electrical resistivities have been made, mostly on the alkali metals and the noble metals (Bardeen 1937; Bailyn 1958, 1960; Collins and Ziman 1961; Joshi and Kashyap 1962; Darby and March 1964; Greene and Kohn 1965; Sundström 1965; Bonsignori and Bortolani 1966; Wiser 1966; Dickey, Meyer, and Young 1967; Robinson 1968; Robinson and Dow 1968; Hasegawa and Kasuya 1970). However, the agreement between these calculations and experimental results often leaves much to be desired; it is probably not too unfair to say that most of these calculations give more information about how to construct sensible pseudopotentials than about the reliability or otherwise of the theory of transport properties.

So far we have regarded the quantum-mechanical transition probabilities that go to make up the scattering cross sections for various processes as quantities to be determined *ab initio* by perturbation theory, using the appropriate wave functions and the correct formal expression for the potential representing the interaction. However, it is possible to proceed in a slightly more empirical manner by thinking in terms of the relaxation time $\tau_{\mathbf{k}}$ for the scattering of an electron with wave vector $\mathbf{k}$ by some given process into any other available state. $\tau_{\mathbf{k}}$ is related to the mean free path $\lambda_{\mathbf{k}}$ and the velocity $\mathbf{v}_{\mathbf{k}}$ of the conduction electron by

$$\tau_{\mathbf{k}} = \frac{\lambda_{\mathbf{k}}}{|\mathbf{v}_{\mathbf{k}}|}. \tag{6.6.13}$$

Although we did not stress this aspect of the situation in Chapter 3, because at that stage our interests were mostly topological, $\tau_{\mathbf{k}}$ and $\lambda_{\mathbf{k}}$ are quantities about which a considerable amount of information can be obtained experimentally. The principal methods used in the determination of relaxation times or mean free paths include the amplitudes in the de Haas–van Alphen effect (Shoenberg 1969*b*), the magnetoresistance, particularly the longitudinal magnetoresistance (Pippard 1965), the Hall effect (Coles 1956), the ultrasonic attenuation, with or without a magnetic field (Pippard 1955; Dooley and Tepley 1969), and various surface and size effects, particularly the amplitudes of radio-frequency size effects (Chambers 1969*b*). Whereas there are several methods that enable Fermi surface dimensions to be determined directly, the same is not true of $\tau_{\mathbf{k}}$ and $\lambda_{\mathbf{k}}$. Relaxation times and mean free paths have to be unravelled indirectly from the experimental data, unless one makes the usually unjustified assumption that $\tau_{\mathbf{k}}$ and $\lambda_{\mathbf{k}}$ are independent of $\mathbf{k}$. The whole field of work on the variation of relaxation times and mean free paths over the Fermi surface is in a much less developed state than work on the topology of Fermi surfaces (Sondheimer 1969). There are also one or two methods available for the determination of the Fermi velocity of the conduction electrons in a metal (Eckstein 1966; Baraff and Phillips 1970). One of the more interesting of these involves measuring the transit time for heat pulses generated in a specimen by the absorption of pulses of optical radiation from a dye laser (von Gutfeld and Nethercot 1967; Kubicar and Krempasky 1970). The object of determining $\tau_{\mathbf{k}}$ experimentally is that since it is inversely proportional to the transition probability or scattering cross section, these experimental determinations of $\tau_{\mathbf{k}}$ enable values of $(\partial f_{\mathbf{k}}/\partial t)_{\text{scatt}}$ to be determined.

In our discussion of transport properties so far in this

section we have concentrated exclusively on the electrical conductivity of a metal. Although some of the details would be different, the general outlines of this discussion can readily be adapted to the consideration of various other transport properties such as the thermal conductivity, the Hall effect, the thermal Hall effect (the Righi-Leduc effect), the thermoelectric power, and the electrical and thermal conductivities in the presence of a magnetic field (that is, the electrical and thermal magneto-resistances). Except for the thermoelectric power, we have already given a qualitative discussion of most of these properties in Section 3.7. For a quantitative discussion of these properties we would refer the reader to the more detailed works of Ziman (1960, 1961, 1964*a*, 1969*a*) and Lifshitz and Kaganov (1965) on the theoretical side, or Rosenberg (1963) on the experimental side. We have also not questioned the validity of Boltzmann's equation in describing transport theory for particles or quasi-particles in a quantum-mechanical system. The condition for the validity of the treatment we have outlined is the same as the condition noted in the previous section in connection with the use of perturbation theory for electron–phonon interactions:

$$k\lambda \gg 2\pi, \tag{6.5.19}$$

so the mean free paths of the particles or quasiparticles are much longer than their de Broglie wavelengths (Hansen 1965).

Before leaving the subject of transport properties it is appropriate to mention the thermoelectric power of a metal. A general expression for the thermoelectric power (or the thermopower), S, can be obtained (see, for example, Ziman 1960, p. 397)

$$S = -\frac{\pi^2}{3}\frac{k^2 T}{|e|}\left[\frac{\partial \ln \sigma(E)}{\partial E}\right]_{E=E_F} \tag{6.6.14}$$

where $|e|$ is the magnitude of the electronic charge. The sign of S is important and can be either positive or negative, being opposite to the sign of $\partial \ln \sigma(E)/\partial E$. By $\sigma(E)$ we mean the electrical conductivity that the metal would possess if the Fermi energy E_F were to take the value E. For the particular case of a cubic metal we can use eqn (3.9.2) for σ so that

$$\frac{\partial \ln \sigma(E)}{\partial E} = \frac{\partial \ln \lambda}{\partial E} + \frac{\partial \ln A}{\partial E}. \tag{6.6.15}$$

Since a more energetic electron is less likely to be scattered than one with lower energy, we should expect λ to increase with energy, i.e. the first term is positive. The sign of the second term will depend on the shape of the Fermi surface. From the detailed descriptions in Chapters 4 and 5 of the Fermi surfaces of numerous cubic metals, it is obvious that $\partial \ln A/\partial E$ may be either positive or negative and consequently S can be expected to have either sign. Moreover τ, and therefore λ, will vary with temperature. The temperature dependence of τ often takes the form

$$\tau^{-1} = a + bT^3, \tag{6.6.16}$$

(Deaton 1963, 1965; Häussler and Welles 1966; Moore 1966; Naberezhnȳkh and Tsymbal 1967), where the first term corresponds to the scattering by static imperfections and the second term corresponds to scattering by phonons. At very low temperatures the scattering by phonons becomes dominated by small-angle scattering which leads to a T^5 rather than a T^3 behaviour (see Ziman 1964*a*, p. 191). Because λ is temperature-dependent the thermoelectric power S will also be temperature-dependent and, for a given metal, it may even change sign as T varies. Some examples of the results of experimental determinations of S are shown in Fig. 6.12 for Li, Na, and K and in Fig. 6.13 for Cu. Results for Ag and Au are very similar to those for Cu as one might expect (MacDonald, Pearson, and Templeton 1958; Gold, MacDonald, Pearson, and Templeton 1960). The Fermi surfaces of all the alkali metals except Li have been described in Section 4.2 as being very slightly distorted from the free-electron sphere. We should therefore expect the thermoelectric powers of Na and K to be negative, as shown in Fig. 6.12. The positive thermoelectric power of Li indicates that its Fermi surface probably touches the Brillouin zone

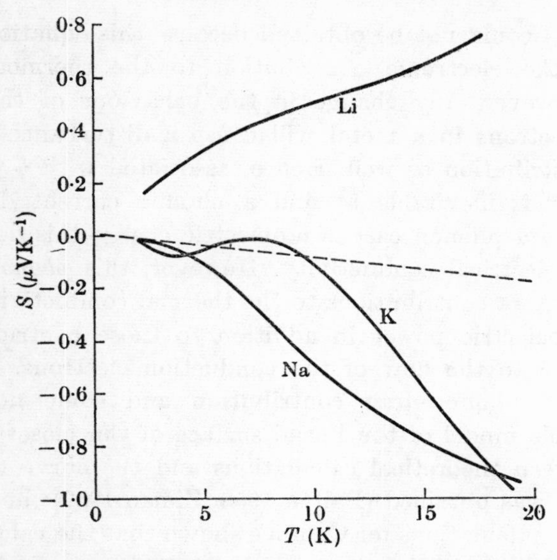

FIG. 6.12. The thermoelectric power of Li, Na, and K; the free-electron thermoelectric power is represented by the broken line (after MacDonald, Pearson, and Templeton 1958).

boundary (so that $\partial \ln A/\partial E$ is negative). However, this feature of the Fermi surface of Li is still an unresolved problem, because of the experimental difficulties involved, see Section 4.2.

Quantitative agreement between experimental results of the type shown in Figs. 6.12 and 6.13 and the formula for S in

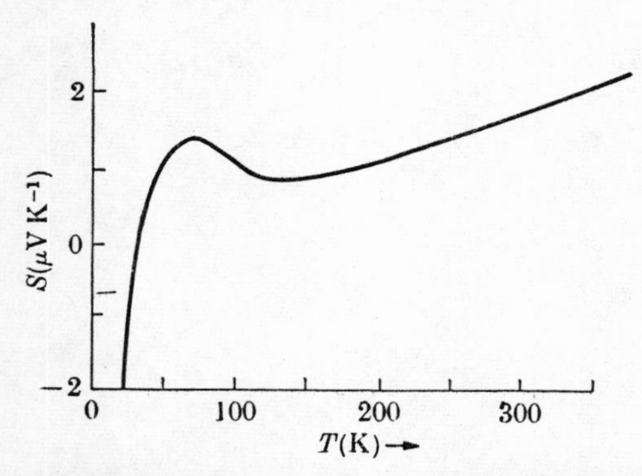

FIG. 6.13. The thermoelectric power of Cu (Gold, MacDonald, Pearson, and Templeton 1960).

eqn (6.6.14) could not be obtained because this equation only describes the electronic contribution to the thermoelectric power. However, any change in the behaviour of the conduction electrons in a metal will cause a disturbance in the phonon distribution as well. Hence, associated with a flow of electrons it is inevitable to find a phonon current dragged along. Since a phonon carries no electric charge this does not affect the electrical conductivity. However, this *phonon drag* will cause extra contributions to the thermal conductivity and the thermoelectric power in addition to those contributions directly due to the flow of the conduction electrons. By including the phonon-drag contribution and some not too unreasonable model of the Fermi surface of Cu, closer agreement between theoretical calculations and the curve of Fig. 6.13 for Cu was obtained (Bailyn 1958; Ziman 1959); however, subsequent calculations for Cu have shown that the calculated thermoelectric power is very sensitive to quite small changes in the assumed band structure and Fermi surface for the metal (Abarenkov and Vedernikov 1966; Hasegawa and Kasuya 1968, 1970; Williams and Davis 1968). Further theoretical considerations of the phonon-drag contributions to the thermoelectric powers of metals have been made by Bailyn (1967).

7
ALLOYS AND METALLIC COMPOUNDS

IN CHAPTERS 2 and 3 we have described in outline the principles which underlie the experimental and theoretical methods which can be used in the determination of the shape of the Fermi surface of a metal. In Chapters 4 and 5 we have also discussed in some detail the results which have accumulated over a number of years giving the details of the shapes of the Fermi surfaces of most of the metallic elements. Now that a large amount of work has been performed on those metals which are chemical elements the attention of workers is turning towards alloys and metallic compounds. In considering alloys it is convenient to consider separately very dilute alloys (with solute concentrations of up to about one per cent), non-dilute random alloys, and ordered alloys. Of these three cases it is the ordered alloys which are the easiest to consider; these alloys can be treated together with metallic compounds, that is, compounds which have metallic properties but contain at least one non-metallic element. Although relatively little work has actually been performed on the band structures and Fermi surfaces of ordered alloys and metallic compounds, the indications are that there are no particularly significant physical differences between the behaviour of electrons in these materials and in metallic elements. In the past it has often been difficult to obtain pure enough samples of these materials for use in Fermiological measurements but substantial progress has now been made in this connection. The mapping of the Fermi surfaces of the ordered alloys and the metallic compounds can therefore be expected to proceed by a fairly routine application of the theoretical and experimental methods which have already been so successful in connection with the mapping of the Fermi surfaces of the metallic elements. To a first approximation the very dilute alloys can be treated on the basis of the rigid-band model (see Section 5.1 and 5.77). However, in the remaining case of non-dilute random alloys there is a fundamental difference from the other cases in that even though the alloy may have a regular periodic arrangement of sites that may be occupied by metallic atoms, there will be a random distribution of the different species of atoms over all these sites. The use of the wave vector **k** as a sensible discriminator for labelling the states of electrons in metals is based on the application of Bloch's theorem, which in turn is based on the existence of a regular periodic potential $V(\mathbf{r})$. However, this periodicity no longer exists in a random alloy and therefore one might feel justified in questioning whether any meaning can actually be attached to the concept of a Fermi surface for such an alloy (see Section 7.2).

7.1. Metallic compounds

The principles involved in the study of the electronic properties of an ordered alloy or of a metallic compound, in which the atoms of different elements are regularly arranged on a lattice with some recognizable crystallographic structure, are not very different from those involved in the case of a crystalline metallic element. The theoretical and experimental methods involved in studying the band structure and Fermi surface of an ordered alloy or of a metallic compound are the same as those used in connection with metallic elements and they were discussed in considerable detail in Chapters 2 and 3. It appears that the first observations of the de Haas–van Alphen effect in a metallic compound were made with BiIn (Thorsen and Berlincourt 1961*a*; Saito 1962). In this section we shall not attempt to describe systematically all the work which has been performed so far on the determination of the shapes of the Fermi surfaces of ordered alloys and of metallic compounds because this is a field which is developing rapidly and any such account would soon become out of date. Instead we shall just describe the results of work which has been performed on a few typical representative materials.

7.1.1. CuZn (*β-brass*) *and related ordered alloys*, AgZn, PdIn

It appears that no band structure calculation had been performed on either an ordered alloy or a metallic compound until the work on β-brass (β'-CuZn) by Johnson and Amar (1965) and Arlinghaus (1965, 1967*b*, 1969) using the Green's function and A.P.W. methods, respectively. β-brass has an ordered alloy structure shown in Fig. 7.1 which is based on a primitive cubic Bravais lattice with, for example, Cu atoms at the corners of the unit cell and a Zn atom at the centre of the unit cell; there is therefore one formula unit of CuZn per unit cell. The unit cell of this superlattice is twice as large as the fundamental unit cell that would exist if the structure consisted

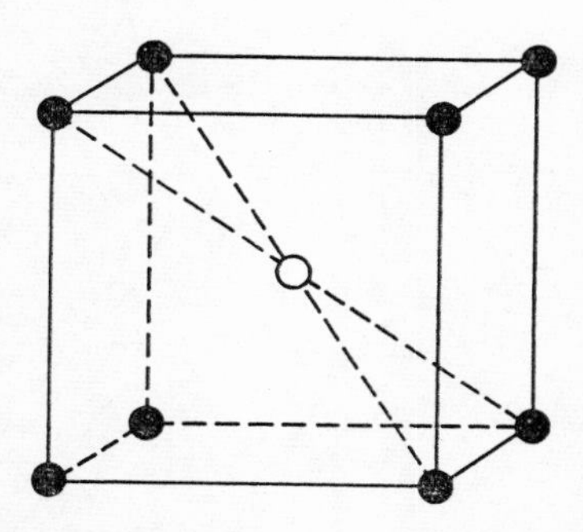

Fig. 7.1. Crystal structure of ordered β-brass (CuZn). The black spheres represent atoms of one element and the white spheres represent atoms of the other element.

entirely of one kind of atom arranged on the same crystallographic sites and which would possess the b.c.c. structure. The volume of the Brillouin zone of the ordered structure is therefore only half the volume of the Brillouin zone of that related b.c.c. structure (see, e.g., Slater 1951*b*). Remembering that in metallic Cu and Zn the 3d bands are full and therefore that Cu has one conduction electron per atom and Zn has two conduction electrons per atom, it is not unreasonable to suppose that β-brass has three conduction electrons per unit cell. In the free-electron Fermi surface for a primitive cubic crystal with three electrons per unit cell, band 1 is full except for small isolated pockets of holes at the corners, R, of the Brillouin zone. The Fermi surface in band 2 consists of a multiply-connected set of lenses centred at X.

The bands obtained by Arlinghaus (1969) for β-brass are shown in Fig. 7.2. There are two regions in which there is a large number of flat bands very close together. The lower of these two regions is very narrow ($\sim$0·05 Ry) and can be considered to originate from the atomic 3d states of the Zn, while

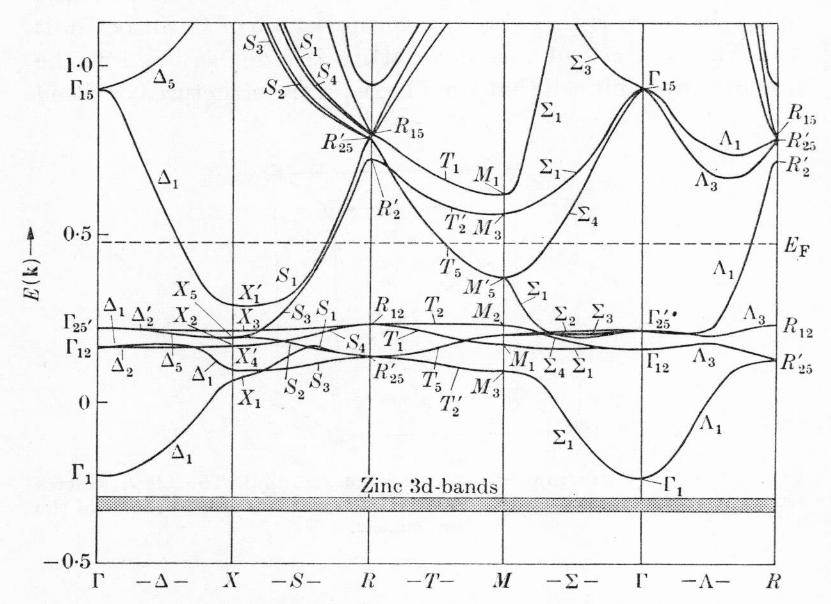

FIG. 7.2. Energy bands for ordered β-brass (Arlinghaus 1969).

the higher is not quite so narrow ($\sim$0·2 Ry) and can be considered to originate from the atomic 3d states of the Cu. In the remaining regions the bands are more typical of bands derived mainly from atomic s- and p-states which are encountered so often in simple metallic elements. The Fermi surfaces calculated by Arlinghaus (1965) and by Johnson and Amar (1965) agreed with each other with respect to all qualitative features and were also in fairly good quantitative agreement with each other, see Fig. 7.3. Closer agreement with the experimental results was obtained in later calculations (Arlinghaus 1969; Taylor 1969). The calculated Fermi surface resembles quite closely the Fermi surface that would be obtained for a free-electron metal with a primitive cubic Bravais lattice and with three conduction electrons per unit cell, which is just the number of conduction electrons that one would expect from one atom of Cu and one atom of Zn. The calculations indicated that the pockets of holes at R in band 1 were slightly smaller than predicted by the free-electron model. From Fig. 7.2 it can be seen that band 1 and

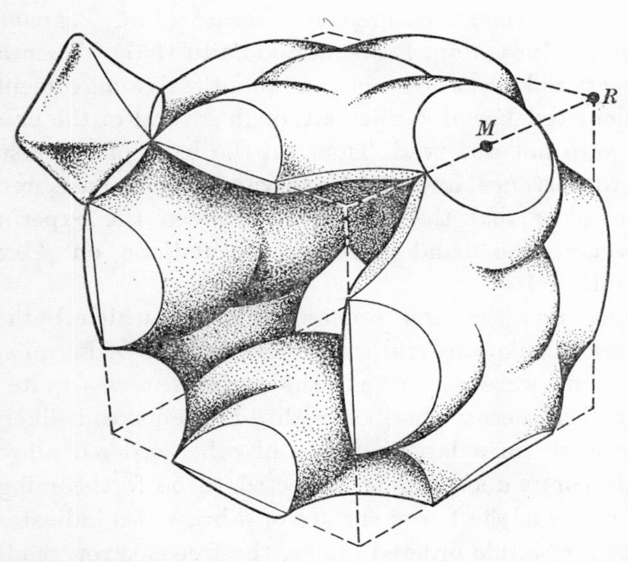

FIG. 7.3. The Fermi surface of ordered β-brass in the repeated zone scheme. The pocket centred on R is the hole surface in band 1 and the multiply-connected surface encloses electrons in band 2 (Sellmyer, Ahn, and Jan 1967).

band 2 form one twofold degenerate (T_5) band along MR in the region of E_F; consequently the pieces of Fermi surface in band 1 and band 2 must remain in contact at some point along MR. In the presence of an applied magnetic field this feature could give rise to orbits involving both bands. Evidence for the existence of the pockets of holes in band 1 is provided by the very low value of the Hall coefficient for β-brass (Frank 1955) which can be assumed to arise as a result of partial cancellation between the contributions from the holes in band 1 and the electrons in band 2 (Arlinghaus 1967*b*). The existence of a multiply-connected sheet of Fermi surface in β-brass was verified by the observation of the effects of open orbits in the results of magnetoresistance measurements by Sellmyer, Ahn, and Jan (1967). However, complete agreement with the magnetoresistance results could only be obtained by making some minor alterations to the calculated Fermi surface. A number of de Haas–van Alphen periods have been measured in β-brass (Jan, Pearson, and Springford, 1964; Jan, Pearson, and Saito 1967). The measured cross-sectional areas were in semiquantitative agreement with the calculated Fermi surface although several of the predicted orbits were not observed. Thus, on the basis of the results of magnetoresistance and de Haas–van Alphen measurements, it seems clear that the agreement between the experimental results and the band structure calculations on β-brass is fairly satisfactory.

β-brass was the first ordered alloy on which both band structure calculations and two different types of Fermi surface experiments were performed; the agreement was quite satisfactory. The success obtained with β-brass makes it likely that similar work on a large number of other ordered alloys and metallic compounds can be expected to be forthcoming. The actual shape of the Fermi surface of β-brass also indicates that, at least for simple ordered alloys, the free-electron model still provides a good first approximation to the shape of the Fermi surface. The work of Amar and Johnson (1965) and Jan, Pearson, and Saito (1967) indicated that, apart from one or two details, there is a close similarity between the Fermi surface of β-brass and the Fermi surfaces of ordered AgZn and PdIn which have the same crystal structure as β-brass.

7.1.2. $AuSb_2$

$AuSb_2$ exhibits the pyrites structure which is illustrated in Fig. 7.4 and which belongs to the space group $Pa3(T_h^6)$. The unit cell shown in Fig. 7.4 contains four formula units and the structure is based on a primitive cubic Bravais lattice; the corresponding Brillouin zone is illustrated in Fig. 7.5.

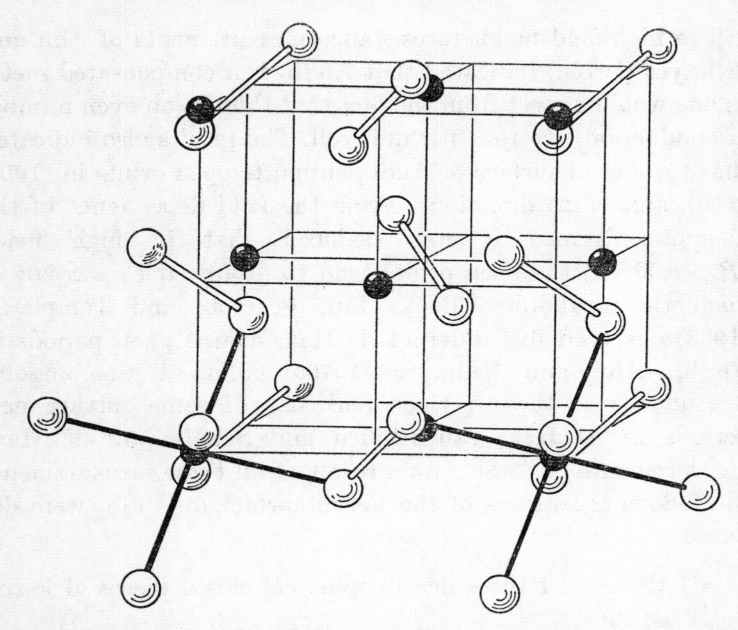

FIG. 7.4. Crystal structure of $AuSb_2$. The black spheres represent Au atoms and the white spheres Sb atoms (Ahn and Sellmyer 1970*a*).

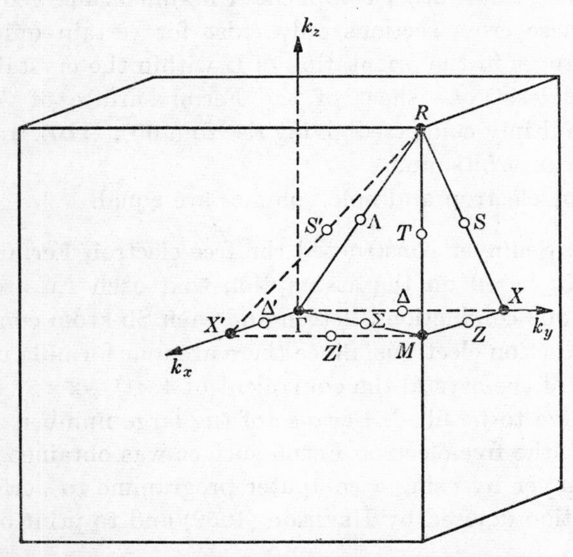

FIG. 7.5. The Brillouin zone for $AuSb_2$, structure $Pa3$ (T_h^6) (Bradley, Wallis, and Cracknell 1970).

The high-field magnetoresistance measurements of Ahn and Sellmyer (1970*a*) indicated that $AuSb_2$ is a compensated metal as one would expect from the fact that there is an even number of conduction electrons per unit cell. The results also indicated that the Fermi surface of $AuSb_2$ supports open orbits in $\langle 100 \rangle$, $\langle 110 \rangle$, and $\langle 112 \rangle$ directions. From the field dependence of the magnetoresistance it was deduced that in high fields ($B > 100$ kG) the open orbits tend to disappear as a result of magnetic breakdown. Beck, Jan, Pearson, and Templeton (1963) observed five different de Haas–van Alphen periods in $AuSb_2$. Ahn and Sellmyer (1970*b*) followed the angular dependence of these periods and also of some further new periods in de Haas–van Alphen and Schubnikow–de Haas measurements of $AuSb_2$. As a result of all these measurements the following features of the Fermi surface of $AuSb_2$ were deduced:

(i) there exist three nearly spherical closed pieces of Fermi surface,
(ii) there exist three extremal cross sections which apparently arise from rather complicated Fermi-surface geometries; these cross sections only exist for certain quite small ranges in the orientation of **B** within the crystal,
(iii) at least one sheet of the Fermi surface of $AuSb_2$ is multiply-connected giving rise to $\langle 100 \rangle$, $\langle 110 \rangle$, and $\langle 112 \rangle$ open orbits, and
(iv) the electron and hole volumes are equal.

Ahn and Sellmyer constructed the free-electron Fermi surface for $AuSb_2$ based on the assumption that each Au atom contributes one conduction electron and each Sb atom contributes five conduction electrons. Since there are four formula units per unit cell of the crystal the equivalent of $4 \times (1+2 \times 5) \times \frac{1}{2} = 22$ bands have to be filled. Because of the large number of bands involved, the free-electron Fermi surface was obtained by Ahn and Sellmyer by using a computer programme to perform the construction devised by Harrison (1959) and to print out cross sections of the intersections of various planes with the free-electron Fermi surface. The free-electron Fermi surface was found to exhibit the following features:

(i) pieces of Fermi surface exist in 14 bands. Of these the lowest three bands contain hole sheets and the remaining higher bands contain electron sheets; some of these sheets of the Fermi surface are multiply-connected;

(ii) on the basis of this Fermi surface one can predict 18 extremal cross sections for **B** parallel to ⟨100⟩ and 16 extremal cross sections for **B** parallel to ⟨110⟩,

(iii) many of the Fermi surface pieces in neighbouring bands are congruent in {100} and {110} planes; that is, extremal areas for **B** parallel to ⟨100⟩ and ⟨110⟩ predicted for many neighbouring bands are the same.

If a comparison is made between the free-electron Fermi surface and the experimental results it appears that the free-electron model does not give even a good first approximation to the band structure and Fermi surface of $AuSb_2$.

In all the successful applications of the free-electron model to pure metallic elements and to metallic compounds, it was always assumed that all the valence electrons go into the conduction bands, where the valence of an element was just determined by the number of the column in which that element occurs in the periodic table. For the cases in which it is successful this assumption can be justified by demonstrating that the experimental results agree with the predictions based on the free-electron Fermi surface. However, there is no *a priori* reason why this assumption should be valid and so, in order to try to obtain better agreement with the experimental results Ahn and Sellmyer (1970*b*) also constructed another possible free-electron Fermi surface for $AuSb_2$ in which it was assumed that each Au atom contributes one 6s-electron and each Sb atom contributes three 5p-electrons to the conduction bands. The remaining two 5s-electrons of each of the Sb atoms were regarded as tightly bound core electrons which are not in the conduction bands and therefore are not involved in the free-electron Fermi surface construction. However, this alternative free-electron Fermi surface was in no better agreement with the experimental results than the previous one. $AuSb_2$ was the first of the metallic compounds that did not contain transition metals which had a Fermi surface that was found to deviate significantly from the free-electron model based on the normal valences of the constituent elements.

7.1.3. Other ordered alloys and metallic compounds

We have so far described one simple ordered alloy structure (that of β-brass) for which the free-electron model gives a good first approximation to the Fermi surface of the metal, and one more complicated ordered alloy or metallic compound ($AuSb_2$) for which the free-electron model appears to give not even a rough first approximation to the Fermi surface of the metal. One could catalogue the results that have been obtained for a number of other ordered alloys and metallic compounds but the results available at the time of writing are fragmentary in most cases and at this stage no clear systematic account is possible. We simply note that materials on which some work relevant to the determination of the shape of the Fermi surface has so far been performed include $AuAl_2$, $AuGa_2$, $AuIn_2$, AuSn, BiIn, Cd_3As_2, CoFe, ReO_3, RuO_2, SiP_2, V_3Si, and several other metallic compounds.

7.2. Alloys

In this section we shall restrict our discussion to the consideration of binary alloys, that is alloys composed of just two chemical elements. We shall at first only consider very dilute alloys in which the concentration of one component, the host or solvent, is very much higher than the concentration of the other component, the impurity or solute. Typical concentrations of solute are probably less than one per cent and certainly less than about five per cent. The problem of more concentrated alloys is much more difficult to handle and only a brief mention of this will be given towards the end of the section. We are, of course, only concerned with alloys which exist as homogeneous solid solutions and we assume that the material possesses a regular crystallographic distribution of sites occupied by atoms. The material would therefore still possess the symmetry of one of the 230 space groups if no distinction were made between the two different species of atom that are present; however, the atoms of the two different species are distributed randomly over the sites and therefore the space-group symmetry is destroyed. We shall not consider the further complication presented by an amorphous material. There are three main factors that contribute to determining the structure that will be exhibited by an alloy of given concentration (see, e.g., Hume-Rothery 1931, 1936, 1966). These are, in roughly decreasing order of importance: the relative atomic sizes of the two constituents, the electro-negativity difference between the constituents, and the electron/atom ratio of the alloy. The atomic size can be considered either in terms of the volume per atom in the alloy or in terms of the distance between nearest-neighbour atoms in the alloy; the volume per atom is often a more basic quantity than the separation between nearest neighbours (see Massalski and King 1961; Massalski 1962; Nevitt 1966). Exchange interactions between the closed shells of two ion cores will lead to a repulsion between the ion cores, sometimes called the *Born–Mayer repulsion* (references are cited in Chapter 4 of Mott and Jones 1936); the magnitude of this repulsion can be expected to vary if the distance between the ion cores is altered. In the early days of work on alloy structures very great importance was attached to the electron/atom ratio, e/a, with particular significance being found in the special values of e/a of 3/2 (= 21/14), 21/13, and 7/4 (= 21/12). A considerable number of alloy structures were found to obey the celebrated rule given by Hume-Rothery (1931) based on these special values of e/a. It would not be appropriate in this book to enter into a lengthy discussion of the detailed application of Hume-Rothery's rule for alloy structures; for a detailed discussion we refer the reader, for example, to the relevant sections of the book by Barrett and Massalski (1966).

However, many exceptions to this rule have also been found, which is not really surprising since the electron/atom ratio is only one of several factors affecting the structure of an alloy.

In dealing with metallic compounds and ordered alloys in the previous section we were still concerned with materials which ideally possess perfectly regular crystal structures, that is, the crystal possesses the full translational symmetry of one of the fourteen Bravais lattices. Therefore, as a result of Bloch's theorem, each eigenstate of an electron in such a crystal can still be characterized by a unique value of $\mathbf{k}$, where the allowed values of $\mathbf{k}$ are determined by the periodicity of the crystal structure and are completely independent of the detailed forces acting on the electrons. From a purely formal point of view, even if only a very small number of impurity atoms are introduced into a crystalline specimen of a pure host metal, the perfectly regular periodic structure of the ideal pure crystal of the host metal will be destroyed. Consequently the wave vector $\mathbf{k}$ is no longer a good quantum number of the system and all the concepts of Brillouin zones and Fermi surfaces that are used so extensively for a pure metal are no longer strictly relevant. Any given one-electron state is characterized not just by a single wave vector $\mathbf{k}$ but by a spread of wave vectors $\mathbf{k}$; some evidence of this can be seen in positron annihilation measurements on dilute alloys (Stern 1968). Of course, even for a single crystal of the pure host metal there is always in practice a number of structural defects, so that the crystal will not possess a perfectly regular periodic structure because of the presence, at any given temperature T, of an equilibrium population of point defects and of dislocations. The presence of this quite small number of structural defects does not seriously affect the motions of the conduction electrons and therefore does not seriously affect the validity of $\mathbf{k}$ as a good quantum number for the conduction electrons. Similarly, for a very dilute alloy, the atoms of the solute will be sufficiently rare that the potential $V(\mathbf{r})$ experienced by a conduction electron in the metal will not be very significantly different from $V(\mathbf{r})$ in a pure specimen of the solvent. In terms of scattering powers and pseudopotentials, the pseudopotential that will be required to provide an adequate description of the scattering of a conduction electron in a dilute alloy will be only very slightly different from the corresponding pseudopotential for the pure host metal. Consequently the energy bands $E(\mathbf{k})$ will not be altered very significantly from the energy bands of the pure host metal. That this is not at all unreasonable can be justified by an adiabatic approximation argument allowing the concentration of the impurity atoms to tend continuously to zero. It is also not unreasonable to suppose that for a given atomic percentage of impurity concentration the magnitude of the difference between the pseudopotential for the alloy and that for the pure host metal will depend on the degree of chemical or electronic dissimilarity between the atoms of the host and impurity elements. Among the factors that we expect to be important are the relative sizes of the atoms of the solute and solvent and the electronegativities and valences of the solute and solvent. Thus, we should expect to find only very small changes in the shapes of the energy bands and of the Fermi surface of a pure noble metal (Cu, Ag, or Au) when a small percentage of one of the other noble metals is introduced. An extensive account of the earlier work on alloys of the noble metals is given by Massalski and King (1961).

Suppose that one considers an alloy in which the valence of the solute is different from that of the solvent but for which the atoms of the solute are roughly the same size as the atoms of the solvent. It is then reasonable to suppose that the principal effect of the presence of the solute atoms will be to change the average concentration of electrons per atom, that is the electron/atom ratio e/a, but that there will be only very small changes in the shapes of the energy bands $E(\mathbf{k})$. In these circumstances the rigid-band model can be expected to give a good first approximation to the band structure, and hence the shape of the Fermi surface, of a dilute alloy. This is particularly likely to be true if the solute and solvent both belong to the same row of the periodic table, when the electronic configurations of the ion cores of the two elements in the metallic form will be the same. We have already mentioned the rigid-band model in Chapter 5 in connection with the band structures of the transition metals. In that connection one was concerned with pure specimens of metallic elements and the electron/atom ratio was (usually) assumed to be an integer. As one moves from one element to another so e/a changes; consequently the position of the Fermi level, E_F, moves up or down through the energy bands and the shape of the Fermi surface changes accordingly. The rigid-band model can be extended for the description of the band structures of dilute alloys by allowing e/a to vary continuously instead of just taking values that are integers. If the valence of the solute is different from the valence of the solvent, the presence of the impurity atoms will change the value of e/a by a small percentage from its value for the pure solvent and this will cause the Fermi level, E_F, to move by a corresponding small amount. By using controlled amounts of an impurity of known valence and measuring the behaviour of one or other of the appropriate properties of a metal, one can often extract some information about the band structure of the pure solvent near the Fermi level, E_F. This is a well-established technique and it can be particularly useful in studying pieces of Fermi surface surrounding very small pockets of carriers; its use dates back at least to the very early work by Jones on the determination of the sizes of the pockets of the carriers in Bi (see Section 4.6).

The effects of the inclusion of impurities in a metal can be observed either in quantities such as the electrical resistivity, which depends among other things on the number of carriers, and the electronic contribution to the specific heat, which depends on $n(E_F)$, or in measurements of caliper dimensions and extremal areas of cross section of the Fermi surface. In principle, the effect of the addition of impurities could be observed in any property of a metal that depends on the behaviour of the conduction electrons (see Chapters 3 and 6). Thus, for example, experimental measurements of the paramagnetic susceptibility and of the electronic contribution to the specific heat of a dilute alloy have been made for numerous pairs of different solute and solvent metals, while de Haas–van Alphen measurements have been made for various solutes in hosts of, for example, Al

(Shepherd and Gordon 1968), As (Ishizawa 1968*b*), Bi (Brandt, Lyubutina, and Kryukova 1967), Cd (Verkin, Kuzmicheva, and Svechkaryov 1968), Cu (Chollet and Templeton 1968), Sb (Ishizawa 1968*a*), and Zn (Hedgecock and Muir 1963; Higgins, Marcus, and Whitmore 1964; Higgins and Marcus 1966, 1967). Other experimental work which has been performed includes measurements on photo-emission, positron annihilation, galvanomagnetic effects, and magnetothermal oscillations.

In some superconducting metals the presence of impurities has an interesting effect on the behaviour of the superconducting transition temperature, T_c, as a function of pressure and this can be used to give some information about the topology of the Fermi surface of the metal. If there is a discontinuity in $\partial T_c/\partial P$, as a function of impurity concentration, this indicates a change in the number of depressions (or concavities) on the Fermi surface as the average number of conduction electrons per atom is changed by varying the amounts of impurity present (Makarov and Bar'yakhtar 1965). This effect has been observed in In (Makarov and Volynskiĭ 1966, 1969) and Tl (Lazarev, Lazareva, Ignat'eva, and Makarov 1965; Ignat'eva, Makarov, and Tereshina 1968). In the case of In the topological changes were assigned to changes in the Fermi surface in band 3 and it was assumed that the change involved was the breaking up of the toriods, which exist in pure In, into separated ellipsoids, see Fig. 7.6.

Let us suppose, for a moment at least, that the rigid-band model does give a completely accurate description of the band structure of a dilute alloy. Any of the topological properties which are only related to the shape of the Fermi surface will then be able to be explained completely by following the change in the electron/atom ratio with impurity concentration, and thence determining the corresponding movement of the Fermi level and the resulting changes in the shape of the Fermi surface. However, for the 'indirect' or 'macroscopic' properties the situation will be more complicated. The behaviour of such a property depends not only on which states, **k**, are actually at the Fermi surface, but also on the relaxation times $\tau_{\mathbf{k}}$ of these states

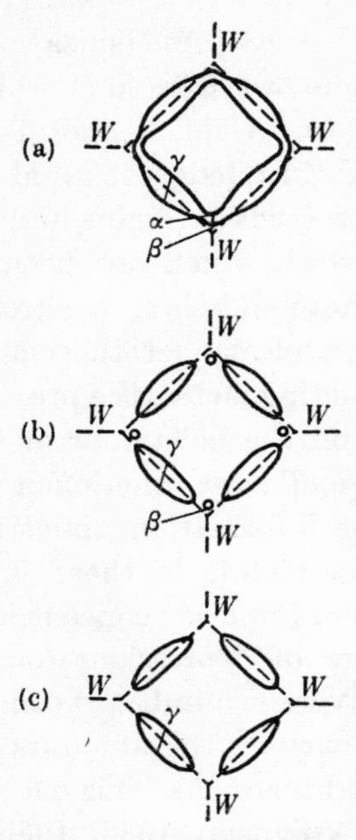

FIG. 7.6. Cross sections of the Fermi surface of In in band 3, β arms: (a) pure, (b) with 1·2 atomic per cent Cd, and (c) with 1·9 atomic per cent Cd (Makarov and Volynskiĭ 1966).

for the process in question; the theoretical determination of values of $\tau_{\mathbf{k}}$ would involve the calculation of the appropriate matrix elements $\int \psi^*_{\mathbf{k}'}(\mathbf{r})X\psi_{\mathbf{k}}(\mathbf{r})\,d\mathbf{r}$ for transitions from the states **k** under the influence described by the operator X. Even within the rigid-band model there is no reason to suppose that such matrix elements should remain constant while **k** is allowed to vary corresponding to the changes in the shape of the Fermi surface. Therefore, even within the rigid-band model, a complete explanation of the behaviour of the indirect or macroscopic properties of an alloy is far from simple. Since the rigid-band model involves using for a dilute alloy the matrix elements or the relaxation times that apply to the pure host metal, it is clear that it takes no account of the extra scattering of the conduction electrons, by collisions with impurity atoms, which is present in an alloy. This additional scattering which is not present in the case of the pure host metal can be expected to reduce the value of the relaxation times $\tau_{\mathbf{k}}$. This can be looked at in a slightly different way. The use of **k** to provide a meaningful label to discriminate between different allowed states of an electron in a specimen of a pure metallic element is a consequence of the use of Bloch's theorem which, in turn, applies because the crystal has a regular periodic structure. Although from a formal point of view the introduction of a small percentage of randomly distributed solute atoms destroys the periodic nature of the lattice, substantial traces of the periodic structure of the pure solvent will still survive in a dilute alloy. Consequently in practice we can still talk meaningfully about the wave vector **k** of an electron, although it is perhaps slightly 'smudged' or 'blurred'; therefore we can also still talk meaningfully about the concept of a Fermi surface in a dilute alloy although it too will be slightly smudged or blurred (for a further discussion of the validity of the rigid-band model for dilute alloys see Stern 1967, 1968, 1969, 1970).

Although the rigid-band model is clearly only a first approximation to the description of the behaviour of the electrons in a disordered alloy it has met with some considerable qualitative success, at least when the atoms of the solute are not too dissimilar from the atoms of the solvent and when the concentration of the solute is very low. Nevertheless there comes a stage at which it is necessary to adopt a rather more sophisticated approach than that of the rigid-band model. For example, the rigid-band model does not provide a completely satisfactory explanation of the electronic specific heat of dilute alloys of the noble metals (Stern 1967). There are two important aspects of the model that need to be improved. One is to allow for changes in the shapes of the bands, $E(\mathbf{k})$, and in the density of states, $n(E)$, when solute atoms are introduced. The other is to allow for changes in the relaxation times $\tau_{\mathbf{k}}$ as a consequence of the extra scattering which is caused by the impurity atoms. Related to the problem of changes in the relaxation time, $\tau_{\mathbf{k}}$, there is also the difficult problem of determining the extent to which the various electrons in a disordered alloy are localized (Anderson

1958; Ziman 1968, 1969*c*). The firstof these modifications will affect both the topological and the macroscopic properties of the metal but the second will only affect the macroscopic properties. One of the first serious attempts to consider the variation of the shapes of the bands $E(\mathbf{k})$ in dilute alloys, apart from a rather formal treatment by Lifshitz (1963), was made by Cohen and Heine (1958) in connection with the noble metals and their

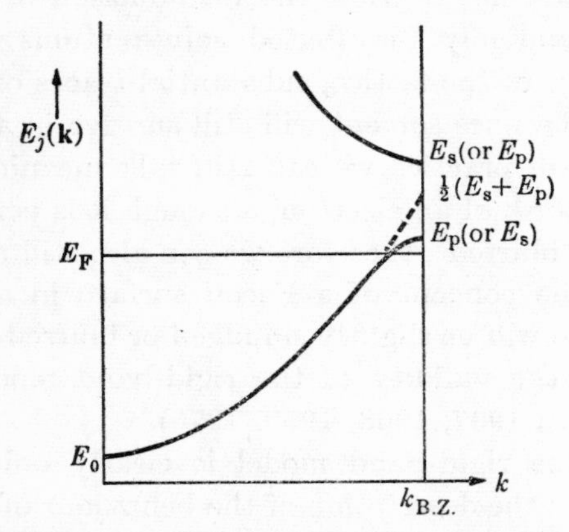

FIG. 7.7. Separation of energy bands at a Brillouin zone boundary (Cohen and Heine 1958).

α-alloys. Cohen and Heine suggested that for monovalent metals the wave functions $\psi_{\mathbf{k}}(\mathbf{r})$ at the surface of the Brillouin zone are either s-like or p-like in character. According to their model, the band gap at the point L on the boundary of the Brillouin zone can be expressed as the difference $|E_s-E_p|$, see Fig. 7.7. For a monovalent metal in the free-electron approximation the Fermi surface would be a sphere that does not reach the Brillouin zone boundary at all. In the free-electron model the band gap at the point L on the surface of the Brillouin zone would vanish and so the size of this band gap can be used to give an indication of the extent of the departure from the free-electron approximation that occurs in the band structure of a given real monovalent metal. Therefore the distortion of the Fermi surface from a sphere will be related to the magnitude of the band gap at L and Cohen and Heine introduced a parameter d_{sp} as a measure of the distortion of the spherical free-electron Fermi surface that occurs in a real metal where

$$d_{sp} = \frac{\frac{1}{2}(E_s - E_p)}{\frac{1}{2}(E_s + E_p) - E_F}. \quad (7.2.1)$$

When $|d_{sp}| \geqslant 1$, either E_s or E_p is depressed below the Fermi level so that the Fermi surface bulges out enough to touch the surface of the Brillouin zone. Cohen and Heine proposed that the band gap $|E_s - E_p|$ for a pure noble metal could be estimated by

$$E_s - E_p = -(\Delta_{sp})_{solvent} + [(BC)_s - (BC)_p] + \tfrac{5}{4}k_F^2, \quad (7.2.2)$$

where Δ_{sp} is the s-p excitation energy of a free atom and $(BC)_s$ and $(BC)_p$ are boundary correction terms for the Schrödinger equation for the s- and p-states respectively. $(E_s - E_p)$ was calculated to be 0·104, 0·007, and −0·055 Ry respectively for Cu, Ag, and Au; that is, it was predicted that for Cu and Ag $E_p < E_s$ whereas for Au $E_p > E_s$. The corresponding values of d_{sp} are then 0·40, 0·04, and −0·27 respectively for Cu, Ag, and Au. Later experimental evidence showed that the Fermi surfaces of the noble metals are sufficiently distorted that they touch the Brillouin zone boundary at L (see Section 5.2) and that these values of d_{sp} are too small. Nevertheless, the degree of distortion from the free-electron Fermi surface, as measured, for example, by the area of cross section of the necks at L, was found experimentally to be greatest in Cu and least in Ag, which is in agreement with the trend exhibited by the values of d_{sp} estimated by Cohen and Heine.

If impurity atoms are now introduced into a specimen of the pure metal and if we relax the requirements of the rigid-band model, then E_s and E_p will vary as the concentration of the solute is varied. The term *soft-band model* is sometimes applied to any treatment in which the shapes of the bands are allowed to change as the concentration of the solute is varied. For an alloy $|E_s - E_p|$ is derived from an average of the values of Δ_{sp} for the solute and solvent elements and so the addition of solute atoms to a pure specimen of a noble metal introduces a change in the band gap. These changes in E_s and E_p lead to a change in the value of d_{sp} and this corresponds to a change in the shape of the constant-energy surfaces. Therefore, even if the valence of the solvent atoms is the same as the valence of the solute atoms so that the value of e/a is not altered, there will be a change in the shape of the Fermi surface on alloying. If the valence of the solvent atoms is different from the valence of the solute atoms there will be changes in the shape of the Fermi surface as a result of changes in both the shape of the bands $E(\mathbf{k})$ and the electron/atom ratio. Cohen and Heine introduced another parameter, η_{sp}, which enables one to compare at a given electron concentration the relative effect of a particular solute element on the band gap of the solvent, where

$$\eta_{sp} = \left(1+\frac{1}{(e/a)-1}\right)[(\Delta_{sp})_{\text{solute}}-(\Delta_{sp})_{\text{solvent}}]. \qquad (7.2.3)$$

For alloys in which the solvent is one of the noble metals and the solute is one of the B-subgroup elements from the four columns that follow the noble metals in the periodic table, the solute always has a larger value of Δ_{sp} than the solvent so that on adding the solute to a specimen of the pure solvent, E_p will be raised relative to E_s. Hence, the addition of solute atoms to Cu should decrease the band gap and cause the Fermi surface to become more spherical. For Au, on the other hand, the addition of solute atoms should cause the band gap to be increased still further resulting in an even more distorted Fermi surface. Although no mention has been made of them so far in the above discussion it must, of course, be remembered that in each of the noble metals there is an occupied d-band just below the conduction band and this may have appreciable effects on the detailed shape of the conduction band.

Although the quantitative results of the work of Cohen and Heine (1958) have now been superseded, their work marked the beginning of serious attempts to study the band structures of dilute alloys without using the assumptions involved in the rigid-band model. Several more sophisticated theoretical schemes have now been proposed for the calculation of the electronic structures of disordered alloys (Anderson 1958; Harrison 1958; Johnson and Amar 1965; Amar, Johnson, and Wang 1966; Soven 1966, 1969; Amar, Johnson, and Sommers 1967; Stocks and Young 1968; Velický, Kirkpatrick, and Ehrenreich 1968; Ziman 1968, 1969*c*). An extensive account of recent developments, particularly in terms of the use of pseudopotential theory, has been given by Heine and Weaire (1970). These developments are of some general interest since they are not necessarily restricted to very dilute alloys. As a starting point one can suppose that each of the sites in the alloy is occupied by a fictitious 'average' ion for which the pseudopotential is given by the weighted mean

$$\bar{v}(k) = cv_B(k)+(1-c)v_A(k), \tag{7.2.4}$$

where $v_A(k)$ and $v_B(k)$ are the pseudopotentials of the A and B ions and c is the concentration of the B atoms in the alloy. Since we are assuming throughout this section that these sites form a crystallographic array, each eigenstate of the electrons for this 'average' crystal will be characterized by a wave vector $\mathbf{k}$ where there is no approximation involved in using $\mathbf{k}$ as a good quantum number. It is then possible to use the average pseudopotential $\bar{v}(k)$ given in eqn (7.2.4) to calculate a band structure $E(\mathbf{k})$, the wave functions $\psi_{\mathbf{k}}(\mathbf{r})$ and the Fermi surface. This is sometimes called the *virtual crystal approximation.* One can then use many-body perturbation theory (see, e.g., March, Young, and Sampanthar 1967; Ziman 1969*b*), carried in principle to infinite order, to determine the effect of introducing (or 'turning on') as a perturbation the differences $(v_A(k)-\bar{v}(k))$ and $(v_B(k)-\bar{v}(k))$ on the sites occupied by A and B atoms respectively. The pseudopotential of the alloy can then be written as the sum of a mean pseudopotential given by eqn (7.2.4) and a residual part due to the difference between the A and B atoms

$$V(\mathbf{k}) = [\mathrm{S}_A(\mathbf{k})+\mathrm{S}_B(\mathbf{k})]\bar{v}(k)+[c\mathrm{S}_A(\mathbf{k})-(1-c)\mathrm{S}_B(\mathbf{k})]v_{\mathrm{diff}}(k), \tag{7.2.5}$$

where

$$\mathrm{S}_A(\mathbf{k}) = \frac{1}{N}\sum_i \exp(\mathrm{i}\mathbf{k}\,.\,\mathbf{R}_i), \tag{7.2.6}$$

where the summation is over A atom sites,

$$S_B(\mathbf{k}) = \frac{1}{N}\sum_i \exp(i\mathbf{k}\,.\,\mathbf{R}_i), \tag{7.2.7}$$

where the summation is over B atom sites, and

$$v_{\text{diff}}(k) = v_A(k) - v_B(k). \tag{7.2.8}$$

The ground state of the perturbed system, which represents the real alloy, can be expanded in terms of determinants of the unperturbed Bloch wave functions $\psi_{\mathbf{k}}(\mathbf{r})$. We can then define rigorously a weight function $w(\mathbf{k})$ involving the number of these determinants in which the Bloch function $\psi_{\mathbf{k}}(\mathbf{r})$ is included and the coefficients of these determinants in the expansion. The weight function $w(\mathbf{k})$, which indicates the extent to which the

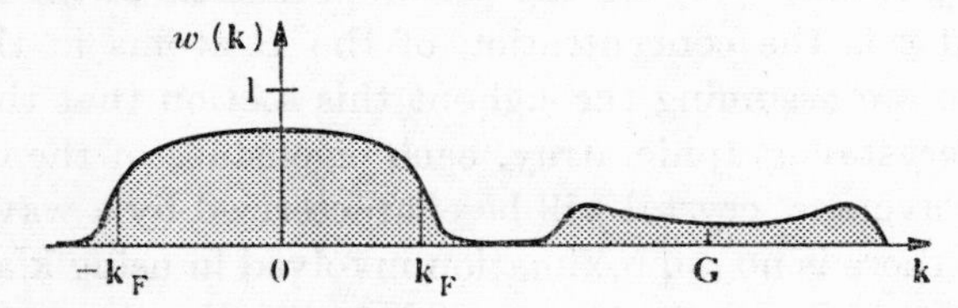

FIG. 7.8. Weight function $w(\mathbf{k})$ for a disordered alloy. $\mathbf{G}$ is a reciprocal lattice vector (after Heine and Weaire 1970).

Bloch function $\psi_{\mathbf{k}}(\mathbf{r})$ is included in the ground state of the alloy, will take values between 0 and 1 and it will look similar to the Fermi–Dirac distribution function. The form of $w(\mathbf{k})$ is sketched in Fig. 7.8; the main part of the function is centred on the origin with satellites around the other reciprocal lattice points $\mathbf{G}_n$.The Fermi level E_F in the virtual crystal lies at $\mathbf{k}_F$ which is very close to the value of $\mathbf{k} = \mathbf{k}_{\frac{1}{2}}$ at which $w(\mathbf{k})$ drops to half of its maximum value. This value of $\mathbf{k}_{\frac{1}{2}}$ could be taken as defining the Fermi surface of an alloy since the usual definition of the Fermi surface is no longer strictly valid once the wave vector $\mathbf{k}$ ceases to be a good quantum number for the eigenstates of the electrons in the material.

The concept of the Fermi surface of an alloy, now rigorously defined in terms of $w(\mathbf{k})$, can be expected still to be useful for a very dilute alloy as we have already seen earlier in this section. If the mean free path of an electron with wave vector **k** is $\lambda_{\mathbf{k}}$, the Fermi surface in the region of **k** will be blurred by an amount $\Delta\mathbf{k}$ which is of the order $1/\lambda_{\mathbf{k}}$. For a very dilute alloy the mean free path $\lambda_{\mathbf{k}}$ will not be very much reduced from the value in the pure solvent and the blurring of the Fermi surface will be quite small. As the concentration of the solvent is increased, so $\tau_{\mathbf{k}}$ and $\lambda_{\mathbf{k}}$ will decrease, and the blurring of the Fermi surface will be increased; in such circumstances the usefulness of the concept of the Fermi surface may well be correspondingly reduced. Of course, $\lambda_{\mathbf{k}}$ depends strongly on temperature so that while

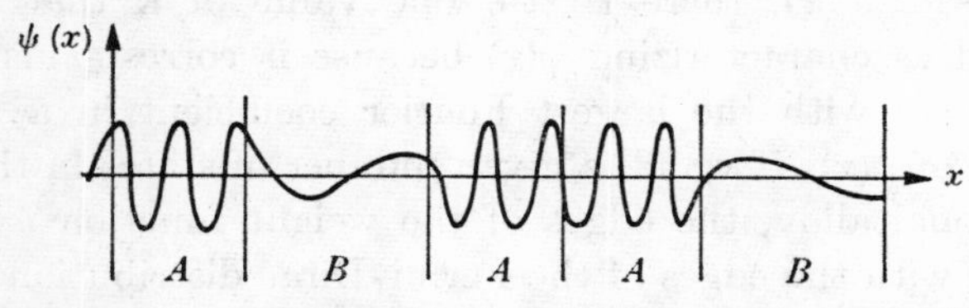

FIG. 7.9. Schematic wave function for an electron in a one-dimensional alloy showing high electron density at A atoms and low electron density at B atoms (after Heine and Weaire 1970).

impurity scattering is usually dominant in a concentrated alloy, nevertheless an alloy at a very low temperature near absolute zero may be a closer approximation to a perfect crystal than a pure metal at room temperature. Whereas it is perhaps fashionable to question the usefulness of the concept of the Fermi surface for concentrated alloys, it should also be borne in mind that similar objections can be made to the use of the concept of the Fermi surface of a pure metal at high temperatures (see also Section 1.3).

That the definition of the Fermi surface of an alloy based on the use of the weight function $w(\mathbf{k})$ is not unreasonable and approximates more and more closely, as the concentration of solute is reduced, to the conventional definition of the Fermi surface of a pure metallic element or compound, can be illustrated by considering a hypothetical one-dimensional alloy. The wave function $\psi(x)$ of a single electron in a disordered alloy is illustrated schematically in Fig. 7.9 in which the wavelength has been reduced to highlight the variation, and the long-range variations of the amplitude of $\psi(x)$ have been ignored. For such a one-dimensional system there is a theorem which states that the $(n+1)$th wave function has n nodes. The principal Fourier component of $\psi(x)$ will be determined by the mean distance between these nodes and this will be the same as the wavelength of the plane wave with n nodes which is the $(n+1)$th solution of the Schrödinger equation for the free-electron problem corresponding to a linear chain of atoms of the same length as our one-dimensional alloy and with the potential $V(x)$ set equal to zero. Therefore, although $\mathbf{k}$ is no longer strictly a good quantum number for $\psi(x)$, there is still one value of $\mathbf{k}$ that can be regarded as characterizing $\psi(x)$ because it corresponds to the plane wave with the largest Fourier coefficient in a Fourier analysis of $\psi(x)$. If we fill a given number of states in this one-dimensional alloy, the edges of the weight function $w(\mathbf{k})$ will coincide with the edges of the Fermi–Dirac distribution for the same number of occupied states in the corresponding one-dimensional free-electron problem. The existence of the satellite structure (see Fig. 7.8) in $w(\mathbf{k})$ reduces the amplitude of $w(\mathbf{k})$ in the central region but does not affect the width of the distribution. These arguments can be extended to the three-dimensional case but the mathematical details will be more complicated.

Although the form of the wave function for a conduction electron in a random alloy A_xB_y will be different according to whether the electron is close to an A atom or to a B atom, one can expect that, at least to a good first approximation, the form of the wave function near to an A or B atom in the alloy will resemble quite closely the corresponding wave function for a pure specimen of metal A or B respectively (Friedel 1954; Kittel 1963*b*). The argument behind this is analogous to the classical thermodynamic argument that leads to the conclusion that the energy density of black-body radiation in a box is independent of the configuration or physical properties of the walls of the box. Experimental evidence in favour of this behaviour of the electrons in an alloy can be obtained from measurements of the soft X-ray spectra of the atoms in an alloy

(Appleton and Curry 1965; Rooke 1968). In particular the bandwidths of the spectra of the two atoms were very close to the bandwidths for the two respective elements whereas on the basis of the rigid-band model only one common bandwidth would be expected for the spectra from both atoms in the alloy. Therefore we see that for an alloy there are some properties, which can be described as *atomic properties*, which because they depend on the local behaviour of the wave function near to an A or B atom will be very similar to those properties for a pure specimen of metal A or metal B respectively. However, there will be other properties of an alloy which will depend on the long-range variation of the wave function and we can refer to such properties as *band properties* (Cohen 1963).

7.3. The metal–insulator transition

In the first six chapters of this book we confined our attention to pure crystalline specimens of metals consisting of a single metallic element. In the first two sections of this chapter we have extended our discussion to include alloys and metallic compounds. It seemed to be appropriate in conclusion to make some mention of the possibility that certain materials may exhibit transitions between the metallic and non-metallic states. The study of the metal-insulator transition is an area of considerable research activity at present and, therefore, the brief account given in this section may quickly become obsolete.

Although the conventional description of the difference between metals and insulators in terms of a model of non-interacting electrons (Wilson 1931) has been very successful, it does not always work (Verwey and de Boer 1936, de Boer and Verwey 1937). For example, in a transition metal oxide such as NiO the d-bands are not full and this would suggest that NiO should be a metal, which is not in fact the case. We have encountered a related difficulty in connection with the position and occupation of the 4f-levels in the rare-earth metals (see p. 383). Wigner (1938) showed that when the Coulomb repulsion $e^2/4\pi\varepsilon_0|\mathbf{r}_{12}|$ between the electrons is considered, a free-electron gas at low densities would form a non-conducting phase.

Suppose that we consider an array of atoms in an arbitrary crystal structure in which each atom consists of a well-defined ion core with net charge $+e$ and one outer electron. Because of the presence of the ion cores this is a more complicated system than the ideal gas of free electrons considered by Wigner (1938) in which the existence of a transition between conducting and non-conducting phases was predicted. We now investigate the behaviour of the electrons under the combined influence of the Coulomb electrostatic forces between them and the background of positive charges when we make imaginary changes in $\bar{a}$, the average interatomic spacing. This is not as futile an exercise as it might appear to be at first sight. For a monatomic solid some relatively small changes in $\bar{a}$ can be achieved by variation of the external pressure. However, the model also gives a good first approximation to the behaviour of the impurity atoms in a doped semiconductor: if the valence of the impurity atoms is different from the valence of the host material, the host material acts as an inert background which dilutes the system of the impurity atoms. A very wide range of values of $\bar{a}$, the mean separation of the impurity atoms, can be studied by varying the impurity concentration and, in particular, very large values of $\bar{a}$ can be achieved by using very low levels of doping. Another example of a system of metal atoms in which wide variations in $\bar{a}$ can be achieved is the case of a solution of an alkali metal in ammonia (Thompson 1968; Catterall and Mott 1969).

If we consider the ground state of a one-electron atom in free space the electron will be in an atomic orbital with a well-defined mean radius, $\bar{r}$. If $\bar{a} \gg \bar{r}$ it is reasonable to suppose that the electron in any given atom will be only very slightly perturbed by the presence of the other atoms in the material. Kohn (1964) has given a formal proof that an array of atoms of this kind is non-conducting at low densities and there is considerable experimental evidence for this, provided both by the transition metal oxides and by doped semiconductors. However, if the density of the material is increased, that is $\bar{a}$ is decreased, the wave functions of the electrons in the different atoms will overlap so that the electrons are no longer completely localized on particular atoms in the crystal. The atomic energy levels of the electrons will spread out into bands and, depending on the detailed nature of the band structure, the material may become a metal at low values of $\bar{a}$. Mott (1949, 1956, 1958, 1961, 1967) gave reasons for supposing that if we make these imaginary changes in $\bar{a}$ at $T = 0$, this change from a non-metal to a metal would not just happen gradually but would be a sharp change of character from a non-metallic phase with no free carriers when $\bar{a}$ is large to a metallic phase with a large density of carriers when $\bar{a}$ is small. This transition between the metallic and non-metallic states has come to be called the *Mott transition;* a recent review is given by Mott and Zinamon (1970). The material may be either a metal, with unoccupied states immediately above the Fermi level, or an insulator with the Fermi level at the top of the conduction band and a gap immediately above the Fermi level; however small this gap may be, the restriction to $T = 0$ eliminates the possibility of semiconducting behaviour which might complicate matters at non-zero temperatures.

The argument given by Mott which leads to the conclusion that there is a sharp transition between the metallic and non-metallic states is roughly as follows. In order to produce a pair of carriers—an electron and a hole—in the non-metallic state it is necessary to provide an activation energy of at least $2E$, where $2E$ is the gap between the conduction band and the valence band. The value of $2E$ can be estimated by a simple argument because the electrostatic force of attraction between an electron and a hole is capable of leading to the existence of a bound state, an *exciton.* The energy required for this cannot be less than a quantity of the order of $m^*e^4/32\hbar\pi^2\varepsilon_0^2\varepsilon^2$,[2] where ε is the dielectric constant of the surroundings and m^* is the reduced effective mass given by $1/m^* = 1/m_1+1/m_2$ where m_1 and m_2 are the effective masses of the electron and of the hole. This gives an estimate of a lower limit for the value of the band gap $2E$:

$$2E \sim \frac{m^*e^4}{32\hbar^2\pi^2\varepsilon_0^2\varepsilon^2}. \tag{7.3.1}$$

The tendency to form a bound state of an electron-hole pair will be reduced because the electrostatic force between the electron and the hole will be screened by any other free carriers that may be present. A first-order transition can be expected to occur, in which n, the number of free carriers at $T = 0$, changes from zero to a finite value n_0; n_0 must be sufficiently large to give a screening constant q such that the screened Coulomb potential

$$V(r) = (e^2/4\pi\varepsilon_0\varepsilon r)\exp(-qr) \tag{7.3.2}$$

does not give rise to bound states. If this condition is applied to the transition to the metallic state that occurs in heavily doped Ge and Si it gives

$$n_0^{\frac{1}{3}}a_{\mathrm{H}} \sim 0{\cdot}2 \tag{7.3.3}$$

where a_{H} is the hydrogenic radius given by

$$a_{\mathrm{H}} = \frac{\hbar^2 4\pi\varepsilon_0\varepsilon}{m^*e^2}\,. \tag{7.3.4}$$

Equation (7.3.3) was found to give quite a satisfactory estimate of the concentration of carriers at which metallic conductivity occurs in doped Ge and Si (Mott 1961, Mott and Twose 1961).

We now turn to the consideration of the other experimental evidence for the existence of the metal–non-metal transition. First, one can alter $\bar{a}$ by varying the pressure although the changes that can be achieved in this way are bound to be quite small. The metals that are most likely to exhibit a transition to the non-metallic state are those metals which are compensated and for which the Fermi surface consists of small isolated pockets of holes in the conduction band and small isolated pockets of electrons in the valence band. From Chapter 4 we see that metals that satisfy this condition include Ca, Sr, graphite, As, Sb, and Bi. The effect of pressure on the conductivity of most of these metals has been investigated either directly or by determining the variation of the sizes of the pockets of carriers as a function of pressure. The electrical conductivity of Ca decreases steeply under the application of increasing pressure, reaching a minimum at about 400 kbar; Sr behaves similarly but reaches its minimum at a much lower pressure, namely at about 35 kbar (Stager and Drickamer 1963; Drickamer 1965). By calculating the band structure of Ca for various different values of the lattice constant, Altmann and Cracknell (1964) showed that if the pressure is increased the sizes of the pockets of holes in band 1 and of electrons in band 2 will diminish and eventually the pockets will vanish completely. Were it not for a degeneracy at one point along the line LW in the Brillouin zone there would be a complete separation between bands 1 and 2 and a gap would appear thereby making Ca into an intrinsic semiconductor or, at $T = 0$, an insulator. However, because of this isolated degeneracy there is never actually a gap between bands 1 and 2 as the pressure is raised and there always remains this isolated point of contact so that Ca is a semi-metal rather than a semiconductor. The effect of pressure on the Schubnikow–de Haas oscillations in Bi has been studied experimentally with pressures up to 15 kbar (Itskevich and Fisher 1967*b*; Itskevich, Krechetova, and Fisher 1967); all the measured cross sections, both of the electron and of the hole ellipsoids, were found to decrease with increasing pressure. The effect of pressure on the electrical conductivity of Bi was measured for pressures up to 25 kbar by Balla and Brandt (1964) and it was predicted that at a pressure of about 26 kbar the overlap between the valence and conduction bands in Bi should disappear completely so that at low temperatures Bi would become an insulator (see also Fal'kovskiĭ 1967, and Brandt, Minina, and Pospelov 1968). Under pressure the carrier density in As is thought to decrease (Brandt and Minina 1968) as occurs in Bi. A calculation by Arkhipov, Kechin, Likhter, and Pospelov (1963) based on the Slonczewski–Weiss model predicted a 23 per cent increase in the total number of carriers in graphite at a pressure of 10 kbar. Good agreement with this theory was obtained in the measurements of the resistance, Hall effect, and magnetoresistance made by Likhter and Kechin (1963). Measurements of the Schubnikow–de Haas oscillations in graphite at pressures up to 8 kbar by Itskevich and Fisher (1967*a*) gave results in agreement with the results of the calculations of Arkhipov, Kechin, Likhter, and Pospelov (1963). Investigations of the pressure dependence of the cross-sectional area of the Fermi surface of Sb by Minomura, Tanuma, Fujii, Nishizawa, and Nagano (1969) using the de Haas–van Alphen effect showed that the cross-sectional area increases with pressure. The corresponding increase in carrier density with pressure was observed by Brandt and Minina (1968), although according to the galvanomagnetic work of Kechin (1967) at pressures up to 10 kbar the ellipsoids in Sb become elongated but thinner so that the total number of carriers is very nearly constant (see also Kechin, Likhter, and Pospelov 1965).

The simple idea of the Mott transition involves a non-metallic state for large values of $\bar{a}$ and a metallic state for small values of $\bar{a}$. Since the materials mentioned in the previous paragraph are already metals at atmospheric pressure this means that if the pressure is increased, and therefore $\bar{a}$ is decreased, the sizes of the pockets of carriers, and therefore also the conductivity, should be expected to increase. However, we have seen that for the four materials Ca, Sr, As, and Bi the electrical conductivity decreases with increasing pressure which is the reverse of the behaviour predicted on the basis of simple ideas of the Mott transition. On the other hand, for graphite and Sb the electrical conductivity and carrier density do increase with increasing pressure as predicted, but even at zero pressure $\bar{a}$ is still too small for the transition to the non-metallic state to occur and these materials remain metals. Although it now seems quite likely that no examples of the Mott transition can be found in pure monatomic metals, one might expect to be able to observe the transition in a monatomic insulator or intrinsic semiconductor by the application of a sufficiently high pressure. However, this too is thought to be unlikely (Mott 1968) because it would be accompanied by a lattice distortion.

While it seems unlikely that there are any examples of pure monatomic solids in which the Mott transition can be observed there is still the possibility that examples of the transition may be observed in ionic crystals such as transition metal oxides, in doped semiconductors, or in solutions of alkali metals in ammonia. For systems of this type it is possible to find several examples of sharp transitions between conducting and non-conducting phases for which the mechanism involved in the Mott transition, namely the effect of the Coulomb electrostatic interactions between the electrons, can be invoked as one possible explanation of the transition. We mention one or two examples and give references to sources in which more complete discussions of these systems can be found. The conductivity of V_2O_3, in common with that of VO, shows a jump by a factor of 10^5 at a temperature near 200 K and this transition temperature drops with increasing pressure; one of the several explanations which have been advanced is that it is an example of the Mott transition (Mott 1967). We have already mentioned the case of a transition to metallic conductivity in heavily doped Ge and Si (see p. 510) for which it appeared that Mott's theoretical estimates of the concentrations of carriers at which this occurs were quite close to the experimental values (for further details see Mott 1961; Mott and Twose 1961).

BIBLIOGRAPHY

ABARENKOV, I. V. and VEDERNIKOV, M. V. (1966) *Fizika tverd. Tela* **8,** 236. English trans: *Soviet Phys. Solid St.* **8,** 186.

ABELÈS, B. and MEIBOOM, S. (1956) *Phys. Rev.* **101,** 544.

ABRIKOSOV, A. A. (1969) *Zh. éksp. teor. Fiz.* **56,** 1391. English trans: *Soviet Phys. JETP* **29,** 746.

—— and FAL'KOVSKIĬ, L. A. (1962) *Zh. éksp. teor. Fiz.* **43,** 1089. English trans: *Soviet Phys. JETP* **16,** 769.

—— GOR'KOV, L. P., and DZYALOSHINSKIĬ, I. YE. (1965) *Quantum field theoretical methods in statistical physics*. Pergamon Press, Oxford.

ADAMS, E. N. (1952*a*) *Phys. Rev.* **85,** 41.

—— (1952*b*) *Phys. Rev.* **86,** 427.

AHN, J. and SELLMYER, D. J. (1970*a*) *Phys. Rev.* B **1,** 1273.

—— —— (1970*b*) *Phys. Rev.* B **1,** 1285.

AIGRAIN, P. (1960) *Proc. Int. Conf. Semicond. Phys., Prague (1960)*. p. 224.

ALEKSEEVSKIĬ, N. E. and EGOROV, V. S. (1963*a*) *Zh. éksp. teor. Fiz.* **45,** 388. English trans: *Soviet Phys. JETP* **18,** 268.

—— —— (1963*b*) *Zh. éksp. teor. Fiz.* **45,** 448. English trans: *Soviet Phys. JETP* **18,** 309.

—— —— (1964) *Zh. éksp. teor. Fiz.* **46,** 1205. English trans: *Soviet Phys. JETP* **19,** 815.

—— —— KARSTENS, G. É., and KAZAK, B. N. (1962) *Zh. éksp. teor. Fiz.* **43,** 731. English trans: *Soviet Phys. JETP* **16,** 519.

—— and GAĬDUKOV, YU. P. (1958) *Zh. éksp. teor. Fiz.* **35,** 554. English trans: *Soviet Phys. JETP* **8,** 383.

—— —— (1959*a*) *Zh. éksp. teor. Fiz.* **36,** 447. English trans: *Soviet Phys. JETP* **9,** 311.

—— —— (1959*b*) *Zh. éksp. teor. Fiz.* **37,** 672. English trans: *Soviet Phys. JETP* **10,** 481.

—— —— (1961) *Zh. éksp. teor. Fiz.* **41,** 354. English trans: *Soviet Phys. JETP* **14,** 256.

—— —— (1962) *Zh. éksp. teor. Fiz.* **43,** 2094. English trans: *Soviet Phys. JETP* **16,** 1481.

—— KARSTENS, G. É., and MOZHAEV, V. V. (1964) *Zh. éksp. teor. Fiz.* **46,** 1979. English trans: *Soviet Phys. JETP* **19,** 1333.

ALERS, G. A. and SWIM, R. T. (1963) *Phys. Rev. Lett.* **11,** 72.

ALLEN, P. B. and COHEN, M. L. (1969) *Phys. Rev.* **187,** 529.

—— —— FALICOV, L. M., and KASOWSKI, R. V. (1968) *Phys. Rev. Lett.* **21,** 1794.

ALLEN, W. D., DAWTON, R. H., BÄR, M., MENDELSSOHN, K. and OLSEN, J. L. (1950) *Nature, Lond.* **166,** 1071.

—— —— LOCK, J. M., PIPPARD, A. B., and SHOENBERG, D. (1950) *Nature, Lond.* **166,** 1071.

ALTMANN, S. L. (1958*a*) *Proc. R. Soc.* A **244,** 141.

—— (1958*b*) *Proc. R. Soc.* A **244,** 153.

—— and BRADLEY, C. J. (1962) *Phys. Lett.* **1,** 336.

—— —— (1963) *Phil. Trans. R. Soc.* A **255,** 193.

—— —— (1964) *Phys. Rev.* **135,** A1253.

—— —— (1965) *Rev. mod. Phys.* **37,** 33.

—— —— (1967) *Proc. phys. Soc.* **92,** 764.

—— and CRACKNELL, A. P. (1964) *Proc. phys. Soc.* **84,** 761.

—— —— (1965) *Rev. mod. Phys.* **37,** 19.

—— DAVIES, B. L., and HARFORD, A. R. (1968) *J. Phys. C* **1,** 1633.

—— HARFORD, A. R. and BLAKE, R. G. (1971) *J. Phys. F* **1,** 791.

AMAR, H. and JOHNSON, K. H. (1965) *Proceedings of the International Colloquium on Optical Properties and Electronic Structure of Metals and Alloys. Paris (1965).* (ed. F. Abelès), p. 586. North-Holland, Amsterdam.

—— —— and SOMMERS, C. B. (1967) *Phys. Rev.* **153,** 655.

—— —— and WANG, K. P. (1966) *Phys. Rev.* **148,** 672.

AMUNDSEN, T. (1968) *Phil. Mag.* **17,** 1303.

—— (1969) *Phil. Mag.* **20,** 687.

ANDERSEN, O. K. and LOUCKS, T. L. (1968) *Phys. Rev.* **167,** 551.

—— and MACKINTOSH, A. R. (1968) *Solid State Commun.* **6,** 285.

ANDERSON, A. C., PETERSON, R. E., and ROBICHAUX, J. E. (1968) *Phys. Rev. Lett.* **20,** 459.

ANDERSON, J. R. and GOLD, A. V. (1963) *Phys. Rev. Lett.* **10,** 227.

—— —— (1965) *Phys. Rev.* **139,** A1459.

—— MCCAFFREY, J. W., and PAPACONSTANTOPOULOS, D. A. (1969) *Solid State Commun.* **7,** 1439.

ANDERSON, P. W. (1958) *Phys. Rev.* **109,** 1492.

ANDREW, E. R. (1949) *Proc. phys. Soc.* A **62,** 77.

ANIMALU, A. O. E., BONSIGNORI, F., and BORTOLANI, V. (1966) *Nuovo Cim.* **44,** 159.

—— and HEINE, V. (1965) *Phil. Mag.* **12,** 1249.

ANTONCIK, E. (1959) *J. Phys. Chem. Solids* **10,** 314.

APPELBAUM, J. A. (1966) *Phys. Rev.* **144,** 435.

APPLETON, A. and CURRY, C. (1965) *Phil. Mag.* **12,** 245.

ARKHIPOV, R. G., KECHIN, V. V., LIKHTER, A. I., and POSPELOV, Yu. A. (1963) *Zh. éksp. teor. Fiz.* **44,** 1964. English trans: *Soviet Phys. JETP* **17,** 1321.
ARKO, A. J. and MARCUS, J. A. (1964) *Proc. 9th Int. Conf. low Temp. Phys.* p. 748.
—— —— and REED, W. A. (1966) *Phys. Lett.* **23,** 617.
—— —— —— (1968) *Phys. Rev.* **176,** 671.
—— —— —— (1969) *Phys. Rev.* **185,** 901.
ARLINGHAUS, F. J. (1965) M.I.T. Solid-state and molecular theory group quarterly progress report no. 56, p. 4.
—— (1967*a*) *Phys. Rev.* **153,** 743.
—— (1967*b*) *Phys. Rev.* **157,** 491.
—— (1969) *Phys. Rev.* **186,** 609.
ARROTT, A. and COLES, B. R. (1961) *J. appl. Phys.* **32,** 51S.
ASANO, S. and YAMASHITA, J. (1967) *J. phys. Soc. Japan,* **23,** **714.**
ASHCROFT, N. W. (1963*a*) *Phys. Lett.* **4,** 202.
—— (1963*b*). *Phil. Mag.* **8,** 2055.
—— (1965) *Phys. Rev.* **140,** A935.
—— and LAWRENCE, W. E. (1968) *Phys. Rev.* **175,** 938.
—— and WILKINS, J. W. (1965) *Phys. Lett.* **14,** 285.
AUBREY, J. E. and CHAMBERS, R. G. (1957) *J. Phys. Chem. Solids* **3,** 128.
AUSTIN, B. J., HEINE, V., and SHAM, L. J. (1962) *Phys. Rev.* **127,** 276.
AZBEL', M. Ya. (1954) *Dokl. Akad. Nauk. SSSR* **98,** 519.
—— (1958) *Zh. éksp. teor. Fiz.* **34,** 969. English trans: *Soviet Phys. JETP* **7,** 801.
—— and KANER, É. A. (1956) *Zh. éksp. teor. Fiz.* **30,** 811. English trans: *Soviet Phys. JETP* **3,** 772.
—— —— (1957) *Zh. éksp. teor. Fiz.* **32,** 896. English trans: *Soviet Phys. JETP* **5,** 730.
—— —— (1958) *J. Phys. Chem. Solids* **6,** 113.
BABISKIN, J. and SIEBENMANN, P. G. (1957) *Phys. Rev.* **107,** 1249.
BACON, G. E. (1962) *Neutron diffraction.* Clarendon Press, Oxford.
BAER, Y., HEDÉN, P. F., HEDMAN, J., KLASSON, M., NORDLING, C., and SIEGBAHN, K. (1970) *Physica Scripta* **1,** 55.
BAILYN, M. (1958) *Phys. Rev.* **112,** 1587.
—— (1960) *Phys. Rev.* **120,** 381.
—— (1967) *Phys. Rev.* **157,** 480.
BAK, T. A. (1964) *Phonons and phonon interactions.* Benjamin, New York.
BALAIN, K. S., GRENIER, C. G., and REYNOLDS, J. M. (1960) *Phys. Rev.* **119,** 935.
BALLA, D. and BRANDT, N. B. (1964) *Zh. éksp. teor. Fiz.* **47,** 1653. English trans: *Soviet Phys. JETP* **20,** 1111.

BARAFF, G. A. and PHILLIPS, T. G. (1970). *Phys. Rev. Lett.* **24**, 1428.
BARDEEN, J. (1937) *Phys. Rev.* **52**, 688.
—— COOPER, L. N., and SCHRIEFFER, J. R. (1957*a*) *Phys. Rev.* **106**, 162.
—— —— —— (1957*b*) *Phys. Rev.* **108**, 1175.
BARKHAUSEN, H. (1919) *Phys. Z.* **20**, 401.
BARRETT, C. S. (1955) *J. Inst. Metals* **84**, 43.
—— (1956) *Acta crystallogr.* **9**, 671.
—— (1957) *Acta crystallogr.* **10**, 58.
—— and MASSALSKI, T. B. (1966) *Structure of metals.* McGraw-Hill, New York.
BASSANI, F. and CELLI, V. (1961) *J. Phys. Chem. Solids* **20**, 64.
BATALLAN, G. and ROSENMAN, I. (1970). *Proc. 12th Int. Conf. low Temp. Phys.* p. 595.
BATTERMAN, B. W., CHIPMAN, D. R., and DEMARCO, J. J. (1961) *Phys. Rev.* **122**, 68.
BAYNHAM, A. C. and BOARDMAN, A. D. (1971) *Plasma effects in semiconductors: Helicon and Alfvén waves.* Taylor and Francis, London; reprinted from *Adv. Phys.* **19**, 575 (1970).
BEATTIE, A. G. and UEHLING, E. A. (1966) *Phys. Rev.* **148**, 657.
BECK, A., JAN, J. P., PEARSON, W. B., and TEMPLETON, I. M. (1963) *Phil. Mag.* **8**, 351.
BEHRENDT, D. R., LEGVOLD, S., and SPEDDING, F. H. (1958) *Phys. Rev.* **109**, 1544.
BELETSKI, V. I., KOROLYUK, A. P., OBOLENSKI, M. A., and KHOTKEVICH, V. I. (1970) *Solid State Commun.* **8**, 1249.
BELL, D. G. (1954) *Rev. mod. Phys.* **26**, 311.
BELLESSA, G., REICH, R., and MERCOUROFF, W. (1969) *J. Phys., Paris* **30**, 823.
BELSON, H. S. (1966) *J. appl. Phys.* **37**, 1348.
BENNEMANN, K. H. (1967) *Phys. Lett.* A **25**, 233.
BERGWALL, S. (1965) *Phys. Lett.* **19**, 539.
BERINGER, R. and MONTGOMERY, C. G. (1942) *Phys. Rev.* **61**, 222.
BERKO, S., CUSHNER, C., and ERSKINE, J. C. (1968) *Phys. Lett.* A **27**, 668.
—— and HEREFORD, F. L. (1956) *Rev. mod. Phys.* **28**, 299.
—— and ZUCKERMAN, J. (1964) *Phys. Rev. Lett.* **13**, 339a.
BERLINCOURT, T. G. (1955) *Phys. Rev.* **99**, 1716.
—— and STEELE, M. C. (1955) *Phys. Rev.* **98**, 956.
BERRE, B. and OLSEN, T. (1965) *Phys. Status solidi* **11**, 657.
BHARGAVA, R. N. (1967) *Phys. Rev.* **156**, 785.
BIRSS, R. R. (1964) *Symmetry and magnetism.* North-Holland, Amsterdam.
BLACKMAN, M. (1938) *Proc. R. Soc.* A **166**, 1.

BLATT, F. J., BURMESTER, A., and LAROY, B. (1967) *Phys. Rev.* **155,** 611.

BLATT, J. M. (1964) *Theory of superconductivity*. Academic Press, New York.

BLEANEY, B. I. and BLEANEY, B. (1965) *Electricity and Magnetism*. Clarendon Press, Oxford.

BLOEMBERGEN, N. and ROWLAND, T. J. (1953) *Acta metall.* **1,** 731.

BLOUNT, E. I. (1959) *Phys. Rev.* **114,** 418.

—— (1962*a*) *Phys. Rev.* **126,** 1636.

—— (1962*b*) *Solid St. Phys.* **13,** 305.

BOGLE, T. E., COON, J. B., and GRENIER, C. G. (1969) *Phys. Rev.* **177,** 1122.

BOGOD, YU. A. and EREMENKO, V. V. (1965) *Phys. Status solidi* **11,** K51.

BOHM, D. and PINES, D. (1951) *Phys. Rev.* **82,** 625.

—— —— (1953) *Phys. Rev.* **92,** 609.

BOHM, H. V. and EASTERLING, V. J. (1962) *Phys. Rev.* **128,** 1021.

BÖMMEL, H. E. (1955) *Phys. Rev.* **100,** 758.

BONSIGNORI, F. and BORTOLANI, V. (1966) *Nuovo Cim.* B **46,** 113.

BORN, M. and HUANG, K. (1954) *Dynamical theory of crystal lattices*. Clarendon Press, Oxford.

BOUCKAERT, L. P., SMOLUCHOWSKI, R., and WIGNER, E. (1936) *Phys. Rev.* **50,** 58.

BOWERS, R., LEGENDY, C., and ROSE, F. (1961) *Phys. Rev. Lett.* **7,** 339.

BOYLE, D. J. and GOLD, A. V. (1969) *Phys. Rev. Lett.* **22** 461.

BOYLE, W. S., HSU, F. S. L., and KUNZLER, J. E. (1960) *Phys. Rev. Lett.* **4,** 278.

—— and RODGERS, K. F. (1959) *Phys. Rev. Lett.* **2,** 338.

—— and SMITH, G. E. (1963) *Prog. Semicond.* **7,** 1.

BRADLEY, C. J. and CRACKNELL, A. P. (1970) *J. Phys. C* **3,** 610.

—— —— (1972) *The mathematical theory of symmetry in solids. Representation theory for point groups and space groups*. Clarendon Press, Oxford.

—— and DAVIES, B. L. (1968) *Rev. mod. Phys.* **40,** 359.

—— WALLIS, D. E., and CRACKNELL, A. P. (1970) *J. Phys. C* **3,** 619.

BRANDT, G. B. and RAYNE, J. A. (1965) *Phys. Lett.* **15,** 18.

—— —— (1966) *Phys. Rev.* **148,** 644.

BRANDT, N. B. (1960) *Zh. éksp. teor. Fiz.* **38,** 1355. English trans: *Soviet Phys. JETP* **11,** 975.

—— DUBROVSKAYA, A. E. and KYTIN, G. A. (1959) *Zh. éksp. teor. Fiz.* **37,** 572. English trans: *Soviet Phys. JETP* **10,** 405.

—— LYUBUTINA, L. G., and KRYUKOVA, N. A. (1967) *Zh. éksp. teor. Fiz.* **53,** 134. English trans: *Soviet Phys. JETP* **26,** 93.

—— and MININA, N. YA. (1968) *Zh. éksp. teor. Fiz., Pis'ma* **7,** 264. English trans: *Soviet Phys. JETP Lett.* **7,** 205.

BRANDT, N. B., MININA, N. YA., and POSPELOV, YU. A. (1968) *Fizika tverd. Tela*, **10**, 1268. English trans: *Soviet Phys. Solid St.* **10**, 1011.
—— and VENTTSEL', V. A. (1958) *Zh. éksp. teor. Fiz.* **35**, 1083. English trans: *Soviet Phys. JETP* **8**, 757.
BREWER, L. (1968) *Science* **161**, 115.
BRISCOE, C. V., BEARDSLEY, G. M., and STEWART, A. T. (1966) *Phys. Rev.* **141**, 379.
BROCKHOUSE, B. N., RAO, K. R., and WOODS, A. D. B. (1961) *Phys. Rev. Lett.* **7**, 93.
BROOKS, H. and HAM, F. S. (1958) *Phys. Rev.* **112**, 344.
BROWN, R. D., HARTMAN, R. L., and KOENIG, S. H. (1968) *Phys. Rev.* **172**, 598.
BRUST, D. (1969) *Phys. Rev. Lett.* **23**, 1232.
—— (1970) *Phys. Lett.* A **31**, 289.
BRYANT, C. A., and KEESOM, P. H. (1961) *Phys. Rev.* **123**, 491.
BURDICK, G. A. (1961) *Phys. Rev. Lett.* **7**, 156.
—— (1963) *Phys. Rev.* **129**, 138.
BURSTEIN, E. and LUNDQVIST, S. (1969) *Tunneling phenomena in solids.* Plenum, New York.
BUTLER, F. A., and BROWN, E. (1968) *Phys. Rev.* **166**, 630.
CABLE, J. W., MOON, R. M., KOEHLER, W. C., and WOLLAN, E. O. (1964) *Phys. Rev. Lett.* **12**, 553.
—— and WOLLAN, E. O. (1968) *Phys. Rev.* **165**, 733.
CALLAWAY, J. (1958) *Solid St. Phys.* **7**, 99.
—— (1961) *Phys. Rev.* **124**, 1824.
—— (1964) *Energy band theory*. Academic Press, New York.
CAPLIN, A. D. and SHOENBERG, D. (1965) *Phys. Lett.* **18**, 238.
CAPOCCI, F. A., HOLTHAM, P. M., PARSONS, D., and PRIESTLEY, M. G. (1970) *J. Phys. C* **3**, 2081.
CARBOTTE, J. P. (1966) *Phys. Rev.* **144**, 309.
—— (1967) *Phys. Rev.* **155**, 197.
—— and DYNES, R. C. (1967) *Phys. Lett.* A **25**, 685.
—— —— (1968) *Phys. Rev.* **172**, 476.
—— and SALVADORI, A. (1967) *Phys. Rev.* **162**, 290.
—— and KAHANA, S. (1965) *Phys. Rev.* **139**, A213.
CATTERALL, R. and MOTT, N. F. (1969) *Adv. Phys.* **18**, 665.
CHAMBERS, R. G. (1950) *Proc. R. Soc.* A **202**, 378.
—— (1952) *Proc. R. Soc.* A **215**, 481.
—— (1956a) *Proc. R. Soc.* A **238**, 344.
—— (1956b) *Can. J. Phys.* **34**, 1395.
—— (1960) *Proc. 7th Int. Conf. low Temp. Phys.* p. 243.
—— (1965) *Proc. phys. Soc.* **86**, 305.
—— (1966) *Proc. phys. Soc.* **89**, 695.

CHAMBERS, R. G. (1969a) In *The physics of metals. I. Electrons.* (ed. J. M. Ziman) p. 175. University Press, Cambridge.
—— (1969b) *Phys. kondens. Materie* **9,** 171.
CHATTERJEE, S. and CHAKRABORTI, D. K. (1970) *J. Phys. C* **3,** S120.
CHODOROW, M. I. (1939) *Phys. Rev.* **55,** 675.
CHOLLET, L. F. and TEMPLETON, I. M. (1968) *Phys. Rev.* **170,** 656.
CLAUS, H. and ULMER, K. (1963) *Z. Phys.* **173,** 462.
CLEVELAND, J. R. and STANFORD, J. L. (1970) *Phys. Rev. Lett.* **24,** 1482.
COCHRAN, W. (1965). In *Phonons in perfect lattices and in lattices with point imperfections* (ed. R. W. H. Stevenson). p. 53. Oliver and Boyd, Edinburgh.
—— and COWLEY, R. A. (1967) In *Handbuch der Physik* (ed. S. Flügge) vol XXV/2a Light and Matter Ia (ed. L. Genzel) p. 59. Springer, Berlin.
COHEN, M. H. (1961) *Phys. Rev.* **121,** 387.
—— (1963) In *Alloying behaviour and effects in concentrated solid solutions* (ed. T. B. Massalski) p. 1. Gordon and Breach, New York.
—— and BLOUNT, E. I. (1960). *Phil. Mag.* **5,** 115.
—— and FALICOV, L. M. (1960) *Phys. Rev. Lett.* **5,** 544.
—— —— (1961) *Phys. Rev. Lett.* **7,** 231.
—— —— and GOLIN, S. (1964) *I.B.M. Jl. Res. Dev.* **8,** 215.
—— HARRISON, M. J., and HARRISON, W. A. (1960) *Phys. Rev.* **117,** 937.
—— and HEINE, V. (1958) *Adv. Phys.* **7,** 395.
—— —— (1961) *Phys. Rev.* **122,** 1821.
COHEN, M. L. and HEINE, V. (1970) *Solid St. Phys.* **24,** 37.
COLERIDGE, P. T. (1965) *Phys. Lett.* **15,** 223.
—— (1966*a*) *Phys. Lett.* **22,** 367.
—— (1966*b*) *Proc. R. Soc.* A **295,** 458.
—— (1969) *J. low Temp. Phys.* **1,** 577.
COLES, B. R. (1956) *Phys. Rev.* **101,** 1254.
COLLINS, J. G. (1966) *Ann. Acad. Sci. Fennicae* AVI, no. 210, 239.
—— and ZIMAN, J. M. (1961) *Proc. R. Soc.* A **264,** 60.
COMBLEY, F. H. (1968) *Acta crystallogr.* B **24,** 142.
COMPTON, A. H. and ALLISON, S. K. (1967) *X-rays in theory and experiment.* Van Nostrand, Princeton.
CONDON, J. H. (1964) *Bull. Am. phys. Soc.* **9,** 239.
—— (1966*a*) *Phys. Rev.* **145,** 526.
—— (1966*b*) *Proc. 10th Int. Conf. low Temp. Phys.* volume 3, p. 289.
—— and MARCUS, J. A. (1964) *Phys. Rev.* **134,** A446.
—— and WALSTEDT, R. E. (1968) *Phys. Rev. Lett.* **21,** 612.
CONNOLLY, J. W. D. (1967) *Phys. Rev.* **159,** 415.
COOK, J. R. and DATARS, W. R. (1970) *Phys. Rev.* B **1,** 1415.

COON, J. B., GRENIER, C. G., and REYNOLDS, J. M. (1967) *J. Phys. Chem. Solids* **28,** 301.

COOPER, B. R., KREIGER, E. L., and SEGALL, B. (1969) *Phys. Lett.* A **30,** 333.

COOPER, D. G. (1968) *The Periodic Table*. Butterworth, London.

COOPER, L. N. (1956) *Phys. Rev.* **104,** 1189.

COOPER, M. and LEAKE, J. A. (1967) *Phil. Mag.* **15,** 1201.

CORNWELL, J. F. (1961) *Proc. R. Soc.* A **261,** 551.

—— (1969) *Group theory and electronic energy bands in solids*. North-Holland, Amsterdam.

—— HUM, D. M. and WONG, K. C. (1968) *Phys. Lett.* A **26,** 365.

COTTI, P. (1963) *Phys. Lett.* **4,** 114.

—— (1964) *Phys. kondens. Materie* **3,** 40.

—— FRYER, E. M. and OLSEN, J . L. (1964) *Helv. phys. Acta*, **37,** 585.

COULSON, C. A. (1947) *Nature, Lond.* **159,** 265.

COUSINS, J. E. and DUPREE, R. (1965) *Phys. Lett.* **19,** 464.

COWLEY, R. A., WOODS, A. D. B., and DOLLING, G. (1966) *Phys. Rev.* **150,** 487.

CRACKNELL, A. P. (1967*a*) *Contemp. Phys.* **8,** 459.

—— (1967*b*) *Phys. Lett.* A **24,** 263.

—— (1968) *Applied group theory*. Pergamon Press, Oxford.

—— (1969*a*) *J. Phys. C* **2,** 1425.

—— (1969*b*) *Rep. Prog. Phys.* **32,** 633.

—— (1969*c*) *Adv. Phys.* **18,** 681.

—— (1970) *Phys. Rev.* B **1,** 1261.

—— (1971*a*) *Adv. Phys.* **20,** 1.

—— (1971*b*) *The Fermi surfaces of metals*. Taylor and Francis, London; reprinted from *Adv. Phys.* **18,** 681 (1969) and **20,** 1 (1971).

—— (1971*c*) *Adv. Phys.* **20,** 747.

—— and JOSHUA, S. J. (1970) *Proc. Camb. phil. Soc. math. phys. Sci.* **67,** 647.

—— and WONG, K. C. (1967) *Aust. J. Phys.* **20,** 173.

CRAVEN, J. E. (1969) *Phys. Rev.* **182,** 693.

—— and STARK, R. W. (1968) *Phys. Rev.* **168,** 849.

CUSACK, N. E. (1958) *The electrical and magnetic properties of solids*. Longmans, London.

DARBY, J. K. and MARCH, N. H. (1964) *Proc. phys. Soc.* **84,** 591.

DAVIS, L. C. and LIU, S. H. (1967) *Phys. Rev.* **163,** 503.

DEATON, B. C. (1963) *Phys. Lett.* **7,** 7.

—— (1965) *Phys. Rev.* **140,** A2051.

DE BENEDETTI, S., COWAN, C. E., KONNEKER, W. R., and PRIMAKOFF, H. (1950) *Phys. Rev.* **77,** 205.

DE BOER, J. H. and VERWEY, E. J. W. (1937) *Proc. phys. Soc.* **49,** 59.

DEBYE, P. P. (1912) *Annln Phys.* **39,** 789.

DEBYE, P. P. and HÜCKEL, E. (1923) *Phys. Z.* **24,** 185.
DEDERICHS, P. H. (1972) *Solid St. Phys.* **27,** 135.
DEEGAN, R. A. (1968) *J. Phys. C* **1,** 763.
—— and TWOSE, W. D. (1967) *Phys. Rev.* **164,** 993.
DE HAAS, W. J. and VAN ALPHEN, P. M. (1930*a*) *Proc. Sect. Sci. K. ned. Akad. Wet.* **33,** 680. Reprinted: *Communs phys. Lab. Univ. Leiden* no. 208d.
—— —— (1930*b*) *Proc. Sect. Sci. K. ned. Akad. Wet.* **33,** 1106. Reprinted: *Communs phys. Lab. Univ. Leiden* no. 212a.
—— —— (1932) *Proc. Sect. Sci. K. ned. Akad. Wet.* **35,** 454. Reprinted: *Communs phys. Lab. Univ. Leiden* no. 220d.
DEKHTYAR, I. YA. (1968) *Czech. J. Phys.* B **18,** 1509.
—— (1969) *Phys. Lett.* A **28,** 771.
DEVILLERS, M. A. C. and DE VROOMEN, A. R. (1969) *Phys. Lett.* A **30,** 159.
D'HAENENS, J. P., LIBCHABER, A., LAROCHE, C., and LE HERICY, J. (1968) *Phys. Lett.* A **28,** 312.
DHILLON, J. S. and SHOENBERG, D. (1955) *Phil. Trans. R. Soc.* A **248,** 1.
DICK, B. J. and OVERHAUSER, A. W. (1958) *Phys. Rev.* **112,** 90
DICKE, D. A. and GREEN, B. A. (1967) *Phys. Rev.* **153,** 800.
DICKEY, J. M., MEYER, A., and YOUNG, W. H. (1967) *Proc. phys. Soc.* **92,** 460.
DINGLE, R. B. (1950) *Proc. R. Soc.* A **201,** 545.
—— (1952) *Proc. R. Soc.* A **211,** 517.
—— (1953*a*) *Physica, 's Grav.* **19,** 311.
—— (1953*b*) *Physica, 's Grav.* **19,** 348.
—— (1953*c*) *Physica, 's Grav.* **19,** 729.
—— and SHOENBERG, D. (1950) *Nature, Lond.* **166,** 652.
DIRAC, P. A. M. (1930) *The principles of quantum mechanics.* Clarendon Press, Oxford.
DITLEFSON, E. and LOTHE, J. (1966) *Phil. Mag.* **14,** 759.
DIXON, M., HOARE, F. E., HOLDEN, T. M., and MOODY, D. E. (1965) *Proc. R. Soc.* A **285,** 561.
DMITRENKO, I. M., VERKIN, B. I., and LAZAREV, B. G. (1958) *Zh. éksp. teor. Fiz.* **35,** 328. English trans: *Soviet Phys. JETP* **8,** 229.
DONAGHY, J. J. and STEWART, A. T. (1967*a*) *Phys. Rev.* **164,** 391.
—— —— (1967*b*) *Phys. Rev.* **164,** 396.
—— —— ROCKMORE, D. M., and KUSMISS, J. H. (1964) *Proc. 9th Int. Conf. low Temp. Phys.* p. 835.
DONOVAN, B. and ANGRESS, J. F. (1971) *Lattice vibrations.* Chapman and Hall, London.
DOOLEY, J. W. and TEPLEY, N. (1969) *Phys. Rev.* **181,** 1001.
DRESSELHAUS, G. (1969) *Solid State Commun.* **7,** 419.
—— and MAVROIDES, J. G. (1964*a*) *Solid State Commun.* **2,** 297.
—— —— (1964*b*) *Carbon* **1,** 263.

DRESSELHAUS, G. and MAVROIDES, J. G. (1964c) *I.B.M. Jl. Res. Dev.* **8,** 262.

—— —— (1966) *Carbon,* **3,** 465.

DRICKAMER, H. G. (1965) *Solid St. Phys.* **17,** 1.

DUNCANSON, W. E. and COULSON, C. A. (1945) *Proc. phys. Soc.* **57,** 190.

DUNIFER, G., SCHULTZ, S., and SCHMIDT, P. H. (1968) *J. appl. Phys.* **39,** 397.

EASTERLING, V. J. and BOHM, H. V. (1962) *Phys. Rev.* **125,** 812.

EASTMAN, D. E. (1969*a*) *Solid State Commun.* **7,** 1697.

—— (1969*b*) *J. appl. Phys.* **40,** 1387.

ECKSTEIN, Y. (1966) *Phys. Lett.* **20,** 142.

—— KETTERSON, J. B. and ECKSTEIN, S. G. (1964) *Phys. Rev.* **135,** A 740.

—— —— and PRIESTLEY, M. G. (1966) *Phys. Rev.* **148,** 586.

ÉDEL'MAN, V. S. and KHAĬKIN, M. S. (1965) *Zh. éksp. teor. Fiz.* **49,** 107. English trans: *Soviet Phys. JETP* **22,** 77.

EDWARDS, L. R. and LEGVOLD, S. (1968) *Phys. Rev.* **176,** 753.

EHRENREICH, H., PHILIPP, H. R., and OLECHNA, D. J. (1963) *Phys. Rev.* **131,** 2469.

EINSTEIN, A. (1907) *Annln Phys.* **22,** 180.

ELIASHBERG, G. M. (1962) *Zh. éksp. teor. Fiz.* **43,** 1005. English trans: *Soviet Phys. JETP* **16,** 780.

ELLIOTT, R. J. (1954) *Phys. Rev.* **96,** 280.

—— and WEDGWOOD, F. A. (1963) *Proc. phys. Soc.* **81,** 846.

—— —— (1964) *Proc. phys. Soc.* **84,** 63.

ESAKI, L. and STILES, P. J. (1965) *Phys. Rev. Lett.* **14,** 902.

EUWEMA, R. N., STUKEL, D. J., COLLINS, T. C., DE WITT, J. S., and SHANKLAND, D. G. (1969) *Phys. Rev.* **178,** 1419.

EVENSON, W. E. and LIU, S. H. (1968) *Phys. Rev. Lett.* **21,** 432.

—— —— (1969) *Phys. Rev.* **178,** 783.

FABIAN, D. J. (1968) *Soft X-ray spectra and the electronic structure of metals and materials.* Academic Press, London.

FADDEYEV, D. K. (1964) *Tables of the principal unitary representations of Fedorov groups.* Pergamon Press, Oxford.

FALICOV, L. M. (1962) *Phil. Trans. R. Soc.* A **255,** 55.

—— and COHEN, M. H. (1963) *Phys. Rev.* **130,** 92.

—— and HEINE, V. (1961) *Adv. Phys.* **10,** 57.

—— and LIN, P. J. (1966) *Phys. Rev.* **141,** 562.

—— and RUVALDS, J. (1968) *Phys. Rev.* **172,** 498.

—— and STARK, R. W. (1967) *Prog. low Temp. Phys.* **5,** 235.

—— and ZUCKERMANN, M. J. (1967) *Phys. Rev.* **160,** 372.

FAL'KOVSKIĬ, L. A. (1967) *Zh. éksp. teor. Fiz.* **53,** 2164. English trans: *Soviet Phys. JETP* **26,** 1222.

—— and RAZINA, G. S. (1965) *Zh. éksp. teor. Fiz.* **49,** 265. English trans: *Soviet Phys. JETP* **22,** 187.

FATE, W. A. (1968) *Phys. Rev.* **172,** 402.
FAWCETT, E. (1956) *Phys. Rev.* **103,** 1582.
—— (1961*a*) *Phys. Rev. Lett.* **7,** 370.
—— (1961*b*) *J. Phys. Chem. Solids* **18,** 320.
—— (1962) *Phys. Rev.* **128,** 154.
—— and GRIFFITHS, D. (1962) *J. Phys. Chem. Solids* **23,** 1631.
—— and REED, W. A. (1962) *Phys. Rev. Lett.* **9,** 336.
—— —— (1963) *Phys. Rev.* **131,** 2463.
—— —— (1964*a*) *Phys. Rev.* **134,** A723.
—— —— (1964*b*) *Proc. 9th Int. Conf. low Temp. Phys.* p. 782.
FERREIRA, L. G. (1967) *J. Phys. Chem. Solids* **28,** 1891.
—— (1968) *J. Phys. Chem. Solids* **29,** 357.
FERRELL, R. A. (1956) *Rev. mod. Phys.* **28,** 308.
FESENKO, E. P. (1969) *Fizika tverd. Tela* **11,** 2649. English trans: *Soviet Phys. Solid St.* **11,** 2137.
FETTER, A. L. and WALECKA, J. D. (1971) *Quantum theory of many-particle systems.* McGraw-Hill, New York.
FLEMING, G. S., LIU, S. H., and LOUCKS, T. L. (1968) *Phys. Rev. Lett.* **21,** 1524.
—— and LOUCKS, T. L. (1968) *Phys. Rev.* **173,** 685.
FLETCHER, G. C. (1969) *J. Phys. C* **2,** 1440.
—— (1971) *The electron band theory of solids.* North-Holland, Amsterdam.
FLIPPEN, R. B. (1964) *J. appl. Phys.* **35,** 1047.
FOLDY, L. L. (1968) *Phys. Rev.* **170,** 670.
FOMIN, N. V. (1966) *Fiz. tverd. Tela* **8,** 3613. English trans: *Soviet Phys. Solid St.* **8,** 2885.
FØRSVOLL, K. and HOLWECH, I. (1962) *Phys. Lett.* **3,** 66.
FRANK, V. (1955) *Kgl. Danske Videnskab. Selskab, Mat-Fys. Medd.* **30,** no. **4.**
FREEMAN, A. J., DIMMOCK, J. O., and WATSON, R. E. (1966) *Phys. Rev. Lett.* **16,** 94.
FREI, V. (1966) *Czech. J. Phys.* **16,** 207.
FRIEDEL, J. (1954) *Adv. Phys.* **3,** 446.
FROBENIUS, F. G. and SCHUR, I. (1906) *Sitzber. kgl. preuss. Akad. Wiss.* 186; 209.
FRÖHLICH, H. (1950) *Phys. Rev.* **79,** 845.
—— (1952) *Proc. R. Soc.* A **215,** 291.
FUCHS, K. (1938) *Proc. Camb. phil. Soc. math. phys. Sci.* **34,** 100.
FUJIWARA, K. and SUEOKA, O. (1966) *J. phys. Soc. Japan,* **21,** 1947.
FUKUMOTO, A. and STRANDBERG, M. W. P. (1967) *Phys. Rev.* **155,** 685.
GAĬDUKOV, YU. P. (1959) *Zh. éksp. teor. Fiz.* **37,** 1281. English trans: *Soviet Phys. JETP* **10,** 913.
—— and ITSKEVICH, E. S. (1963) *Zh. éksp. teor. Fiz.* **45,** 71. English trans: *Soviet Phys. JETP* **18,** 51.

GALASIEWICZ, Z. (1970) *Superconductivity and quantum fluids*. Pergamon Press, Oxford.

GALLINA, V. and OMINI, M. (1966) *J. Phys. Chem. Solids* **27**, 1479.

GALT, J. K., MERRITT, F. R., and KLAUDER, J. R. (1965) *Phys. Rev.* **139**, A823.

GANTMAKHER, V. F. (1962*a*) *Zh. éksp. teor. Fiz.* **42**, 1416. English trans: *Soviet Phys. JETP* **15**, 982.

—— (1962*b*) *Zh. éksp. teor. Fiz.* **43**, 345. English trans: *Soviet Phys. JETP* **16**, 247.

—— (1963) *Zh. éksp. teor. Fiz.* **44**, 811. English trans: *Soviet Phys. JETP* **17**, 549.

—— (1964) *Zh. éksp. teor. Fiz.* **46**, 2028. English trans: *Soviet Phys. JETP* **19**, 1366.

—— (1967) *Prog. low Temp. Phys.* **5**, 181.

—— and KANER, É. A. (1963) *Zh. éksp. teor. Fiz.* **45**, 1430. English trans: *Soviet Phys. JETP* **18**, 988.

—— —— (1965) *Zh. éksp. teor. Fiz.* **48**, 1572. English trans: *Soviet Phys. JETP* **21**, 1053.

—— and KRYLOV, I. P. (1964) *Zh. éksp. teor. Fiz.* **47**, 2111. English trans: *Soviet Phys. JETP* **20**, 1418.

—— —— (1965) *Zh. éksp. teor. Fiz.* **49**, 1054. English trans: *Soviet Phys. JETP* **22**, 734.

GARCIA, N., KAO, Y. H. and STRONGIN, M. (1969) *Phys. Lett.* A **29**, 631.

GARCÍA-MOLINER, F. (1958) *Phil. Mag.* **3**, 207.

GARG, J. C. and SARAF, B. L. (1969) *J. phys. Soc. Japan* **27**, 1695.

GARLAND, J. C. and BOWERS, R. (1968) *Phys. Rev. Lett.* **21**, 1007.

GELDART, D. J. W., HOUGHTON, A., and VOSKO, S. H. (1964) *Can. J. Phys.* **42**, 1938.

GERRITSEN, A. N. and DE HAAS, W. J. (1940) *Physica, 's Grav.* **7**, 802.

GIBBONS, D. F. (1961) *Phil. Mag.* **6**, 445.

—— and FALICOV, L. M. (1963) *Phil. Mag.* **8**, 177.

GIRVAN, R. F., GOLD, A. V., and PHILLIPS, R. A. (1968) *J. Phys. Chem. Solids* **29**, 1485.

GIVENS, M. P. (1958) *Solid St. Phys.* **6**, 313.

GOLD, A. V. (1958) *Phil. Trans. R. Soc.* A **251**, 85.

—— (1964) *Proc. Int. Conf. Magnetism, Nottingham* p. 124.

—— MACDONALD, D. K. C., PEARSON, W. B., and TEMPLETON, I. M. (1960) *Phil. Mag.* **5**, 765.

—— and PRIESTLEY, M. G. (1960) *Phil. Mag.* **5**, 1089.

GOLDSTEIN, A. and FONER, S. (1966) *Phys. Rev.* **146**, 442.

GOODRICH, R. G., KHAN, S. A., and REYNOLDS, J. M. (1969) *Phys. Rev. Lett.* **23**, 767.

GRAEBNER, J. E. and MARCUS, J. A. (1966) *J. appl. Phys.* **37**, 1262.
—— —— (1968) *Phys. Rev.* **175**, 659.
GRAHAM, G. M. (1958) *Proc. R. Soc.* A **248**, 522.
GREENE, J. B. and MANNING, M. F. (1943) *Phys. Rev.* **63**, 203.
GREENE, M. P. and KOHN, W. (1965) *Phys. Rev.* **137**, A513.
GRIGSBY, D. L., JOHNSON, D. H., NEUBERGER, M., and WELLES, S. J. (1967) *Electronic properties of materials. A guide to the literature*. Vol. 2. Plenum, New York. See also JOHNSON (1965).
GRIMES, C. C. and KIP, A. F. (1963) *Phys. Rev.* **132**, 1991.
—— —— SPONG, F., STRADLING, R. A. and PINCUS, P. (1963) *Phys. Rev. Lett.* **11**, 455.
GRIMVALL, G. (1969) *Phys. kondens Materie* **9**, 283.
GRODSKI, J. J. and DIXON, A. E. (1969) *Solid State Commun.* **7**, 735.
GSCHNEIDNER, K. A. (1964) *Solid St. Phys.* **16**, 275.
GUBANOV, A. I., and NIKULIN, V. K. (1966) *Phys. Status solidi* **17**, 815.
GUNNERSON, E. M. (1957) *Phil. Trans. R. Soc.* A **249**, 299.
GUPTA, R. P. and LOUCKS, T. L. (1969) *Phys. Rev. Lett.* **22**, 458.
GUREVICH, V. L. (1958) *Zh. éksp. teor. Fiz.* **35**, 668. English trans: *Soviet Phys. JETP* **8**, 464.
—— (1959*a*) *Zh. éksp. teor. Fiz.* **37**, 71. English trans: *Soviet Phys. JETP* **10**, 51.
—— (1959*b*) *Zh. éksp. teor. Fiz.* **37**, 1680. English trans: *Soviet Phys. JETP* **10**, 1190.
—— SKOBOV, V. G., and FIRSOV, YU. A. (1961) *Zh. éksp. teor. Fiz.* **40**, 786. English trans: *Soviet Phys. JETP* **13**, 552.
HAERING, R. R. and MROZOWSKI, S. (1960) *Prog. Semicond.* **5**, 273.
HAGSTRÖM, S. B. M., HEDÉN, P. O., and LÖFGREN, H. (1970) *Solid State Commun.* **8**, 1245.
HAGSTRUM, H. D. (1954) *Phys. Rev.* **96**, 336.
—— (1961) *Phys. Rev.* **122**, 83.
—— (1966) *Phys. Rev.* **150**, 495.
—— and BECKER, G. E. (1967) *Phys. Rev.* **159**, 572.
HALLORAN, M. H., CONDON, J. H., GRAEBNER, J. E., KUNZLER, J. E., and HSU, F. S. L. (1970) *Phys. Rev.* B **1**, 366.
HALSE, M. R. (1969) *Phil. Trans. R. Soc.* A **265**, 507.
HAM, F. S. (1960) in *The Fermi surface. Proceedings of a conference held at Cooperstown, New York, August 1960* (ed. W. A. Harrison and M. B. Webb) p. 9. Wiley, New York.
—— (1962*a*) *Phys. Rev.* **128**, 82.
—— (1962*b*) *Phys. Rev.* **128**, 2524.
—— and SEGALL, B. (1961) *Phys. Rev.* **124**, 1786.
—— —— (1969) *Meth. comput. Phys.* **8**, 251.
HANLON, J. E., and LAWSON, A. W. (1959) *Phys. Rev.* **113**, 472.

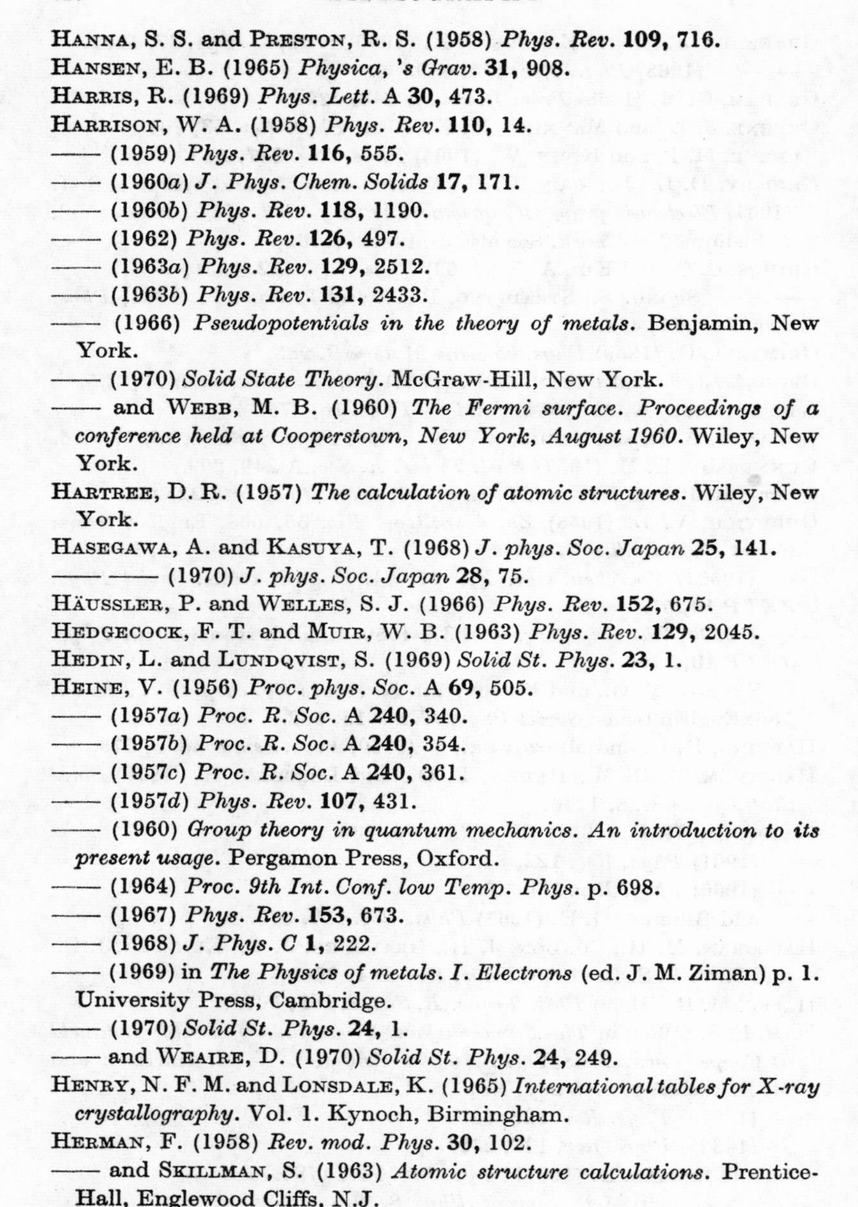

Hanna, S. S. and Preston, R. S. (1958) *Phys. Rev.* **109,** 716.

Hansen, E. B. (1965) *Physica, 's Grav.* **31,** 908.

Harris, R. (1969) *Phys. Lett.* A **30,** 473.

Harrison, W. A. (1958) *Phys. Rev.* **110,** 14.

—— (1959) *Phys. Rev.* **116,** 555.

—— (1960*a*) *J. Phys. Chem. Solids* **17,** 171.

—— (1960*b*) *Phys. Rev.* **118,** 1190.

—— (1962) *Phys. Rev.* **126,** 497.

—— (1963*a*) *Phys. Rev.* **129,** 2512.

—— (1963*b*) *Phys. Rev.* **131,** 2433.

—— (1966) *Pseudopotentials in the theory of metals.* Benjamin, New York.

—— (1970) *Solid State Theory.* McGraw-Hill, New York.

—— and Webb, M. B. (1960) *The Fermi surface. Proceedings of a conference held at Cooperstown, New York, August 1960.* Wiley, New York.

Hartree, D. R. (1957) *The calculation of atomic structures.* Wiley, New York.

Hasegawa, A. and Kasuya, T. (1968) *J. phys. Soc. Japan* **25,** 141.

—— —— (1970) *J. phys. Soc. Japan* **28,** 75.

Häussler, P. and Welles, S. J. (1966) *Phys. Rev.* **152,** 675.

Hedgecock, F. T. and Muir, W. B. (1963) *Phys. Rev.* **129,** 2045.

Hedin, L. and Lundqvist, S. (1969) *Solid St. Phys.* **23,** 1.

Heine, V. (1956) *Proc. phys. Soc.* A **69,** 505.

—— (1957*a*) *Proc. R. Soc.* A **240,** 340.

—— (1957*b*) *Proc. R. Soc.* A **240,** 354.

—— (1957*c*) *Proc. R. Soc.* A **240,** 361.

—— (1957*d*) *Phys. Rev.* **107,** 431.

—— (1960) *Group theory in quantum mechanics. An introduction to its present usage.* Pergamon Press, Oxford.

—— (1964) *Proc. 9th Int. Conf. low Temp. Phys.* p. 698.

—— (1967) *Phys. Rev.* **153,** 673.

—— (1968) *J. Phys. C* **1,** 222.

—— (1969) in *The Physics of metals. I. Electrons* (ed. J. M. Ziman) p. 1. University Press, Cambridge.

—— (1970) *Solid St. Phys.* **24,** 1.

—— and Weaire, D. (1970) *Solid St. Phys.* **24,** 249.

Henry, N. F. M. and Lonsdale, K. (1965) *International tables for X-ray crystallography.* Vol. 1. Kynoch, Birmingham.

Herman, F. (1958) *Rev. mod. Phys.* **30,** 102.

—— and Skillman, S. (1963) *Atomic structure calculations.* Prentice-Hall, Englewood Cliffs, N.J.

Herring, C. (1937*a*) *Phys. Rev.* **52,** 361.

——(1937*b*) *Phys. Rev.* **52,** 365.

HERRING, C. (1940) *Phys. Rev.* **57,** 1169.
—— (1942) *J. Franklin Inst.* **233,** 525.
—— (1966) in *Magnetism* (ed. G. T. Rado and H. Suhl) Vol. 4. Academic Press, New York.
—— and HILL, A. G. (1940) *Phys. Rev.* **58,** 132.
HIGGINS, R. J. and MARCUS, J. A. (1966) *Phys. Rev.* **141,** 553.
—— —— (1967) *Phys. Rev.* **161,** 589.
—— —— and WHITMORE, D. H. (1964) *Proc. 9th Int. Conf. low Temp. Phys.* p. 859.
HO, P. S. and RUOFF, A. L. (1967) *Phys. Status solidi* **23,** 489.
HODGES, L. and EHRENREICH, H. (1965) *Phys. Lett.* **16,** 203.
—— —— (1968) *J. appl. Phys.* **39,** 1280.
—— —— and LANG, N. D. (1966) *Phys. Rev.* **152,** 505.
—— LANG, N. D., EHRENREICH, H., and FREEMAN, A. J. (1966) *J. appl. Phys.* **37,** 1449.
—— STONE, D. R., and GOLD, A. V. (1967) *Phys. Rev. Lett.* **19,** 655.
HOLWECH, I. and RISNES, R. (1968) *Phil. Mag.* **17,** 757.
HÖRNFELDT, S. (1970) *Solid State Commun.* **8,** 673.
HOTZ, H. P., MATHIESEN, J. M., and HURLEY, J. P. (1968) *Phys. Rev.* **170,** 351.
HOUGHTON, J. D. and SMITH, S. D. (1966) *Infra-red physics*, Clarendon Press, Oxford.
HOWARTH, D. J. and JONES, H. (1952) *Proc. phys. Soc.* A **65,** 355.
HUBBARD, J. (1967) *Proc. phys. Soc.* **92,** 921.
—— (1969) *J. Phys. C* **2,** 1222.
—— and DALTON, N. W. (1968) *J. Phys. C* **1,** 1637.
HUGHES, A. J. and CALLAWAY, J. (1964) *Phys. Rev.* **136,** A1390.
—— and LETTINGTON, A. H. (1968) *Phys. Lett.* A **27,** 241.
—— and SHEPHERD, J. P. G. (1969) *J. Phys. C* **2,** 661.
HUM, D. M. and WONG, K. C. (1969) *J. Phys. C* **2,** 833.
HUME-ROTHERY, W. (1931) *The metallic state.* Clarendon Press, Oxford.
—— (1936) *The structure of metals and alloys.* Institute of metals, London.
—— (1966) in *Phase stability in metals and alloys* (ed. P. S. Rudman). McGraw-Hill, New York.
—— and COLES, B. R. (1954) *Adv. Phys.* **3,** 149.
HUNTINGTON, H. B. (1958) *Solid St. Phys.* **7,** 213.
HYGH, E. H. and WELCH, R. M. (1970) *Phys. Rev.* B **1,** 2424.
IGNAT'EVA, T. A., MAKAROV, V. I. and TERESHINA, N. S. (1968) *Zh. éksp. teor. Fiz.* **54,** 1617. English trans: *Soviet Phys. JETP* **27,** 865.
IONA, M. (1941) *Phys. Rev.* **60,** 822.
ISHIZAWA, Y. (1968*a*) *J. phys. Soc. Japan* **25,** 150.
—— (1968*b*) *J. phys. Soc. Japan* **25,** 160.
—— and DATARS, W. R. (1969) *Phys. Lett.* A **30, 463.**

ISHIZAWA, Y. and TANUMA, S. (1965) *J. phys. Soc. Japan*, **20**, 1278.

ITSKEVICH, E. S. (1962) *Zh. éksp. teor. Fiz.* **42**, 1173. English trans: *Soviet Phys. JETP* **15**, 811.

—— and FISHER, L. M. (1967*a*) *Zh. éksp. teor. Fiz., Pis'ma* **5**, 141. English trans: *Soviet Phys. JETP Lett.* **5**, 114.

—— —— (1967*b*) *Zh. éksp. teor. Fiz.* **53**, 98. English trans: *Soviet Phys. JETP* **26**, 66.

—— KRECHETOVA, I. P. and FISHER, L. M. (1967) *Zh. éksp. teor. Fiz.* **52**, 66. English trans: *Soviet Phys. JETP* **25**, 41.

—— and VORONOVSKIĬ, A. N. (1966) *Zh. éksp. teor. Fiz., Pis'ma* **4**, 226. English trans: *Soviet Phys. JETP Lett.* **4**, 154.

JACKSON, C. (1969) *Phys. Rev.* **178**, 949.

—— and DONIACH, S. (1969) *Phys. Lett.* A **30**, 328.

JACOBS, R. L. (1968) *J. Phys. C* **1**, 492.

JAIN, A. L. and KOENIG, S. H. (1962) *Phys. Rev.* **127**, 442.

JAN, J. P. (1957) *Solid St. Phys.* **5**, 1.

—— (1968) *Helv. Phys. Acta* **41**, 957.

—— PEARSON, W. B. and SAITO, Y. (1967) *Proc. R. Soc.* A **297**, 275.

—— —— and SPRINGFORD, M. (1964) *Proc. 9th Int. Conf. low Temp. Phys.* p. 776.

JAYARAMAN, A. (1964) *Phys. Rev.* **135**, A1056.

—— KLEMENT, W. and KENNEDY, G. C. (1963) *Phys. Rev.* **132**, 1620.

JOHANSEN, G. (1969) *Solid State Commun.* **7**, 731.

—— and MACKINTOSH, A. R. (1970) *Solid State Commun.* **8**, 121.

JOHNSON, H. T. (1965) *Electronic properties of materials. A guide to the literature*. Plenum, New York. See also GRIGSBY et al. (1967).

JOHNSON, K. H. (1966) *Phys. Rev.* **150**, 429.

—— and AMAR, H. (1965) *Phys. Rev.* **139**, A760.

JOHNSON, L. E., CONKLIN, J. B., and PRATT, G. W. (1963) *Phys. Rev. Lett.* **11**, 538.

JONES, C. K. and RAYNE, J. A. (1965*a*) *Phys. Lett.* **14**, 13.

—— —— (1965*b*) *Phys. Rev.* **139**, A1876.

JONES, D. and LETTINGTON, A. H. (1967) *Proc. phys. Soc.* **92**, 948.

JONES, E. P. and WILLIAMS, D. L. (1964) *Can. J. Phys.* **42**, 1499.

JONES, H. (1934) *Proc. R. Soc.* A **147**, 396.

—— (1949) *Physica, 's Grav.* **15**, 13.

—— (1955) *Proc. phys. Soc.* A **68**, 1191.

—— (1960) *The theory of Brillouin zones and electronic states in crystals*. North-Holland, Amsterdam.

JOSEPH, A. S. and GORDON, W. L. (1962) *Phys. Rev.* **126**, 489.

—— —— REITZ, J. R. and ECK, T. G. (1961) *Phys. Rev. Lett.* **7**, 334.

—— and THORSEN, A. C. (1963) *Phys. Rev. Lett.* **11**, 554.

JOSEPH, A. S. and THORSEN, A. C. (1964) *Phys. Rev.* **133**, A1546.

JOSHI, S. K. and KASHYAP, B. M. S. (1962) *Phys. Rev.* **126,** 936.

KAGANOV, M. I., LIFSHITZ, I. M. and SINEL'NIKOV, K. D. (1957) *Zh. éksp. teor. Fiz.* **32,** 605. English trans: *Soviet Phys. JETP* **5,** 500.

KAHANA, S. (1963) *Phys. Rev.* **129,** 1622.

KAHN, A. H. and FREDERIKSE, H. P. R. (1959) *Solid St. Phys.* **9,** 257.

KALINKINA, I. N. and STRELKOV, P. G. (1958) *Zh. éksp. teor. Fiz.* **34** 616. English trans: *Soviet Phys. JETP* **7,** 426.

KANER, É. A. (1958) *Dokl. Akad. Nauk SSSR* **119,** 471. English trans: *Soviet Phys. Dokl.* **3,** 314.

—— (1967) *Physics* **3,** 285.

—— and FAL'KO, V. L. (1964) *Zh. éksp. teor. Fiz.* **46,** 1344. English trans: *Soviet Phys. JETP* **19,** 910.

—— and SKOBOV, V. G. (1971) *Plasma effects in metals: Helicon and Alfvén waves.* Taylor and Francis, London; reprinted from *Adv. Phys.* **17,** 605 (1968).

KAO, Y. H. (1965) *Phys. Rev.* **138,** A1412.

KAPLAN, J. I. (1969) *Phys. Lett.* A **29,** 552.

KARO, A. M. (1959) *J. chem. Phys.* **31,** 1489.

—— (1960) *J. chem. Phys.* **33,** 7.

KASPER, J. S. and ROBERTS, B. W. (1956) *Phys. Rev.* **101,** 537.

KASUYA, T. (1956) *Prog. theor. Phys.* **16,** 58.

KECHIN, V. V. (1967) *Fizika tverd. Tela,* **9,** 3595. English trans: *Soviet Phys. Solid St.* **9,** 2828.

—— (1969) *Fizika tverd. Tela* **11,** 1788. English trans: *Soviet Phys. Solid St.* **11,** 1448.

—— LIKHTER, A. I., and POSPELOV, YU. A. (1965) *Zh. éksp. teor. Fiz.* **49,** 36. English trans: *Soviet Phys. JETP* **22,** 66.

KEETON, S. C. and LOUCKS, T. L. (1966*a*) *Phys. Rev.* **146,** 429.

—— —— (1966*b*) *Phys. Rev.* **152,** 548.

—— —— (1968) *Phys. Rev.* **168,** 672.

KELLERMANN, E. W. (1940) *Phil. Trans. R. Soc.* A **238,** 513.

KETTERSON, J. B. and ECKSTEIN, Y. (1966) *Rev. scient. Instrum.* **37,** 44.

—— MUELLER, F. M., and WINDMILLER, L. R. (1969) *Phys. Rev.* **186,** 656.

—— PRIESTLEY, M. G., and VUILLEMIN, J. J. (1966) *Phys. Lett.* **20,** 452.

—— and STARK, R. W. (1967) *Phys. Rev.* **156,** 748.

KHAĬKIN, M. S. (1961) *Zh. éksp. teor. Fiz.* **41,** 1773. English trans: *Soviet Phys. JETP* **14,** 1260.

—— and MINA, R. T. (1962) *Zh. éksp. teor. Fiz.* **42,** 35. English trans: *Soviet Phys. JETP* **15,** 24.

—— —— and ÉDEL'MAN, V. S. (1962) *Zh. éksp. teor. Fiz.* **43,** 2063. English trans: *Soviet Phys. JETP* **16,** 1459.

KILBY, G. E. (1965) *Proc. phys. Soc.* **86,** 1037.

KIMBALL, J. C., STARK, R. W., and MUELLER, F. M. (1967) *Phys. Rev.* **162,** 600.

KIP, A. F., LANGENBERG, D. N., and MOORE, T. W. (1961) *Phys. Rev.* **124,** 359.

KITTEL, C. (1956) *Introduction to solid state physics.* Wiley, New York.

—— (1963*a*) *Phys. Rev. Lett.* **10,** 339.

—— (1963*b*) *Quantum theory of solids.* Wiley, New York.

KJELDAAS, T. (1959) *Phys. Rev.* **113,** 1473.

KLEMENS, P. G. (1954) *Aust. J. Phys.* **7,** 70.

KOCH, J. F., STRADLING, R. A., and KIP, A. F. (1964) *Phys. Rev.* **133,** A240.

KOEHLER, W. C., MOON, R. M., TREGO, A. L., and MACKINTOSH, A. R. (1966) *Phys. Rev.* **151,** 405.

KOENIGSBERG, E. (1953) *Phys. Rev.* **91,** 8.

KOHN, W. (1959*a*) *Phys. Rev. Lett.* **2,** 393.

—— (1959*b*) *Phys. Rev.* **115,** 1460.

—— (1964) *Phys. Rev.* **133,** 171.

—— and ROSTOKER, J. (1954) *Phys. Rev.* **94,** 1111.

KONSTANTINOV, O. V. and PEREL', V. I. (1960) *Zh. éksp. teor. Fiz.* **38,** 161. English trans: *Soviet Phys. JETP* **11,** 117.

KOOPMANS, T. (1934) *Physica, 's Grav.* **1,** 104.

KOROLYUK, A. P. and PRUSHCHAK, T. A. (1961) *Zh. éksp. teor. Fiz.* **41,** 1689. English trans: *Soviet Phys. JETP* **14,** 1201.

KORRINGA, J. (1947) *Physica, 's Grav.* **13,** 392.

KOSTER, G. F. (1957) *Solid St. Phys.* **5,** 173.

—— (1962) *Phys. Rev.* **127,** 2044.

——, DIMMOCK, J. O., WHEELER, R. G., and STATZ, H. (1963) *Properties of the thirty-two point groups.* M.I.T. Press, Cambridge, Mass.

KOVALEV, O. V. (1965) *Irreducible representations of the space groups.* Gordon and Breach, New York.

KREBS, K. (1963) *Phys. Lett.* **3,** 31.

KUBICAR, L. and KREMPASKY, J. (1970) *Phys. Status solidi* A **2,** 739.

KUHN, T. S. and VAN VLECK, J. H. (1950) *Phys. Rev.* **79,** 382.

KUNZLER, J. E. and HSU, F. S. L. (1960) in *The Fermi surface. Proceedings of a conference held at Cooperstown, New York, August 1960* (ed. W. A. Harrison and M. B. Webb) p. 88. Wiley, New York.

—— —— and BOYLE, W. S. (1962) *Phys. Rev.* **128,** 1084.

KUPER, C. G. (1968) *An introduction to the theory of superconductivity.* Clarendon Press, Oxford.

LANDAU, L. D. (1930) *Z. Phys.* **64,** 629.

—— (1939) *Proc. R. Soc.* A **170,** 341.

—— (1956) *Zh. éksp. teor. Fiz.* **30,** 1058. English trans: *Soviet Phys. JETP* **3,** 920.

Landau, L. D. (1957) *Zh. éksp. teor. Fiz.* **32**, 59. English trans: *Soviet Phys. JETP* **5**, 110.

—— (1958) *Zh. éksp. teor. Fiz.* **35**, 97. English trans: *Soviet Phys. JETP* **8**, 70.

Lang, G., de Benedetti, S., and Smoluchowski, R. (1955) *Phys. Rev.* **99**, 596.

Langbein, D. (1969) *Phys. Rev.* **180**, 633.

Langenberg, D. N. (1968) *Am. J. Phys.* **36**, 777.

Larson, C. O. and Gordon, W. L. (1967) *Phys. Rev.* **156**, 703.

Lax, B. and Mavroides, J. G. (1960) *Solid St. Phys.* **11**, 261.

Lazarev, B. G., Lazareva, L. S., Ignat'eva, T. A., and Makarov, V. I. (1965) *Dokl. Akad. Nauk SSSR* **163**, 74. English trans: *Soviet Phys. Dokl.* **10**, 620.

Leaver, G. and Myers, A. (1969) *Phil. Mag.* **19**, 465.

Lee, E. W. and Asgar, M. A. (1969) *Phys. Rev. Lett.* **22**, 1436.

Lee, M. J. G. (1966) *Proc. R. Soc.* A **295**, 440.

—— (1969) *Phys. Rev.* **178**, 953.

—— and Falicov, L. M. (1968) *Proc. R. Soc.* A **304**, 319.

Leibfried, G. and Ludwig, W. (1961) *Solid St. Phys.* **12**, 275.

Lenham, A. P. and Treherne, P. M. (1964) *Proc. phys. Soc.* **83**, 1059.

—— —— (1966) *J. opt. Soc. Am.* **56**, 683.

Lerner, L. S. (1962) *Phys. Rev.* **127**, 1480.

—— (1963) *Phys. Rev.* **130**, 605.

Lettington, A. H. (1964) *Phys. Lett.* **9**, 98.

Lifshitz, I. M. (1963) *Zh. éksp. teor. Fiz.* **44**, 1723. English trans: *Soviet Phys. JETP* **17**, 1159.

—— (1964) *Adv. Phys.* **13**, 483.

—— Azbel', M. Ya. and Kaganov, M. I. (1956) *Zh. éksp. teor. Fiz.* **31**, 63. English trans: *Soviet Phys. JETP* **4**, 41.

—— and Kaganov, M. I. (1959) *Usp. fiz. Nauk* **69**, 419. English trans: *Soviet Phys. Usp.* **2**, 831.

—— —— (1965) *Usp. fiz. Nauk* **87**, 389. English trans: *Soviet Phys. Usp.* **8**, 805.

—— and Pogorelov, A. V. (1954) *Dokl. Akad. Nauk SSSR* **96**, 1143.

Likhter, A. I. and Kechin, V. V. (1963) *Fizika tverd. Tela* **5**, 3066. English trans: *Soviet Phys. Solid St.* **5**, 2246.

Likhter, A. I. and Venttsel', V. A. (1962) *Fizika tverd. Tela* **4**, 485. English trans: *Soviet Phys. Solid St.* **4**, 352.

Lin, P. J. and Falicov, L. M. (1966) *Phys. Rev.* **142**, 441.

Lin, P. J. and Phillips, J. C. (1965) *Adv. Phys.* **14**, 257.

—— —— (1966) *Phys. Rev.* **147**, 469.

Lipson, S. G. (1964) *Proc. 9th Int. Conf. low Temp. Phys.* p. 814.

Lomer, W. M. (1962*a*) *Proc. phys. Soc.* **80**, 489.

—— (1962*b*) *J. Phys., Paris* **23**, 716.

Lomer, W. M. (1964*a*) *Proc. phys. Soc.* **84,** 327.
—— (1964*b*) *Proc. Int. Conf. Magnetism, Nottingham*, p. 127.
—— and Gardner, W. E. (1969) *Prog. Mater. Sci.* **14,** 143.
London, H. (1940) *Proc. R. Soc.* A **176,** 522.
Loucks, T. L. (1964) *Phys. Rev.* **134,** A1618.
—— (1965*a*) *Phys. Rev.* **139,** A1181.
—— (1965*b*) *Phys. Rev. Lett.* **14,** 693.
—— (1965*c*) *Phys. Rev. Lett.* **14,** 1072.
—— (1965*d*) *Phys. Rev.* **139,** A1333.
—— (1966*a*) *Phys. Rev.* **143,** 506.
—— (1966*b*) *Phys. Rev.* **144,** 504.
—— (1967*a*) *Augmented plane wave method*. Benjamin, New York.
—— (1967*b*) *Phys. Rev.* **159,** 544.
—— and Cutler, P. H. (1964) *Phys. Rev.* **133,** A819.
Loudon, R. (1964) *Adv. Phys.* **13,** 423.
Lounasmaa, O. V. (1963) *Phys. Rev.* **129,** 2460.
Lovell, A. C. B. (1936) *Proc. R. Soc.* A **157,** 311.
—— (1938) *Proc. R. Soc.* A **166,** 270.
Löwdin, P. O. (1956) *Adv. Phys.* **5,** 1.
Lukirskiĭ, A. P. and Brȳtov, I. A. (1964) *Fizika tverd. Tela*, **6,** 43. English trans: *Soviet Phys. Solid St.* **6,** 33.
Lüthi, B. (1960) *Helv. Phys. Acta* **33,** 161.
Lutskii, V. N. and Fesenko, E. P. (1968) *Fizika tverd. Tela* **10,** 3661. English trans: *Soviet Phys. Solid St.* **10,** 2902.
—— Korneev, D. N. and Elinson, M. I. (1966) *Zh. éksp. teor. Fiz., Pis'ma* **4,** 267. English trans: *Soviet Phys. JETP Lett.* **4,** 179.
Luttinger, J. M. (1951) *Phys. Rev.* **84,** 814.
Lynton, E. A. (1962) *Superconductivity*. Methuen, London.
McClure, J. W. (1957) *Phys. Rev.* **108,** 612.
—— (1964) *I.B.M. Jl. Res. Dev.* **8,** 255.
—— (1971) *J. Phys. Chem. Solids* **32,** Suppl. 1, 127.
—— and Spry, W. J. (1968) *Phys. Rev.* **165,** 809.
MacDonald, D. K. C. (1949) *Nature, Lond.* **163,** 637.
—— and Barron, T. H. K. (1958) *Physica, 's Grav.* **24,** 102.
—— Pearson, W. B., and Templeton, I. M. (1958) *Proc. R. Soc.* A **248,** 107.
—— and Sarginson, K. (1950) *Proc. R. Soc.* A **203,** 223.
McEwen, K. A. (1969) *Phys. Lett.* A **30,** 77.
—— (1971) *Proc. R. Soc.* A **322,** 509.
Machlin, E. S., Petralia, S., Desalvo, A., Missiroli, G. F., and Zignani, F. (1968) *Nuovo Cim.* B **55,** 263.
Mackinnon, L. (1966) *Experimental physics at low temperatures. An introductory survey.* Wayne State University Press, Detroit.
Mackinnon, L., Taylor, M. T., and Daniel, M. R. (1962) *Phil. Mag.* **7,** 523.

MACKINTOSH, A. R. (1962) *Phys. Rev. Lett.* **9,** 90.

—— (1963) *Scientific American* **209**(1), 110.

—— SPANEL, L. E., and YOUNG, R. C. (1963) *Phys. Rev. Lett.* **10,** 434.

MAJUMDAR, C. K. (1965*a*) *Phys. Rev.* **140,** A 227.

—— (1965*b*) *Phys. Rev.* **140,** A237.

MAKAROV, V. I. and BAR'YAKHTAR, V. G. (1965) *Zh. éksp. teor. Fiz.* **48,** 1717. English trans: *Soviet Phys. JETP* **21,** 1151.

—— and VOLYNSKIĬ, I. YA. (1966) *Zh. éksp. teor. Fiz., Pis'ma* **4,** 369. English trans: *Soviet Phys. JETP Lett.* **4,** 249.

—— —— (1969) *Zh. éksp. teor. Fiz.* **57,** 3. English trans: *Soviet Phys. JETP* **30,** 1.

MANNING, M. F. (1943) *Phys. Rev.* **63,** 190.

MARADUDIN, A. A. and VOSKO, S. H. (1968) *Rev. mod. Phys.* **40,** 1.

MARCH, N. H., YOUNG, W. H., and SAMPANTHAR, S. (1967) *The many-body problem in quantum mechanics.* University Press, Cambridge.

MARSHALL, W. and LOVESEY, S. W. (1971) *Theory of thermal neutron scattering.* Clarendon Press, Oxford.

MARTIN, D. L. (1965) *Phys. Rev.* **139,** A 150.

MASE, S. (1958) *J. phys. Soc. Japan* **13,** 434.

—— (1959*a*) *J. phys. Soc. Japan* **14,** 584.

—— (1959*b*) *J. phys. Soc. Japan* **14,** 1538.

MASSALSKI, T. B. (1962) *J. Phys., Paris* **23,** 607.

—— and KING, H. W. (1961) *Prog. Mater. Sci.* **10,** 1.

MATSUMOTO, T., SAMBONGI, T., and MITSUI, T. (1969) *J. phys. Soc. Japan* **26,** 209.

MATTHEISS, L. F. (1964*a*) *Phys. Rev.* **133,** A1399.

—— (1964*b*) *Phys. Rev.* **134,** A970.

—— (1965) *Phys. Rev.* **139,** A1893.

—— (1966) *Phys. Rev.* **151,** 450.

—— (1970) *Phys. Rev.* B **1,** 373.

—— and WATSON, R. E. (1964) *Phys. Rev. Lett.* **13,** 526.

—— —— (1965) *Phys. Rev.* **139,** A1893.

MATTHIESSEN, A. (1862) *Rep. Br. Ass. Advmt Sci.* **32,** 144.

MATTIS, D. C. and DRESSELHAUS, G. (1958) *Phys. Rev.* **111,** 403.

MATTUCK, R. D. (1967) *A guide to Feynman diagrams in the many-body problem.* McGraw-Hill, London.

MAVROIDES, J. G., LAX, B., BUTTON, K. J., and SHAPIRA, Y. (1962) *Phys. Rev. Lett.* **9,** 451.

MAXWELL, E. (1950) *Phys. Rev.* **78,** 477.

MEADEN, G. T., RAO, K. V., LOO, H. Y., and SZE, N. H. (1969) *J. phys. Soc. Japan* **27,** 1073.

MELNGAILIS, J. and DE BENEDETTI, S. (1966) *Phys. Rev.* **145,** 400.

MERCOUROFF, W. (1967) *La surface de Fermi des metaux.* Masson, Paris.

MERZ, H. and ULMER, K. (1966) *Phys. Lett.* **22,** 251.

MIĄSEK, M. (1957) *Phys. Rev.* **107,** 92.

—— and SUFFCZYŃSKI, M. (1961*a*) *Bull. Acad. pol. Sci., Sér. sci. math. astr. phys.* **9,** 477.

—— —— (1961*b*) *Bull. Acad. pol. Sci., Sér. sci. math. astr. phys.* **9,** 483.

MIHALISIN, T. W. and PARKS, R. D. (1966) *Phys. Lett.* **21,** 610.

—— —— (1967) *Phys. Rev. Lett.* **18,** 210.

—— —— (1969) *Solid State Commun.* **7,** 33.

MIJNARENDS, P. E. and HAMBRO, L. (1964) *Phys. Lett.* **10,** 272.

MILLER, S. C. and LOVE, W. F. (1967) *Tables of irreducible representations of space groups and co-representations of magnetic space groups.* Pruett, Boulder, Col.

MILLIKEN, J. C. and YOUNG, R. C. (1966) *Phys. Rev.* **148,** 558.

MINA, R. T. and KHAĬKIN, M. S. (1963) *Zh. éksp. teor. Fiz.* **45,** 1304. English trans: *Soviet Phys. JETP* **18,** 896.

—— —— (1965) *Zh. éksp. teor. Fiz.* **48,** 111. English trans: *Soviet Phys. JETP* **21,** 75.

—— —— (1966) *Zh. éksp. teor. Fiz.* **51,** 62. English trans: *Soviet Phys. JETP* **24,** 42.

MINOMURA, S., TANUMA, S., FUJII, G., NISHIZAWA, M., and NAGANO, H. (1969) *Phys. Lett.* A **29,** 16.

MIWA, H. (1963) *Prog. theor. Phys.* **29,** 477.

MIZIUMSKI, C. and LAWSON, A. W. (1969) *Phys. Rev.* **180,** 749.

MONTALVO, R. A. and MARCUS, J. A. (1964) *Phys. Lett.* **8,** 151.

MOON, R. M., CABLE, J. W., and KOEHLER, W. C. (1964) *J. appl. Phys.* **35,** 1041.

MOORE, T. W. (1966) *Phys. Rev. Lett.* **16,** 581.

MORSE, R. W. (1959) *Prog. Cryogen.* **1,** 226.

—— (1960) in *The Fermi surface. Proceedings of a conference held at Cooperstown, New York, August 1960* (ed. W. A. Harrison and M. B. Webb) p. 214. Wiley, New York.

—— MYERS, A., and WALKER, C. T. (1960) *Phys. Rev. Lett.* **4,** 605.

—— —— —— (1961) *J. acoust. Soc. Am.* **33,** 699.

MOSS, J. S. and DATARS, W. R. (1967) *Phys. Lett.* A **24,** 630.

MOTT, N. F. (1935) *Proc. phys. Soc.* **47,** 571.

—— (1949) *Proc. phys. Soc.* A **62,** 416.

—— (1956) *Can. J. Phys.* **34,** 1356.

—— (1958) *Nuovo Cim.* Suppl. 7, Ser. X, 312.

—— (1961) *Phil. Mag.* **6,** 287.

—— (1962) *Rep. Prog. Phys.* **25,** 218.

—— (1964) *Adv. Phys.* **13,** 325.

—— (1967) *Adv. Phys.* **16,** 49.

—— (1968) *Rev. mod. Phys.* **40,** 677.

—— and JONES, H. (1936) *The theory of the properties of metals and alloys.* Clarendon Press, Oxford.

MOTT, N. F. and TWOSE, W. D. (1961) *Adv. Phys.* **10**, 107.
—— and ZINAMON, Z. (1970) *Rep. Prog. Phys.* **33**, 881.
MUELLER, F. M. (1966) *Phys. Rev.* **148**, 636.
—— (1967) *Phys. Rev.* **153**, 659.
—— FREEMAN, A. J., DIMMOCK, J. O., and FURDYNA, A. M. (1970) *Phys. Rev.* B **1**, 4617.
—— and PRIESTLEY, M. G. (1966) *Phys. Rev.* **148**, 638.
MUKHOPADHYAY, G. and MAJUMDAR, C. K. (1969) *J. Phys. C* **2**, 924.
MÜLLER, W. E. (1965) *Phys. Lett.* **17**, 82.
—— (1966) *Solid State Commun.* **4**, 581.
MUNARIN, J. A. and MARCUS, J. A. (1964) *Proc. 9th Int. Conf. low Temp. Phys.* p. 743.
—— —— and BLOOMFIELD, P. E. (1968) *Phys. Rev.* **172**, 718.
MYRON, H. W. and LIU, S. H. (1970) *Phys. Rev.* B **1**, 2414.
NABEREZHNYKH, V. P. and TSYMBAL, D. T. (1967) *Zh. éksp. teor. Fiz., Pis'ma* **5**, 319. English trans: *Soviet Phys. JETP Lett.* **5**, 263.
NAKAJIMA, S. (1967) *Prog. theor. Phys.* **38**, 23.
NATHANS, R., SHULL, C. G., SHIRANE, G., and ANDRESEN, A. (1959) *J. Phys. Chem. Solids* **10**, 138.
NELSON, K. S., STANFORD, J. L., and SCHMIDT, F. A. (1968) *Phys. Lett.* A **28**, 402.
NEVITT, M. V. (1966) in *Phase stability in metals and alloys* (ed. P. S. Rudman) p. 281. McGraw-Hill, New York.
NIKULIN, V. K. (1966) *Phys. Status solidi* **16**, K125.
—— and TRZHASKOVSKAYA, M. B. (1968) *Phys. Status solidi* **28**, 801.
NORDBERG, R., HEDMAN, J., HEDÉN, P. F., NORDLING, C., and SIEGBAHN, K. (1968) *Arkiv Fysik* **37**, 489.
NOZIÈRES, P. (1964) *Theory of interacting Fermi systems.* Benjamin, New York.
NUSSBAUM, A. (1966) *Solid St. Phys.* **18**, 165.
NYE, J. F. (1957) *Physical properties of crystals.* Clarendon Press, Oxford.
OGRIN, Yu. F., LUTSKII, V. N. and ELINSON, M. I. (1966) *Zh. éksp. teor. Fiz., Pis'ma* **3**, 114. English trans: *Soviet Phys. JETP Lett.* **3**, 71.
ÖKTÜ, Ö. and SAUNDERS, G. A. (1967) *Proc. phys. Soc.* **91**, 156.
OKUMURA, K. and TEMPLETON, I. (1965) *Proc. R. Soc.* A **287**, 89.
ONODERA, Y. and OKAZAKI, M. (1966*a*) *J. phys. Soc. Japan* **21**, 1273.
—— —— (1966*b*) *J. phys. Soc. Japan* **21**, 2400.
ONSAGER, L. (1952) *Phil. Mag.* **43**, 1006.
OPECHOWSKI, W. and GUCCIONE, R. (1965) in *Magnetism* (ed. G. T. Rado and H. Suhl) Vol. 2A, p. 105. Academic Press, New York.
ORCHARD-WEBB, J. H. and COUSINS, J. E. (1968) *Phys. Lett.* A **28**, 236.
O'SULLIVAN, W. J. and SCHIRBER, J. E. (1965) *Phys. Lett.* **18**, 212.
OVERHAUSER, A. W. (1962) *Phys. Rev.* **128**, 1437.

Overhauser, A. W. (1967) *Phys. Rev.* **156**, 844.

Paskin, A. and Weiss, R. J. (1962) *Phys. Rev. Lett.* **9**, 199.

Patterson, J. (1901) *Proc. Camb. phil. Soc. math. phys. Sci.* **11**, 118.

Peierls, R. (1933*a*) *Z. Phys.* **80**, 763.

—— (1933*b*) *Z. Phys.* **81**, 186.

—— (1934) *Helv. Phys. Acta* **7** (Suppl.) 24.

—— (1955) *Quantum theory of solids*. Clarendon Press, Oxford.

Perz, J. M., Hum, R. H., and Coleridge, P. T. (1969) *Phys. Lett.* A **30**, 235.

Pettifor, D. G. (1969) *J. Phys. C* **2**, 1051.

—— (1970) *J. Phys. C* **3**, 367.

Phillips, F. C. (1963) *An introduction to crystallography*. Longmans, Green, London.

Phillips, J. C. (1961) *Phys. Rev.* **123**, 420.

—— (1964) *Phys. Rev.* **133**, A1020.

—— (1966) *Solid St. Phys.* **18**, 55.

—— (1968*a*) *J. appl. Phys.* **39**, 755.

—— (1968*b*) *Adv. Phys.* **17**, 79.

—— and Kleinman, L. (1959) *Phys. Rev.* **116**, 287.

Phillips, W. C. and Weiss, R. J. (1968) *Phys. Rev.* **171**, 790.

Pick, R. (1967) *J. Phys., Paris* **28**, 539.

Pincherle, L. (1960) *Rep. Prog. Phys.* **23**, 355.

—— (1971) *Electronic energy bands in solids*. Macdonald, London.

Pines, D. (1953) *Phys. Rev.* **92**, 626.

——(1955) *Solid St. Phys.* **1**, 367.

—— and Bohm, D. (1952) *Phys. Rev.* **85**, 338.

Pippard, A. B. (1947) *Proc. R. Soc.* A **191**, 385.

—— (1950) *Proc. R. Soc.* A **203**, 98.

—— (1954) *Adv. Electronics Electron Phys.* **6**, 1.

—— (1955) *Phil. Mag.* **46**, 1104.

—— (1957*a*) *Phil. Mag.* **2**, 1147.

—— (1957*b*) *Phil. Trans. R. Soc.* A **250**, 325.

—— (1960*a*) *Rep. Prog. Phys.* **23**, 176.

—— (1960*b*) *Proc. R. Soc.* A **257**, 165.

—— (1960*c*) in *The Fermi surface. Proceedings of a conference held at Cooperstown, New York, August 1960* (ed. W. A. Harrison and M. B. Webb) p. 330. Wiley, New York.

—— (1965) *The dynamics of conduction electrons*. Blackie, London.

—— (1966) *Phil. Mag.* **13**, 1143.

—— and Chambers, R. G. (1952) *Proc. phys. Soc.* A **65**, 955.

—— Reuter, G. E. H., and Sondheimer, E. H. (1948) *Phys. Rev.* **73**, 920.

Platzman, P. M. and Tzoar, N. (1965) *Phys. Rev.* **139**, A410.

Platzman, P. M. and Walsh, W. M. (1967) *Phys. Rev. Lett.* **19,** 514 and Erratum: *Phys. Rev. Lett.* **20,** 89.
—— —— and Foo, E. N. (1968) *Phys. Rev.* **172,** 689.
—— and Wolff, P. A. (1967) *Phys. Rev. Lett.* **18,** 280.
Prange, R. E. and Kadanoff, L. P. (1964) *Phys. Rev.* **134,** A566.
Priestley, M. G. (1960) *Phil. Mag.* **5,** 111.
—— (1966) *Phys. Rev.* **148,** 580.
—— Windmiller, L. R., Ketterson, J. B., and Eckstein, Y. (1967) *Phys. Rev.* **154,** 671.
Quinn, J. J. (1963) *Phys. Rev. Lett.* **11,** 316.
Rayne, J. A. (1963) *Phys. Rev.* **129,** 652.
Reed, W. A. (1969) *Phys. Rev.* **188,** 1184.
—— and Fawcett, E. (1964) *J. appl. Phys.* **35,** 754.
—— and Marcus, J. A. (1962) *Phys. Rev.* **126,** 1298.
Reitz, J. R. (1955) *Solid St. Phys.* **1,** 1.
Reneker, D. H. (1958) *Phys. Rev. Lett.* **1,** 440.
Reuter, G. E. H. and Sondheimer, E. H. (1948) *Proc. R. Soc.* A **195,** 336.
Reynolds, C. A., Serin, B., Wright, W. H., and Nesbitt, L. B. (1950) *Phys. Rev.* **78,** 487.
Rice, T. M. (1968) *Phys. Rev.* **175,** 858.
Rickayzen, G. (1965) *Theory of superconductivity.* Interscience, New York.
Roaf, D. J. (1962) *Phil. Trans. R. Soc.* A **255,** 135.
Roberts, B. W. (1961) *Phys. Rev. Lett.* **6,** 453.
—— (1968) in *Physical acoustics* (ed. W. P. Mason) volume 4B, p. 1. Academic Press, New York.
Robinson, J. E. (1968) *Nuovo Cim.* Suppl. **6,** 745.
—— and Dow, J. D. (1968) *Phys. Rev.* **171,** 815.
Rooke, G. A. (1968) in *Soft X-ray band spectra and the electronic structure of metals and materials* (ed. D. J. Fabian) p. 3; 185. Academic Press, New York.
Rose, M. E. (1961) *Relativistic electron theory.* Wiley, New York.
Rosenberg, H. M. (1963) *Low temperature solid state physics.* Clarendon Press, Oxford.
Roth, L. M. (1962) *J. Phys. Chem. Solids,* **23,** 433.
—— Zeiger, H. J. and Kaplan, T. A. (1966) *Phys. Rev.* **149,** 519.
Ruderman, M. A. and Kittel, C. (1954) *Phys. Rev.* **96,** 99.
Ruvalds, J. and Falicov, L. M. (1968) *Phys. Rev.* **172,** 508.
Saito, Y. (1962) *J. phys. Soc. Japan,* **17,** 716.
—— (1963) *J. phys. Soc. Japan,* **18,** 1845.
Schiff, L. I. (1955) *Quantum mechanics.* McGraw-Hill, New York.
Schirber, J. E. (1963) *Phys. Rev.* **131,** 2459.
Schneider, T. and Stoll, E. (1966*a*) *Phys. kondens. Materie* **5,** 331.

SCHNEIDER, T. and STOLL, E. (1966*b*) *Phys. kondens. Materie* **5**, 364.
—— —— (1967*a*) *Phys. kondens. Materie* **6**, 135.
—— —— (1967*b*) *Phys. Lett.* A **24**, 258.
SCHRIEFFER, J. R. (1964) *Theory of superconductivity*. Benjamin, New York.
SCHROEDER, P. R., DRESSELHAUS, M. S., and JAVAN, A. (1968) *Phys. Rev. Lett.* **20**, 1292.
SCHUBNIKOW, L. and DE HAAS, W. J. (1930*a*) *Proc. Sect. Sci. K. ned. Akad. Wet.* **33**, 130. Reprinted: *Communs phys. Lab. Univ. Leiden* no. 207a.
—— —— (1930*b*) *Proc. Sect. Sci. K. ned. Akad. Wet.* **33**, 350. Reprinted: *Communs phys. Lab. Univ. Leiden* no. 207c.
—— —— (1930*c*) *Proc. Sect. Sci. K. ned. Akad. Wet.* **33**, 363. Reprinted: *Communs phys. Lab. Univ. Leiden* no. 207d.
—— —— (1930*d*) *Proc. Sect. Sci. K. ned. Akad. Wet.* **33**, 418. Reprinted: *Communs phys. Lab. Univ. Leiden* no. 210a.
SCHÜLER, C. C. (1965) in *Proceedings of the international colloquium on optical properties and electronic structure of metals and alloys, Paris (1965)* (ed. F. Abelès) p. 221. North-Holland, Amsterdam.
SCHULTZ, S. and DUNIFER, G. (1967) *Phys. Rev. Lett.* **18**, 283.
—— and SHANABARGER, M. R. (1966) *Phys. Rev. Lett.* **16**, 178.
SCHULTZ, T. D. (1964) *Quantum field theory and the many-body problem*. Gordon and Breach, New York.
SCOTT, G. B. and SPRINGFORD, M. (1970) *Proc. R. Soc.* A **320**, 115.
—— —— and STOCKTON, J. R. (1968*a*) *Phys. Lett.* A **27**, 655.
—— —— —— (1968*b*) *Proc. 11th Int. Conf. low Temp. Phys.* p. 1129.
SEDOV, V. L. (1968) *Usp. fiz. Nauk.* **94**, 417. English trans: *Soviet Phys. Usp.* **11**, 163.
SEGALL, B. (1957) *Phys. Rev.* **105**, 108.
—— (1961*a*) *Phys. Rev.* **124**, 1797.
—— (1961*b*) *Phys. Rev. Lett.* **7**, 154.
—— (1962) *Phys. Rev.* **125**, 109.
—— (1963) *Phys. Rev.* **131**, 121.
—— and HAM, F. S. (1968) *Meth. comput. Phys.* **8**, 251.
SEITZ, F. (1936) *Ann. Math.* **37**, 17.
SELLMYER, D. J., AHN, J. and JAN, J. P. (1967) *Phys. Rev.* **161**, 618.
SHAM, L. J. and ZIMAN, J. M. (1963) *Solid St. Phys.* **15**, 221.
SHAPIRA, Y. (1964) *Phys. Rev. Lett.* **13**, 162.
—— (1968) in *Physical acoustics* (ed. W. P. Mason) Vol. 5, p. 1. Academic Press, New York.
—— and LAX, B. (1965) *Phys. Rev.* **138**, A1191.
—— and NEURINGER, L. J. (1967) *Phys. Rev. Lett.* **18**, 1133.
SHARP, R. I. (1969*a*) *J. Phys. C* **2**, 421.

SHARP, R. I. (1969*b*) *J. Phys. C* **2,** 432.

SHARVIN, Yu. V. (1965) *Zh. éksp. teor. Fiz.* **48,** 984. English trans: *Soviet Phys. JETP* **21,** 655.

—— and FISHER, L. M. (1965) *Zh. éksp. teor. Fiz., Pis'ma,* **1** (5), 54. English trans: *Soviet Phys. JETP Lett.* **1,** 152.

SHAW, R. W. (1969) *J. Phys. C* **2,** 2335.

SHEPHERD, J. P. G. and GORDON, W. L. (1968) *Phys. Rev.* **169,** 541.

SHKLIAREVSKIĬ, I. N. and PADALKA, V. G. (1959) *Optika Spektrosk.* **6,** 776. English trans: *Optics Spectrosc.* **6,** 505.

SHOCKLEY, W. (1937) *Phys. Rev.* **52,** 866.

SHOENBERG, D. (1939) *Proc. R. Soc.* A **170,** 341.

—— (1952*a*) *Phil. Trans. R. Soc.* A **245,** 1.

—— (1952*b*) *Superconductivity.* University Press, Cambridge.

—— (1957) *Prog. low Temp. Phys.* **2,** 226.

—— (1959) *Nature, Lond.* **183,** 171.

—— (1960*a*) *Phil. Mag.* **5,** 105.

—— (1960*b*) in *The Fermi surface. Proceedings of a conference held at Cooperstown, New York, August 1960* (ed. W. A. Harrison and M. B. Webb) p. 74. Wiley, New York.

—— (1962) *Phil. Trans. R. Soc.* A **255,** 85.

—— (1964) *Proc. 9th Int. Conf. low Temp. Phys.* p. 680.

—— (1968) *Can. J. Phys.* **46,** 1915.

—— (1969*a*) in *The physics of metals. I. Electrons* (ed. J. M. Ziman) p. 62. University Press, Cambridge.

—— (1969*b*) *Phys. kondens. Materie* **9,** 1.

—— and STILES, P. J. (1963) *Phys. Lett.* **4,** 274.

—— —— (1964) *Proc. R. Soc.* A **281,** 62.

—— and TEMPLETON, I. M. (1968) *Can. J. Phys.* **46,** 1925.

—— and UDDIN, M. Z. (1936) *Proc. R. Soc.* A **156,** 701.

SHULL, C. G. (1963) in *Electronic structure and alloy chemistry of the transition elements* (ed. P. A. Beck) p. 69. Wiley, New York.

—— and WILKINSON, M. K. (1953) *Rev. mod. Phys.* **25,** 100.

—— and WOLLAN, E. O. (1956) *Solid St. Phys.* **2,** 137.

SIDGWICK, N. V. (1950) *The chemical elements and their compounds.* Clarendon Press, Oxford.

SINHA, S. K., BRUN, T. O., MUHLESTEIN, L. D., and SAKURAI, J. (1970) *Phys. Rev.* B **1,** 2430.

SIROTA, N. N. and OLEKHNOVICH, N. M. (1961) *Dokl. Akad. Nauk SSSR* **139,** 844. English trans: *Soviet Phys. Dokl.* **6,** 704.

SKINNER, H. W. B. (1938) *Rep. Prog. Phys.* **5,** 257.

—— (1940) *Phil. Trans. R. Soc.* A **239,** 95.

—— BULLEN, T. G., and JOHNSTON, J. E. (1954) *Phil. Mag.* **45,** 1070.

—— and JOHNSTON, J. E. (1937) *Proc. R. Soc.* A **161,** 420.

SLATER, J. C. (1934) *Phys. Rev.* **45,** 794.
—— (1937) *Phys. Rev.* **51,** 846.
—— (1951*a*) *Phys. Rev.* **81,** 385.
—— (1951*b*) *Phys. Rev.* **84,** 179.
—— (1953) *Phys. Rev.* **92,** 603.
—— (1960) *Quantum theory of atomic structure*, Vol. I. McGraw-Hill, New York.
—— (1965) *Quantum theory of molecules and solids*, volume 2. *Symmetry and energy bands in crystals*. McGraw-Hill, New York.
—— (1966) *Phys. Rev.* **145,** 599.
—— (1967) *Quantum theory of molecules and solids*, volume 3. *Insulators, semiconductors, and metals*. McGraw-Hill, New York.
—— and KOSTER, G. F. (1954) *Phys. Rev.* **94,** 1498.
—— —— and WOOD, J. H. (1962) *Phys. Rev.* **126,** 1307.
SLONCZEWSKI, J. C. and WEISS, P. R. (1958) *Phys. Rev.* **109,** 272.
SMITH, D. A. (1967) *Proc. R. Soc.* A **297,** 205.
SMITH, G. E. (1959) *Phys. Rev.* **115,** 1561.
—— (1961) *J. Phys. Chem. Solids* **20,** 168.
SMITH, N. V. (1969*a*) *Phys. Rev. Lett.* **23,** 1452.
—— (1969*b*) *Phys. Rev.* **183,** 634.
SMITHELLS, C. J. (1967) *Metals reference book*. Butterworth, London.
SNOW, E. C. and WABER, J. T. (1967) *Phys. Rev.* **157,** 570.
—— —— (1969) *Acta metall.* **17,** 623.
SONDHEIMER, E. H. (1950) *Phys. Rev.* **80,** 401.
—— (1952) *Adv. Phys.* **1,** 1.
—— (1969) *Phys. kondens. Materie* **9,** 208.
SOULE, D. E., MCCLURE, J. W. and SMITH, L. B. (1964) *Phys. Rev.* **134,** A453.
SOVEN, P. (1965*a*) *Phys. Rev.* **137,** A1706.
—— (1965*b*) *Phys. Rev.* **137,** A1717.
—— (1966) *Phys. Rev.* **151,** 539.
—— (1969) *Phys. Rev.* **178,** 1136.
SPARLIN, D. M. and MARCUS, J. A. (1966) *Phys. Rev.* **144,** 484.
SPECTOR, H. N. (1960) *Phys. Rev.* **120,** 1261.
—— (1962) *Phys. Rev.* **125,** 1192.
—— (1966) *Solid St. Phys.* **19,** 291.
SPICER, W. E. and BERGLUND, C. N. (1964) *Phys. Rev. Lett.* **12,** 9.
SPONG, F. W. and KIP, A. F. (1965) *Phys. Rev.* **137,** A431.
SRIVASTAVA, R. S. and SINGH, K. (1970) *Phys. Status solidi* **39,** 25.
STAFLEU, M. D. and DE VROOMEN, A. R. (1965) *Phys. Lett.* **19,** 81.
—— (1966) *Phys. Lett.* **23,** 179.
—— (1967*a*) *Phys. Status solidi* **23,** 675.
—— (1967*b*) *Phys. Status solidi* **23,** 683.
STAGER, R. A. and DRICKAMER, H. G. (1963) *Phys. Rev.* **131,** 2524.

STARK, R. W. (1967) *Phys. Rev.* **162,** 589.
—— and FALICOV, L. M. (1967) *Phys. Rev. Lett.* **19,** 795.
—— and TSUI, D. C. (1968) *J. appl. Phys.* **39,** 1056.
STEELE, M. C. and BABISKIN, J. (1955) *Phys. Rev.* **98,** 359.
STERN, E. A. (1963) *Phys. Rev. Lett.* **10,** 91.
—— (1967) *Phys. Rev.* **157,** 544.
—— (1968) *Phys. Rev.* **168,** 730.
—— (1969) *Phys. Rev.* **188,** 1163.
—— (1970) *Phys. Rev.* B **1,** 1518.
STERN, R. M. and TAUB, H. (1970) *Crit. Rev. solid St. Sci.* **1,** 221.
STEWART, A. T. and ROELLIG, L. O. (1965) *Positron annihilation. Proceedings of the conference held at Wayne State University July 1965.* Academic Press, New York.
STOCKS, G. M. and YOUNG, W. H. (1968) *Phil. Mag.* **18,** 895.
STOLZ, H. (1963*a*) *Phys. Status solidi* **3,** 1153.
—— (1963*b*) *Phys. Status solidi* **3,** 1493.
STONE, I. (1898) *Phys. Rev.* **6,** 1.
STONER, E. C. (1936) *Phil. Mag.* **21,** 145.
STREITWOLF, H. W. (1971) *Group theory in solid-state physics.* Macdonald, London.
STROUD, D. and EHRENREICH, H. (1968) *Phys. Rev.* **171,** 399.
SUFFCZYŃSKI, M. (1960) *J. Phys. Chem. Solids* **16,** 174.
—— (1961) *Bull. Acad. pol. Sci., Sér. Sci. math. astr. phys.* **9,** 489.
SUNDSTRÖM, L. J. (1965) *Phil. Mag.* **11,** 657.
SYBERT, J. R., GRENIER, C. G., and REYNOLDS, J. M. (1962) *Bull. Am. phys. Soc.* **7,** 74.
TANUMA, S., DATARS, W. R., DOI, H., and DUNSWORTH, A. (1970) *Solid State Commun.* **8,** 1107.
TAYLOR, R. (1969) *Proc. R. Soc.* A **312,** 495.
TERRELL, J. H. (1964) *Phys. Lett.* **8,** 149.
—— (1966) *Phys. Rev.* **149,** 526.
TESTARDI, L. R. and CONDON, J. H. (1970) *Phys. Rev.* B **1,** 3928.
—— and SODEN, R. R. (1967) *Phys. Rev.* **158,** 581.
THOMPSON, J. C. (1968) *Rev. mod. Phys.* **40,** 704.
THOMSON, J. J. (1901) *Proc. Camb. phil. Soc. math. phys. Sci.* **11,** 120.
THORSEN, A. C. and BERLINCOURT, T. G. (1961*a*) *Nature, Lond.* **192,** 959.
—— —— (1961*b*) *Phys. Rev. Lett.* **7,** 244.
—— and JOSEPH, A. S. (1963) *Phys. Rev.* **131,** 2078.
—— —— and VALBY, L. E. (1966) *Phys. Rev.* **150,** 523.
—— —— —— (1967) *Phys. Rev.* **162,** 574.
TINKHAM, M. (1964) *Group theory and quantum mechanics.* McGraw-Hill, New York.
TOBIN, P. J., SELLMYER, D. J., and AVERBACH, B. L. (1969) *Phys. Lett.* A **28,** 723.

Townes, C. H., Herring, C. and Knight, W. D. (1950) *Phys. Rev.* **77,** 852.

Toya, T. (1958) *J. Res. Inst. Catalysis Hokkaido Univ., Sapporo* **6,** 183.

Tripp, J. H. (1970) *Phys. Rev.* B **1,** 550.

Tsui, D. C. (1967) *Phys. Rev.* **164,** 669.

—— and Stark, R. W. (1966) *Phys. Rev. Lett.* **17,** 871.

Ulmer, K. (1961) *Z. Phys.* **162,** 254.

van der Hoeven, B. J. C., and Keesom, P. H. (1965) *Phys. Rev.* **137,** A 103.

VanderVen, N. S. (1968) *Phys. Rev.* **168,** 787.

van Goor, J. M. N. (1968) *Phys. Lett.* A **26,** 490.

van Haeringen, W. and Junginger, G. (1969) *Solid State Commun.* **7,** 1723.

van Hove, L. (1953) *Phys. Rev.* **89,** 1189.

van Kranendonk, J. and van Vleck, J. H. (1958) *Rev. mod. Phys.* **30,** 1.

Vasvari, B. (1968) *Rev. mod. Phys.* **40,** 776.

—— Animalu, A. O. E., and Heine, V. (1967) *Phys. Rev.* **154,** 535.

—— and Heine, V. (1967) *Phil. Mag.* **15,** 731.

Velický, B., Kirkpatrick, S., and Ehrenreich, H. (1968) *Phys. Rev.* **175,** 747.

Verkin, B. I. and Dmitrenko, I. M. (1958) *Zh. éksp. teor. Fiz.* **35,** 201. English trans: *Soviet Phys. JETP* **8,** 200.

—— —— (1959) *Dokl. Akad. Nauk SSSR* **124,** 557. English trans: *Soviet Phys. Dokl.* **4,** 118.

—— Kuzmicheva, L. B., and Svechkaryov, I. V. (1968) *Zh. éksp. teor. Fiz.* **54,** 74. English trans: *Soviet Phys. JETP* **27,** 41.

—— Pelikh, L. N., and Eremenko, V. V. (1965) *Dokl. Akad. Nauk SSSR* **159,** 771. English trans: *Soviet Phys. Dokl.* **9,** 1076.

Verwey, E. J. W. and de Boer, J. H. (1936) *Recl. Trav. chim. Pays-Bas. Belg.* **55,** 531.

Visscher, P. B. and Falicov, L. M. (1972) *Phys. Status solidi* b **54,** 9.

Vol'skiĭ, E. P. (1964) *Zh. éksp. teor. Fiz.* **46,** 123. English trans: *Soviet Phys. JETP* **19,** 89.

von der Lage, F. C. and Bethe, H. A. (1947) *Phys. Rev.* **71,** 612.

von Gutfeld, R. J. and Nethercot, A. H. (1967) *Phys. Rev. Lett.* **18,** 855.

Vosko, W. H., Taylor, R., and Keech, G. H. (1965) *Can. J. Phys.* **43,** 1187.

Vuillemin, J. J. (1966) *Phys. Rev.* **144,** 396.

—— and Priestley, M. G. (1965) *Phys. Rev. Lett.* **15,** 307.

Wakoh, S. and Yamashita, J. (1964) *J. phys. Soc. Japan* **19,** 1342.

—— —— (1966) *J. phys. Soc. Japan,* **21** 1712.

—— —— (1970) *J. phys. Soc. Japan,* **28** 1151.

WALKER, C. B. (1956) *Phys. Rev.* **103,** 547.
WALLACE, D. C. (1968) *Phys. Rev.* **176,** 832.
WALLACE, D. C. (1970) *Solid St. Phys.* **25,** 301.
—— (1972) *Thermodynamics of crystals.* Wiley, New York.
WALLACE, P. R. (1947) *Phys. Rev.* **71,** 622.
—— (1960) *Solid St. Phys.* **10,** 1 (1960).
WALLACE, W. D. and BOHM, H. V. (1968) *J. Phys. Chem. Solids* **29,** 721.
WALMSLEY, R. H. (1962) *Phys. Rev. Lett.* **8,** 242.
WALSH, W. M. and GRIMES, C. C. (1964) *Phys. Rev. Lett.* **13,** 523.
—— and PLATZMAN, P. M. (1965) *Phys. Rev. Lett.* **15,** 784.
WANNIER, G. H. (1962) *Rev. mod. Phys.* **34,** 645.
WARREN, J. L. (1968) *Rev. mod. Phys.* **40,** 38.
WATSON, R. E., FREEMAN, A. J., and DIMMOCK, J. O. (1968) *Phys. Rev.* **167,** 497.
WATTS, B. R. (1963) *Phys. Lett.* **3,** 284.
—— (1964*a*) *Phys. Lett.* **10,** 275.
—— (1964*b*) *Proc. R. Soc.* A **282,** 521.
—— (1964*c*) *Proc. 9th Int. Conf. low Temp. Phys.* p. 779.
WEISS, R. J. (1970) *Phys. Rev. Lett.* **24,** 883.
—— and DE MARCO, J. J. (1958) *Rev. mod. Phys.* **30,** 59.
—— —— (1965) *Phys. Rev.* **140,** A1223.
—— and FREEMAN, A. J. (1959) *J. Phys. Chem. Solids* **10,** 147.
—— HARVEY, A., and PHILLIPS, W. C. (1968) *Phil. Mag.* **17,** 241.
WEISZ, G. (1966) *Phys. Rev.* **149,** 504.
WERNER, S. A., ARROTT, A. and KENDRICK, H. (1966) *J. appl. Phys.* **37,** 1260.
WHITE, G. K. and TAINSH, R. J. (1967) *Phys. Rev. Lett.* **19,** 165.
WHITTEN, W. B. and PICCINI, A. (1966) *Phys. Lett.* **20,** 248.
WIGNER, E. P. (1932) *Nachr. Ges. Wiss. Göttingen* p. 546.
—— (1934) *Phys. Rev.* **46,** 1002.
—— (1938) *Trans. Faraday Soc.* **34,** 678.
—— and SEITZ, F. (1933) *Phys. Rev.* **43,** 804.
—— —— (1934) *Phys. Rev.* **46,** 509.
—— —— (1955) *Solid St. Phys.* **1,** 97.
WILDING, M. D. and LEE, E. W. (1965) *Proc. phys. Soc.* **85,** 955.
WILLIAMS, G. A. (1965) *Phys. Rev.* **139,** A771.
WILLIAMS, R. W. and DAVIS, H. L. (1968) *Phys. Lett.* A **28,** 412.
WILLIAMSON, S. J., SURMA, M., PRADDAUDE, H. C., PATTEN, R. A., and FURDYNA, J. K. (1966) *Solid State Commun.* **4,** 37.
WILSON, A. H. (1931) *Proc. R. Soc.* A **133,** 458.
WINDMILLER, L. R. (1966) *Phys. Rev.* **149,** 472.
WISER, N. (1966) *Phys. Rev.* **143,** 393.
WOHLFARTH, E. P. (1953) *Proc. phys. Soc.* A **66,** 889.
—— and CORNWELL, J. F. (1961) *Phys. Rev. Lett.* **7,** 342.

Woll, E. J. and Kohn, W. (1962) *Phys. Rev.* **126,** 1693.

Wood, J. H. (1962) *Phys. Rev.* **126,** 517.

—— (1966) *Phys. Rev.* **146,** 432.

Woods, A. D. B., Brockhouse, B. N., March, R. H., and Bowers, R. (1962) *Proc. phys. Soc.* **79,** 440.

—— Cochran, W., and Brockhouse, B. N. (1960) *Phys. Rev.* **119,** 980.

Yaqub, M. and Cochran, J. F. (1965) *Phys. Rev.* **137,** A1182.

Yosida, K. (1957) *Phys. Rev.* **106,** 893.

—— and Watabe, A. (1962) *Prog. theor. Phys.* **28,** 361.

Young, R. C. (1967) *Phys. Rev.* **163,** 676.

Zak, J. (1968) *Phys. Rev.* **168,** 686.

—— Casher, A., Glück, M., and Gur, Y. (1969) *The irreducible representations of space groups*. Benjamin, New York.

Zebouni, N. H., Hamburg, R. E., and Mackey, H. J. (1963) *Phys. Rev. Lett.* **11,** 260.

Zener, C. (1951*a*) *Phys. Rev.* **81,** 440.

—— (1951*b*) *Phys. Rev.* **82,** 403.

—— (1951*c*) *Phys. Rev.* **83,** 299.

—— (1952) *Phys. Rev.* **85,** 324.

Ziman, J. M. (1959) *Proc. R. Soc.* A **252,** 63.

—— (1960) *Electrons and phonons*. Clarendon Press, Oxford.

—— (1961) *Adv. Phys.* **10,** 1.

—— (1964*a*) *Principles of the theory of solids*. University Press, Cambridge.

—— (1964*b*) *Adv. Phys.* **13,** 89.

—— (1965) *Proc. phys. Soc.* **86,** 337.

—— (1968) *J. Phys. C* **1,** 1532.

—— (1969*a*) in *The physics of metals. I. Electrons*. (ed. J. M. Ziman). p. 250. University Press, Cambridge.

——(1969*b*) *Elements of advanced quantum theory*. University Press, Cambridge.

——(1969*c*) *J. Phys. C* **2,** 1230.

Zitter, R. N. (1962) *Phys. Rev.* **127,** 1471.

Zornberg, R. I. and Mueller, F. M. (1966) *Phys. Rev.* **151,** 557.

AUTHOR INDEX

The page numbers in italic type refer to tabular material or figures.

SUBJECT INDEX

The page numbers in italic type refer to tabular material or figures.